T0396708

MATHEMATICAL GEOPHYSICS

MODERN APPROACHES IN GEOPHYSICS

formerly Seismology and Exploration Geophysics

Mathematical Geophysics

A Survey of Recent Developments in Seismology and Geodynamics

edited by

N. J. VLAAR

G. NOLET

M. J. R. WORTEL

and

S. A. P. L. CLOETINGH

University of Utrecht, The Netherlands

D. Reidel Publishing Company

A MEMBER OF THE KLUWER ACADEMIC PUBLISHERS GROUP 

Dordrecht / Boston / Lancaster / Tokyo

Library of Congress Cataloging in Publication Data

Mathematical geophysics.

(Modern approaches in geophysics)
Includes index.
1. Seismology—Mathematics. 2. Geodynamics—Mathematics.
I. Vlaar, N. J., 1933– II. Series.
QE539.M33 1987 551.2′2′0151 87–23462

ISBN-13: 978-94-010-7785-9 e-ISBN-13: 978-94-009-2857-2
DOI: 10.1007/978-94-009-2857-2

Published by D. Reidel Publishing Company,
P.O. Box 17, 3300 AA Dordrecht, Holland.

Sold and distributed in the U.S.A. and Canada
by Kluwer Academic Publishers,
101 Philip Drive, Assinippi Park, Norwell, MA 02061, U.S.A.

In all other countries, sold and distributed
by Kluwer Academic Publishers Group,
P.O. Box 322, 3300 AH Dordrecht, Holland.

Softcover reprint of the hardcover 1st edition 1988

CONTENTS

II. Convection and Lithospheric Processes

Geomagnetism

Mantle convection

Postglacial rebound

Thermomechanical processes in the lithosphere

PREFACE

The contributions to this book follow a topical trend. In several geophysical fields evidence is accumulating concerning the deviation of the earth's structure from radial symmetry. Seismology provides the most adequate resolution for revealing the earth's lateral inhomogeneity on a global to local scale. Lateral structure in the density distribution is also manifest in the earth's gravity field and in the geoid. Asphericity in physical parameters, generally supposed only to vary with the vertical coordinate, has a profound influence on geodynamics. The effects of these deviations from spherical symmetry concern in particular convection theory, post-glacial rebound and the dynamics of the lithosphere and upper mantle in general.

At the 16th International Conference on Mathematical Geophysics which was held in Oosterbeek, the Netherlands, in 1986, the need was felt to present the state of the art. Several prospective authors were found interested to contribute to the present book. This Oosterbeek conference was one in a long series of topical conferences starting with the Upper Mantle Project Symposia on Geophysical Theory and Computers in the 1960s, and thence their successors, the conferences on Mathematical Geophysics, until the present.

Without exaggeration it can be said that these conferences were at the heart of the main developments in solid earth geophysics. They were by their very character dominated by the application of mathematical and computational physics to the problems connected with studying the earth's interior structure and dynamics. The main developments in global seismology have, since the beginnings, formed the hard core of the symposia. Emphasis has been placed on developments in the theory of propagation and excitation of seismic waves, free oscillations of the earth, data processing, and inverse problems. Whereas much of this work was connected with a radially symmetric, rotating or non-rotating earth model, the last years have seen a growing interest in lateral deviations from a standard earth. This is witnessed by the increase of contributions in the fields of seismic tomography, the effect of lateral structure on free oscillations, and on the propagation of surface waves.

As the resolution of tomographic imaging of the earth increases rapidly, the time is approaching when geological structures on lithospheric and crustal scales can be mapped with great precision, thus providing an important data base for unravelling the history of the earth's crust. Therefore it was felt to be justified to lay extra emphasis on this topic, also for the purpose of searching for connections between the fundaments of seismology and those of seismic exploration methods.

Another topic of growing interest is the evolution of both oceanic and continental lithosphere. This is particularly the case given the greater resolution mentioned above, increasing knowledge of the earth's gravity field, and deeper insight into rheological properties of the earth's materials. Greater understanding of plate tectonic mechanisms, particularly their connection with deeper seated mantle dynamics, is required to grasp hold of geological reality.

Traditionally, the conferences on mathematical geophysics were firmly based within a rather closed system of the mathematical description of mostly observable geophysical fields. Present day geodynamics is becoming more and more multi-disciplinary. Mathematical geophysics may offer the geophysical framework for eliminating speculation on the earth's structure, dynamics, and history.

All chapters in this book, deal, one way or another, with geophysical expressions and effects of lateral heterogeneity in the earth.

The chapters on seismology deal with different bands in the seismic spectrum and different approaches for resolving lateral heterogeneity from the inversion of imperfect data. Low frequency seismology develops towards a high sophistication and appears to be at the threshold of unravelling the large scale 3-D structure of the earth. Intermediate frequency seismology, as represented by the study of surface waves, employs horizontal asymptotic ray theory for resolving intermediate scale structure, and even incorporates off great circle path effects by means of a scattering Born approximation approach. High frequency body waves are used for tomographic studies which reveal strong lateral heterogeneity, particularly in the upper mantle. Attention is given to the problem of large scale inversion and the resolving power of seismological data sets. Seismology is the most important source of information on the earth's structure and the relevant chapters are indicative of the high standards required in frontline research in this field.

Geomagnetic observations may be inverted into the fluid velocity field near the core surface. This type of study is relevant in the light of emerging seismological evidence for lateral core structure. The interpretation of seismological findings in terms of the solid earth's internal and surface dynamics which should find its expression in the gravity field, surface heat flux, and vertical and horizontal motion of the earth's surface appears to be a promising field of research. In this context is knowledge of the earth's internal rheology of primary importance. The earth's rheological behaviour is manifest in the refined studies of post-glacial rebound. The rheological structure, still considered to be only vertically varying is fundamental to the problem of convective motion in the earth's mantle. The question arises in how far laterally varying rheology, as is clearly revealed by lithosphere and upper mantle structure may invalidate some of the present approach to the earth's internal dynamics.

An interesting development is the attempt to connect the plate tectonic velocity field, poloidal and toroidal, to lateral heterogeneity and its geoidal expression. The toroidal velocity field of the lithosphere indicates that spherically symmetric viscosity structure may not be warranted.

The most obvious laterally varying structure is the earth's strong lithosphere. Understanding the processes of its creation, evolution and destruction would impose very strong constraints on internal dynamics. Whereas long standing traditions treat the lithosphere as circumstantial to mantle flow, thorough knowledge of lithospheric processes may give way to better understanding deeper seated dynamics. Moreover, geology and geophysics present many observables which still await explanation. Here a quantitative approach to lithospheric evolution and its thermo-mechanical properties might provide the setting for the unravelling of the earth's evolution and dynamics.

The editors wish to express their great appreciation to the authors for the high level of their achievements as laid down in the chapters of this volume, and also for their constructive cooperation. Thanks are due to Maarten Remkes and Hong-Kie Thio for their enthusiastic assistence to the editors.

Utrecht, July 1987

N.J. Vlaar

Chapter 1

Low frequency seismology and three-dimensional structure — observational aspects

Guy Masters and Michael Ritzwoller

The availability of a large quantity of long period digital seismic data has played a major role in recent advances in our understanding of the large scale aspherical structure of the Earth. Many new models exist which have been designed to fit these data. They succeed to various degrees but it is indisputable that much of the signal in the data remains to be explained.

We illustrate the kinds of signals that models of aspherical structure should reproduce and then indicate the theoretical framework in which these observations may be interpreted. Finally, we discuss the "first generation" of aspherical Earth models which demonstrate that we need to be more sophisticated in our interpretative techniques if we are to obtain consistent models of structure.

1. Introduction

If the Earth were truly spherically symmetric, it is probably fair to say that the low frequency seismologist would now be out of business. Compact expressions for an observed seismogram on a spherical Earth as a sum of decaying cosinusoids have been around since the early 1970s (Gilbert, 1970). Array processing techniques based on this theory were introduced in the mid 1970s and were successfully applied to hand-digitized data (Gilbert and Dziewonski, 1975). With the vast improvement in the quantity and quality of digital data since that time, one might reasonably expect that we would, by now, fully understand the long period spectrum of seismic motion. Needless to say, this has not

N.J. Vlaar, G. Nolet, M.J.R. Wortel & S.A.P.L. Cloetingh (eds), Mathematical Geophysics, 1-30.

happened. Better data and more sophisticated ways of processing them have revealed a plethora of signals that cannot be modeled by spherically symmetric structure alone. The main purpose of this paper is to acquaint the reader with the types of signals that the low frequency seismologist is now trying to explain. Some of the signals are unsurprising, e.g., the effect of rotation of the Earth on the longest period part of the spectrum has been observed and understood for over 25 years (Backus and Gilbert, 1961; Pekeris et al., 1961; Ness et al., 1961; Benioff et al., 1961). Others are completely unexpected and, as of the time of writing, unexplained.

The complexities in the low frequency spectrum arise because rotation and lateral structure remove the spherical symmetry. On a spherically symmetric Earth, modes of oscillations would be degenerate, i.e., a group of modes or "singlets" would have the same frequency. Such groups of singlets are called "multiplets." The removal of spherical symmetry completely removes the degeneracy so that each singlet within a multiplet has a slightly different frequency. This "splitting" dominantly affects spectra by producing phase shifts relative to the spherically symmetric reference state. Coupling between singlets, either within a multiplet or between multiplets, can also produce further phase and amplitude perturbations. These effects are now observationally well documented and are slowly becoming better understood.

The next section of this paper gives a brief look at the low frequency digital data and attempts to highlight some of the observational problems we encounter. This is followed by an overview of the theoretical background required to model the data and we close with an evaluation of some recently proposed three-dimensional models.

2. Long period data

The main sources of long period data are the IDA array (Agnew et al., 1976, 1986), the SRO and ASRO components (Peterson et al., 1980) of the Global Digital Seismic Network (GDSN) and, more recently, GEOSCOPE (Romanowicz et al., 1984). The original IDA instrumentation consists of an electrostatically fed-back Lacoste-Romberg gravimeter. Electrostatic feedback has the disadvantage of being relatively weak but has the advantage that only modest amplification is needed to get a usable signal. (The development of low noise electronic amplifiers has allowed the use of electromagnetic feedback which is capable of a much larger force.) The IDA instruments are therefore susceptible to signal distortion of the early part of recordings of large events and the first Rayleigh wave is usually lost for events greater than $M_S = 6.5$. Nevertheless, we are usually interested in many hours of recording for free oscillation work so the loss of the initial part of the record is not as severe as it might seem. Another drawback of the IDA data is that only the vertical component is recorded. Many modes of oscillation are dominantly horizontally polarized so three component recording is desirable. This led to the use of the SRO and ASRO components of the GDSN. The difficulty of making long time series from SRO data has meant that few investigators have used this data source though useful signal exists to periods longer than 500 seconds for the largest events. Figure 1 shows the acceleration response of several seismic systems. The ASRO system has a long period response intermediate between the IDA and SRO systems. Note the very poor low frequency response of the DWWSSN and RSTN components of the GDSN. The DWWSSN is also extremely noisy so neither the DWWSSN nor RSTN system have seen much use in very

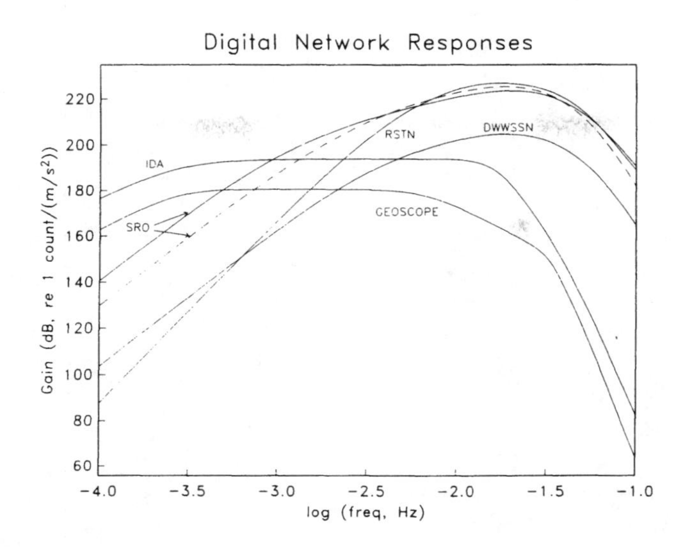

Figure 1. Response to acceleration of some currently operating global digital networks. The long period channels of the components of the GDSN are shown (i.e., SRO, RSTN, and DWWSSN). The mode channel for IDA and the very long period channel for GEOSCOPE are also shown. Recent modification of the SRO has resulted in an improved long period response.

long period seismology. Even the SRO and ASRO networks were not initially designed for long period work and so calibration of the long periods is relatively poor. It appears that calibration signals provided with the data are not reliable enough to constrain the very long period response. This leads to difficulties when performing analyses which mix GDSN and IDA data. New sensor and electronics design have led to improvements in long period instrumentation (Wielandt and Streckheisen, 1982) and new arrays (e.g., GEOSCOPE) are now providing easy to use, high quality three component long period data. Upgrading of WWSSN and IDA stations will hopefully soon result in a large, high quality global network.

Figure 2 shows some typical waveforms from the IDA, SRO and ASRO networks which have been filtered so that only periods longer than 150 seconds are visible. Also shown are synthetic seismograms constructed for a spherically symmetric Earth model. While visual agreement is very good there are many details which are not modeled and which provide strong constraints on aspherical structure. Figure 3 shows some Fourier amplitude spectra of Hanning tapered recordings. The fundamental mode oscillations are clearly visible along with some lower amplitude overtones. Note that individual mode peak amplitudes are quite poorly predicted by the spherical Earth synthetics even though the time domain fits of Figure 2 are very good. These examples are for events with moments of about 10^{27} dyne cm and signal with periods in excess of 750 seconds is clearly visible. We primarily look at the data in the frequency domain so the method of estimating the Fourier spectra merits some discussion. Until recently, a direct estimate using a single, *ad-hoc* data window has been the method of choice. Different windows give different

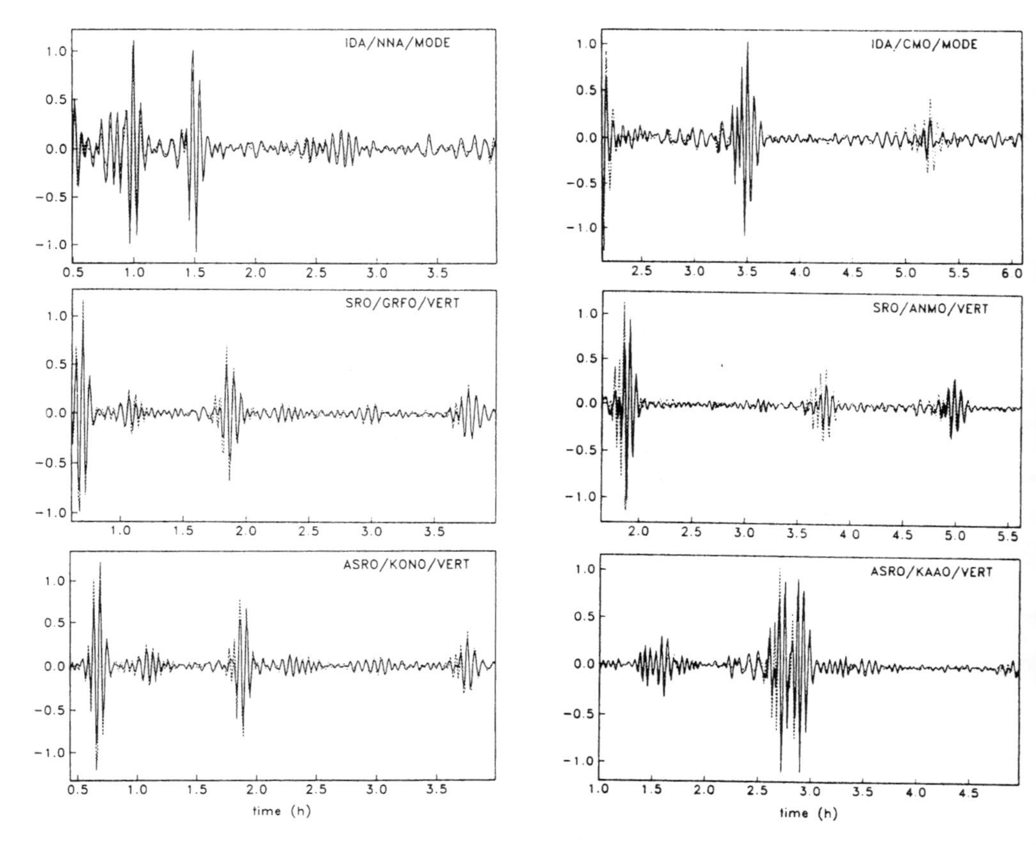

Figure 2a (left panel). A comparison of long period waveforms (T>150 secs) with spherical Earth synthetic seismograms for a) a large, deep event (Banda Sea 1982/173). Examples for the IDA, SRO and ASRO networks are shown. Note that fundamental mode Rayleigh wave packets dominate even the recordings of very deep events.
Figure 2b (right panel). As for Figure 2a except for a large shallow event (Iran 1978/259).

compromises between resolution and spectral leakage and we show the effect of several different data windows in Figure 4. If we have a harmonic signal in noise, it is possible to find a taper which minimizes spectral leakage for a particular resolution width. Such tapers are relatively easy to calculate (Slepian, 1978) and are called "discrete prolate spheroidal sequences" (or just "prolate windows"). For our example in Figure 4, the 2π prolate window gives a reasonable compromise between peak width and side band level. The familiar Hanning window gives similar results and is, of course, much easier to compute and so has seen the most usage. The Blackman-Harris 4-term taper (Harris, 1978) gives similar results to the 4π prolate window but the peaks are sufficiently broadened that spectral details are lost. (On the other hand spectral leakage is almost negligible.) We need not limit ourselves to one taper as families of tapers can be constructed for a given resolution width. These have progressively worsening spectral leakage properties but the

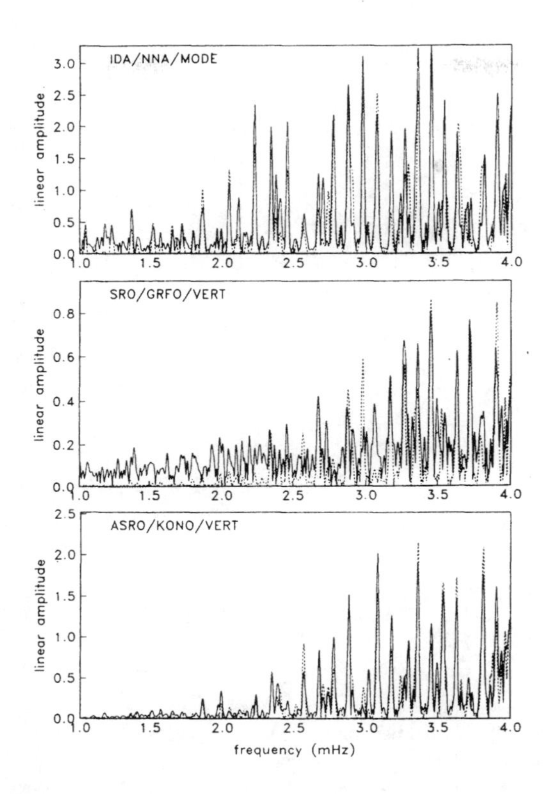

Figure 3. Spectra of 30-hour recordings in the frequency band 1 to 4 mHz. The recordings have been Hanning tapered to reduce spectral leakage. Examples from the IDA, SRO and ASRO networks are shown. The SRO instruments are generally of slightly inferior quality in this frequency band. The dashed lines are spectra of spherical Earth model synthetic seismograms. Note that the detailed fit in the frequency domain looks quite poor despite the excellent fit in the time domain (Figure 2a).

first few can be used to obtain useful spectral estimates. This multiple-taper method has been described by Thomson (1982) and extended for use with decaying sinusoids by Lindberg (1986). The method allows reliable line detection and it may be possible to detect low signal spectral lines that would otherwise be missed. Multiple-taper analyses are relatively new so examples presented in the remaining part of this section will use only the Hanning taper. It is clearly apparent from Figure 4 that some taper must be used as the distortion due to spectral leakage in the absence of a taper is sufficient to give grossly erroneous estimates of center frequencies, apparent attenuation, etc. Further discussion of this point can be found in Dahlen (1982) and Masters and Gilbert (1983).

As noted in the introduction, departures of the Earth from sphericity lead to splitting and coupling of free oscillations. These effects manifest themselves in the data in a variety of ways. At the lowest frequencies, the rotation of the Earth causes strong splitting of modes of oscillation (Figure 5) and leads dominantly to a linear distribution of the singlet frequencies in azimuthal order. The rotational bulge of the Earth is an axisymmetric

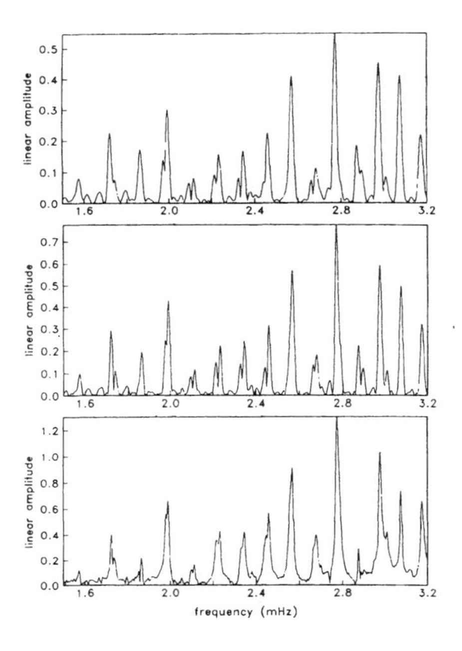

Figure 4. Direct spectral estimates of a 35-hour time series using a boxcar taper (lower panel), a 2π prolate (middle panel) and a 4π prolate (upper panel). In the lower panel, the distortion caused by spectral leakage is so great that the spectral estimate using a boxcar taper is useless. In the upper panel the peak widths are so broad that some multiplets appear to overlap. The 2π prolate gives a reasonable compromise between resolution and spectral leakage.

feature which is dominated by harmonic degree 2 and leads to a parabolic distribution of singlet frequencies. The joint effect is a quadratic in azimuthal order (Dahlen, 1968). Careful processing of many records allows the individual singlets to be isolated (see below) and the results are often in good agreement with the predictions of a rotating Earth in hydrostatic equilibrium. For many modes (in particular high Q, low l modes which sample into the Earth's core), the rotating, hydrostatic Earth model explains little of the observed signal. Some modes can have singlets distributed over a frequency band up to two and a half times greater than theoretically predicted (Figure 6). This "anomalous splitting" of modes suggests the presence of deep seated large scale aspherical structure and will be discussed further below.

As one goes to higher frequency, higher l modes it is impossible to directly observe splitting as in the above examples and a multiplet may often appear to be a single resonance function. In fact, the model of a single decaying cosinusoid can usually explain the observations though the inferred center frequency and attenuation rate vary from recording to recording (Figure 7). A discussion of the many methods that can be used for estimating frequencies and attenuation rates can be found in Masters and Gilbert (1983). In practice, the techniques require the use of high signal spectral peaks that can be unambiguously

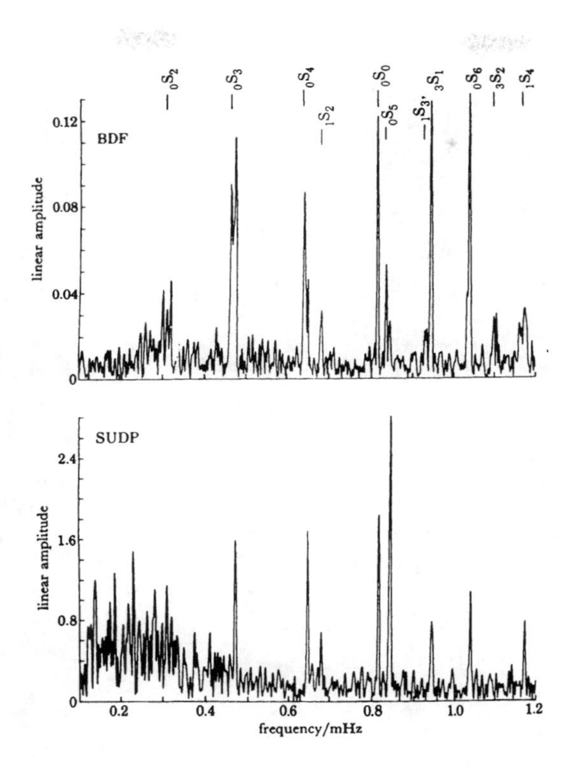

Figure 5. Two spectra of 120-hr records from the Sumbawa earthquake of 1977. In the upper panel, the effects of rotation and ellipticity cause obvious splitting at the IDA station at Brasilia. In the lower panel, the spectrum of the UCLA South Pole recording shows no obvious splitting though it is noisy at frequencies below 0.4 mHz. The lack of splitting at a pole is to be expected if rotation and axisymmetric aspherical structure are the dominant perturbing influences.

identified so the analysis of single recordings by this method is mainly limited to periods longer than about 150 seconds and to the highly excited fundamental modes. Some high Q overtones can be analyzed by using the Earth as an attenuation filter, i.e., the data window used in spectral estimation is started several hours after the origin time of the event so that low Q modes have been significantly attenuated.

The effects of splitting of a multiplet in the low frequency spectra are obvious. More subtle is the effect of multiplet/multiplet coupling. Detection of such coupling is often difficult as closely spaced multiplets may or may not be coupled depending upon the nature of aspherical structure. One type of coupling which will obviously be present is Coriolis coupling between $_0S_l$ and $_0T_{l\pm1}$ multiplets if they are sufficiently close in complex frequency (Dahlen, 1969; Luh, 1973, 1974; Woodhouse, 1980; Masters et al., 1983; Fichler et al., 1986). Conditions for observable amounts of Coriolis coupling actually exist for $_0S_8/_0T_9$ through $_0S_{22}/_0T_{23}$ (a frequency range of 1.7 to 3.1 mHz). This coupling means that we observe spectral lines which are hybrid spheroidal/toroidal motion and thus spectral

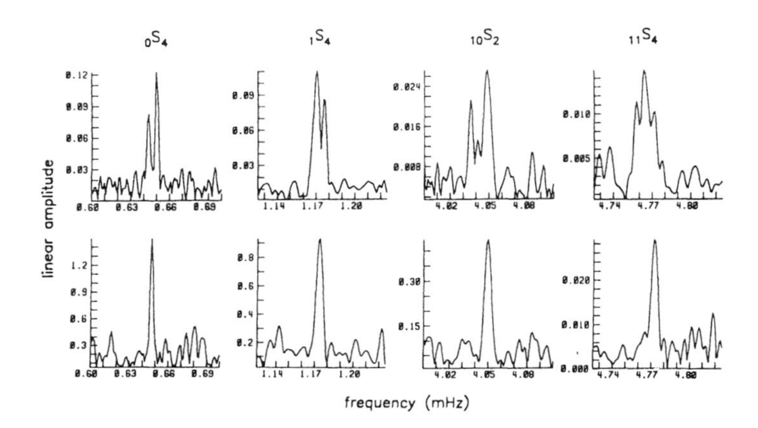

Figure 6. Detailed spectra of four low harmonic degree multiplets recorded at nonpolar latitudes (upper figures) and polar latitudes (lower figures). Each nonpolar spectrum shows clear evidence of splitting though the polar spectra are not obviously split. This is surprising for $_{10}S_2$ and $_{11}S_4$ as they are insensitive to the rotation of the Earth. These latter two modes sample into the core and are very strongly split. They both span frequency bands in excess of 13 μHz though are predicted to be less than 7 μHz wide for a rotating hydrostatic Earth.

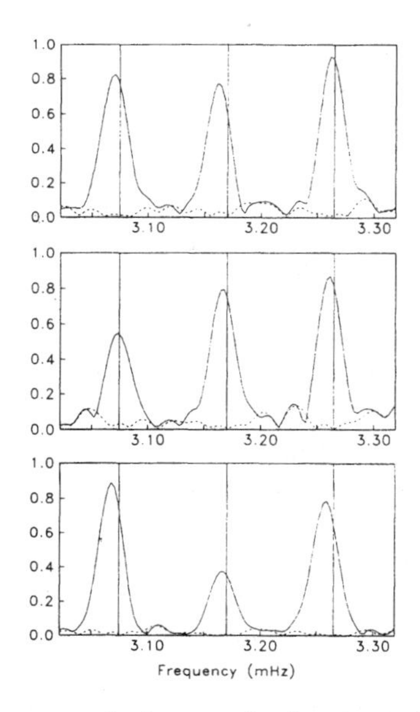

Figure 7. Linear amplitude spectra of a frequency band encompassing the fundamental modes $_0S_{22} \rightarrow {_0S_{24}}$. The solid line is the original spectrum while the dashed line is the residual after the best fitting decaying cosinusoid model has been fit to the data. The data are for the Iran 1978/259 event recorded at CMO (top), HAL (middle), and PFO (bottom). Also shown are the theoretical degenerate multiplet frequencies as predicted by Model 1066A. A decaying cosinusoid is a good model for the data though the apparent center frequencies are quite variable.

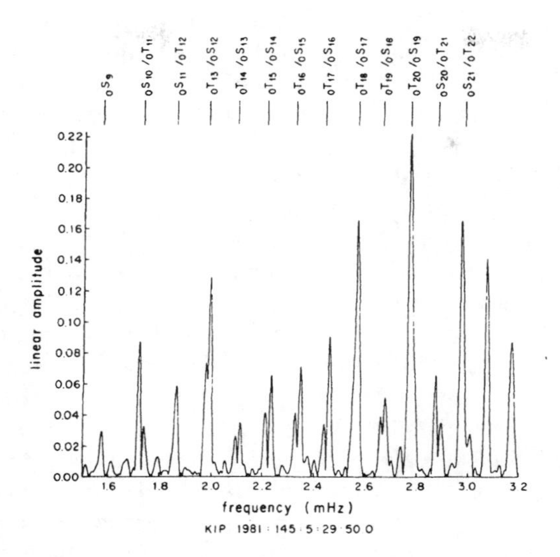

Figure 8. A 35-hour Hanning-tapered spectrum of a vertical component recording made at KIP of an event south of New Zealand 1981/145. The source mechanism of this event was particularly good at exciting toroidal modes and the Coriolis coupling of these modes to nearby spheroidal modes is particularly obvious.

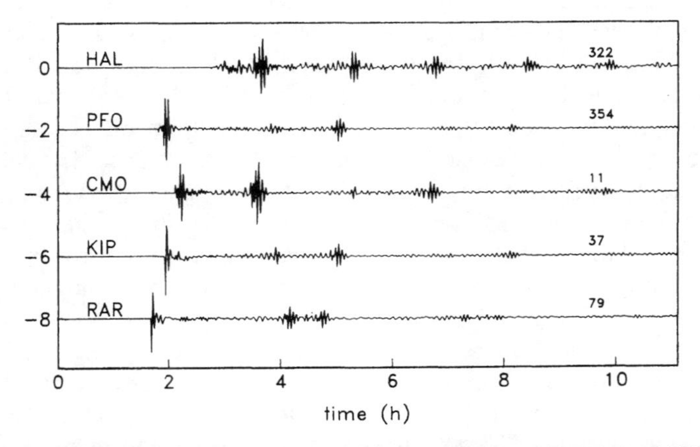

Figure 9. IDA recordings of the Iran 1978/259 event low passed so that only periods of 150 sec or longer are visible. The numbers on the right are the azimuths to each receiver in degrees. Note that PFO, CMO and KIP are all at a very similar azimuth but show very different Rayleigh wave amplitude behavior. This example of surface wave amplitude anomalies was first discussed by Niazi and Kanamori (1981)

peaks can occur on vertical component recordings near the frequencies of toroidal modes. An example of this is shown in Figure 8 where a whole suite of spectral peaks near toroidal mode frequencies can be seen. On vertical component recordings, two clumps of energy are visible with one (closest to the original uncoupled spheroidal mode) being dominant. If the properties of this spectral peak are interpreted as being those of an uncoupled spheroidal mode then significant bias can result. For example, the mean frequency of a dominantly spheroidal hybrid mode can be two or three microhertz from the true spheroidal mode frequency. This constitutes a bias of several standard deviations. Apparent attenuation rates are similarly affected. Another form of coupling which should also provide measurable signals is so called "along-branch" coupling; e.g., the coupling of ${}_0S_l$ to ${}_0S_{l\pm1}$ modes. This coupling provides some sensitivity to odd-order structure (see below) and is therefore of great interest. Because of the relatively large frequency spacing of fundamental modes (~100 μHz), this kind of coupling causes only a weak signal which cannot be illustrated by direct inspection of the spectrum. It may be more apparent in the time domain where the expected effect of coupling between neighboring fundamental modes is to produce a phase shifted wave packet though theoretical calculations with existing models of long wavelength structure show that the effect is small (Park, 1986).

Inspection of the time domain fits of spherical Earth point source synthetic seismograms to data often reveals large surface wave amplitude anomalies. These anomalies are most apparent when a Rayleigh wave packet R_n is smaller than the next packet, R_{n+1} (Figure 9). For the largest events, it is possible that some of the observed anomaly is due to source finiteness. A more likely cause is a focusing/defocusing effect of aspherical structure (Lay and Kanamori, 1985; Woodhouse and Wong, 1986). The amplitude anomalies are visible for nearly every event, regardless of size, which makes an explanation based on source finiteness untenable.

At periods shorter than about 150 seconds, it is extremely difficult to identify individual peaks in the spectra of single recordings and it becomes more natural to work in the time domain. (Note that stacking of the spectra of many recordings still allows individual modes to be identified at much higher frequencies.) Individual surface wave packets consisting dominantly of fundamental modes can be isolated in the data and great circle or single station phase velocity or group velocity measurements performed (see, e.g., Nakanishi and Anderson, 1983, 1984). Great circle phase velocity analysis reduces basically to a transfer function estimation problem for mapping R_n into R_{n+2} and several quite sophisticated techniques have been developed to do this (e.g., Nakanishi, 1979). Unfortunately the analysis is restricted to relatively short data lengths so that the transfer function at periods longer than 250 secs is unreliable. Traditional transfer function estimation techniques use frequency domain averaging to estimate errors (e.g., Jenkins and Watts, 1968) though we are unaware of a formal error analysis in a seismological context. Single station measurements are much more difficult to perform and require corrections for source mechanism, instrumental response and higher mode contamination and are also sensitive to errors in epicenter location. These difficulties have led some authors to abandon the estimation of intermediate quantities from the data and to directly model waveforms in terms of Earth structure (Dziewonski and Steim, 1982; Woodhouse and Dziewonski, 1984; Tanimoto, 1987). This is intuitively very satisfying but suffers from some practical drawbacks. Perhaps the most contentious problem is the assessment of goodness of fit and the consequent determination of errors on the model parameters. Assigning an "error" to a waveform is not a trivial task.

Returning to the main theme of this section, i.e., the signal from aspherical structure, we can summarize as follows. Low frequency time series are dominated by fundamental mode energy (even for very deep events). In the frequency domain, the spectral peaks of fundamental modes are clearly shifted or observably split and individual peak amplitudes are not well fit by spherical Earth synthetics. A consequence of the splitting is that measured Q values are extremely variable. We can also observe Coriolis coupling of toroidal/spheroidal multiplets. In the time domain we observe phase shifting of surface wave packets and strong amplitude anomalies. If we use extremely long records we can enhance high Q overtones relative to the low Q fundamental modes and the spectra of many of these overtones show strong anomalous splitting.

3. Theoretical review

We consider first the case of uncoupled multiplets. A complete description of the seismogram of an uncoupled multiplet on a slightly aspherical Earth is given by Woodhouse and Dahlen (1978) and by Woodhouse and Girnius (1982). We summarize the results for the kth multiplet where k is shorthand for (n,l,q) where n is the radial order, l the harmonic degree, and q the mode type (spheroidal or toroidal).

A component of the displacement field at position $\mathbf{r}=(r,\theta,\phi)$ excited by a point source at $\mathbf{r}_0$ with moment rate tensor $\mathbf{M}$ can be written as the inner product

$$s(\mathbf{r},t) = \mathrm{Re}\left\{ \sum_k \boldsymbol{\sigma}_k^T(\mathbf{r}) \cdot \mathbf{a}_k(\mathbf{r}_0,t) e^{i\omega_k t} \right\} \tag{1}$$

where the complex envelope function vector $\mathbf{a}_k(t)$ is given by

$$\mathbf{a}_k(t) = \mathbf{P}_k(t)\mathbf{a}_k\,, \quad \mathbf{a}_k(0) = \mathbf{a}_k \tag{2}$$

$\mathbf{P}_k(t)=\exp(i\mathbf{H}_k t)$ is the matrizant or propagator matrix of the following first-order propagator equation with initial condition given in (2):

$$\frac{d}{dt}\mathbf{a}_k(t) = i\mathbf{H}_k \mathbf{a}_k(t) \tag{3}$$

The receiver vector $\boldsymbol{\sigma}_k$ in (1) is composed of the $2l+1$ singlet eigenfunctions:

$$\sigma_k^m(r) = [\hat{\mathbf{r}}U_k(r)Y_l^m(\theta,\phi) + V_k(r)\nabla_1 Y_l^m(\theta,\phi) - W_k(r)\hat{\mathbf{r}}\times \nabla_1 Y_l^m(\theta,\phi)]\cdot\hat{\mathbf{n}} \tag{4}$$

where $\nabla_1=\hat{\boldsymbol{\theta}}\partial_\theta+(\sin\phi)^{-1}\hat{\boldsymbol{\phi}}\partial_\phi$ and where the unit vector $\hat{\mathbf{n}}$ determines which component is represented by s in (1).

The excitation vector $\mathbf{a}_k$ comprises the $2l+1$ excitation coefficients $a_k^m=\mathbf{M}:\boldsymbol{\varepsilon}_k^{m*}(\mathbf{r}_0)$. The term $\boldsymbol{\varepsilon}_k^{m*}$ is the complex conjugate of the strain tensor for the azimuthal order m singlet and is expressible in terms of the multiplet scalars U_k, V_k, and W_k in (4) (Gilbert and Dziewonski, 1975). The complex spherical harmonics, Y_l^m, are normalized according to the convention of Edmonds (1960).

The dependence of displacement on aspherical structure in (1) can be made explicit by considering the components of the $(2l+1)\times(2l+1)$ complex splitting matrix $\mathbf{H}_k$:

$$H_{mm'}^k = \omega_k(a_k+mb_k+m^2c_k)\delta_{mm'} + \sum_{\substack{s=2\\ \text{even}}}^{2l} \gamma_s^{mm'}{}_k c_s^{m-m'} \tag{5}$$

where

$$ {}_k c_s^t = \int_0^a \delta \mathbf{m}_s^t(r) \cdot {}_k\mathbf{G}_s(r) r^2 dr - \sum_d^D r_d^2 h_{sd}^t \, {}_kB_s \tag{6} $$

The first term on the right hand side of (5) represents the contribution by rotation and hydrostatic ellipticity of figure. Here ω_k is the degenerate frequency and numerical values of the splitting parameters a_k, b_k, and c_k evaluated for some multiplets can be found in Dahlen and Sailor (1979) and Ritzwoller et al. (1986). We call the model containing only these contributions to $\mathbf{H}_k$ the RH model (standing for rotating, hydrostatic earth model). The second term contains the additional effect of general even-order aspherical volumetric ($\delta \mathbf{m}_s^t(r)$) and boundary (h_{sd}^t) perturbations. Aspherical perturbations are represented with spherical harmonic basis functions of degree s and order t:

$$ \delta \mathbf{m}(\mathbf{r}) = \sum_{s,t} \delta \mathbf{m}_s^t(r) Y_s^t(\theta,\phi) \qquad h_d(\theta,\phi) = \sum_{s,t} h_{sd}^t Y_s^t(\theta,\phi) \tag{7} $$

for each boundary d. The model vector is

$$ \delta \mathbf{m}_s^t(r) = (\delta\rho_s^t(r)\,,\, \delta\kappa_s^t(r)\,,\, \delta\mu_s^t(r))^T $$

and the integral kernel vector is

$$ \mathbf{G}_s(r) = (R_s(r), K_s(r), M_s(r))^T $$

where R_s, K_s, and M_s can be computed using the formulas given by Woodhouse and Dahlen (1978). Each multiplet k possesses a unique set of complex structure coefficients ${}_k c_s^t$ whose amplitude and phase are functions of the amount and distribution of heterogeneity in the earth and of the manner in which the multiplet samples this heterogeneity. The γ function in equation (5) can be written in terms of Wigner $3j$ symbols which may be computed using the recurrence relations given by Schulten and Gordon (1975). Thus, estimation of the c_s^t fully determines the splitting matrix. Note that, for an isolated multiplet, only those components of $\delta\mathbf{m}$ with s even contribute to the splitting matrix. Coupling between nearby multiplets can give some sensitivity to odd-order structure but the coupling is usually weak so that low frequency seismograms are dominantly sensitive only to even-order structure. This fact should be borne in mind when looking at models of aspherical structure derived from low frequency data.

Insight into the way aspherical structure affects the displacement field is gained by considering the spectral decomposition of $\mathbf{H}_k$:

$$ \mathbf{H}_k \mathbf{U}_k = \mathbf{U}_k \mathbf{\Omega}_k \tag{8} $$

Here $\mathbf{U}_k$ is the unitary matrix whose columns are the eigenvectors of $\mathbf{H}_k$ and $\Omega_{mm'}^k = \delta\omega_k^m \delta_{mm'}$ is the diagonal matrix of eigenvalues. Equation (1) can then be written as

$$
\begin{aligned}
s(\mathbf{r},t) &= \mathrm{Re}\left\{ \sum_k \boldsymbol{\sigma}_k^T(\mathbf{r}) e^{i\mathbf{H}_k t} \mathbf{a}_k(\mathbf{r}_0,0) e^{i\omega_k t} \right\} \\
&= \mathrm{Re}\left\{ \sum_k (\boldsymbol{\sigma}_k^T \mathbf{U}_k) e^{i(\boldsymbol{\Omega}_k+\omega_k)t} (\mathbf{U}_k^{-1} \mathbf{a}_k(\mathbf{r}_0,0)) \right\}
\end{aligned} \tag{9}
$$

Equation (9) demonstrates that aspherical structure splits the singlet frequencies within each multiplet: $\omega_k^m = \omega_k + \delta\omega_k^m$. For the RH model, $\omega_k^m = \omega_k + \omega_k(a_k + mb_k + m^2 c_k)$ and $\mathbf{U}_k = \mathbf{I}$. If $\mathbf{U}_k = \mathbf{I}$, the mth singlet frequency is uniquely associated with the mth element of the receiver and excitation vectors. In this case, each envelope function $a_k^m(t)$ in (1) is a pure harmonic time function. Additional aspherical structure further splits the singlet frequencies and perturbs $\mathbf{U}_k$ from $\mathbf{I}$, producing cross-azimuthal coupling which associates more than one element of the receiver and excitation vectors with each singlet frequency. Thus, $a_k^m(t)$ becomes a sum of single harmonics displaying a more complicated temporal behaviour caused by the interchange of energy among azimuthal orders. Since the apparent period of the envelope function is controlled by the singlet frequency perturbations in $\boldsymbol{\Omega}_k$, which are relatively small, $a_k^m(t)$ will be very slowly varying in time.

Provided that a multiplet is approximately uncoupled to any other multiplet, (1) can be used in a variety of ways to model the data. A convenient way to proceed is to linearize equation (1) with respect to the structure coefficients which are themselves linear functionals of aspherical structure (Woodhouse and Giardini, 1985; Ritzwoller et al., 1986). Thus we write

$$
s(\mathbf{r},t) = s_0(\mathbf{r},t) + \sum_{s,t} \frac{\partial s_0(r,t)}{\partial c_s^t} \delta c_s^t + \frac{\partial s_0(\mathbf{r},t)}{\partial \omega_k} \delta\omega_k \tag{10}
$$

Equation (10) is solved iteratively for the δc_s^t in a small frequency band surrounding the mode(s) of interest until a set of c_s^t which adequately predict the spectra are found. These c_s^t may later be used to constrain aspherical structure through inversion of (6). This iterative spectral fitting technique has recently been applied to spectra of resolvably split multiplets from a dataset of recordings from large events (see below). The technique could also be applied to unresolvably split multiplets but we have already seen that rather few parameters are actually required to model the spectra of such multiplets. In fact, for isolated, highly excited fundamental modes, a single resonance function appears to be useful approximation to the observed spectra. To see why this might be so we write equation (1) as

$$
s(\mathbf{r},t) = A(t)\exp(i\omega_k t)
$$

where the real part is understood and

$$
A(t) = \boldsymbol{\sigma}^T \cdot \mathbf{a}(t)
$$

It is straightforward to show that

$$A(t) = A_0 \exp\left[i \int_0^t \frac{\sigma^T \cdot \mathbf{H} \cdot \mathbf{a}(t)}{\sigma^T \cdot \mathbf{a}(t)} \, dt \right] \tag{11}$$

where A_0 is the initial excitation of the multiplet as it would be on a spherical Earth. If we now write

$$\lambda(t) = \frac{\sigma^T \cdot \mathbf{H} \cdot \mathbf{a}(t)}{\sigma^T \cdot \mathbf{a}(t)} \tag{12}$$

we see that $\lambda_0 = \lambda(0)$ is Jordan's location parameter. Equation (1) may now be written

$$s(\mathbf{r},t) = A_0 \exp\left[i \int_0^t \lambda(t) dt \right] \exp(i \omega_k t) \tag{13}$$

If $\lambda(t)$ is only weakly dependent on time, i.e., $\lambda(t) \approx \lambda_0$ then

$$A(t) \approx A_0 \exp(i \lambda_0 t)$$

and (13) becomes

$$s(\mathbf{r},t) \approx A_0 \exp[i(\omega_k + \lambda_0)t] \tag{14}$$

This corresponds to a peak shift in the spectrum which is the asymptotic result of Jordan (1978). Woodhouse and Girnius (1982) give a very elegant analysis of λ_0 and demonstrate that it can be written as a weighted surface integral of aspherical structure. For high l modes, the kernel in the integral is highly oscillatory except in the region of the great circle path joining source and receiver. Thus, in the case of very long wavelength structure, off-path contributions to the integral cancel and λ_0 just becomes sensitive to structure under the great circle path. In fact it is straightforward to show that, in the limit $l \to \infty$ all time derivatives of $\lambda(t)$ are zero so λ is a constant (Dahlen, 1979) and in this limit, the peak shift, λ_0 becomes (Jordan, 1978)

$$\lambda_0 = \frac{\sigma^T \cdot \mathbf{H} \cdot \mathbf{a}(0)}{\sigma^T \cdot \mathbf{a}(0)} \approx \sum_s P_s(0) \sum_t c_s^t \, Y_s^t(\theta, \Phi) \tag{15}$$

where θ, Φ is the location of the pole of the great circle joining the source and receiver. Equation (15) is not very precise in practice, particularly near nodes in the radiation pattern and several authors have suggested improvements. Davis and Henson (1986) and Romanowicz and Roult (1986) show that to $O(1/l)$ where l is the harmonic degree of a multiplet

$$\lambda_0 = A(l) + \frac{B(l)}{l} \tan((l + 1/2)\Delta - \pi/4 + z) \tag{16}$$

$A(l), B(l)$ and z are source dependent but for an isotropic source $A(l)$ is given by (15) and the second term is the desired correction (Δ is epicentral distance). Romanowicz and Roult (1986) claim that the tangent fluctuations can be sufficiently well measured to provide additional constraints on aspherical structure though it is our experience that the results are sensitive to the peak estimation technique. Davis (1987) avoids using any of the

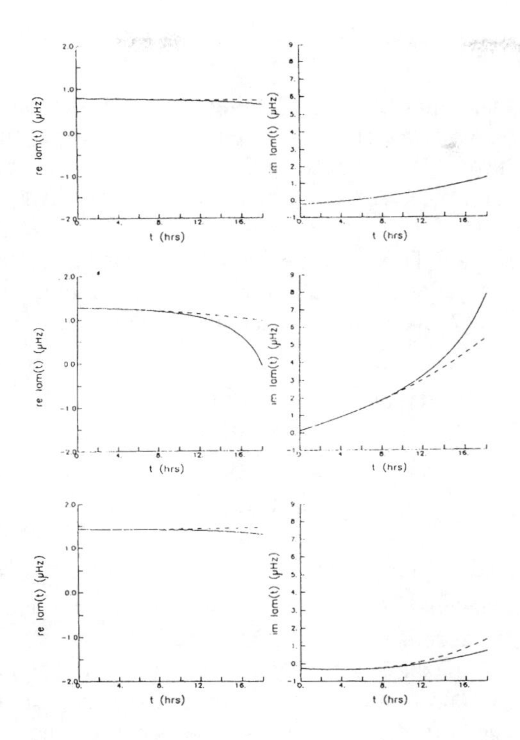

Figure 10. **The generalized location parameter, $\lambda(t)$ for ${}_0S_{23}$ (top), ${}_0S_{24}$ (middle) and ${}_0S_{25}$ (bottom) computed for an event in Tonga and recorded at GUMO. The time duration of these plots is what one would typically use in an analysis of these multiplets. Note that $\lambda(t)$ can have quite a strong time dependence with a significant imaginary part. The dashed line is a cubic approximation to $\lambda(t)$. The computations are performed with model M84A.**

approximations to λ_0 and uses the definition of $\lambda(0)$ (equation (12)) to relate mode peak measurements to aspherical structure. (This procedure was originally suggested by Jordan (1978)). He finds that degree two structure dominates the peak shifting pattern of fundamental spheroidal modes and otherwise only part of the degree 6 structure is reliably determined. Perhaps one of the reasons for this is that, to order $(1/l)$, more terms appear in the expression for the displacement field because λ cannot be regarded as constant to this level of approximation. A better approximation to (13) might therefore be to let λ be a slowly varying function of time. We write

$$\lambda(t) = \lambda_0 + \frac{d\lambda_0}{dt}\, t + \tfrac{1}{2}\, \frac{d^2\lambda_0}{dt^2}\, t^2 + \cdots \tag{17}$$

so (13) becomes

$$s(\mathbf{r},t) = A_0 \exp\left[i(\lambda_0+\omega_k)t + i\,\frac{d\lambda_0}{dt}\,\frac{t^2}{2} + i\,\frac{d^2\lambda_0}{dt^2}\,\frac{t^3}{6} + \cdots \right] \tag{18}$$

That (18) might be a reasonable basis for representing the data is demonstrated in Figure 10 where we plot $\lambda(t)$ along with an approximation to it for some fundamental modes calculated using model M84A of Woodhouse and Dziewonski (1984). Note that if rotation and aspherical anelastic structure are neglected, λ_0 is purely real, $d\lambda_0/dt$ is purely imaginary and so on.

It is not immediately apparent how to relate the time domain representation given above to the apparent frequency and attenuation rate measured when fitting a resonance function to the data. Smith et al. (1987) have found that time averages of $\lambda(t)$ give an adequate representation, i.e.,

$$s(\mathbf{r},t) = A_0 \exp(i(\omega_k+\overline{\Lambda})t) \tag{19}$$

where

$$\overline{\Lambda} = \frac{1}{T}\int_0^T w(t)\lambda(t)dt$$

where $w(t)$ is a normalized weight function and T is record length. The measured variation in $\overline{\Lambda}$ can then be related to the splitting matrix and so the c_s^t can be constrained by both measurements of apparent center frequency and attenuation rate.

The above analysis of uncoupled multiplets is relatively complete but a proper treatment of low frequency data requires that we extend the analysis to coupled modes. Loosely speaking, all we need to do is to redefine the vectors in equation (1) to include more than one multiplet and to include in the splitting matrix blocks which describe the interactions of the multiplets. For example if we couple the k and k' multiplets and write

$$\sigma = \begin{bmatrix} \sigma_k \\ \cdots\cdots \\ \sigma_{k'} \end{bmatrix} \qquad \mathbf{a} = \begin{bmatrix} \mathbf{a}_k \\ \cdots\cdots \\ \mathbf{a}_{k'} \end{bmatrix}$$

and

$$\mathbf{H} = \begin{bmatrix} \mathbf{H}_{kk} & \vdots & \mathbf{H}_{kk'} \\ \cdots\cdots\cdots & \vdots & \cdots\cdots\cdots \\ \mathbf{H}_{k'k} & \vdots & \mathbf{H}_{k'k'} \end{bmatrix}$$

then all the foregoing analysis remains valid. Woodhouse (1980) has provided expressions for the interaction blocks $\mathbf{H}_{kk'}$ in the above formula. These interaction blocks can be written in a similar form to (5) so that the spectrum depends upon an extended set of structure coefficients which can be estimated by iterative spectral fitting. The interaction blocks may have terms depending upon odd-order structure so that the seismograms of coupled multiplets can be weakly dependent on odd-order structure. Recently, expressions for the interaction blocks in $\mathbf{H}$ have been extended to include anisotropy and it is now possible to calculate the effect of a small general anisotropic perturbation to a transversely isotropic reference Earth (Mochizuki, 1986; Tanimoto, 1986b; I. H. Henson, personal communication).

Synthetic seismogram construction including coupling between multiplets has been treated in detail by Tanimoto and Bolt (1983), Morris et al. (1984), and Park and Gilbert (1986). The Galerkin calculation of Park and Gilbert allows probably the most complete treatment of seismic motion on an aspherical Earth and has been used to compare the predictions of some current Earth models with the data (Park, 1986, and below). This calculation is capable of properly coupling modes with very different attenuation rates. This effect is particularly important when coupling spheroidal and toroidal modes together as was first demonstrated by Woodhouse (1980). Along-branch coupling involves modes with very similar Qs so that the Galerkin technique can be replaced with the computationally faster Rayleigh-Ritz technique. The similarity of the radial eigenfunctions of adjacent modes along a branch also allows analytical approximations to be made so that asymptotic forms for coupled mode seismograms can be obtained (Park, 1987; Romanowicz, 1987).

4. Modeling long period data

At frequencies lower than about 6 mHz existing spherically averaged models of elastic structure (e.g., 1066A, Gilbert and Dziewonski, 1975; PREM, Dziewonski and Anderson, 1981), and attenuation structure (Masters and Gilbert, 1983; Masters et al., 1983) are sufficiently accurate to allow retrieval of a reliable source mechanism. We have found that, at these frequencies, the source can be adequately characterized by the six elements of the "seismic moment tensor" and a source time function. This latter feature is necessary to account for the fact that the low frequency centroid time of the event is not necessarily the same as the origin time deduced from body wave arrivals (Backus, 1977; Dziewonski and Woodhouse, 1983). The exact form of the source time function is not important as far as the quality of the fit to the data in this frequency band is concerned. We have also found that mislocation of the spatial centroid of the event does not significantly degrade the fit to the data in this frequency band though it is possible to find an optimum location (Dziewonski et al., 1981). Source retrieval is, in principle, a straightforward operation (e.g., Buland and Gilbert, 1976) though linear fitting for the elements of the seismic moment tensor does not always succeed. The main problem is phase mismatching due to the presence of aspherical structure leading to a bias toward low moment. This bias can be reduced by modifying the source orientation to predict, as best as possible, the observed distribution of power at all stations recording the event. The moment tensors used to calculate the synthetic seismograms of Figure 2 were obtained with this technique.

Despite the perturbing influence of aspherical structure, our fit to the long period data is sufficiently good that we are encouraged to use the multiplet stacking and stripping techniques described by Gilbert and Dziewonski (1975) to improve our estimates of multiplet degenerate frequencies. Such measurements can then be used to improve our models of spherically averaged structure (the "terrestrial monopole"). We rewrite equation (1) in the frequency domain, i.e.,

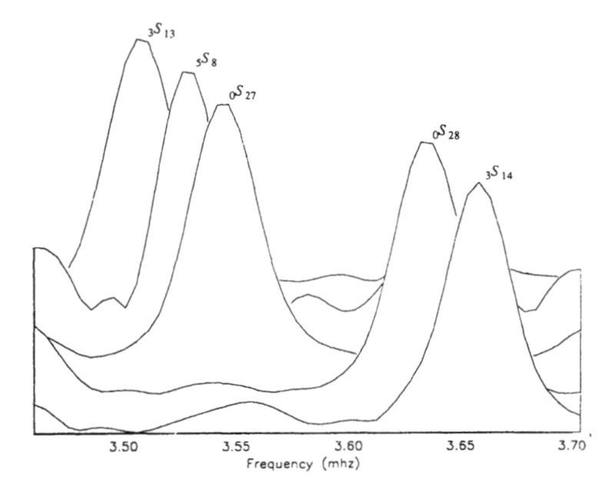

Figure 11. The results of stripping 800 recordings in a small frequency band which includes the fundamental spheroidal modes $_0S_{27}$ and $_0S_{28}$. Each "row" of the figure is the amplitude spectrum of a single $c_k(\omega)$. Note that overtones such as $_3S_{14}$ are clearly separated from the highly excited fundamental modes.

$$s_j(\omega) = \sum_{k=1}^{K} A_{jk}(\omega)\, C_k(\omega) \tag{20}$$

where $C_k(\omega)$ is the resonance function of the kth mode (i.e., the spectrum of a decaying cosinusoid). A_{jk} is the complex amplitude of the kth mode recorded at the jth station which may be estimated if we have a model of the source and a provisional model of Earth structure. Equation (20) can be solved for the $C_k(\omega)$ in a small frequency band incorporating K modes if we have K or more recordings. Typically we will have several hundred recordings so the multiplet resonance functions can be recovered (Figure 11). Frequency and attenuation can now be estimated from the isolated resonance functions.

Multiplet stacking and stripping only gives unbiased degenerate frequencies when the multiplet in question has a relatively uniform frequency distribution of singlets. When a multiplet has a highly nonuniform distribution of singlets or its singlets are nonuniformly excited, stripping fails by giving a peak at a subset of the singlets which may have a very different frequency from the true degenerate frequency. Diagnosis of such a failure may be difficult but a common symptom is for the strip to have multiple peaks. For example, multiplet stacking and stripping is incapable of producing accurate results for broadly split modes such as $_1S_4$. The situation is even worse for anomalously split low l, high Q modes such as $_{10}S_2$ where multiplet stripping produces a single peak at an extreme end of the multiplet leading to systematically biased degenerate frequencies.

Ritzwoller et al. (1986, 1987) demonstrate that accurate degenerate frequencies for broadly split multiplets can be found by either stripping for individual singlets or by directly modeling the effect of aspherical structure on the data spectra. Equation (20) can be modified in a straightforward way to strip for individual singlets within a multiplet provided that a reasonably accurate estimate of the singlet eigenvectors is known. For example, it is reasonable to suppose that, for rotationally dominated modes like $_1S_4$, the splitting matrix is diagonally dominant. If this is true, $\mathbf{U}$ in equation (9) is the unit matrix so that (9) can be written in exactly the same form as (1) and we are able to calculate the

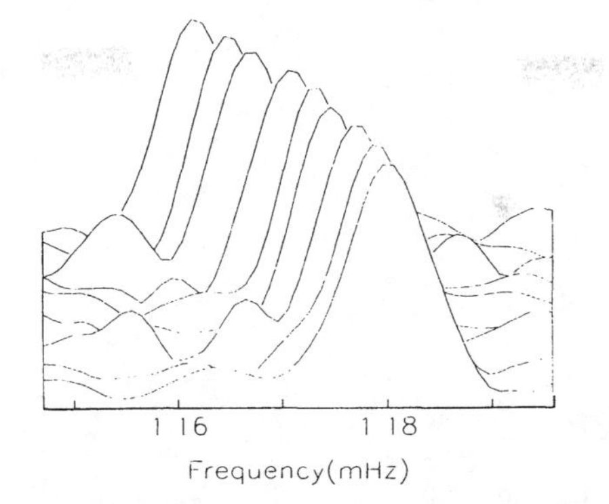

Figure 12. This figure has the same format as Figure 11 but now each row is the amplitude spectrum of a singlet of ${}_1S_4$. All nine singlets are recovered and follow a quadratic in azimuthal order close to that predicted for a rotating, hydrostatic Earth model.

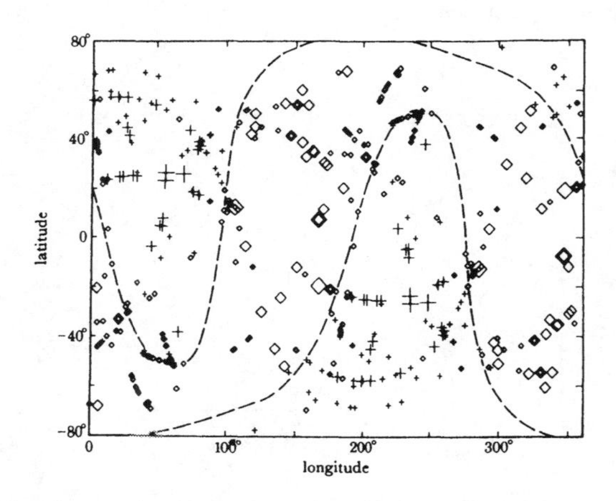

Figure 13. The effect of aspherical structure on center frequency can be seen by plotting the shift in frequency as a function of the pole position of the great circle joining source and receiver. Each symbol is plotted at a pole position: a + corresponds to a positive frequency shift and a ◇ corresponds to a negative frequency shift. The size of the symbol is indicative of the magnitude of the shift: the smallest symbols correspond to a 0–0.1% shift and the largest to a 0.3–0.4% shift. A degree-two spherical harmonic pattern accounts for most of the structure in the observations; the nodal lines of this pattern are shown by the dashed lines. This example is a combination of measurements for the modes ${}_0S_{21} - {}_0S_{23}$.

excitations of individual singlets. We may then stack or strip recordings to enhance desired singlets. It turns out that we are able to recover all the singlet resonance functions of ${}_1S_4$ (Figure 12) and so estimate a precise degenerate frequency (i.e., by employing the diagonal sum rule [Gilbert, 1971]).

Even when singlets within a multiplet are uniformly distributed, multiplet stripping can still give biased measurements due to inadequate geographical sampling of the Earth. The exact location of the center frequency of a recovered resonance function will depend upon

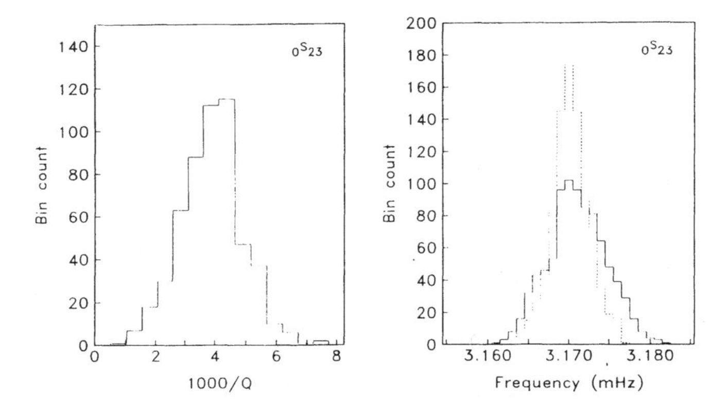

Figure 14. Histograms of 1000/Q and center frequency measurements obtained by fitting a decaying cosinusoid model to many spectra of the multiplet $_0S_{23}$. The effect of removing the frequency shifts due to a dominant degree 2 elastic aspherical structure is shown by the dashed line in the right-hand figure. No large scale structure is visible in the attenuation data.

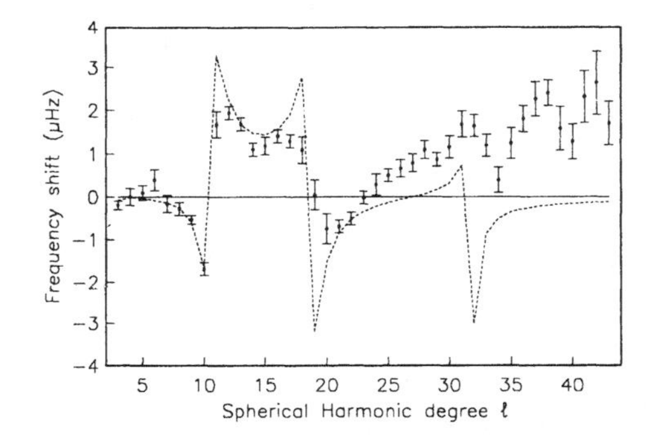

Figure 15. Observed degenerate frequencies of fundamental spheroidal modes presented as frequency shifts from the predictions of model 1066A. The discontinuous nature of the observations is mimicked by a calculation of the perturbing effect of Coriolis coupling to nearby toroidal modes (dashed line). Model 1066A systematically misfits the observations for $l \gtrsim 25$.

the choice of records employed in the analysis. If a mode is sufficiently well isolated and of sufficiently high amplitude that its properties can be measured from individual recordings, we may analyze the geographical peak shifting effect directly. This was the approach taken by Masters et al. (1982) and, more recently (and with less assumptions), by Davis (1987). Equation (15) suggests that, if the asymptotic limit is valid, the magnitude of a peak frequency shift should be a smooth function of the pole position of the great circle joining the source and receiver. If we plot the raw measurements in this way (Figure 13), we find that a simple pattern emerges from the data which is dominantly of harmonic degree two. This pattern persists for fundamental spheroidal modes over a large range of l though by $l \geq 43$ (i.e., periods shorter than 200 secs) the degree 2 pattern is no longer as dominant. It should be noted that, if we use the asymptotic formula (15) to interpret peak shift data we can only reliably determine the large degree 2 structure and even this is subject to some bias. The signal from higher order structure is sufficiently small (at least at periods longer than about 200 seconds) that we must use the full expression for λ which includes off-path propagation effects. It is unlikely that the great circle approximation improves at shorter periods. Histograms of the results of single station peak analysis are shown in Figure 14.

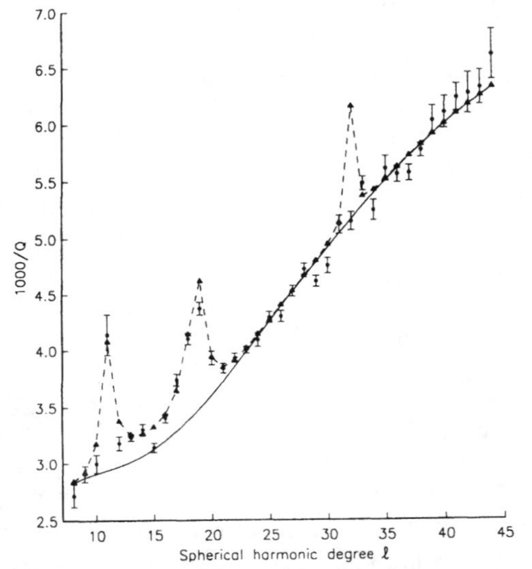

Figure 16. **Observed mean attenuation rates of fundamental spheroidal modes. The dashed line again shows the predicted effect of Coriolis coupling and follows the data for $l \lesssim 25$. The predicted strong coupling at $l \approx 32$ is absent in the data (see text).**

The spread in center frequency estimates is dominantly controlled by large-scale aspherical structure. Removal of the effect of a dominant degree 2 elastic structure greatly reduces the variance in the frequency observations. The spread in apparent attenuation measurements is too large to be ascribed to aspherical attenuation structure and is dominantly due to interference effects. Davis (1985) shows that the observed spread in apparent attenuation can be generated solely by elastic aspherical structure and it is likely that true aspherical attenuation structure constitutes only a small part of the signal. Davis (1987) reports that part of the degree 6 structure can also be reliably determined from peak shift data (as well as the degree 2 structure) provided the full expression for $\lambda(0)$ is used. He also notes that there is still signal remaining in the peak shift data which may be partly due to the fact that $\lambda(0)$ is not precisely the same as the observed peak shift (see equation (18) for a better approximation). The analysis also neglects multiplet/multiplet coupling which certainly contributes to the observed signal where Coriolis coupling is known to occur but may also contribute because of along-branch coupling (see below).

Accurate estimates of fundamental spheroidal mode degenerate frequencies from the analysis of peak shift patterns are shown in Figure 15. The frequencies are not a smooth function of harmonic degree and the "tears" observed at $l = 11$ and 19 are due to strong

Coriolis coupling to ${}_0T_{12}$ and ${}_0T_{20}$ respectively. A coupling calculation using the Galerkin technique described by Park and Gilbert (1986) shows that, unless coupling is extremely strong, the singlets of a coupled ${}_0S_l - {}_0T_{l+1}$ pair fall into two groups, one with dominantly spheroidal characteristics, the other with dominantly toroidal characteristics. The frequencies of the singlets within each group are repelled so that, on average, the multiplets appear to be further apart than when they are uncoupled. The attenuations, on the other hand, are averaged so that coupled singlets which are dominantly spheroidal are more quickly attenuated than uncoupled spheroidal mode singlets. This latter feature is illustrated in Figure 16 where mean attenuation rates for spheroidal modes (or the dominantly spheroidal part of a spheroidal/toroidal coupled pair) are shown along with a prediction of the effect of coupling. The strong coupling calculated at harmonic degree 32 occurs because ${}_0S_{32}$ and ${}_0T_{31}$ have frequencies separated by only 0.6 μHz in model 1066A of Gilbert and Dziewonski (1975). This predicted strong coupling is a feature of most modern models but is absent in the data because ${}_0S_{32}$ and ${}_0T_{31}$ are actually separated by about 10 μHz in degenerate frequency which is sufficient to significantly reduce the strength of coupling. This sensitivity of coupling calculations to the frequency separation of multiplets highlights the importance of having more accurate spherically averaged Earth models than are currently available.

It is difficult to extend the peak shift analysis to higher frequencies unless measurements of the center frequencies of several adjacent modes are made simultaneously. Another way to proceed is to use surface wave dispersion analysis. This has been the approach of Nakanishi and Anderson (1983, 1984) who have estimated phase velocity from Rayleigh and Love wave packets which have traveled through a great circle and have also made single station phase and group velocity measurements between 100- and 300-sec periods. These authors have performed a spherical harmonic analysis of their data assuming that the observed phase shifts represent line integral averages over the minor arc from source to receiver and over great circles. Inspection of their results shows that odd-order structure is poorly resolved by the data and only a few of their spherical harmonic coefficients are significant at the 95% confidence level. A formal resolution analysis by Tanimoto (1985) demonstrates that this dataset constrains little of the geographical variation in phase velocity at the 95% confidence level. This result is a little disappointing, especially if one remembers that no off-path propagation has been accounted for in the analysis and that the assessment of data errors is quite subjective. The neglect of off-path propagation has been justified by invoking Fermat's principle which tells us that errors in the phase will be of second order if we ignore path perturbations and great circle wave propagation has also been assumed by Woodhouse and Dziewonski (1984) in their waveform fitting experiment. They showed that the difference in phase anomaly accumulated along a great circle and minor arc can be represented by a fictitious shift in the epicentral distance. This clever observation allowed them to efficiently construct differential seismograms for both even- and odd-order structure components. Their differential seismograms included overtone contributions but the dominance of the fundamental modes in the data means that this is largely the signal that is being fit.

The availability of a model like M84A allows us to test the various approximations that are used in modeling aspherical structure. Park (1986) has used this model to construct synthetic seismograms using various coupling schemes. He finds that the addition of

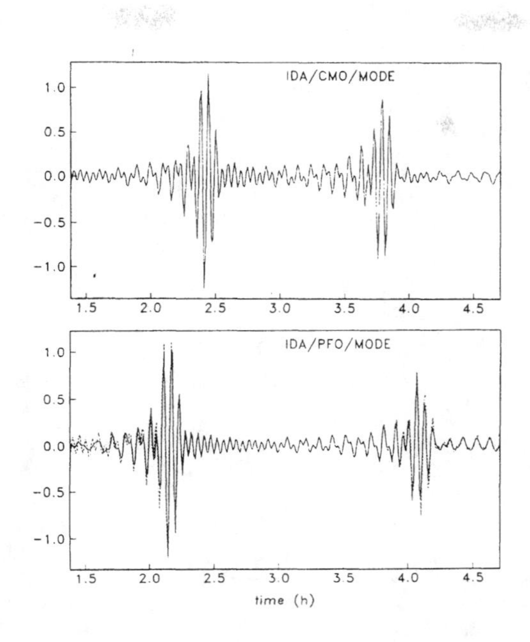

Figure 17. Comparisons of along-branch fundamental spheroidal mode coupled synthetic seismograms (solid line) with uncoupled synthetic seismograms (dashed line). All fundamental modes with frequencies less than 6 mHz are included and nearest neighbor coupling is calculated using model M84A for the event in Iran, 1978/259. The difference between the synthetics is the effect of odd-order structure. In most cases the differences are extremely small but occasionally, amplitude and phase anomalies are visible (e.g., PFO).

along-branch coupling (so giving sensitivity to odd-order structure) causes very small additional perturbations to the seismograms at long periods. In fact the phase shift induced by odd-order structure is of the same order of magnitude as the error in the phase shift incurred by neglecting off-path propagation. We illustrate the additional signal from the odd-order structure of M84A by comparing uncoupled and along-branch coupled fundamental mode synthetics in Figures 17 and 18. The calculations include only nearest-neighbor fundamental spheroidal mode coupling up to a frequency of 6 mHz. The synthetics are computed for an Iranian earthquake (1978/259) which was extremely shallow and so is dominated by fundamental mode energy. A typical comparison is given by the station CMO with almost no difference in the time or frequency domain between uncoupled and coupled synthetics. One of the largest computed differences is shown by station PFO in which R_2 is phase shifted by several seconds and the amplitude spectra are strongly perturbed. Comparison of such coupled mode synthetic seismograms with the data demonstrates that much of the signal is not modeled. In particular, mode amplitudes (and surface wave amplitude anomalies in the time domain) are not well fit. The general impression is that the current models underpredict the magnitude of the anomalies and higher-order structure will probably be required to match the data. Another conclusion that

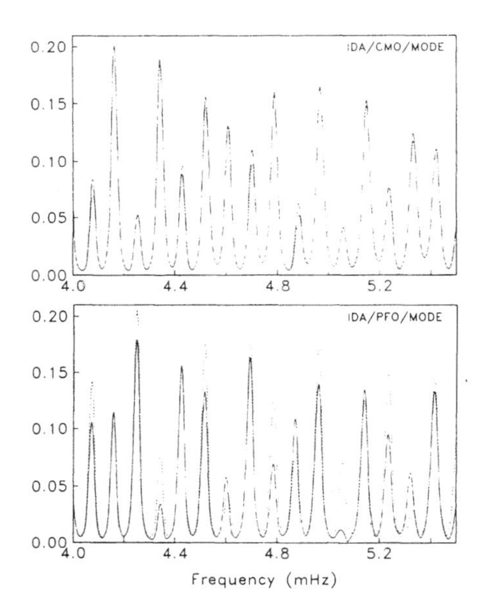

Figure 18. Comparisons of Hanning tapered spectra of 25-hour recordings with and without the effects of along-branch coupling (see Figure 17 caption for details). Most spectral comparisons are like CMO where large differences are only seen near excitation nodes. PFO shows quite large amplitude perturbations.

we can draw is that the current models predict basically no effect of odd-order structure via along-branch coupling on low-frequency peak shifts or apparent attenuation rates (Figure 19) so it is unlikely that this will account for our present inability to model these data.

Clearly, we do not yet have an aspherical model which gives a reliable prediction of surface wave propagation. In any case, the models we have been discussing are dominantly controlled by fundamental modes which are only sensitive to structure in the upper mantle. The depth resolution of fundamental modes is notoriously poor and a formal analysis by Tanimoto (1986a) shows that vertical resolving lengths are of the order of two-hundred kilometers or more. To improve this we must add overtone information or, if we are performing waveform fits, the overtone wave packets must be given additional weight to compete with the larger amplitude fundamental modes (e.g., Tanimoto, 1987). As noted earlier in this paper we do have some observations of split overtone multiplets and we now discuss the current interpretation of these data. Ritzwoller et al. (1987) and Giardini et al. (1987) have both analyzed a similar set of about 40 multiplets and recovered the c_s^t structure coefficients for harmonic degrees $s = 2$ and 4. An example of the ability of iterative spectral fitting to fit the data is shown in Figure 20. Roughly ten of these multiplets are anomalously split, all of which sample the core. For most multiplets, Ritzwoller et al. (1987) find that only the c_2^t are significant and simple aspherical structures in the mantle are capable of explaining most of the observations. What these models do not explain are the c_2^0 coefficients of the anomalously split multiplets. We require some perturbation in core structure to fit these data. Improvements can be achieved by the inclusion of any of the following structures: isotropic perturbations in the inner or outer

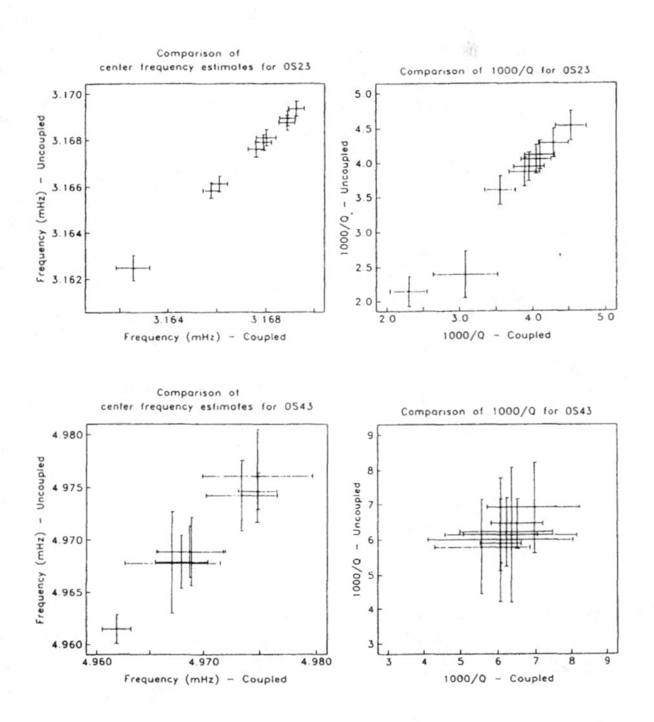

Figure 19. A comparison of center frequency and attenuation estimates for ${}_0S_{23}$ (upper panels) and ${}_0S_{43}$ (lower panels) made from synthetic seismograms with and without along-branch coupling (see Figure 17 caption). These results indicate that this type of coupling has little effect on such estimates.

core, perturbations to the boundaries of the fluid outer core and also, apparently, anisotropic perturbations to the inner core (Woodhouse et al., 1986). This last structure is the only one on this list which appears to allow the very anomalously split but poorly excited multiplet ${}_3S_2$ to be fit. However, the amount of anomalous structure in these models is extremely large and a definitive solution has yet to be found.

5. Conclusions

Aspherical Earth structure manifests itself in long period seismic data in many ways through the splitting and coupling of multiplets. Nearly all research to date has concentrated on either the highly excited fundamental modes or on low harmonic degree, high Q overtones. Fundamental modes show peak shifting and anomalous attenuation rates in the frequency domain and phase perturbation and amplitude anomalies of surface wave

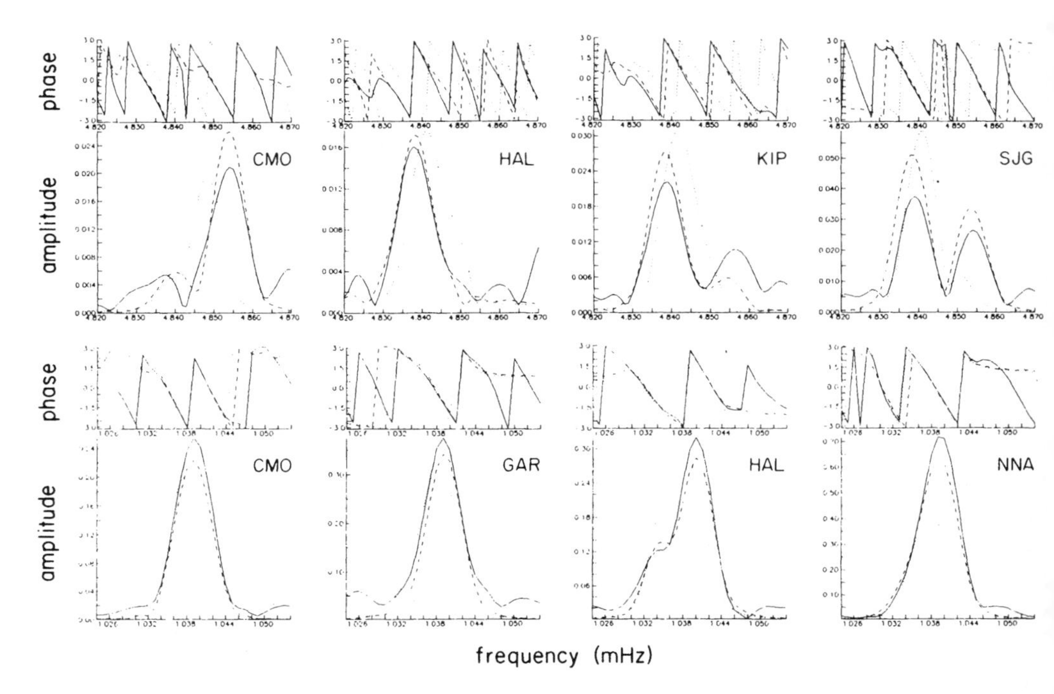

Figure 20. Amplitude and phase spectra showing the fit to the data (solid line) of a rotating hydrostatic Earth model (dotted line) and of a model which includes c_2^t and c_4^t coefficients (dashed line). The rotating hydrostatic Earth model is incapable of fitting the data (note that agreement in phase can only occur when peaks are above the noise). The upper figures show spectra of the mode ${}_{13}S_2$ computed using 60-hour recordings of the Banda Sea event (1982/173). The lower figures show spectra of the mode ${}_0S_6$ computed using 90-hour recordings of the Sumbawa event (1977/231). All time series were recorded on the IDA array.

packets in the time domain. Also clearly visible in the frequency band 1.5–3 mHz is Coriolis coupling between spheroidal and toroidal modes. Peak shifting measurements are mainly restricted to fundamental modes of periods longer than 200 sec because of the difficulty of identifying isolated multiplets on single recordings at shorter periods. The peak shifting pattern is dominated by a degree 2 pattern with some resolvable degree 6 structure. Current analyses of the peak shift data include nonasymptotic effects though neglect multiplet/multiplet coupling. Synthetic seismogram experiments indicate that along-branch coupling has little effect on peak frequency measurements. Surface wave phase velocity perturbations have, until now, been interpreted in a pure-path framework where off-path propagation is ignored. Synthetic experiments indicate that the error induced by this approximation can rival the weak signal from odd-order structure so that the odd-order part of the model may be suspect. Woodhouse and Dziewonski's (1984) models predict that the power in degrees 2 and 6 structure exceeds that in degrees 4 and 8 showing some consistency with the peak shifting results though the agreement of the structure coefficients is quite poor. Agreement of the peak shifting and waveform fitting experiments with the dispersion models of Nakanishi and Anderson (1983, 1984) is also poor. This is particularly true at long periods suggesting that the dispersion analyses are unreliable at periods longer than about 250 sec.

The lack of agreement of results described above is probably not surprising given the fact that these models are the first real attempts at constraining global aspherical structure. These models provide a first chance to construct accurate synthetic seismograms and see if classical approximations (such as the pure path approximation) are useful. The results indicate that more accurate theoretical formulations must be used in the inversion process if we are to recover reliable models of aspherical structure. It is probably also true that we will need to be more sophisticated in our analysis of surface wave dispersion if we are to get precise enough data to usefully constrain large-scale structure. In particular, ways of reliably assigning observational errors to dispersion data are needed so that we can have confidence in the resulting models.

The analysis of overtones, including that of the anomalously split multiplets, has provided some reliable estimates of structure coefficients, c_s^t, which in principle are sensitive to structure throughout the Earth. Unfortunately there are insufficient observations to independently constrain deep Earth structure though some progress appears to have been made by combining these data with constraints from the ISC dataset (Giardini et al., 1987). One thing is clear, the retrieval of aspherical Earth structure from low frequency seismic data will remain a fertile area of research for many years to come.

6. Acknowledgments

We gratefully acknowledge the help of Freeman Gilbert, Ivan Henson, Mark Smith, and Rudolf Widmer. Rudy's help in making pictures went above and beyond the call of duty. We also wish to thank those people who responded to our request for information. The revised scope of this paper made it impossible to properly survey the field and we hope to do this in a later contribution. Thanks are due to the operators of the IDA, GDSN and GEOSCOPE networks for their continued efforts to maintain the high quality of digital data and a special thanks to Cecil H. and Ida Green for their past and continued support of the IDA network. This research was supported by National Science Foundation grants EAR-84-10369 and EAR-84-18471.

References

Agnew, D., J. Berger, R. Buland, W. Farrell, and F. Gilbert (1976). International deployment of accelerometers: A network for very long period seismology, *Eos*, **57**, 180-188.

Agnew, D. C., J. Berger, W. E. Farrell, J. F. Gilbert, G. Masters, and D. Miller (1986). Project IDA: A Decade in review, *Eos*, **67**, 203-212.

Backus, G. E. (1977). Seismic sources with observable glut moments of spatial degree two, *Geophys. J. R. Astron. Soc.*, **51**, 27-45.

Backus, G. E., and F. Gilbert (1961). Rotational splitting of the free oscillations of the earth, *Proc. Natl. Acad. Sci.*, **47**, 362-371.

Benioff, H., F. Press, and S. W.Smith (1961). Excitation of the free oscillations of the earth by earthquakes, *J. Geophys. Res.*, **66**, 605-619.

Buland, R., and F. Gilbert (1976). Matched filtering for the seismic moment tensor, *Geophys. Res. Lttrs.*, **3**, 205-

206.

Dahlen, F. A. (1968). The normal modes of a rotating, elliptical earth, *Geophys. J. R. Astron. Soc.*, **16**, 329-367.

Dahlen, F. A. (1969). The normal modes of a rotating, elliptical earth, II, Near resonance multiplet coupling, *Geophys, J. R. Astron. Soc.*, **18**, 397-436.

Dahlen, F. A. (1979). The spectra of unresolved split normal mode multiplets, *Geophys. J. R. Astron. Soc.*, **58**, 1-33.

Dahlen, F. A. (1982). The effect of data windows on the estimation of free oscillation parameters, *Geophys. J. R. Astron. Soc.*, **69**, 537-549.

Dahlen, F. A., and R. V. Sailor (1979). Rotational and elliptical splitting of the free oscillations of the earth, *Geophys. J. R. Astron. Soc.*, **58**, 609-623.

Davis, J. P. (1985). Variation in apparent attenuation of the earth's normal modes due to lateral heterogeneity, *Geophys. Res. Lett.*, **12**, 141-143.

Davis, J. P. (1987). Local eigenfrequency and its uncertainty inferred from fundamental spheroidal mode frequency shifts, *Geophys. J. R. Astron. Soc.*, **88**, 693-722.

Davis, J. P., and I. H. Henson (1986). Validity of the great circular average approximation for inversion of normal mode measurements, *Geophys. J. R. Astron. Soc.*, **85**, 69-92.

Dziewonski, A. M., and D. Anderson (1981). Preliminary reference Earth model, *Phys. Earth Planet. Inter.*, **25**, 297-356.

Dziewonski, A. M., and J. Steim (1982). Dispersion and attenuation of mantle waves through wave-form inversion, *Geophys. J. R. Astron. Soc.*, **70**, 503-527.

Dziewonski, A. M., T. A. Chou, and J. H. Woodhouse (1981). Determination of earthquake source parameters from waveform data for studies of global and regional seismicity, *J. Geophys. Res.*, **36**, 2825-2831.

Dziewonski, A. M., and J. H. Woodhouse (1983). Studies of the seismic source using normal mode theory, in *Earthquakes: Observation, Theory and Interpretation*, pp45-137. edited by E. Boschi and H. Kanamori.

Edmonds, A. R. (1960). *Angular Momentum and Quantum Mechanics*, Princeton University Press, Princeton, NJ.

Fichler, C., M. Grünewald, and W. Zürn (1986). Observation of spheroidal-toroidal mode coupling, *Ann. Geoph.*, **4B**, 251-260.

Giardini, D., X.-D. Li, and J. H. Woodhouse (1987). Three-dimensional structure of the earth from splitting in free oscillation spectra, *Nature*, **325**, 405-409.

Gilbert, F. (1970). Excitation of the normal modes of the Earth by earthquake sources, research note, *Geophys. J. R. Astron. Soc.*, **22**, 223-226.

Gilbert, F. (1971). The diagonal sum rule and averaged eigenfrequencies, *Geophys. J. R. Astron. Soc.*, **23**, 119-123.

Gilbert, F., and A. Dziewonski (1975). An application of normal mode theory to the retrieval of structural parameters and source mechanisms from seismic spectra, *Philos. Trans. R. Soc. London Ser. A*, **278**, 187-269.

Harris, F. (1978) On the use of windows for harmonic analysis with the discrete Fourier transform, *Proc. IEEE*, **66**, 51-83.

Jenkins, G. M., and D. G. Watts (1968). *Spectral Analysis and its Applications*, Holden-Day, pp. 525.

Jordan, T. H. (1978). A procedure for estimating lateral variations from low-frequency eigenspectra data, *Geophys. J. R. Astron. Soc.*, **52**, 441-455.

Lay, T., and H. Kanamori (1985). Geometric effects of global lateral heterogeneity on long-period surface wave propagation, *J. Geophys. Res.*, **90**, 605-621.

Lindberg, C. L. (1986). Multiple taper spectral analysis of terrestrial free oscillations, Ph.D. Thesis, Univ. of California, San Diego.

Luh, P. C. (1973). Free oscillations of the laterally inhomogeneous earth: Quasi degenerate multiplet coupling, *Geophys. J. R. Astron. Soc.*, **32**, 187-202.

Luh, P. C. (1974). The normal modes of the rotating self gravitating inhomogeneous earth, *Geophys. J. R. Astron. Soc.*, **38**, 187-224.

Masters, G., T. H. Jordan, P. G. Silver, and F. Gilbert (1982). Aspherical earth structure from fundamental spheroidal-mode data, *Nature*, **298**, 609-613.

Masters, G., J. Park, and F. Gilbert (1983). Observations of coupled spheroidal and toroidal modes, *J. Geophys. Res.* **88**, 10,285-10,298.

Masters, G., and F. Gilbert (1983). Attenuation in the earth at low frequencies, *Philos. Trans. R. Soc. London Ser. A*, **308**, 479-522.

Morris, S. P., H. Kawakatsu, R. J. Geller, and S. Tsuboi (1984). A calculation of the normal modes of realistic laterally heterogeneous, anelastic earthmodels at 250 sec period, *Eos Trans AGU*, **65**, 1002.

Nakanishi, I. (1979). Phase velocity and Q of mantle Rayleigh waves, *Geophys. J. Roy. Astron. Soc.*, **58**, 35-59.

Nakanishi, I., and D. Anderson (1983). Measurements of mantle wave velocities and inversion for lateral heterogeneity and anisotropy. I. Analysis of great-circle phase velocities, *J. Geophys. Res.*, **88**, 10-267.

Nakanishi, I., and D. Anderson (1984). Measurements of mantle wave velocities and inversion for lateral heterogeneity and anisotropy. II. Analysis by the singlet station method, *Geophys. J. R. Astron. Soc.*, **78**, 573-617.

Nataf, H.-C., I. Nakanishi, and D. L. Anderson (1986). Measurements of mantle wave velocities and inversion for lateral heterogeneities and anisotropy, 3. Inversion, *J. Geophys. Res.*, **91**, 7261-7307.

Ness, N. F, C. J. Harrison, and L. J. Slichter (1961). Observations of the free oscillations of the earth, *J. Geophys. Res.*, **66**, 621-629.

Niazi, M., and H. Kanamori (1981). Source parameters of 1978 Tabas and 1979 Quainat, Iran, earthquakes from long period surface waves, *Bull. Seismol. Soc. Am.*, **71**, 1201-1213.

Park, J. (1986). Synthetic seismograms from coupled free oscillations: The effects of lateral structure and rotation, *J. Geophys. Res.*, **91**, 6441-6464.

Park, J. (1987). Asymptotic coupled-mode expressions for multiplet amplitude anomalies and frequency shifts on a laterally heterogeneous Earth, *Geophys. J. R. Astron. Soc.* (in press).

Park, J., and F. Gilbert (1986). Coupled free oscillations of an aspherical dissipative rotation earth: Galerkin theory, *J. Geophys. Res.*, **91**, 7241-7260.

Pekeris, C. L., A. Alterman, and M. Jarosch (1961). Rotational multiplets in the spectrum of the earth, *Phys. Rev.*, **122**, 1692-1700.

Peterson, J., C. R. Hutt, and L. G. Holkomb (1980). Test and calibration of the SRO observations, U.S.G.S. report.

Ritzwoller, M., G. Masters, and F. Gilbert (1986). Observations of anomalous splitting and their interpretation in terms of aspherical structure, *J. Geophys. Res.*, **91**, 10,203-10,228.

Ritzwoller, M., G. Masters, and F. Gilbert (1987). Constraining aspherical structure with low frequency interaction coefficients: Application to uncoupled multiplets, *J. Geophys. Res.*, in review.

Romanowicz, B. (1987). Multiplet-multiplet coupling due to lateral heterogeneity: asymptotic effects on the amplitude and frequency of the Earth's normal modes, *Geophys. J. R. Astron. Soc.* (submitted).

Romanowicz, B., and G. Roult (1986). First order asymptotics for the eigenfrequencies of the Earth and application to the retrieval of large scale lateral variations of structure, *Geophys. J. Roy. Astron. Soc.*, **87**, 209-239.

Romanowicz, B., M. Cara, J. F. Fels, and G. Roult (1984). GEOSCOPE; a French initiative in long period three component global seismic networks, *Eos Trans. AGU*, **65**, 753-756.

Schulten, K., and R. Gordon (1975). Exact recursive evaluation of $3-j$ and $6-j$ coefficients for quantum mechanical coupling of angular momenta, *J. Math. Phys.*, **16**, 1961-1870.

Slepian, D. (1978). Prolate spheroidal wave functions, Fourier analysis, and uncertainty. V: The discrete case, *Bill Systems Tech. J.*, **57**, 1371-1430.

Smith, M. F., G. Masters, and M. Ritzwoller (1987). Constraining aspherical structure with normal mode frequency and attenuation measurements, *Eos Trans. AGU*, **68**, 358.

Tanimoto, T. (1985). The Backus-Gilbert approach to the three-dimensional structure in the upper mantle, I., Lateral variation of surface wave phase velocity with its error and resolution, *Geophys. J. R. Astron. Soc.*, **82**, 105-124.

Tanimoto, T. (1986a). The Backus-Gilbert approach to the three-dimensional structure in the upper mantle, II, SH and SV velocity, *Geophys. J. R. Astron. Soc.*, **84**, 49-70.

Tanimoto, T. (1986b). Free oscillations in a slightly anisotropic earth, *Geophys. J. R. Astron. Soc.*, **87**, 493-517.

Tanimoto, T. (1987). The three-dimensional shear wave structure in the mantle by overtone waveform inversion — 1. Radial seismogram inversion, *Geophys. J. R. Astron. Soc.*, **89**, 713-740.

Tanimoto, T., and B. A. Bolt (1983). Coupling of torsional modes in the earth, *Geophys. J. R. Astron. Soc.*, **74**, 83-96.

Thomson, D. J. (1982). Spectrum estimation and harmonic analysis, *IEEE Proc.*, **70**, 1055-1096.

Wielandt, E., and G. Streckheisen (1982). The leaf spring seismometer: design and performance, *Bull. Seismol. Soc. Am.*, **72**, 2349-2368.

Woodhouse, J. H. (1980). The coupling and attenuation of nearly resonant multiplets in the earth's free oscillation spectrum, *Geophys. J. R. Astron. Soc.*, **61**, 261-283.

Woodhouse, J. H., and F. A. Dahlen (1978). The effect of a general aspherical perturbation on the free oscillations of the earth, *Geophys. J. R. Astron. Soc.*, **53**, 335-354.

Woodhouse, J. H., and A. M. Dziewonski (1984). Mapping of the upper mantle: three-dimensional modeling of earth structure by inversion of seismic waveforms, *J.Geophys. Res.*, **89**, 5953-5986.

Woodhouse, J. H., and D. Giardini (1985). Inversion for the splitting function of isolated low order normal mode multiplets, *Eos Trans. AGU*, **66**, 300.

Woodhouse, J. H., and T. P. Girnius (1982). Surface waves and free oscillations in a regionalized earth model, *Geophys. J. R. Astron. Soc.*, **68**, 653-673.

Woodhouse, J. H., and Y. K. Wong (1986). Amplitude, phase and ray anomalies of mantle waves, *Geophys. J. Roy. Astron. Soc.*, **87**, 753-773.

Woodhouse, J. H., D. Giardini, and X.-D. Li (1986). Evidence for inner core anisotropy from free oscillations, *Geophys. Res. Lett.*, **13**, 1549-1552.

G. MASTERS and M. RITZWOLLER, Institute of Geophysics and Planetary Physics, University of California, San Diego, La Jolla CA 92093, USA.

Chapter 2

Free-oscillation coupling theory

Jeffrey Park

The free oscillations of an elastic, spherical earth model can be combined using variational or Galerkin techniques to form hybrid oscillations that represent the free oscillations of a rotating, anelastic and laterally-variable earth model. With supercomputers and/or array processors, it is feasible to synthesize realistic long-period seismograms from sums of these coupled free oscillations. Such calculations demonstrate pervasive coupling between toroidal (SH, Love) and spheroidal (P-SV, Rayleigh) motion in surface wave with periods $T > 240s$ due to the earth's rotation, and show these effects to be especially prevalent on north-south propagation paths. Surface-wave propagation effects, such as wavepacket amplitude anomalies, can also be modelled. Theoretical techniques have been developed to represent functionals of the seismogram (e.g. modal amplitude and frequency shift in the frequency domain, the oscillation envelope of an isolated multiplet in the time domain) in the presence of mode-mode coupling. These techniques promise, along with recent work on the coupling induced by anisotropy, a significant improvement in the accuracy of future inversions of the global long-period seismic data set.

1. Introduction

Starting with the first confirmed observations that the long-period seismic response of the earth consists of discrete spectral peaks corresponding to elastogravitational free oscillations, seismologists have attempted to use their frequencies and amplitudes to constrain the density and elastic properties of the earth's interior. To a first approximation the earth is spherically-symmetric, making it possible to constrain useful models of spherically-averaged density and seismic velocities $\rho(r)$, $\alpha(r)$, $\beta(r)$, functions of radius r ,

N.J. Vlaar, G. Nolet, M.J.R. Wortel & S.A.P.L. Cloetingh (eds), Mathematical Geophysics, 31-52.

without knowing the lateral variability in these quantities (often called 'lateral structure' or 'asphericity'). The first systematic attempts to understand the effects of the earth's lateral structure on low frequency free oscillations were theoretical, e.g. Dahlen (1968), who calculated the effect of the earth's ellipticity of figure, assumed to be in hydrostatic equilibrium. See Dahlen (1980) for a review of efforts up to 1979. Observational breakthroughs in aspherical structure studies awaited the advent of global digitally-recording low frequency seismic arrays, such as IDA (International Deployment of Accelerometers, see Agnew, et al (1976)) and GDSN (Global Digital Seismic Network, see Engdahl, Peterson and Orsini (1982)). IDA and GDSN began recording in 1975, but did not accumulate a substantial dataset until this decade. The first lateral structure inversions using free oscillation observations (Silver and Jordan, 1981; Masters, Jordan, Silver and Gilbert, 1982) were based on a 'pure-path' asymptotic representation of the response of seismic free oscillations to lateral structure. Much of the seismograms remained unexplained. This has spurred research towards understanding better the physics of coupled free oscillations. Since 1980 progress has been made both in developing higher-order asymptotic representations of free oscillation behavior and in the enterprise of calculating synthetic seismograms from coupled free oscillations. This paper summarizes these recent developments.

We must first introduce some terminology. Seismic motion within a given earth model that is everywhere frictionless is governed by self-adjoint equations of motion. Such equations have a countably-infinite set of eigensolutions $\mathbf{s}_k(\mathbf{r})$ that represent small oscillations within the earth model's volume with eigenfrequencies ω_k. Secular motions (e.g. rigid rotations) are represented by modes with zero frequency. Any seismic motion $\mathbf{u}(\mathbf{r},t)$ on a frictionless model can be represented by a sum of free oscillations. Each free oscillation is an independent vibrational degree of freedom of the chosen earth model, and is called a 'singlet.' The functional form of singlets on a general earth model is often complicated. Coupled-mode theory attempts to simplify this representation by expressing the singlets of a general earth model in terms the singlets of an earth model with simple geometry and hence more tractable equations of motion.

A typical reference earth model is spherically-symmetric, non-rotating, elastic, and isotropic (SNREI). The free-oscillation particle motions $\mathbf{s}_k(\mathbf{r})$ of such a model can be represented in terms of one or two scalar functions of r and a single complex fully-normalized spherical harmonic $Y_l^m(\theta,\phi)$, of angular degree l and azimuthal order m. The angle θ is colatitude measured from the North Pole, and ϕ is longitude east of Greenwich. There are two types of SNREI free oscillation singlets. The first type corresponds to spheroidal particle motion:

$$\mathbf{s}_k(\mathbf{r}) = {}_nU_l(r)\hat{\mathbf{r}}\, Y_l^m(\theta,\phi) + {}_nV_l(r)\, \nabla_1\, Y_l^m(\theta,\phi) \tag{1}$$

where ${}_nU_l(r)$, ${}_nV_l(r)$ are real-valued scalar functions, $\nabla_1 = \hat{\boldsymbol{\theta}}\partial_\theta + (\sin\theta)^{-1}\hat{\boldsymbol{\phi}}\,\partial_\phi$ is the surface gradient, and $\hat{\mathbf{r}}$, $\hat{\boldsymbol{\theta}}$, and $\hat{\boldsymbol{\phi}}$ are unit coordinate vectors. The index n is the overtone number of the oscillation: $n = 0$ is the 'fundamental' mode, $n = 1$ is the first overtone, etc. Spheroidal oscillations, indexed by the notation ${}_nS_l^m$, correspond to Rayleigh surface waves and the P-SV system of body waves. The second type of free oscillation, indexed by the notation ${}_nT_l^m$, corresponds to toroidal particle motion:

$$\mathbf{s}_k(\mathbf{r}) = {}_nW_l(r)\,\hat{\mathbf{r}} \times \nabla_1 Y_l^m(\theta,\phi) \tag{2}$$

where ${}_nW_l(r)$ is a real-valued scalar function. Toroidal oscillations correspond to Love surface waves and SH body waves. The eigenfunction solutions (1) and (2) are often referred to as 'wavefunctions,' similar to the electron wavefunctions of quantum mechanics.

If $\mathbf{s}_k(\mathbf{r})$ and $\mathbf{s}_j(\mathbf{r})$ are vibrational singlets of a SNREI earth, then suitable normalization factors will render them orthonormal:

$$\int_V \rho(r)\mathbf{s}_j^* \cdot \mathbf{s}_k d\mathbf{r} = \delta_{jk} \tag{3}$$

where the asterisk denotes complex conjugation and the integral is over the earth's volume V. δ_{jk} is the Kronecker delta. The vibrational singlets $\mathbf{s}_k(\mathbf{r})$ form an orthonormal basis of the infinite-dimensional vector space of vector-valued seismic particle motions $\mathbf{u}(\mathbf{r},t)$ that satisfy the equations of motion with stress-free boundary conditions at the surface. Hence they are referred to as 'normal modes.'

A spherically-symmetric earth model cannot distinguish between singlets of equal type, overtone number n and angular degree l, so that the vibrational frequencies of the singlets ${}_nS_l^m$ and ${}_nT_l^m$ do not depend on the azimuthal number m. The $2l+1$ singlets $(m = l, -l+1, \cdots l+1, l)$ comprise a degenerate 'multiplet,' with notations ${}_nS_l$ and ${}_nT_l$. Each singlet within a degenerate multiplet has equal frequency ${}_n\omega_l^S$ or ${}_n\omega_l^T$. (We will drop the superscripts in future use.) Free oscillation singlets group into degenerate multiplets only on a non-rotating, spherically-averaged earth model (e.g. models 1066A, 1066B of Gilbert and Dziewonski (1975), and model PREM of Dziewonski and Anderson (1981)). This restriction is looser than SNREI, as degeneracy is preserved in the presence of spherically-averaged anelasticity and a limited type of anisotropy in which the stress-strain relation for vertical strains differs from that for horizontal strains (model PREM contains this type of anisotropy in the upper 220km of the mantle). On the real earth the deviations from spherical symmetry are small. Free oscillations are observed typically as nearly-degenerate multiplets rather than as singlets.

Anelasticity is incorporated into the equations of motion usually by adding a small imaginary component to the bulk modulus κ and shear modulus μ. Nonzero frequency-dependence in the real-part modulus ('physical dispersion') is necessary in order to satisfy causality (e.g. Hudson, 1980, chap. 9). The most tractable way to model modal attenuation is to apply Rayleigh's Principle: treat the imaginary portions of the moduli as small perturbations and calculate an imaginary-part perturbation to the singlet eigenfrequencies ω_k. If one takes a non-perturbative approach and solves the radial systems of equations explicitly using complex bulk and shear moduli, the radial dependence of the singlet particle motion (${}_nU_l(r)$, ${}_nV_l(r)$ and ${}_nW_l(r)$) becomes complex-valued. Buland, et al (1985) show that this leads to temporal phase delays in the anelastic free oscillations. The source time for an earthquake beneath a low Q layer in the mantle would be biased late. If the anelasticity is small ($Q_\kappa(r)$, $Q_\mu(r) \geq 100$ for all r), they show the potential bias to be observationally negligible ($\leq 1^o$ in the phase of the oscillation for the toroidal modes used in their calculations).

The earth's rotation, azimuthally-dependent anisotropy, and lateral structure in ρ, κ, μ, Q_κ and Q_μ break the symmetry of the equations of motion. The free oscillations of an aspherical earth model can be represented (not always exactly) by hybrid singlets that are infinite sums of coupled spherical-earth singlets. This paper discusses the wave propagation effects that can be represented by the coupling of spherical-earth free oscillations. We provide background for this discussion in sections 2 and 3, which describe, respectively, coupling selection rules and the variational/Galerkin formalism for the calculation of hybrid singlets. Section 4 describes recent advances in understanding the coupling of singlets within an isolated multiplet. Section 5 discusses coupling among multiplets along a single dispersion branch, which leads to WKBJ phase integrals for the observables of surface-wave-equivalent modes. Section 6 covers recent work on coupling between spheroidal and toroidal modes due to rotational Coriolis force. Section 7 speculates on future directions in free oscillation studies.

2. Selection rules

In free oscillation studies we express the earth's isotropic lateral structure in a spherical harmonic expansion

$$\begin{aligned} \rho(\mathbf{r}) &= \rho_o(r) + \sum_{s=0}^{s_{max}} \sum_{t=-s}^{s} \delta\rho_s^t(r) Y_s^t(\theta,\phi) \\ \kappa(\mathbf{r}) &= \kappa_o(r) + \sum_{s=0}^{s_{max}} \sum_{t=-s}^{s} \delta\kappa_s^t(r) Y_s^t(\theta,\phi) \\ \mu(\mathbf{r}) &= \mu_o(r) + \sum_{s=0}^{s_{max}} \sum_{t=-s}^{s} \delta\mu_s^t(r) Y_s^t(\theta,\phi) \end{aligned} \tag{4}$$

with maximum degree s_{max}. This expansion is more flexible than tectonic regionalizations, but can fall prey to spurious features due to the truncation e.g. a 'ringing' in the model with wavelength $\lambda \approx 2\pi a / [s_{max} + \frac{1}{2}]$, where $a = 6371$km is the earth's radius.

The interaction of spherical-earth singlets $\mathbf{s}_k(\mathbf{r})$ through the aspherical structure governs the composition of hybrid singlets. This interaction is subject to a set of selection rules. Given spherical-earth singlets of like type with angular degrees l, l' and azimuthal numbers m, m', interaction through lateral structure with Y_s^t-dependence will occur if the integral

$$\int_0^\pi d\theta \, \sin\theta \int_0^{2\pi} d\phi \, (Y_l^m(\theta,\phi))^* \; Y_s^t(\theta,\phi) \; Y_{l'}^{m'}(\theta,\phi) \neq 0 \tag{5}$$

The ϕ-integral

$$\int_0^{2\pi} d\phi \, e^{-im\phi} e^{it\phi} e^{im'\phi} \tag{6}$$

can be isolated, and is nonzero only if $-m + t + m' = 0$. Axisymmetric perturbations, such as rotation and hydrostatic ellipticity, have azimuthal number $t = 0$ and so restrict interaction to cases where $m = m'$. The integral in (5) will vanish if the integrand has odd parity; therefore $l + s + l'$ must be an even number. As a consequence, the singlets within

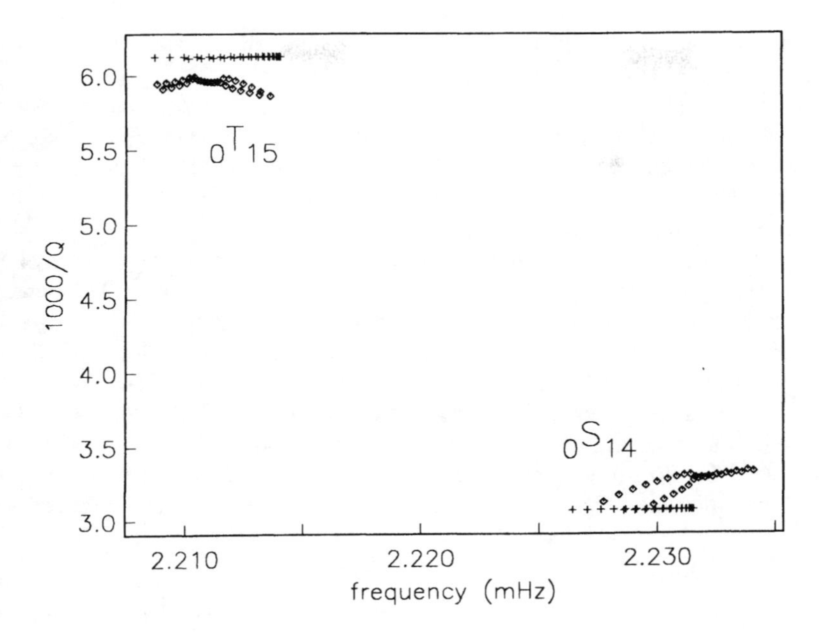

Figure 1. Hybrid singlet frequencies for ${}_0S_{14}$-${}_0T_{15}$. The imaginary portion of the eigenfrequency is plotted as $1000/Q$. The plus symbols indicate singlet eigenfrequencies in the isolated multiplet approximation, in which the attenuation parameter remains uniform within each multiplet. The diamonds plot hybrid eigenfrequencies for mixed-mode coupling due to rotation, hydrostatic ellipticity and the $s=2$ transition zone model suggested by Masters, et al (1982). Note that the separation of the averaged real-part hybrid multiplet frequencies increases due to coupling, while the averaged imaginary-part frequencies are brought closer.

any degenerate multiplet (i.e. $l=l'$) can interact only through even-degree lateral structure. Finally there is the triangle rule, which requires that a triangle can be made with sides of l, s and l' length units. The triangle rule restricts the lateral structure through which the two singlets can interact to have angular degree: $|l-l'| \leq s \leq l+l'$. For instance, the coupling among the five singlets of ${}_{10}S_2$ is influenced only by $s=0,2,4$ asphericity. Likewise, interaction between the neighboring multiplets ${}_0S_{33}$ and ${}_1S_{21}$ is forbidden for lateral structure with $s<12$. Second-order interaction is possible by means of coupling through intermediary modes, but this effect is usually negligible unless the asphericity is large.

The selection rules for mixed-type interaction are similar, but the different tangential gradients of $Y_l^m(\theta,\phi)$ in (1) and (2) alter the parity selection rule, so that $l+s+l'$ must be an *odd* number. For instance, ${}_0S_2$ and ${}_0T_2$ can couple through lateral structures with angular degree $s=1,3$.

Like-type rotational interaction is possible only if the two singlets have identical values of l and m – it is this behavior that makes complex spherical harmonics Y_l^m the natural functions to use to describe spherical-earth singlets. Mixed-type rotational interaction between singlets ${}_nS_l^m$ and ${}_{n'}T_{l'}^{m'}$ is possible only if $m=m'$ and $l=l'\pm 1$. The selection rules for interaction on an anisotropic earth are not yet available in a compact form. For coupling within an isolated multiplet, Tanimoto (1986) and Mochizuki (1986a) present general coupling formulas from which they can be derived.

There are not radial selection rules *per se* , but intuition suggests (and experience confirms) that interaction is maximized when both coupling partners possess substantial vibrational energy in the depth range where lateral structure is prevalent. Lateral structure in the lithosphere strongly influences surface-wave-equivalent modes ${}_0S_l$, ${}_0T_l$ with $l \approx 40$, but has weaker influence on PKP-equivalent modes such as ${}_{13}S_2$.

If other factors are equal, the strength of interaction between two spherical-earth singlets varies roughly inversely with their difference in frequency. The strongest interaction therefore occurs among singlets within degenerate multiplets. The real-part frequency difference of two coupling partners tends to increase, as though the modes were repulsed. On an anelastic earth, the initial frequency difference between coupling partners can be complex-valued, allowing differences in both oscillation frequency and Q. The mixture of a high-Q spheroidal mode with a low-Q toroidal mode produces hybrid singlets with intermediate values of Q (figure 1). These effects are demonstrated in data by the averaged-frequency measurements of ${}_0S_l$, $l = 9-23$ by Masters, Park and Gilbert (1983).

3. Hybrid singlet calculation

A free oscillation $\boldsymbol{\sigma}(\mathbf{r})$ of an anelastic, aspherical earth model with oscillation frequency ω must satisfy an equation of bilinear functionals

$$L = V(\boldsymbol{\sigma}^A, \boldsymbol{\sigma}) + \omega W(\boldsymbol{\sigma}^A, \boldsymbol{\sigma}) - \omega^2 T(\boldsymbol{\sigma}^A, \boldsymbol{\sigma}) = 0. \tag{7}$$

$\boldsymbol{\sigma}^A$ is the 'adjoint' of $\boldsymbol{\sigma}$, necessary because anelasticity renders the equations of motion non-self-adjoint (Park and Gilbert, 1986). V , W and T are the potential energy, Coriolis and kinetic energy interaction functionals, respectively. The kinetic energy functional T is positive definite, and serves as an inner product for the space of possible deformational motions. We substitute $\boldsymbol{\sigma}(\mathbf{r}) = \sum_{i=1}^{N} \alpha_i \mathbf{s}_i(\mathbf{r})$ in (7), where the $\mathbf{s}_i(\mathbf{r})$ are spherical-earth singlets and the coefficients α_i are to be determined. $\boldsymbol{\sigma}^A$ is represented by a sum of the same spherical-earth singlets $\sum_{i=1}^{N} \beta_i \mathbf{s}_i(\mathbf{r})$ with different coefficients β_k . If attenuation is neglected, $\beta_k = \alpha_k$ and $\boldsymbol{\sigma}^A = \boldsymbol{\sigma}$. Note the difference from Park and Gilbert (1986) in the conjugation convention of the adjoint $\boldsymbol{\sigma}^A$.

In practice the number of singlets N is dependent on one's computer and one's patience, but in principle can be arbitrarily large. The infinite-sum representation is exact in the case where the earth is elastic and there are no perturbations to internal and external boundaries. Boundary perturbations can be treated as linearized distortions of the spherical polar coordinate system (Woodhouse, 1976; Dahlen, 1976; Woodhouse and Dahlen, 1978). Such a linearization is most accurate where the distortion of internal layer thicknesses is small e.g. the hydrostatic perturbations associated with the earth's rotational bulge. Some inaccuracy may be expected, for example, if large short-wavelength fluctuations in crustal thickness are modelled, but this has yet to be thoroughly studied. The imprecision associated with anelasticity will likely lead to temporal phase shifts similar to those reported by Buland, et al (1985), but this has not been investigated rigorously. The inability of the infinite sum of spherical-earth singlets to converge to the true anelastic/aspherical-earth singlet is a characteristic of a Galerkin procedure, in which a variational method is used to solve a problem for which the prerequisites for solution - in this case self-adjointness and basis vectors that fit boundary conditions exactly - are not satisfied (see e.g. Mikhlin, 1964).

With these expansions substituted into (7), we obtain a quadratic matrix eigenvalue equation for singlet eigenfrequency ω

$$L = \boldsymbol{\beta}^* \cdot (\mathbf{V} + \omega \mathbf{W} - \omega^2 \mathbf{T}) \cdot \boldsymbol{\alpha} = 0 \tag{8}$$

where $\boldsymbol{\beta}$ and $\boldsymbol{\alpha}$ are N-vectors of summation coefficients, and $\mathrm{V}_{ij} = V(\mathbf{s}_i, \mathbf{s}_j)$, etc. The eigenfrequencies of (8) are complex with nonnegative imaginary part. General formulas for calculating these functionals for spherical-earth singlets are given by Woodhouse and Dahlen (1978). Woodhouse (1980) gives more detailed formulas for interaction through isotropic lateral structure and boundary perturbations, in terms of the radial eigenfunctions ${}_nU_l(r)$, ${}_nV_l(r)$ and ${}_nW_l(r)$. Mochizuki (1986a) and Tanimoto (1986) derive similar formulas for the interaction of singlets within an isolated multiplet through general anisotropy. The term in (8) linear in ω is significant for modes with $f \le 1$mHz that suffer significant rotational splitting e.g. ${}_0S_2$. In other cases we can approximate $\omega\mathbf{W} \approx \omega_o \mathbf{W}$ for a fixed ω_o (typically the degenerate frequency ${}_n\omega_l$ of the multiplet we wish to model). This reduces (8) to a linear eigenvalue problem for ω^2, thus reducing the calculation by a factor of 4 to 8.

Hybrid singlets satisfy Rayleigh's Principle: the first variation of L with respect to small changes in $\boldsymbol{\alpha}$ must vanish. Solving the variational problem $\delta L = 0$ is equivalent to solving (8) for right-eigenvectors $\boldsymbol{\alpha}_k$, $k = 1, \cdots, N$ (which correspond to hybrid singlet displacements at the receiver) and left- or dual-eigenvectors $\boldsymbol{\beta}_k$, $k = 1, \cdots, N$ (which correspond to singlet strains at the earthquake source). For a definition of 'dual' vectors, see Halmos (1974), pp23-24. In solid state physics dual vectors are used as the 'reciprocal lattice vectors' of periodic crystals (Ashcroft and Mermin, 1976, pp84-96). The $\boldsymbol{\alpha}_k$ need not be orthogonal, and are sometimes nearly parallel (Park and Gilbert, 1986). The $\boldsymbol{\beta}_k$ are dual to the $\boldsymbol{\alpha}_k$ with inner product defined by the kinetic energy matrix $\mathbf{T}$, so that $\boldsymbol{\beta}_i^* \cdot \mathbf{T} \cdot \boldsymbol{\alpha}_j = \delta_{ij}$. This is equivalent to the hybrid singlet normalization, analogous to (3),

$$T(\boldsymbol{\sigma}_j^A, \boldsymbol{\sigma}_k) = \int_V \rho(\mathbf{r})(\boldsymbol{\sigma}_j^A)^* \cdot \boldsymbol{\sigma}_k d\mathbf{r} = \delta_{jk} \tag{9}$$

Consider a seismic event that occurs at position $\mathbf{r}_o$ and time $t = 0$. The particle motion $\mathbf{u}(\mathbf{r},t)$, observed at receiver position $\mathbf{r}$, can be expressed as a sum of decaying sinusoids. Each singlet has time dependence, accurate to order Q_k^{-1},

$$C_k(t) = |\omega_k|^{-2} (\cos(\mathrm{Re}\,\omega_k t - \psi_k)\, e^{-\mathrm{Im}\,\omega_k t} - 1) H(t) \tag{10}$$

where $H(t)$ is the stepfunction and $\tan \psi_k = \mathrm{Im}\,\omega_k / \mathrm{Re}\,\omega_k = [2Q_k]^{-1}$. $C_k(t)$ differs from the time dependence used by Gilbert and Dziewonski (1975) by a factor of $|\omega_k|^{-2}$ due to a difference in normalization and by terms of order Q_k^{-1}. The portion of the seismic motion caused by the N hybrid singlets $\boldsymbol{\sigma}_k(\mathbf{r})$ is

$$\mathbf{u}(\mathbf{r},t) = \sum_{i=1}^{N} \sum_{j=1}^{N} \sum_{k=1}^{N} \alpha_{ki}\, \beta_{kj}^* \mathbf{s}_i(\mathbf{r}) (\mathbf{M} : \mathbf{E}_j^*(\mathbf{r}_o)) C_k(t) \tag{11}$$

where $\mathbf{M}$ is the source moment tensor, modelled as a point source in space and time. $\mathbf{E}_j$ is the strain of the jth spherical-earth singlet at the source location $\mathbf{r}_o$. α_{ki} and β_{ki} are the coefficients of the ith spherical-earth singlet in the kth hybrid singlet and its adjoint, respectively. Finite source extent and duration can be modelled by convolving (11) with the source rupture function.

Seismic motion on an aspherical earth model is best approximated when the number N of spherical-earth singlets is very large. Eigenvector-eigenvalue decompositions require $O(N^3)$ scalar operations, with a proportionality constant that varies with the exploitable symmetries of the matrix system. The Galerkin matrix systems are complex-valued with no special symmetry, so the computation required can be formidable. Vectorizable supercomputers can reduce the burden to $O(N^2)$ machine cycles. The author expended one minute of CPU time on a Cray 1S computer to solve a coupled system involving 330 singlets. A similar calculation required 30 seconds on a Hitachi S810 machine (Morris, et al, 1987). Even with supercomputer speed, the accumulation of roundoff error in the calculations limits $N \leq 500$ for full matrix decomposition using 64-bit floating point arithmetic. It is prudent to assess what wave propagation effects can be modelled with the fewest number of spherical-earth singlets. In many cases of interest the desirable $N > 1000$ and approximation methods are necessary.

Henson (1987) demonstrates an accurate approximation to the 'brute-force' coupled-mode sum (11). The hybrid multiplets corresponding to ${}_nS_l$ and ${}_nT_l$ are bandlimited, with $2l+1$ singlet frequencies grouped within a splitting width $\Delta\omega \ll {}_n\omega_l$. It is not usually necessary to include $2l+1$ terms in the coupled-mode sum, especially when the record duration T is comparable to $2\pi/\Delta\omega$. Instead one uses a basis set of prolate spheroidal wavefunctions to represent the spectrum of the hybrid multiplet. If $\Delta\omega \approx 2\pi p/T$, with p an integer, only p basis functions are necessary to represent the hybrid multiplet accurately.

Similar computational savings can be had in the time domain as well as the frequency domain. Woodhouse and Girnius (1982) rewrite the sum (11) for an isolated split multiplet ${}_nS_l$ or ${}_nT_l$ in the form, good to order ${}_nQ_l^{-1}$,

$$\mathbf{u}(\mathbf{r},t) = \sum_{m=-l}^{l} |{}_n\omega_l|^{-2}\, {}_n\mathbf{s}_l^m(\mathbf{r})(\mathrm{Re}({}_n\mathrm{a}_l^m(t)\, e^{i_n\omega_l t - i_n\psi_l}) - 1)\, H(t) \tag{12}$$

where the ${}_n\mathbf{s}_l^m$ are spherical-earth singlets, ${}_n\omega_l$ is complex-valued to model modal attenuation, and $\tan {}_n\psi_l = \mathrm{Im}_n\omega_l/\mathrm{Re}_n\omega_l = [2\,{}_nQ_l]^{-1}$. All effects of coupling within the multiplet are contained in the complex-valued coefficients ${}_n\mathrm{a}_l^m(t)$, which modulate the 'carrier' frequency ${}_n\omega_l$. Since the multiplet spectrum is bandlimited, the ${}_n\mathrm{a}_l^m$ are slowly-varying in time. Form the modified interaction matrix $\mathbf{\Lambda}$ for the multiplet:

$$2\,{}_n\omega_l\mathbf{\Lambda} = \mathbf{V} + {}_n\omega_l\,\mathbf{W} - {}_n\omega_l^2(\mathbf{T} - \mathbf{I}). \tag{13}$$

The vector $\mathbf{a}(t) = {}_n\mathrm{a}_l^{-l}(t), \cdots, {}_n\mathrm{a}_l^l(t))$ of singlet coefficients is governed by a system of first-order differential equations

$$\frac{d}{dt}\mathbf{a}(t) = i\mathbf{\Lambda}\cdot\mathbf{a}(t) \tag{14}$$

with $\mathbf{a}(0)$ equal to the initial excitation coefficients of the spherical-earth singlets. The formal solution of (14) is $\mathbf{a}(t) = \mathbf{P}(t)\cdot\mathbf{a}(0)$, where $\mathbf{P}(t)$ is the propagator matrix

$$\mathbf{P}(t) = e^{i\Lambda t} = \mathbf{I} + it\boldsymbol{\Lambda} - \frac{t^2}{2}\boldsymbol{\Lambda}\cdot\boldsymbol{\Lambda} + \cdots \tag{15}$$

The term linear in time is linearly-dependent on asphericity through $\boldsymbol{\Lambda}$. Solving for $\mathbf{a}(t)$ and weighting the components by the singlet displacements ${}_n\mathbf{s}_l^m(\mathbf{r})$ as in (12) obtains the oscillation envelope of the hybrid multiplet at receiver location $\mathbf{r}$.

The interaction matrix $\boldsymbol{\Lambda}$ need not be limited to an isolated multiplet. In fact, Woodhouse (1983) derives the above formulas considering interactions between *all* spherical-earth singlets, so that - formally, at least - $\boldsymbol{\Lambda}$ can be infinite-dimensional. Dahlen (1987) and Tanimoto (1984) truncate (15) after the linear term to derive a time-domain inversion algorithm for lateral structure, analogous to the Born approximation in quantum mechanical scattering studies (Schiff, 1968; pp 324-335). Since higher-order terms in the expansion grow with time, this algorithm is only valid for $t \,||\boldsymbol{\Lambda}||\ll 1$. $||\boldsymbol{\Lambda}|| \approx \frac{1}{2}\Delta\omega$, where $\Delta\omega$ is the splitting width of the multiplet. Observations of ${}_0S_l$ near $l = 30$ have splitting widths of roughly 20μHz (Masters, et al, 1982), so that the above approximation is technically valid only for $t \ll 4.4$hr after earthquake onset. Model M84A of Woodhouse and Dziewonski (1984) predicts splitting widths of 45μHz for ${}_0S_{50}$, so that the Born-like representation of 180s surface wave scattering is valid for $t \ll 2.0$hr. It is likely that the truncated form of (15) is not reliable past the second surface wave arrivals R_2 and G_2.

Synthetic seismograms for individual multiplets can be calculated by integrating (14) numerically and substituting into (13). Since $\mathbf{a}(t)$ is slowly-varying, the time step for numerical integration can be relatively large. Giardini, Li and Woodhouse (1987) and Ritzwoller, Masters and Gilbert (1986) follow this strategy to study the splitting of overtones sensitive to lower mantle and core structure. Although these calculations involve relatively small matrix systems, the strategy can be extended to incorporate interactions among many multiplets (Woodhouse, 1983; Dahlen, 1987).

4. First-order splitting of isolated multiplets

In the isolated multiplet approximation, one assumes that the interaction between neighboring multiplets is negligible compared to the interaction among singlets within the multiplets themselves. Angular selection rules require a corollary of this assumption, that odd-degree lateral structure exerts negligible influence on the seismogram, or at least on the aspect of the seismogram we wish to measure. By comparing synthetic seismograms constructed from (11), Park (1986) demonstrates that odd-degree structure exerts a small influence relative to even-degree structure on surface wave amplitude anomalies and higher-orbit wavepacket phase delays. The physics of isolated multiplet interaction must be understood before more general coupling effects can be explored.

In the frequency domain, free oscillation multiplet resonances are composed of the individual resonances of their singlets. Define the zeroth and first moments of the multiplet spectrum about its degenerate spherical-earth frequency ${}_n\omega_l$:

$$\mu_0 = \int_{{}_n\omega_l - \delta\omega}^{{}_n\omega_l + \delta\omega} \hat{\mathbf{x}} \cdot \mathbf{u}(\mathbf{r},\omega)\, d\omega$$

$$\mu_1 = \int_{{}_n\omega_l - \delta\omega}^{{}_n\omega_l + \delta\omega} (\omega - {}_n\omega_l)\hat{\mathbf{x}} \cdot \mathbf{u}(\mathbf{r},\omega)\, d\omega \tag{16}$$

where $\hat{\mathbf{x}}$ is the unit vector of the recorded particle motion component, $\mathbf{u}(\mathbf{r},\omega)$ is the multiplet resonance at receiver position $\mathbf{r}$, and $\delta\omega$ is the half-bandwidth of integration, chosen as large as feasible while avoiding interference from other spectral peaks. Jordan (1978) shows that the ratio of these spectral moments, which he called the 'multiplet location parameter' λ , satisfies

$$\lambda = \frac{\mu_1}{\mu_0} \approx \frac{1}{2\pi} \oint \delta\omega_{local}(\theta,\phi)\, d\xi = \overline{\delta\omega} \tag{17}$$

where $\delta\omega_{local}(\theta,\phi)$ is the local perturbation to multiplet frequency, which is proportional to the local phase velocity perturbation of the travelling wave associated with the mode. The integral is over the great circle connecting source and receiver. $\overline{\delta\omega}$ is the average of $\delta\omega_{local}$ over the great circle. This expression is asymptotically valid for multiplets with $l \gg s_{max}$, the maximum angular degree of the lateral structure. In addition, Dahlen (1979) shows that such multiplets tend to resemble singlet resonances with apparent frequency ${}_n\omega_l + \overline{\delta\omega}$. However, only 15-40% of the observations of ${}_0S_{12} - {}_0S_{38}$ by Masters and Gilbert (1983) and Masters, Jordan, Silver and Gilbert (1982) were considered sufficiently singlet-like to exploit this tendency.

Relation (17) implies that only lateral structure directly beneath the source-receiver great circle influences measurements of λ . This is equivalent to assuming that the travelling waves whose interference pattern forms the multiplet resonance never focus along or refract away from the source-receiver great circle. This assumption is not always tenable, and the observations discarded or poorly-fit by Masters, et al (1982) encouraged higher-order asymptotic representations. Woodhouse and Girnius (1982) show that measurements of the first multiplet moment μ_1 can be interpreted in terms of only three source-dependent data kernels within the earth's volume. The kernels have relative sensitivities of roughly 1, $\frac{s(s+1)}{l(l+1)}$, and $\left[\frac{s(s+1)}{l(l+1)}\right]^2$, so that, for $l \gg s_{max}$, the first kernel dominates. Plots of this kernel demonstrate the sensitivity of μ_1 to off-path structure, especially asphericity near the source, its antipode, the receiver and its antipode.

When μ_1 is calculated for a tectonically-regionalized earth model, the off-path effects on μ_1 appear small enough to neglect (Woodhouse and Girnius, 1982). Davis and Henson (1986) and Romanowicz and Roult (1986), however, demonstrate how off-path structure, even if smooth, can disrupt measurements of λ . Both derive equivalent asymptotic representations of the sensitivity of μ_1 to smooth off-path structure which offer a improved data inversion algorithm. When off-path structure is neglected, μ_0 , μ_1 vary with epicentral distance Δ as a linear combination of $X_l^m(\Delta)$, $m = 0,1,2$, where $Y_l^m(\theta,\phi) = X_l^m(\theta)e^{im\phi}$. The zeroes of the two multiplet moments coincide, so that (17) remains valid there, in the limiting sense. Off-path lateral structure generates terms of order s/l , $(s/l)^2$, etc. that asymptotically depend on the derivatives of $\delta\omega_{local}(\theta,\phi)$ transverse to the source-receiver great circle. These higher-order asymptotic terms shift the zeroes of μ_1, but leave μ_0 unchanged. The formula for the location parameter

$$\lambda = \overline{\delta\omega} + \frac{(a\,\Omega'_o + \frac{1}{2}\Omega''_o)\tan((l+\frac{1}{2})\Delta - \chi) + b\,\Omega'_o}{(l+\frac{1}{2})\sin\Delta} \tag{18}$$

acquires singularities at the zeroes of μ_0. In (18), Ω'_o and Ω''_o are great-circle integrals involving the first and second transverse derivatives, respectively, of $\delta\omega_{local}$. The constants a, b and χ are source-dependent. In numerical experiments comparing (17), (18) and measurements from the spectra of synthetic seismograms calculated with the isolated multiplet approximation, Davis and Henson (1986) show that large fluctuations in λ measured from the synthetics correlated well with the predictions of the asymptotic theory. Romanowicz and Roult (1986) report evidence of such behavior in seismograms from project GEOSCOPE. Davis (1987) shows that the tangent-function fluctuations, along with a tendency for λ to approach its global average for receivers close to the source and its antipode, lead to a bias toward smooth models if (17) is used for lateral structure inversion. Davis (1987) reports an inversion, using (18) rather than (17), of 6297 multiplet location parameter measurements of spheroidal fundamental modes, assembled from 968 IDA records. In addition to the $s = 2$ anomaly seen by Masters, et al (1982), this inversion finds a robust $s = 6$ anomaly in $\delta\omega_{local}$ that correlates above the 95% confidence level with the $s = 6$ portion of model M84A of Woodhouse and Dziewonski (1984), a model derived from time-domain waveform fitting. Correlation between the $s = 4$ and $s = 8$ portions of the models is relatively poor.

Isolated multiplet splitting caused by anisotropy has been studied by Mochizuki (1986a), Tanimoto (1986) and Giardini, Xi and Woodhouse (1987). For general anisotropy, the stress-strain relation has 21 independent components, each of which can generate coupling interactions. This leads to formidable algebraic expressions and little immediate insight. However, the strong evidence for anisotropy in the oceanic upper mantle and its possible relation to convection flow patterns indicates that a suitable simplification of anisotropic coupling effects would aid greatly our understanding of mantle structure and dynamics. Tanimoto (1986) derives asymptotic relations for coupling in the presence of anisotropy. One can generalize $\delta\omega_{local}$ in (17) to include dependence on the propagation azimuth ψ. This leads to an anisotropic multiplet location parameter

$$\lambda = \frac{\mu_1}{\mu_0} \approx \frac{1}{2\pi}\oint \delta\omega_{local}(\theta,\phi,\psi)\, d\xi \tag{19}$$

where ψ is the azimuth of the great circle at point (θ,ϕ). The expression for $\delta\omega_{local}(\theta,\phi,\psi)$ given by Tanimoto (1986) contains 15 independent terms, each calculated with an integration over radius of the structure beneath (θ,ϕ). The 12 terms additional to the three given by Woodhouse and Girnius (1982) include azimuthal dependence $\cos p\psi$ or $\sin p\psi$ for $p = 1,2,3,4$. In the limit of smooth anisotropic structure and $l \gg 1$, the $p = 1,3$ terms can be neglected relative to the $p = 2,4$ terms. In this limit, the expressions for eigenfrequency perturbations of surface-wave-equivalent modes can be reduced, in a flat-layered geometry, to the expressions of Smith and Dahlen (1973) for anisotropic surface-wave phase velocity perturbations.

The hybrid singlets for an aspherical earth are formally given by the nonzero vectors of coefficients $\boldsymbol{\alpha}$, $\boldsymbol{\beta}$ that solve (8). For physical insight into how these singlets behave, Dahlen and Henson (1985) and Henson and Dahlen (1986) are useful. Given an isolated free oscillation multiplet ${}_nS_l$ or ${}_nT_l$, one can form the great circle average $\overline{\delta\omega}$ of the local eigenfrequency perturbation $\delta\omega_{local}$ as a function of all possible great circles on the sphere. $\overline{\delta\omega}$ can be graphed as a function of the colatitude Θ and longitude Φ of its pole of revolution. For the $s=2$ lateral structure model of Masters, et al (1982), $\overline{\delta\omega}(\Theta,\Phi)$ exhibits one maximum and one minimum, each with antipodal mirror images. The great circle about the minimum of $\overline{\delta\omega}$ has the slowest average phase velocity. The singlet eigenfunctions associated with relatively low eigenfrequencies are trapped in the neighborhood of this great circle. These singlets correspond to surface waves whose global trajectories precess slowly, and whose orbits cover densely a belt of finite, irregular width about the maximum-time great circle. The boundary of this belt is a caustic that divides the 'oscillatory' portion from the 'decaying' portion of the singlet wavefunction. Similar behavior occurs among the singlet eigenfunctions with largest eigenfrequencies, which are trapped in the neighborhood of the minimum-time great circle about the maximum of $\overline{\delta\omega}$. The trapped wavefunctions must be periodic around the extremal great circles, possessing $p \leq l$ wavelengths. The number of wavefunction zero-crossings in the direction transverse to the extremal great circle is $l-p$. For each p , the earth's rotation splits the possible motion into two propagating modes that rotate in opposite directions about the extrema of $\overline{\delta\omega}(\Theta,\Phi)$. As $p \approx \frac{1}{2} l$, a 'separatrix' is reached at which the singlets shift affinity between the maximum path and the minimum path.

In addition to these descriptive properties of the hybrid singlets, Dahlen and Henson (1985) provide prescriptive criteria for the determination of their eigenfrequencies and wavefunctions, good to first order in $\varepsilon = \delta\omega_{local} / {}_n\omega_l$. They derive quantum conditions that are similar to semi-classical quantum conditions (e.g. Keller, 1985) used in the asymptotic calculation of the energy states of molecules. Their asymptotic formulation carries over into more general lateral structures, with the complication that one must deal with several local extrema in $\overline{\delta\omega}$, each of which may possess a group of trapped hybrid singlet eigenfunctions. Using the $s_{\max}=8$ model M84A, excellent agreement in singlet eigenfrequencies is found between the asymptotic quantization analysis and degenerate-multiplet splitting calculations, except for the handful of singlets near the transitions among $\overline{\delta\omega}$-extrema.

5. Coupling along a single dispersion branch

Coupling can occur between any two spherical-earth multiplets on a general anisotropic, aspherical, rotating earth. Experiments with recently-proposed mantle structure models indicate the predominance of certain types of multiplet-multiplet coupling. Rotational coupling is a special issue and will be discussed in section 6. In this section we discuss coupling between multiplets ${}_nS_l$ and ${}_nS_{l'}$ (or multiplets ${}_nT_l$ and ${}_nT_{l'}$), with the condition that $l \gg n$ and $|l-l'| \ll l$. Since the coupling partners have identical overtone number n , this is called 'dispersion-branch' coupling. It may seem unlikely that nearest neighbors along a single dispersion branch ${}_nS$ or ${}_nT$ couple strongly, for the coupling partners are typically separated in frequency by 90μHz or more. Two factors favor dispersion-branch

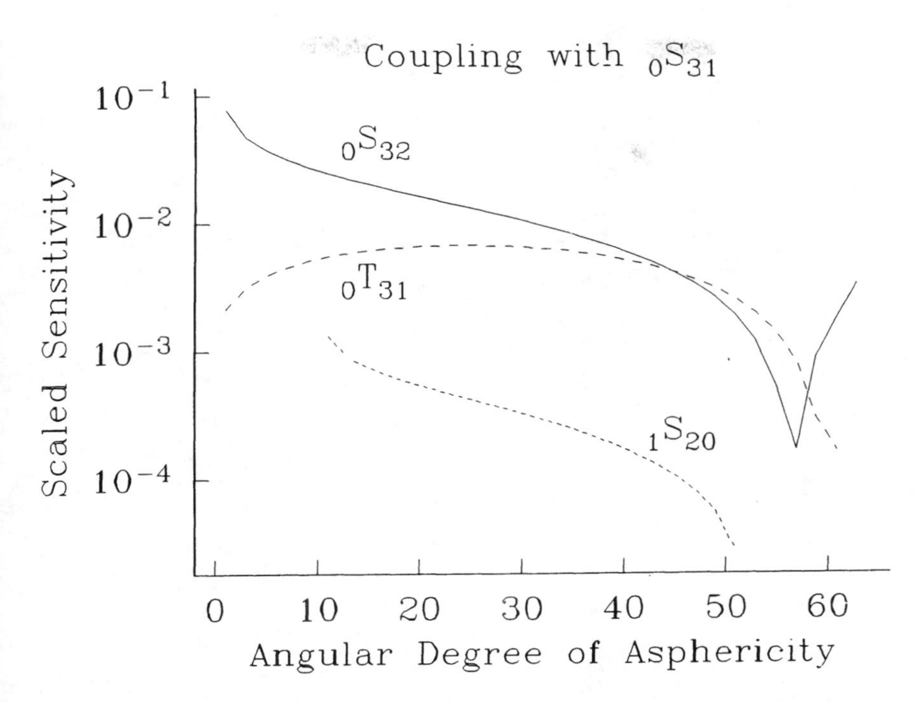

Figure 2. Coupling sensitivity between $_0S_{31}$ and potential coupling partners $_0S_{32}$, $_0T_{31}$ and $_1S_{20}$, plotted versus the angular degree s of lateral structure. Sensitivity to even s vanishes in this example; the curves connect values for odd s .

coupling: one, the radial eigenfunctions for the modes $_nS_l$ and $_nS_{l'}$ are very similar, enhancing the interaction; two, smooth lateral structure will not couple modes for which $|l-l'| > s_{\max}$. For instance, direct coupling between nearest neighbors in frequency on the $_0S$ and $_1S$ branches is not possible using model M84A for $l > 24$. Coupling between spheroidal and toroidal modes with comparable angular degree is inhibited for $l \gg s_{\max}$, though not prohibited. Secondary coupling through intermediary coupling partners is possible but second-order in smallness.

As an example, figure 2 shows the sensitivity of $_0S_{31}$ (f = 3.9031 mHz, Q = 198 on model 1066A with the attenuation model of Masters and Gilbert (1983)) to coupling with $_0S_{32}$ (f = 3.9925 mHz, Q = 194), $_0T_{31}$ (f = 3.9932 mHz, Q = 132) and $_1S_{20}$ (f = 3.9425 mHz, Q = 174), as a function of the angular degree of the lateral structure s . We compare using a unit variance of lateral structure in $\delta\beta/\beta$, constant with radius in the upper mantle (33-670km depth), with angular dependence $Y_s^t(\theta,\phi)$. To calculate the interaction we set

$$\frac{\delta\rho}{\rho} = .4\frac{\delta\beta}{\beta} \qquad \frac{\delta\kappa}{\kappa} = 1.7\frac{\delta\beta}{\beta} \qquad \frac{\delta\mu}{\mu} = 2.4\frac{\delta\beta}{\beta} .$$

The coupling sensitivity parameter ε between a 'target' multiplet with angular degree l_1 and degenerate frequency ω_1 and another multiplet with angular degree l_2 and degenerate frequency ω_2 is given by Park (1987):

$$\varepsilon = \frac{\| \mathbf{H}_{12} \|_F}{| \omega_1^2 - \omega_2^2 | (2l_1 + 1)^{1/2}} \tag{20}$$

where $\mathbf{H}_{12}$ is the $(2l_1 + 1)$ by $(2l_2 + 1)$ matrix of interactions among the singlets, and $\| \mathbf{A} \|_F = \sqrt{\sum_{ij} | A_{ij} |^2}$ is the Frobenius matrix norm. Coupling to the overtone ${}_1S_{20}$ is more than an order of magnitude weaker than that to ${}_0S_{32}$, even though both oscillate mostly within the upper mantle and ${}_0S_{32}$ is twice as far away in frequency. Coupling to ${}_0T_{31}$ is only comparable to coupling to ${}_0S_{32}$ for $s > 30$.

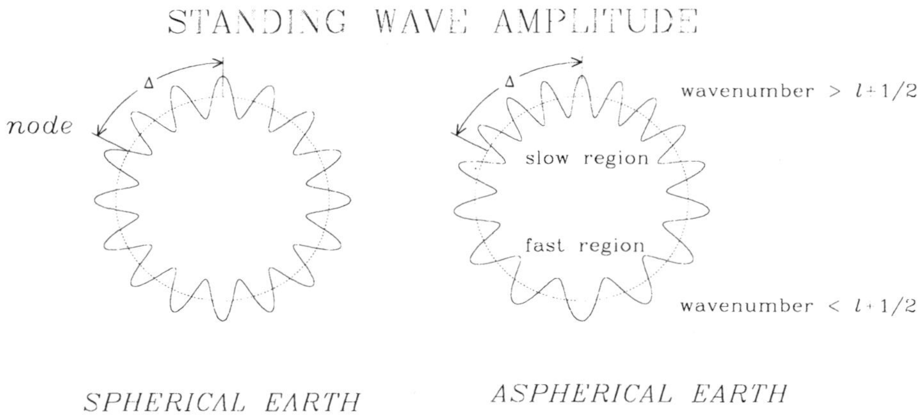

Figure 3. Cartoon illustrating the distortion of the hybrid multiplet wavefunction caused by lateral structure. Although the number of wavelengths around any great circle remains constant, the local wavenumber k varies with $\delta\omega_{local}$. The spatial phase shift at epicentral distance Δ perturbs the observed multiplet amplitude.

Dispersion-branch coupling is easily characterized by asymptotic formulas, as the factors governing the interaction vary slowly from multiplet to multiplet. The frequency spacing along the branch is related to the local group velocity U(ω) of the surface wave associated with the branch: ${}_n\omega_l - {}_n\omega_{l-1} \approx U({}_n\omega_l)/a$, with a the earth's radius. The radial eigenfunctions of the mode ${}_nS_{l+p}$ are similar enough to that of ${}_nS_l$ in the limit $l \gg n,p$ to neglect the difference to first order. Using these approximations, Park (1987), Romanowicz (1987) and Mochizuki (1986b) express the effects of dispersion-branch coupling in a compact form. The most promising result of these studies, from the standpoint of future inversion studies, is the relation of multiplet amplitude, as quantified by the zeroth multiplet moment μ_0, to $\delta\omega_{local}(\theta,\phi)$. In the isolated multiplet approximation, μ_0 is not perturbed. When distinct multiplets couple to form hybrid multiplets, the asymptotic wavenumber $k \approx l + ½$ of the standing wave associated with ${}_nS_l$ can vary. This effect can cause large variations in the recorded multiplet amplitude (figure 3). The total number of standing-wave oscillations around the great circle remains constant, but $k > l + ½$ in regions with relatively slow velocities, and $k < l + ½$ in fast regions. On a spherical earth model, μ_0 varies as a linear combination of

$X_l^m(\Delta)$, $m = 0,1,2$ and their derivatives. On an aspherical earth model, μ_0 can be represented approximately by replacing Δ with $\tilde{\Delta}$, where

$$\tilde{\Delta} = \Delta - \frac{a}{(l + \frac{1}{2})\, U({}_n\omega_l)} \int_0^{\Delta} d\xi \, (\delta\omega_{local} - \overline{\delta\omega}) \tag{21}$$

This WKBJ phase-integral is performed on the source-receiver great circle. The inclusion of the great-circle average $\overline{\delta\omega}$ is necessary for the hybrid multiplet to retain an integer number of wavelengths around the earth. The first multiplet moment μ_1 suffers the same perturbation to its spatial phase along the great circle, so that the formula (18) for the multiplet location parameter λ is minimally altered. (21) is the modal equivalent of the major arc/minor arc decomposition used in long-period surface wave inversions. Estimates of $\overline{\delta\omega}$ from multiplet frequency shifts, like great-circle phase velocity measurements, sense the even part of the lateral structure. Multiplet amplitudes, like minor-arc surface wave phase velocities, are sensitive principally to odd-degree lateral structure. It can be seen from (21) that structure with $\sin p\xi$, $\cos p\xi$ dependence along the source-receiver great circle will affect μ_0 with sensitivity p^{-1}. The perturbation vanishes at the source. The sensitivity of μ_0 to even-degree structure vanishes near the source antipode. Pollitz, Park and Dahlen (1987) show that many seismic records exhibit amplitude anomalies that follow the predictions of (21). They note, however, many discrepancies that may be caused by other types of coupling, short-wavelength structure, and difficulties in obtaining sufficiently accurate measurements of μ_0.

The phase integral (21) dominates dispersion-branch coupling in the presence of long-wavelength lateral structure. Several higher-order terms are derived by Park (1987) and Romanowicz (1987) that express the dependence of μ_0 on the transverse derivatives of $\delta\omega_{local}$ along the source-receiver great circle. Park (1987) has grouped these effects in terms of a focusing $\delta A(\Delta)$ of the radiation pattern and apparent rotations of the receiver and source azimuths Φ_r and Φ_s. These effects are analogous to the focussing and refraction of surface waves. In fact, the formulas are nearly identical to those of Woodhouse and Wong (1986) for the first-arriving surface wavepacket. The effects on μ_0 are relatively small except within 30° of the source antipode: $\frac{\delta A}{A} < .04$, $\delta\Phi_s$, $\delta\Phi_r < 3^o$. At the source antipode the asymptotic formulas break down and the predicted perturbation to μ_0 has a singularity.

Romanowicz (1987) and Mochizuki (1986b) derive the phase-integral (21) from the short time expansion of mode-mode coupling derived by Woodhouse (1983). The derivation of Park (1987) is equivalent, but identifies an intermediate approximation step in which the coupling interaction is represented as a set of matrix multiplications rather than as a line integral of $\delta\omega_{local}$ and its derivatives. This intermediate step is called 'subspace projection,' as the algorithm attempts to isolate the subspace of vibrational motion corresponding to a chosen hybrid multiplet with a projection of the available spherical-earth singlets. The projection itself is approximate, but includes all first-order coupling interactions. No path from source to receiver is specified by the algorithm, so that it remains regular at the antipodal caustic. Subspace projection can be used to check the WKBJ phase integral formula for μ_0, and by extension the major-arc/minor-arc surface wave decomposition. Park (1987) found good agreement between the methods for

$30^o \leq \Delta \leq 150^o$ using model M84A, although the δA , $\delta\Phi_s$ and $\delta\Phi_r$ perturbations are sometimes important. Agreement near the source and its antipode is often poor for general moment tensor sources. The δA , etc. perturbations aid the match in spatial phase between the methods, but the antipodal singularities in these terms degrade the reliability of the phase integral prediction for μ_0 .

6. Rotational effects

The effect of the earth's rotation on free oscillation frequencies was noticed in the observations of the 1960 Chilean event, in which the rotational splitting of the modes ${}_0S_2$ and ${}_0S_3$ was partially resolved (Benioff, Press and Smith, 1961; Ness, Harrison and Slichter, 1961). Backus and Gilbert (1961) showed that rotation splits the singlet frequencies of an isolated multiplet according to the amount of angular momentum they possess about the earth's rotation axis. The singlet eigenfunctions depend on a single spherical harmonic Y_l^m and suffer a frequency perturbation proportional to azimuthal number m . Luh (1974) calculated mixed-type coupling interaction between multiplets ${}_nS_l$ and ${}_nT_{l\pm1}$. Previously, McDonald and Ness (1961) had predicted that mixed-type interaction due to Coriolis force would deflect seismic particle motion and cause toroidal multiplets to appear on vertical-component spectra. Mixed-type coupling, however, was thought to have minor influence on the seismogram for f > 1mHz until Masters, Park and Gilbert (1983) verified these predictions with observations of hybrid multiplets composed of the modes ${}_0S_l$, ${}_nT_{l+1}$, $l = 9, \cdots, 23$.

Masters, et al (1983) find that Coriolis interaction biases averaged measurements of frequency and Q . The existence of significant Coriolis coupling will likely also bias the determination of long-period source mechanisms and lateral structure. The ${}_nS_l - {}_nT_{l-1}$ mode branch crossing occurs at $l = 32$ in models 1066A and PREM, after which the possible Coriolis coupling partners are widely spaced in frequency. Since ${}_0f_{32}^S = 3.993$mHz, 250s surface waves are susceptible to Coriolis coupling. Masters, et al (1983) conclude that existing spherical earth models must be modified to place the ${}_nS_l - {}_nT_{l-1}$ mode branch crossing at l between 33 and 34, but this modification would only shift slightly the period of maximum Coriolis interaction. Park (1986) presents coupled-mode synthetic seismograms that demonstrate the time-domain waveform distortions caused by mixed-type Coriolis coupling. Figure 4 shows coupled-mode synthetic IDA records using model M84A for the June 22, 1977 Tonga event as would be recorded at ESK (Eskdalemuir, Scotland). Only fundamental modes with $l \leq 56$ are used in these synthetics. A lowpass filter is applied to suppress 'ringing' at the cutoff frequency. Coriolis coupling among spheroidal and toroidal modes disrupts the envelope of the seismogram drastically; waveforms appear to arrive with the Love wave velocity. It would be difficult to infer lateral structure from such a record. Park (1986) shows examples of similar distortions in selected IDA records (e.g. figure 5).

Park (1986) shows, by varying selected parameters in synthetic calculations, that mixed-type Coriolis coupling has a simple geometric dependence. Seismic particle motion is deflected by Coriolis force from spheroidal to toroidal polarity and vice versa. In the travelling wave, polarity conversion occurs mostly where the Coriolis force is strongest. Coriolis force on the rotating earth varies from zero at the equator to a maximum at the

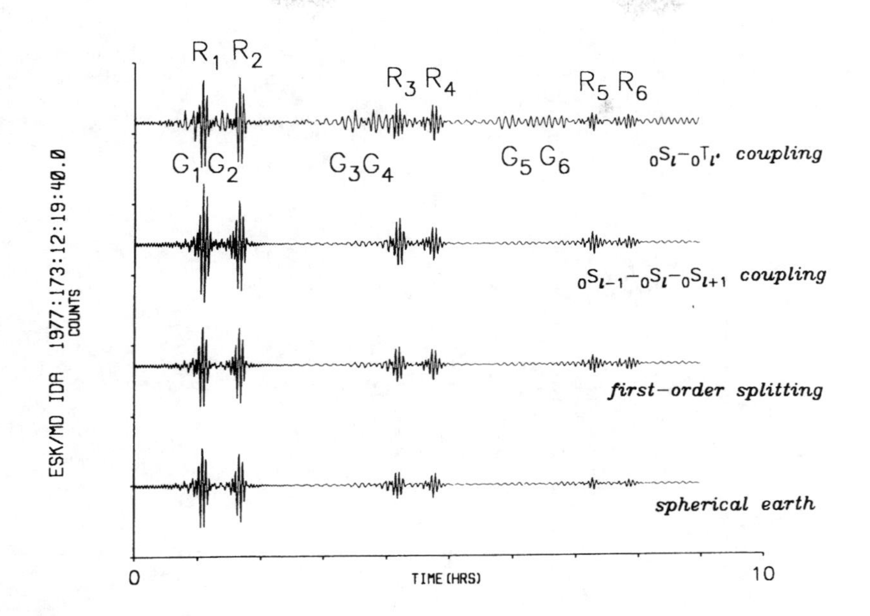

Figure 4. Coupled-fundamental-mode synthetic records using model M84A, rotation and hydrostatic ellipticity for the June 22, 1977 event in Tonga as would be observed at IDA station ESK (Eskdalemuir, Scotland). The figure compares a spherical-earth modal sum with the isolated multiplet approximation, dispersion branch coupling, and mixed-type coupling due principally to Coriolis force. The coupling schemes are described in the text.

North and South Poles. Therefore surface waves which propagate near the poles will suffer the largest effects. On vertical-component records (which normally contain no evidence of toroidal motion), this conversion is most evident for source azimuths corresponding to Rayleigh radiation nodes and Love radiation maxima. The ideal geometry for observing Coriolis coupling effects such as those in figures 4 and 5 on vertical-component records is a high-latitude great-circle path from a strike-slip event with north-south or east-west strike. The same holds true for observing toroidal-to-spheroidal conversion on the radial horizontal component. Optimal conditions for observing spheroidal-to-toroidal conversion occur if the source strike is rotated 45° from the geographic axes, so that the Love wave radiation node aligns north-south. The paucity of records with optimal mode-conversion geometry explains why such effects have not been commonly-recognized. The bias in lateral structure inversions caused by modest amounts of Coriolis coupling has yet to be assessed.

7. Future directions

Despite its potential, the best way to apply coupled-mode calculations to real-earth inverse problems remains an open question. Tsuboi, Geller, and Morris (1985) advocate a full-

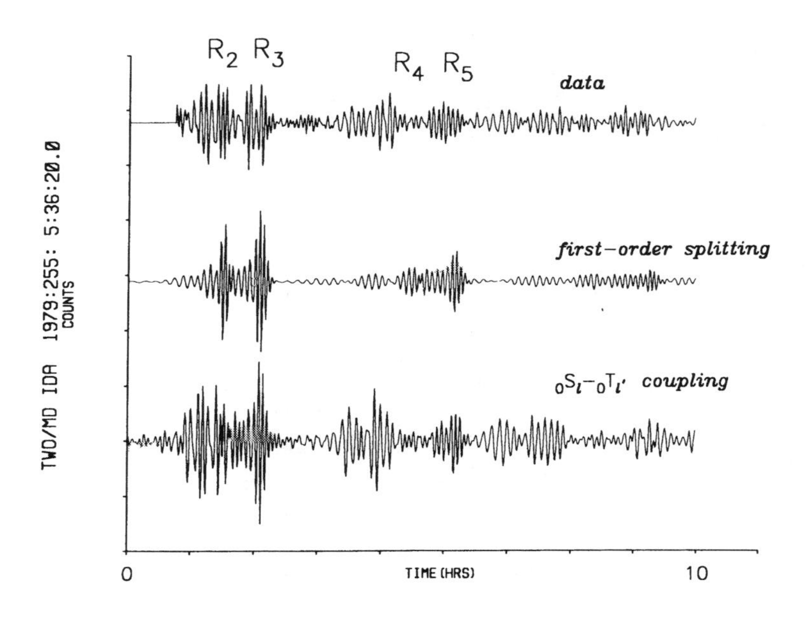

Figure 5. Data-synthetic comparison for the September 12, 1979 New Guinea event (M_S = 7.6, h = 33 km (ISC)) at IDA station TWO. The source-receiver great circle passes within 5° of the rotation axis for this geometry, enhancing the effects of Coriolis coupling between spheroidal and toroidal motion. The first Rayleigh wave arrival R_1 was discarded due to instrument nonlinearity. Note the anomalous wavepackets that precede R_4 on both data and mixed-type synthetic, but not in the record calculated with the isolated multiplet approximation.

blown linearized inversion for structure using the hybrid eigenfrequencies of coupled systems as data. They demonstrate the feasibility of such an enterprise using a system containing over 300 singlets. Their test examples, however, assume that a good first guess for the asphericity is known, and the perturbation is small compared to this first guess. Another difficulty is that the individual hybrid eigenfrequencies are notoriously hard to measure. Ritzwoller, et al (1986) demonstrate the potential for bias in systems of only a few hybrid singlets.

Another application of coupled-mode theory emphasizes linearizing the response of hybrid multiplets to asphericity. Many approaches have been followed, several of which can be traced back to Woodhouse (1983). Time-domain linearizations analogous to the Born approximation have been developed and applied to data. Prescriptions for the oscillation envelopes of hybrid multiplets promise more accurate time-domain parameterizations. Recent detailed investigations of the behavior of the zeroth and first frequency moments μ_0 and μ_1 of hybrid multiplets has led to improved frequency-domain inversion algorithms.

Much work remains to understand the effects of general anisotropy on hybrid multiplets. The effects of rough ($s \approx l$) lateral structure have been neglected in most theoretical investigations, perhaps because powerful asymptotic tools like WKBJ theory are not readily applicable. Such structure, however, is not absent from the upper mantle, and its effects on the seismogram must be estimated even if the modal data set is not large enough to constrain it uniquely. Snieder (1986) develops a scattering formalism for surface waves in which a Born-like approximation incorporates off-path structure explicitly rather than through the derivatives of $\delta\omega_{local}$ transverse to the source-receiver great circle. The 'dyadic Green's function' used by Snieder (1986) to represent the scattering interaction is similar in form to intermediate steps in the asymptotic-coupling derivations of Romanowicz (1987) and Park (1987). The approximation $l \gg s_{max}$ used in the latter studies can be relaxed at the cost of requiring an area integral over the sphere rather than a line integral over the source-receiver great circle. Such a cost is necessary to model accurately mode-type conversion, as spheroidal-toroidal interaction is greatest for $s \geq l$.

Coupled-mode seismograms, although not linearly related to asphericity, can be useful checks on asymptotic schemes and lateral structure inversions. Accuracy is limited, however, by restrictions on the number of spherical-earth singlets available for the coupling calculation, which in turn is limited by computer resources and numerical roundoff. For instance, in the coupled-mode seismograms reported by Park (1986), the hybrid multiplet on model M84A corresponding to ${}_0S_{50}$ was composed of the 303 spherical-earth singlets of ${}_0S_{49}$, ${}_0S_{50}$ and ${}_0S_{51}$. This was near the upper limit of the computing resources then available. The derivations of Park (1987), Romanowicz (1987) and Mochizuki (1986b) make clear, however, that interactions between all multiplets ${}_0S_l$, $l = 42\text{–}58$ must be included to represent the behavior of the phase integral (21) for ${}_0S_{50}$. This calculation would include 1717 singlets, and is not feasible using straightforward eigensystem decomposition routines, at least for the near future. Dahlen (1987) shows how to manipulate the subspace-projection algorithm of Park (1987) to calculate narrow-band synthetic seismograms for individual hybrid multiplets with error uniformly of order ε^2 over a time interval of order $\|\varepsilon\Lambda_{00}\|^{-1}$, where ε is the maximum applicable coupling parameter (see equation (20)) and $\|\Lambda_{00}\|$ is a generalization of (13), roughly equal to half the splitting width $\Delta\omega$ of the multiplet. Since $\varepsilon < .1$ for the vast majority of cases, this approach offers a powerful tool for calculating coupled-mode seismograms.

The coupled mode formalism is more accurate than are pure-path, ray tracing, and other travelling-wave algorithms for calculating long-period seismograms of extended duration and for calculating frequency-domain observables. With the increasing availability of cheap and powerful computers, the study of coupled free oscillations has grown from purely theoretical studies to large-scale numerical experiments. It is now feasible to compare coupled-mode seismograms with long-period data. This will lead to an unprecedented ability to model unusual features of the seismogram in an accurate manner, especially interactions (e.g. Coriolis coupling) that do not have straightforward travelling-wave representations.

Acknowledgements

I thank F.A. Dahlen for a critical reading of the manuscript. I thank Guust Nolet for his invitation to write this article, and his patience with its tardy completion. This work has been supported by an award from the Atlantic Richfield Junior Faculty Program in Science and Engineering, and by NSF grant EAR-8617935.

References

Agnew, D.C., Berger, J., Buland, R., Farrell, W., and Gilbert, F., 1976. International deployment of accelerometers: A network for very long period seismology, *Trans. AGU Eos*, **57**, 180-188.

Ashcroft, N.W., and Mermin, N.D., 1976. *Solid State Physics*, Holt, Rinehart, and Winston, New York.

Backus, G.E., and Gilbert, F., 1961. The rotational splitting of the free oscillations of the earth, *Proc. Nat. Acad. Sci. USA*, **47**, 362-371.

Benioff, H., Press, F., and Smith, S., 1961. Excitation of the free oscillations of the earth by earthquakes, *J. Geophys. Res.*, **66**, 605-619.

Buland, R., Yuen, D.A., Konstanty, K., and Widmer, R., 1985. Source phase shift: A new phenomenon in wave propagation due to anelasticity, *Geophys. Res. Lett.*, **12**, 569-572.

Dahlen, F.A., 1968. The normal modes of a rotating, elliptical earth, *Geophys. J. R. Astr. Soc.*, **16**, 329-367.

Dahlen, F.A., 1976. Reply, *J. Geophys. Res.*, **81**, 4951-4956.

Dahlen, F.A. 1979. The spectra of unresolved split normal mode multiplets, *Geophys. J. R. Astr. Soc.*, **58**, 1-33.

Dahlen, F.A., 1980. Splitting of the free oscillations of the earth, in *Proceedings of the Enrico Fermi International School of Physics*, **78**, A.M. Dziewonski and E. Boschi, eds., North Holland Publishing, Amsterdam, 88-126.

Dahlen, F.A., 1987. Multiplet-multiplet coupling and the calculation of synthetic long-period seismograms, *Geophys. J. R. Astr. Soc.*, in press.

Dahlen, F.A., and Henson, I.H., 1985. Asymptotic normal modes of a laterally heterogeneous earth, *J. Geophys. Res.*, **90**, 12653-12681.

Davis, J.P., 1987. Local eigenfrequency and its uncertainty inferred from fundamental spheroidal mode frequency shifts, *Geophys. J. R. Astr. Soc.*, **88**, 693-722.

Davis, J.P., and Henson, I.H., 1986. Validity of the great circular average approximation for inversion of normal mode measurements, *Geophys. J. R. Astr. Soc.*, **85**, 69-92.

Dziewonski, A.D., and Anderson, D.L., Preliminary reference earth model, *Phys. Earth Plan. Inter.*, **25**, 297-356.

Engdahl, E.R., Peterson, J., and Orsini, N.A., 1982. Global digital networks -- Current status and future directions, *Bull. Seismol. Soc. Am.*, **72**, s243-s259.

Garbow, B.S., Boyle, J.M., Dongarra, J., and Moler, C.B., 1977. *Matrix Eigensystem Routines - EISPACK Guide Extension*, Lecture Notes in Computer Science, Springer-Verlag, New York.

Giardini, D., Li, X., and Woodhouse, J.H., 1987. Three dimensional structure of the Earth from splitting in free oscillation spectra, *Nature*, **325**, 405-411.

Gilbert, F., and Dziewonski, A.M., 1975. An application of normal mode theory to the retrieval of structural parameters and source mechanisms from seismic spectra, *Philos. Trans. R. Soc. London Ser. A*, **278**, 187-269.

Halmos, P.R., 1974. *Finite-Dimensional Vector Spaces*, Springer Verlag, New York.

Henson, I.H., 1987. Computational methods for normal mode synthetic seismograms, *Geophys. J. R. Astr. Soc.*, in press.

Henson, I.H., and Dahlen, F.A., 1986. Asymptotic normal modes of a laterally heterogeneous earth, 2, further results, *J. Geophys. Res.*, **91**, 12467-12481.

Hudson, J.A., 1980. *The Excitation and Propagation of Elastic Waves*, Cambridge University Press, Cambridge.

Jordan, T.H., 1978. A procedure for estimating lateral variations from low-frequency eigenspectra data, *Geophys. J. R. Astr. Soc.* , **52**, 441-455.

Keller, J.B., 1985. Semiclassical mechanics, *SIAM Review*, **27**, 485-504.

Luh, P.C., 1974. The normal modes of a rotating, self-gravitating, inhomogeneous earth, *Geophys. J. R. Astr. Soc.*, **38**, 187-224.

MacDonald, G.J.F., and Ness, N.F., 1961. A study of the free oscillations of the earth, *J. Geophys. Res.*, **66**, 1865-

1911.

Masters, G., and Gilbert, F., 1983. Attenuation in the earth at low frequencies, *Philos. Trans. R. Soc. London Ser. A*, **308**, 479-522.

Masters, G., Jordan, T.H., Silver, P.G., and Gilbert, F., 1982. Aspherical earth structure from fundamental spheroidal-mode data, *Nature*, **298**, 609-613.

Masters, G., Park, J., and Gilbert, F., 1983. Observations of coupled spheroidal and toroidal modes, *J. Geophys. Res.*, **88**, 10285-10298.

Mikhlin, S.G., 1964. *Variational Methods in Mathematical Physics*, MacMillan, New York.

Mochizuki, E., 1986a. The free oscillations of an anisotropic and heterogeneous earth, *Geophys. J. R. Astr. Soc.*, **86**, 167-176.

Mochizuki, E., 1986b. Free oscillations and surface waves of an aspherical earth, *Geophys. Res. Lett.*, **13**, 1478-1481.

Morris, S.P., Geller, R.J., Kawakatsu, H., and Tsuboi, S., 1987. Variational normal mode computations for three laterally heterogeneous earth models, *Phys. Earth Plan. Inter..*, submitted.

Nakanishi, I., and Anderson, D.L., 1984. Measurements of mantle wave velocities and inversion for lateral heterogeneity and anisotropy. I. Analysis by the single station method, *Geophys. J. R. Astr. Soc.*, **78** , 573-617.

Ness, N.F., Harrison, J.C., and Slichter, L.B., 1961. Observations of the free oscillations of the earth, *J. Geophys. Res.*,**66**, 621-629.

Park, J., 1986. Synthetic seismograms from coupled free oscillations: effects of lateral structure and rotation, *J. Geophys. Res.*, **91**, 6441-6464.

Park, J., 1987. Asymptotic coupled-mode expressions for multiplet amplitude anomalies and frequency shifts on an aspherical earth, *Geophys. J. R. Astr. Soc.*, in press.

Park, J., and Gilbert, F., 1986. Coupled free oscillations of an aspherical, dissipative, rotating earth: Galerkin theory, *J. Geophys. Res.*, **91**, 7241-7260.

Pollitz, F., Park, J., and Dahlen, F.A., 1987. Observations of free oscillation amplitude anomalies, *Geophys. Res. Lett.*, submitted.

Ritzwoller, M., Masters, G., and Gilbert, F., 1986. Observations of anomalous splitting and their interpretation in terms of aspherical structure, *J. Geophys. Res.*, **91**, 10203-10228.

Romanowicz, B., 1987. Multiplet-multiplet coupling due to lateral heterogeneity: asymptotic effects on the amplitude and frequency of the earth's normal modes, *Geophys. J. R. Astr. Soc.*, in press.

Romanowicz, B., and Roult, G., 1986. First order asymptotics for the eigenfrequencies of the earth and application to the retrieval of large-scale lateral variations of structure, *Geophys. J. R. Astr. Soc.*, **87**, 209-240.

Schiff, L.I., 1968. *Quantum Mechanics*, 3rd ed., McGraw Hill, New York.

Silver, P.G., and Jordan, T.H., 1981. Fundamental spheroidal mode observations of aspherical heterogeneity, *Geophys. J. R. Astr. Soc.*, **64**, 605-634.

Smith, M.L., and Dahlen, F.A., 1973. The azimuthal dependence of Love and Rayleigh wave propagation in a slightly anisotropic earth, *J. Geophys. Res.*, **78**, 3321-3333.

Snieder, R., 1986. 3-D linearized scattering of surface waves and a formalism for surface wave holography, *Geophys. J. R. Astr. Soc.*, **84**, 581-606.

Tanimoto, T., 1984. A simple derivation of the formula to calculate long-period seismograms in a heterogeneous earth by normal mode summation, *Geophys. J. R. Astr. Soc.*, **77**, 275-278.

Tanimoto, T., 1986. Free oscillations of a slightly anisotropic earth, *Geophys. J. R. Astr. Soc.*, **87**, 493-518.

Tanimoto, T., and Anderson, D.L., 1985. Lateral heterogeneity and anisotropy of the upper mantle 100-200s, *J. Geophys. Res.*, **90**, 1842-1858.

Tanimoto, T. and Bolt, B.A., 1983. Coupling of torsional modes in the earth, *Geophys. J. R. Astr. Soc.*, **74**, 83-96.

Tsuboi, S., Geller, R.J., and Morris, S.P., 1985. Partial derivatives of the eigenfrequencies of a laterally heterogeneous earth model, *Geophys. Res. Lett.*, **12**, 817-820.

Woodhouse, J.H., 1976. On Rayleigh's principle, *Geophys. J. R. Astr. Soc.*, **46**, 11-22.

Woodhouse, J.H., 1980. The coupling and attenuation of nearly-resonant multiplets in the earth's free oscillation spectrum, *Geophys. J. R. Astr. Soc.*, **61**, 261-283.

Woodhouse, J.H., 1983. The joint inversion of seismic waveforms for lateral variations in Earth structure and earthquake source parameters, in *Proceedings of the Enrico Fermi International School of Physics*, **85**, H. Kanamori and E. Boschi, eds., pp366-397, North Holland Publishing, Amsterdam.

Woodhouse, J.H., and Dahlen, F.A., 1978. The effect of a general aspherical perturbation on the free oscillations of the earth, *Geophys. J. R. Astr. Soc.*, **53**, 335-354.

Woodhouse, J.H., and Dziewonski, A.M., 1984. Mapping the upper mantle: Three-dimensional modeling of earth structure by inversion of seismic waveforms, *J. Geophys. Res.*, **89**, 5953-5986.

Woodhouse, J.H., and Girnius, T.P., 1982. Surface waves and free oscillations in a regionalized earth model, *Geophys. J. R. Astr. Soc.*, **68**, 653-673.

Woodhouse, J.H., and Wong, Y.K., 1986. Amplitude, phase and path anomalies of mantle waves, *Geophys. J. R. Astr. Soc.*, **87**, 753-774.

J. PARK, Kline Geology Laboratory, Yale University, PO Box 6666, New Haven Connecticut 06511, USA.

Chapter 3

Surface waves in weakly heterogeneous media

K. Yomogida

The theory of the propagations of surface waves in the media with weak lateral heterogeneities is reviewed. In such media we can represent wavefields as the conventional normal modes in the vertical profiles while the horizontal propagations can be treated as rays spreading on the surface with the phase velocity distributions. The formulations of surface waves share many common features with acoustic or elastic body waves in two-dimensional media. Therefore, the direct applications of the recently developed theories for acoustic or elastic body waves such as the Gaussian beam method and the Maslov method are applicable to the surface wave problems. Because of the simplicity and the capability of the rapid calculations, the present approach can be applied to the inverse problems with the use of Born approximations. We are now in the stage to invert both the amplitude and the phase anomalies for the lateral heterogeneities of the earth.

1. Introduction

The theory on propagation and excitation of surface waves in a vertically heterogeneous but laterally homogeneous medium is now established and routinely used among seismologists (e.g. Chapter 7 of Aki and Richards (1980)). Such a theory is still being used for studies of lateral heterogeneities in the sense of geometrical ray theory: the heterogeneity is assumed to be so smooth that the Earth is considered as a stack of homogeneous areas and then the observed travel time (or phase delay) is a sum of travel times for each homogeneous area. For the actual data, even the most sophisticated studies so far such as Woodhouse and Dziewonski (1984) are based on the above simple approach.

N.J. Vlaar, G. Nolet, M.J.R. Wortel & S.A.P.L. Cloetingh (eds), Mathematical Geophysics, 53-75.

Even though such a method has imaged the gross lateral heterogeneities in the Earth, seismologists have realized some deficiencies. For example, in heterogeneous media surface waves no longer propagate along great circles and the discrepancies in travel times between the great circle path and the exact ray path cannot be neglected in some cases. Also, there are many observations of amplitude anomalies for surface waves (e.g., Lay and Kanamori, 1984) which cannot be explained by the above approach. Thus, we need to develop the theory of surface wave propagations in laterally heterogeneous media.

Although recent improvements of both theory and computational power are remarkable, we are still far from the stage of calculating the exact solutions of surface wave problems in media with general lateral heterogeneities. Instead, since most of the Earth is strongly stratified horizontally and the lateral variations are fairly smooth, we can develop efficient methods to simulate surface wave propagations by some approximations. The essence of such methods is the different treatment of wavefields between vertical and horizontal directions. Vertically the wavefield is expressed by the normal mode theory while surface waves are considered to propagate horizontally in a similar way to ray theory. Such an approach has been popular for a long time in the field of acoustic wave propagations in the ocean (e.g., Pierce, 1965; Burridge and Weinberg, 1976) and we can find some literature related to this approach in seismology back to the early 1960's (e.g., De Noyer, 1961; Babich and Rusakova, 1962). Much clearer formulations were found in the 1970's (e.g., Gjevik, 1973; Woodhouse, 1974; Babich et al., 1976). However, these methods still use conventional geometrical ray theory in the horizontal directions and still require large amounts of computations and in some cases results are unstable. Stimulated by recent theoretical developments of body wave propagations in heterogeneous media, such as the works by Červený and Pšenčík (1983) and Chapman and Drummond (1982), practical methods to calculate surface wavefields in laterally heterogeneous media have been established (e.g., Hudson, 1981; Yomogida, 1985; Saastamoinen, 1986) from the original forms of Kirpichinikova (1969). In this paper we shall look through the above developments although the fundamental approach is based on the works by the author (Yomogida, 1985; Yomogida and Aki, 1985; 1986), that is, the "Gaussian beam method". Since Červený (1985a,b) gave excellent summaries on the Gaussian beam method for body wave problems, here we shall emphasize only its unique character for surface wave problems.

2. Asymptotic ray theory

Even though our purpose is to derive formulations of surface waves in slightly heterogeneous media, we shall start with scalar waves and seismic body waves then compare these results with surface wave problems in order to give easier explanations for the concepts of the final formulations for surface waves. In these processes, it will be easily noticed that the propagation of surface waves shares many common concepts with that of scalar waves and body waves, especially 2-D body waves. This is why we can apply most of the recently developed theories for body wave propagations in heterogeneous media to surface wave problems.

2.a Scalar waves

Let us start with the simple scalar wave equation:

$$\nabla^2 U(\mathbf{x},t) - \frac{1}{C(\mathbf{x})^2}\frac{\partial^2}{\partial t^2} U(\mathbf{x},t) = 0 \tag{1}$$

where U is the displacement as a function of space $\mathbf{x}$ and time t and $C(\mathbf{x})$ is the velocity. Hereafter, we neglect the body force term, i.e. the excitation term. In the asymptotic ray theory, the solutions are expanded by power series of angular frequency ω and we take only a few leading terms as $\omega \rightarrow \infty$, that is, the high frequency limit pioneered by V. M. Babich and J. B. Keller. The trial form of the solutions is called the "Ansatz". Now

$$U(\mathbf{x},t) = \exp[-i\,\omega(t-\tau(\mathbf{x}))] \sum_{k=0}^{\infty} U_k(\mathbf{x})(-i\,\omega)^{-k} \tag{2}$$

where $\tau(\mathbf{x})$ is the travel time. Near the wavefront, the exponential term is fluctuating much more rapidly than the power-series terms. For details, the readers may refer to section 4.4 of Aki and Richards (1980). Substituting (2) into (1), we take the leading terms:

$$(\nabla\,\tau)^2 = \frac{1}{C^2} \quad \text{for } k = 2, \tag{3}$$

which is called the eikonal equation and

$$(\nabla^2\tau)\, U_0 + 2\nabla\tau.\nabla U_0 = 0 \quad \text{for } k = 1, \tag{4}$$

called the transport equation. In this paper we shall investigate terms up to this order. The eikonal equation (3) can be interpreted as follows. The wavefront is the surface where τ = constant. So the direction of $\nabla\tau$ is perpendicular to the wavefront. Therefore, (3) gives

$$\nabla\tau = \frac{1}{C}\mathbf{v} = \mathbf{p}, \tag{5}$$

where $\mathbf{v}$ is the unit vector perpendicular to the wavefront, which is the direction of the "ray path". $\mathbf{p}$ is called the slowness vector. If we define the arclength s along the ray path as

$$ds = \mathbf{v}.d\mathbf{x} \tag{6}$$

then the travel time is expressed by the integration of the slowness along the ray path:

$$\tau(s) = \int_0^s \nabla\tau\cdot d\mathbf{x} = \int_0^s \frac{d\zeta}{C(\zeta)}. \tag{7}$$

From (6) we get

$$\frac{d\mathbf{x}}{ds} = \mathbf{v} = C\,\mathbf{p} \tag{8}$$

Also, differentiating the slowness vector (5) with respect to s,

$$\frac{d\mathbf{p}}{ds} = \frac{d}{ds}(\nabla\tau) = (\mathbf{v}.\nabla)(\nabla\tau) = C\,(\nabla\tau.\nabla)\nabla\tau = \frac{C}{2}\nabla((\nabla\tau)^2) = \nabla(\frac{1}{C}) \tag{9}$$

where we used eikonal equation (3). Equations (8) and (9) determine the ray path, so we call them "ray tracing equations".

While the eikonal equation (3) determines the phase term, the transport equation (4) gives the measure of amplitude. Let us take the ray-coordinate (s, γ_1, γ_2) where s is tangent to the ray path in (6), and γ_1 and γ_2 are perpendicular to the ray path, that is, spanning the plane of the wavefront. Then, (5) gives

$$\frac{\partial \tau}{\partial s} = \frac{1}{C}, \quad \frac{\partial \tau}{\partial \gamma_1} = \frac{\partial \tau}{\partial \gamma_2} = 0 \tag{10}$$

and the Laplacian of τ is expressed as

$$\nabla^2 \tau = \frac{1}{h_1 h_2 h_3} \frac{\partial}{\partial s} \left[\frac{h_2 h_3}{h_1} \frac{\partial \tau}{\partial s} \right] = \frac{1}{h_1 h_2 h_3} \frac{\partial}{\partial s} \left[\frac{h_2 h_3}{h_1} \frac{1}{C} \right] \tag{11}$$

where h_1, h_2 and h_3 are scaling factors of the coordinates s, γ_1, and γ_2, respectively. Therefore, $h_1 = 1$. We define the geometrical spreading

$$J = h_2 h_3 = \left[\frac{\partial \mathbf{x}}{\partial \gamma_1} \times \frac{\partial \mathbf{x}}{\partial \gamma_2} \right] = \frac{\partial (x_2, x_3)}{\partial (\gamma_1, \gamma_2)}. \tag{12}$$

J is the measure of the cross-section of the considered ray tube. (11) and (12) may give

$$\nabla^2 \tau = \frac{1}{J} \frac{d}{ds} \left[\frac{J}{C} \right] \tag{13}$$

Using (13) and (5), transport equation (4) may become

$$\frac{1}{J} \frac{d}{ds} \left[\frac{J}{C} \right] U_0 + \frac{2}{C} \frac{d}{ds} U_0 = 0.$$

That is

$$\frac{d}{ds} \left[\frac{J}{C} U_0^2 \right] = 0. \tag{14}$$

This means

$$\frac{J(s)}{C(s)} U_0(s)^2 = \text{constant along the ray} = \Phi^2.$$

Since $J(s)$ is the measure of the cross-section of the ray tube and the total energy is proportional to the square of amplitude, equation (14) means that the energy flow along the ray tube is constant. The final solution (2) may be written as

$$u(s,t) = \Phi \sqrt{\frac{C(s)}{J(s)}} \exp \left[-i \omega (t - \int_0^s \frac{d\zeta}{C(\zeta)}) \right] \tag{15}$$

where the integration is along the ray path. We shall show how to calculate $J(s)$ in the next chapter.

2.b Elastic body waves

Let us proceed to elastic body waves. Now we deal with the elastodynamic equation

$$\rho\frac{\partial^2 u_i}{\partial t^2}=\frac{\partial}{\partial x_j}\left[C_{ijkl}\frac{\partial u_k}{\partial x_l}\right] \tag{16}$$

where ρ is the density and C_{ijkl} is the elastic constant. In this paper we shall consider only the isotropic case:

$$C_{ijkl}=\lambda\delta_{ij}\delta_{kl}+\mu(\delta_{ik}\delta_{jl}+\delta_{il}\delta_{jk}). \tag{17}$$

where λ and μ are Lamé constants. Since the difference from the scalar wave is that there is a polarization in body waves, the trial form ''Ansatz'' may be written as

$$\mathbf{u}(\mathbf{x},t)=e^{-i\omega(t-\tau(\mathbf{x},t))}\sum_{k=0}^{\infty}\mathbf{U}^{(k)}(-i\omega)^{-k} \tag{18}$$

Substituting this into (16), we consider only the leading terms. For $k=2$,

$$\left[-\rho\delta_{ij}+(\lambda+\mu)\frac{\partial\tau}{\partial x_i}\frac{\partial\tau}{\partial x_j}+\mu\delta_{ij}\frac{\partial\tau}{\partial x_k}\frac{\partial\tau}{\partial x_k}\right]U_j^{(0)}=0 \quad (i=1,2,3). \tag{19}$$

Actually, these are three linear simultaneous equations. For the solutions $U_j^{(0)}$ to be non-zero, the determinant of the coefficient matrix (i.e., det[ ]) has to be zero. This may be written as

$$\left(\frac{\partial\tau}{\partial x_k}\frac{\partial\tau}{\partial x_k}-\frac{1}{\alpha^2}\right)\left(\frac{\partial\tau}{\partial x_k}\frac{\partial\tau}{\partial x_k}-\frac{1}{\beta^2}\right)^2=0, \tag{20}$$

that is,

$$\left[(\nabla\tau)^2-\frac{1}{\alpha^2}\right]\left[(\nabla\tau)^2-\frac{1}{\beta^2}\right]^2=0,$$

where $\alpha^2=\frac{(\lambda+2\mu)}{\rho}$ and $\beta^2=\frac{\mu}{\rho}$, that is, P and S wave velocities, respectively. (20) is the eikonal equation and we notice that there are two cases: one is $(\nabla\tau)^2=\alpha^{-2}$, the P wave and the other is $(\nabla\tau)^2=\beta^{-2}$, the S waves with two different polarizations. Taking the dot and the cross products of $\nabla\tau$ with (19), we get

$$\left[(\nabla\tau)^2-\frac{1}{\alpha^2}\right](\mathbf{U}^{(0)}.\nabla\tau)=0 \tag{21a}$$

and

$$\left[(\nabla\tau)^2-\frac{1}{\beta^2}\right](\mathbf{U}^{(0)}\times\nabla\tau)=0. \tag{21b}$$

Since $\nabla\tau$ is tangent to the ray path, (21a) shows that for P wave the polarization vector is directed along the ray path while (21b) shows that for S wave the particle motion is transverse to the ray path.

The term of the next order, $k = 1$, gives the transport equation:

$$\frac{d\mathbf{U}^{(0)}}{ds} + \frac{1}{2}\mathbf{U}^{(0)}[C\nabla^2\tau + \frac{1}{\rho C^2}\frac{d}{ds}(\rho C^2)] = 0. \tag{22}$$

where C is either α or β. From (13), we get

$$\frac{d}{ds}[J\rho C(\mathbf{U}^{(0)})^2] = 0, \tag{23}$$

that is,

$$J\rho C(\mathbf{U}^{(0)})^2 = \text{constant along the ray path} = \Phi^2.$$

Thus, the final result is

$$\mathbf{u}(s,t) = \frac{\Phi}{\sqrt{\rho(s)C(s)J(s)}}\,\mathbf{r}\exp\left[-i\omega(t-\int_0^s\frac{d\zeta}{C(\zeta)})\right] \tag{24}$$

where $C = \alpha$ and $\mathbf{r} = \mathbf{t}$, a unit vector tangent to the ray path for the P wave, and $C = \beta$ and $\mathbf{r} = \mathbf{n}$, normal to the ray path for the S wave. The result for elastic body waves (24) has a very similar form to (15) for scalar waves.

2.c *Surface waves*

Now let us proceed to surface wave problems, our goal. In this case, the heterogeneities in the horizontal direction are assumed to be much weaker than in the vertical direction. For variations of wavefields in each direction to be of a similar length scale, the original coordinate system (x, y, z, t) is changed into the new system:

$$\eta_1 = X = \varepsilon x, \quad \eta_2 = Y = \varepsilon y, \quad \eta_3 = z \quad \text{and } T = \varepsilon t. \tag{25}$$

Here the medium is semi-infinite with z-axis directed downward. Elastic constants λ and μ, and density ρ are varying slowly in the horizontal directions and the order of variations are

$$\frac{\partial\lambda}{\partial\eta_i} = O(1), \quad \frac{\partial\mu}{\partial\eta_i} = O(1) \text{ and } \frac{\partial\rho}{\partial\eta_i} = O(1) \quad (i = 1, 2, 3). \tag{26}$$

ε is a small parameter indicating the order of lateral heterogeneities. The free surface $z = \zeta(x, y)$ is allowed to have smooth undulations:

$$\frac{\partial\zeta}{\partial\eta_i} = O(1) \qquad (i = 1, 2). \tag{27}$$

Instead of expressing it in the original coordinates, the elastodynamic equation (16) should be written in the re-scaled coordinate system (X, Y, z, T). While we used the expansion series of $(-i\omega^{-1})$ for the previous cases, here we use the expansion of ε and we take the limit of $\varepsilon\to 0$ (i.e., homogeneous case) rather than $\omega\to\infty$. Also, the phase term is not necessarily harmonic in time such as it was in the previous examples: $e^{-i\omega t}$. Thus, in this case the ansatz is

$$\mathbf{u}(x,y,z,t) = e^{i\psi(x,y,t)} \sum_{k=0}^{\infty} \mathbf{U}^{(k)}(X,Y,z,T)\varepsilon^k . \tag{28}$$

(28) has more general features compared with (2) or (18): the phase term $e^{i\psi}$ is not explicitly expressed in time, but we allow the wavenumber vector $\mathbf{k}$ and the angular frequency to vary slowly according to the following definitions:

$$\mathbf{k}(X,Y,T) = \nabla\psi \quad \text{and} \quad \omega(X,Y,T) = -\frac{\partial\psi}{\partial t}. \tag{29}$$

Also, $\mathbf{U}^{(k)}$ is allowed to vary slowly with time and space.

At first, let us consider the eikonal equation for surface waves directly derived from (29):

$$(\nabla\psi)^2 = |\mathbf{k}|^2. \tag{30}$$

For a fixed angular frequency, dividing both sides by ω^2, (30) is reduced to the previous case (3). Thus, (30) is the eikonal equation in a more general form. From (30), we can obtain ray tracing equations for surface waves. Let us take the local dispersion relation which will be shown to satisfy the characteristic equations of Love or Rayleigh waves:

$$\omega = \omega(\mathbf{k},\mathbf{x}). \tag{31}$$

Then, we get the following partial differential equation for ψ:

$$\frac{\partial\psi(\mathbf{x},t)}{\partial t} + \omega(\nabla\psi,\mathbf{x}) = 0. \tag{32}$$

This has a similar form to the Hamilton-Jacobi equation in the classical mechanics. From (29) and (30), we notice the clear analogy between geometrical ray theory and classical mechanics (see section 53 of Landau and Lifshitz, 1975). The following quantities follow similar equations:

$$\begin{array}{llll} \psi & \longleftrightarrow & S & : \textit{action} \\ \omega & \longleftrightarrow & H & : \textit{Hamiltonian} \\ \mathbf{k} & \longleftrightarrow & \mathbf{p} & : \textit{momentum} \\ \mathbf{x} & \longleftrightarrow & \mathbf{q} & : \textit{general coordinate} \end{array} \tag{33}$$

Since $\mathbf{p}$ and q form the pair of canonical equations in the classical mechanics, in the case of geometrical ray theory they may be

$$d\frac{\mathbf{x}}{dt} = \frac{\partial\omega}{\partial\mathbf{k}} \qquad d\frac{\mathbf{k}}{dt} = -\frac{\partial\omega}{\partial\mathbf{x}} \tag{34}$$

which actually correspond to ray tracing equations. If the medium is isotropic or transverse isotropic, by introducing the arclength of the ray path $ds = Udt$ where U is the local group velocity, (34) can be reduced to the same ray tracing equations as the previous ones (8) and (9) (Yomogida and Aki, 1985). Note that the velocity C in these equations corresponds to the phase velocity but not the group velocity. Another implication in the present analogy is the principle of least action expressed by

$$\delta S = \delta\int \mathbf{p}.d\mathbf{s} = 0.$$

In the geometrical ray theory this corresponds to

$$\delta\psi = \delta\int \mathbf{k}.d\mathbf{s} = 0. \tag{35}$$

This means that the phase delay along the actual ray path is extreme, i.e. Fermat's principle. This equation is more general compared to the familiar form which is derived by dividing both sides by the fixed ω:

$$\delta\tau = \delta\int \frac{1}{C} ds = 0, \tag{36}$$

which means that the travel time is extreme. Eq. (35) can be used even for anisotropic media (e.g., Tanimoto, 1986). If the Hamiltonian H does not explicitly depend on time, the total derivative of H with time is zero. Since in our case the angular frequency does not explicitly depend on t, from (34) we may obtain

$$\frac{d\,\omega(\mathbf{x},\mathbf{k})}{dt} = \frac{\partial\omega}{\partial\mathbf{x}}\cdot d\frac{\mathbf{x}}{dt} + \frac{\partial\omega}{\partial\mathbf{k}}\cdot d\frac{\mathbf{k}}{dt} = 0, \tag{37}$$

that is, ω is constant along the ray. In summary, in the present surface wave problems, in the phase term there are many common features with the previous cases.

Let us go back to the original problem and try to obtain the asymptotic solution of surface waves in laterally heterogeneous media. Substituting (28) into the elastodynamic equations with the new coordinates (X, Y, z, T), we may obtain the leading terms of order unity:

$$R_1(U_s{}^0, U_z{}^0) = [\rho\omega^2 - (\lambda+2\mu)k^2 + \frac{\partial}{\partial z}\mu\frac{\partial}{\partial z}]U_s{}^0 + ik[\lambda\frac{\partial}{\partial z} + \frac{\partial}{\partial z}\mu]U_z{}^0 = 0, \tag{38a}$$

$$L(U_n{}^0) = [\rho\omega^2 - \mu k^2 + \frac{\partial}{\partial z}\mu\frac{\partial}{\partial z}]U_n{}^0 = 0, \tag{38b}$$

$$R_2(U_s{}^0, U_z{}^0) = [\rho\omega^2 - \mu k^2 + \frac{\partial}{\partial z}(\lambda+2\mu)\frac{\partial}{\partial z}]U_z{}^0 + ik[\mu\frac{\partial}{\partial z} + \frac{\partial}{\partial z}\lambda]U_s{}^0 = 0. \tag{38c}$$

Here, we take U_s as the component tangent to the ray path and U_n as perpendicular to it on the surface. Also, if we put (28) into the boundary conditions, that is, all the components of traction must vanish at the free surface $z = \zeta$, and take only the leading terms, we may get

$$ikU_z{}^0 + \frac{\partial U_s{}^0}{\partial z} = 0, \tag{39a}$$

$$\mu\frac{\partial U_n{}^0}{\partial z} = 0, \tag{39b}$$

$$(\lambda + 2\mu)\frac{\partial U_z{}^0}{\partial z} + i\lambda k U_s{}^0 = 0. \tag{39c}$$

at $z=\zeta$. Also, the radiation condition should be satisfied:

$$U_s{}^0,\ U_n{}^0,\ U_s{}^0 \to 0 \quad \text{as } z \to \infty. \tag{40}$$

Actually, these are characteristic equations and boundary conditions of Love (U_n) and Rayleigh (U_s and U_z) waves in the laterally homogeneous media. The important point is that Love and Rayleigh waves are decoupled up to the present terms. This is true even for the transverse isotropic case (Yomogida, 1987).

2.c.1 Love waves

Let us take the phase velocity C as that of Love waves which satisfy the characteristic equation $L(U_n^0) = 0$ with (39b) and (40). In general, the phase velocity of Rayleigh waves is different from that of Love waves, thus

$$U_s = U_z = 0. \tag{41}$$

The amplitude U_n is determined by the terms of the next order, the transport equation, as are scalar or body waves. For Love waves, that equation is

$$L(U_n^1) + 2i\rho\omega \frac{\partial U_n^0}{\partial T} + i\rho \frac{\partial \omega}{\partial T} U_n^0 + i\nabla\cdot(\mu \mathbf{k})U_n^0 + 2i\mu\mathbf{k}\cdot(\nabla U_n^0) = 0 \tag{42}$$

and the boundary condition at $z = \zeta$ is

$$\frac{\partial U_n^1}{\partial z} = i\nabla\cdot U_n^0. \tag{43}$$

Hereafter, $\mathbf{x}$ and $\nabla = \frac{\partial}{\partial \mathbf{x}}$ are only horizontal components: x and y but not z. Multiplying the left hand side of (42) by U_n^0 and integrating from the surface $\zeta(\mathbf{x})$ to ∞ with respect to z, we obtain

$$\int_{\zeta(\mathbf{x})}^{\infty} U_n^0 L(U_n^1)dz + i\frac{\partial}{\partial T}\left[\omega\int_{\zeta}^{\infty}\rho(U_n^0)^2 dz\right]$$

$$+ i\nabla\cdot\left[\mathbf{k}\int_{\zeta}^{\infty}\mu(U_n^0)^2 dz\right] + i\nabla\zeta\cdot k\mu(U_n^0)^2|_{z=\zeta} = 0. \tag{44}$$

Using integration by parts, the first term of the left hand side of (44) becomes zero because of (38b), (39b) and (40). The last term (44) evaluated at $z = \zeta$ should vanish because of the boundary condition (43).

Then, we separate $U_n^0(X, Y, z, T)$ into the terms dependent on and independent of depth as $U_n^0 = A(\mathbf{x},t)l_1(\mathbf{x},z)$. $l_1(\mathbf{x},z)$ must satisfy $L(l_1) = 0$ and boundary conditions (39b) and (40), thus corresponding to the eigenfunction of Love waves for laterally homogeneous media defined by chapter 7 of Aki and Richards (1980). Here, we normalize l_1 at the surface so that $l_1(\mathbf{x},z)|_{z=\zeta} = 1$ and $A(\mathbf{x},t)$ becomes the amplitude at the surface. Therefore, (44) may be written as

$$\frac{\partial}{\partial T}(\omega A^2 I_1) + \nabla\cdot(\mathbf{k}A^2 I_2) = 0 \tag{45}$$

where I_1 and I_2 are energy integrals similarly defined by Aki and Richards (1980):

$$I_1(\mathbf{x}) = \frac{1}{2}\int_{\zeta(\mathbf{x})}^{\infty} \rho(\mathbf{x},z) l_1(\mathbf{x},z)^2 dz \tag{46}$$

$$I_2(\mathbf{x}) = \frac{1}{2}\int_{\zeta(\mathbf{x})}^{\infty} \mu(\mathbf{x},z) l_1(\mathbf{x},z)^2 dz$$

Using Rayleigh's principle, the group velocity $\mathbf{U}$ is expressed by

$$\mathbf{U} = \mathbf{k} I_2 / \omega I_1 \tag{47}$$

(see section 7.3 of Aki and Richards (1980)). From this relation and

$$\frac{d\omega}{dT} = \frac{\partial \omega}{\partial T} + \mathbf{U}\cdot\nabla\omega = 0,$$

given by (37), (45) may be written as

$$\frac{\partial}{\partial T}(I_1 A^2) + \nabla\cdot(\mathbf{U} I_1 A^2) = 0. \tag{48}$$

The second term may written as

$$\nabla\cdot(\mathbf{U} I_1 A^2) = \nabla\cdot(\mathbf{v} U I_1 A^2) = \nabla\cdot((\nabla\tau) C U I_1 A^2) = (\nabla^2\tau) C U I_1 A^2 + \nabla\tau\cdot\nabla(C U I_1 A^2)$$
$$= \frac{1}{J}\frac{\partial}{\partial S}\left[\frac{J}{C}\right] C U I_1 A^2 + \frac{1}{C}\frac{\partial}{\partial S}(C U I_1 A^2) = \frac{1}{J}\frac{\partial}{\partial S}(J U I_1 A^2)$$

from (5), (6) and (13). Since the geometrical spreading J does not depend on time implicitly, (48) may be expressed as

$$\frac{\partial}{\partial T}(A^2 J I_1) + \frac{\partial}{\partial S}(U A^2 J I_1) = 0. \tag{49}$$

This is actually the transport equation for Love waves and has a similar form to the continuity equation in fluid mechanics. Therefore, it means that

$$A^2 J I_1 U = \text{constant along the ray} = \Phi^2.$$

$J(s)$ represents the geometrical spreading in the horizontal direction and I_1 gives the vertical energy profile of the eigenfunction. Thus, (49) implies that the energy flow along the vertical ray tube is constant, similar to the previous cases (14) and (23). The final result for the Love wave is

$$\mathbf{u}(s,z,t) = \frac{\Phi}{[J(s)U(s)I_1(s)]^{1/2}} \mathbf{n} l_1(s,z) \exp[i\psi(\mathbf{x},t)] \tag{50}$$

where $\mathbf{n}$ is the unit vector on the surface perpendicular to the ray path and $\partial\psi/\partial t = -\omega$ and $\partial\psi/\partial s = |\mathbf{k}| = \omega/C$. Compared to the body wave case (24), the large difference is that for surface waves the profile of wavefields in the vertical direction is expressed by the eigenfunction of normal mode theory l_1 while the profile in the horizontal direction is followed by 2-D ray theory.

2.c.2 Rayleigh waves

Next, let us take the phase velocity C as that of Rayleigh waves. Then, the characteristic equations (38a) and (38c) and the boundary conditions (39a), (39c) and (40) are satisfied. In this case (39b) are not satisfied unless

$$U_n = 0. \tag{51}$$

The terms of the next order may be written as:

$$R_1(U_s^{\,1}, U_z^{\,1}) + 2i\rho\omega\frac{\partial U_s^{\,0}}{\partial T} + i\rho\frac{\partial\omega}{\partial T}U_s^{\,0} + 2i(\lambda+2\mu)\mathbf{k}\cdot\nabla U_s^{\,0}$$

$$+ \nabla\cdot[(\lambda+2\mu)\mathbf{k}]U_s^{\,0} + k^{-1}\mathbf{k}\cdot[\nabla(\lambda\frac{\partial U_z^{\,0}}{\partial z}) + \frac{\partial}{\partial z}(\mu\nabla U_z^{\,0})] = 0, \tag{52a}$$

$$R_2(U_s^{\,1}, U_z^{\,1}) + 2i\rho\omega\frac{\partial U_z^{\,0}}{\partial T} + i\rho\frac{\partial\omega}{\partial T}U_z^{\,0} + 2i\mu\mathbf{k}\cdot\nabla U_z^{\,0}$$

$$+ \nabla\cdot(\mu\mathbf{k})U_z^{\,0} + k^{-1}\mathbf{k}\cdot[\nabla(\mu\frac{\partial U_s^{\,0}}{\partial z}) + \frac{\partial}{\partial z}(\lambda\nabla U_s^{\,0})] = 0, \tag{52b}$$

and the boundary conditions at $z = \zeta$ are

$$i\mu k U_z^{\,1} + \mu\frac{\partial U_s^{\,1}}{\partial z} - \mu k^{-1}\mathbf{k}\cdot\nabla U_z^{\,0} + \nabla\zeta\cdot[(\lambda+2\mu)i\mathbf{k}U_s^{\,0} + \lambda k^{-1}\mathbf{k}\frac{\partial U_z^{\,0}}{\partial z}] = 0, \tag{53a}$$

$$i\lambda k U_s^{\,1} + (\lambda+2\mu)\frac{\partial U_z^{\,1}}{\partial z} - \lambda k^{-1}\mathbf{k}\cdot\nabla U_s^{\,0} + \nabla\zeta\cdot\mu(i\mathbf{k}U_z^{\,0} + k^{-1}\mathbf{k}\frac{\partial U_s^{\,0}}{\partial z}) = 0. \tag{53b}$$

Multiplying (52a) and (52b) by $U_s^{\,0}$ and $U_z^{\,0}$, respectively, and integrating from $\zeta(s)$ to ∞ with respect to z, we may get their sum:

$$\frac{\partial}{\partial T}(\omega A^2 I_1) + \nabla\cdot[(I_2 + I_3/2k)\mathbf{k}A^2] = 0 \tag{54}$$

by using (53a) and (53b). Here we used the eigenfunctions r_1 and r_2: $U_s^{\,0} = A(\mathbf{x},t)r_1(\mathbf{x},z)$ and $U_z^{\,0} = iA(\mathbf{x},t)r_2(\mathbf{x},z)$ and normalized $r_2(\mathbf{x},z)|_{z=\zeta} = 1$, following the notation of Aki and Richards (1980). Similarly, the energy integrals are defined as

$$I_1 = \frac{1}{2}\int_\zeta^\infty \rho(r_1^2 + r_2^2)dz$$

$$I_2 = \frac{1}{2}\int_\zeta^\infty [(\lambda+2\mu)r_1^2 + \mu r_2^2]dz \tag{55}$$

$$I_3 = \int_\zeta^\infty (\lambda r_1\frac{\partial r_2}{\partial z} - \mu r_2\frac{\partial r_1}{\partial z})dz.$$

Since $\mathbf{U} = (I_2 + I_3/2k)\mathbf{k}/\omega I_1$ from (7.76) of Aki and Richards (1980), we may obtain the same transport equation as (49). The final result may be written as

$$\mathbf{u}(s,z,t) = \frac{\Phi}{[J(s)U(s)I_1(s)]^{1/2}}[\mathbf{t}r_1(s,z) + i\mathbf{z}r_2(s,z)]\exp[i\psi(\mathbf{x},t)]. \tag{56}$$

Here we have followed the straightforward derivations such as Babich et al. (1976). Woodhouse (1974) used the variational principle for the present problem and obtained similar results by much more elegant and compact procedures. Once again, it should be emphasized that there are many common characteristics in asymptotic ray theory among scalar waves (15), elastic body waves (24) and surface waves (50) and (56). However, there are still several problems in the above results. One of the largest deficiencies is that we can evaluate wavefields only at a point on the ray path. For heterogeneous media, it may sometimes take a large amount of computation to get a ray path connecting the source and the receiver by conventional shooting or bending methods (e.g., Julian and Gubbins, 1977).

3. Paraxial ray approximation

To overcome the difficulties of geometrical ray theory, several powerful methods have been proposed recently. Probably, the Gaussian beam method (e.g., Červený, 1985a,b) and the Maslov method (e.g., Chapman and Drummond, 1982) are the most popular among seismologists. Some researchers (e.g., Madariaga, 1984; Klimes, 1984; Thomson and Chapman, 1985) have found close relationship between these two approaches, and the reader may consult these references for details. Here, we shall explain the paraxial ray approximation, which is directly related to the above methods.

The results in the previous section give the solutions at a point on the ray path. To extend the solution in the neighborhood of the ray the natural approach is to expand the wavefield by a Taylor series with respect to distance from the ray path. For convenience the ray-centered coordinates (s, q_1, q_2) are introduced by Popov and Pšenčík (1978) to expand the wavefields: s corresponds to the arclength along the ray, q_1 and q_2 form a 2-D Cartesian coordinate system in the plane perpendicular to the ray path. Then, the travel time τ is expanded around the point on the ray $(s, 0, 0)$:

$$\tau(s,q_1,q_2) = \tau(s) + \nabla\tau.\mathbf{q} + \frac{1}{2}\mathbf{q}^T M(s)\mathbf{q} \tag{57}$$

where $M_{ij} = \partial^2\tau/\partial q_i \partial q_j$ evaluated at $q_1 = q_2 = 0$. Since $\nabla\tau$ evaluated on the ray $(q_1 = q_2 = 0)$ is tangent to the ray path (5), it is perpendicular to $\mathbf{q}$. Thus, $\nabla\tau\cdot\mathbf{q} = 0$. If we truncate the expansion (57) at the quadratic form of $\mathbf{q}$, the wavefront around the point $(s,0,0)$ is approximated by the parabola as shown in Fig. 1. This is similar to the 15 degree equation in zero-offset migration (e.g., Claerbout, 1985). From the geometry, $CM_{ij} = C\partial^2\tau/\partial q_i \partial q_j$ is the curvature matrix of the wavefront at $(s,0,0)$.

Let us show the relationship between $\mathbf{M}$ and geometrical spreading J. From (57),

$$\frac{\partial\tau}{\partial s} = \frac{1}{C} + \frac{1}{2}\mathbf{q}^T\frac{d\mathbf{M}}{ds}\mathbf{q},$$

$$\frac{\partial\tau}{\partial q_1} = M_{11}q_1 + M_{12}q_2, \tag{58}$$

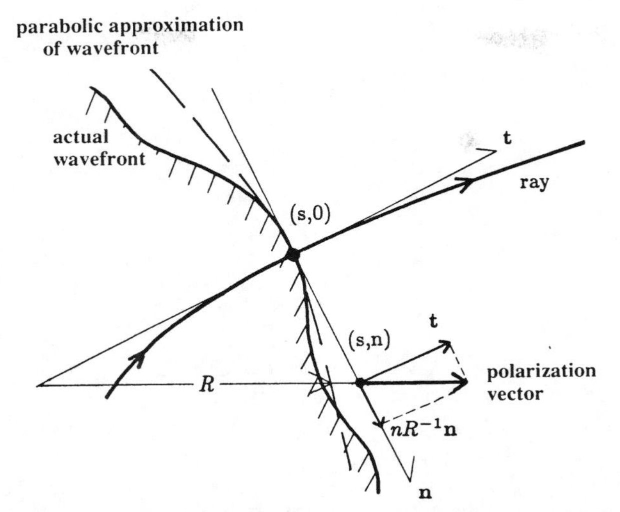

Figure 1. Approximation of the wavefront by parabolic form and the additional transverse component nR^{-1} of P waves. R is the radius of curvature of wavefront at $(s,0)$.

$$\frac{\partial \tau}{\partial q_2} = M_{21}q_1 + M_{22}q_2,$$

Because of (13), we may get

$$\frac{1}{J}\frac{d}{ds}\left[\frac{J}{C}\right] = \nabla^2\tau\,|_{q_1=q_2=0} = \left[\frac{\partial^2\tau}{\partial s^2} + \frac{\partial^2\tau}{\partial {q_1}^2} + \frac{\partial^2\tau}{\partial {q_2}^2}\right]_{q_1=q_2=0}$$

$$= \frac{d}{ds}\left(\frac{1}{C}\right) + M_{11} + M_{22}.$$

Thus, finally the following relation is noticed:

$$\text{trace of } \mathbf{M} = \frac{1}{JC}\frac{dJ}{ds}. \tag{59}$$

The remaining problem is how to calculate these variables. Following Červený and Hron (1980) we introduce the following two variables:

$$Q_{ij} = \frac{\partial q_i}{\partial \gamma_j}, \tag{60a}$$

$$P_{ij} = \frac{\partial p_i}{\partial \gamma_j} = \frac{\partial^2\tau}{\partial \gamma_j \partial q_i} = \frac{\partial^2\tau}{\partial q_i \partial q_k}\frac{\partial q_k}{\partial \gamma_j} = M_{ik}Q_{kj}. \tag{60b}$$

Here we used (5). From the definition of the geometrical spreading (12), Q_{ij} is related to the geometrical spreading J(s) by

$$J = Q_{11}Q_{22} - Q_{12}Q_{21} = \det\mathbf{Q}. \tag{61}$$

Differentiating ray tracing equations (8) and (9) with respect to γ_j, we may obtain the differential equations for Q_{ij} and P_{ij} :

$$\frac{dQ_{ij}}{ds} = C\frac{\partial p_i}{\partial \gamma_j} = CP_{ij}, \tag{62}$$

$$\frac{dP_{ij}}{ds} = -C^{-2}\frac{\partial^2 C}{\partial q_i \partial q_k}Q_{kj} \tag{63}$$

or in matrix form:

$$\frac{d\mathbf{Q}}{ds} = C\mathbf{P},$$

$$\frac{d\mathbf{P}}{ds} = -C^{-2}\mathbf{V}\mathbf{Q}$$

where $V_{ij} = [\partial^2 C / \partial q_i \partial q_j]_{q_1=q_2=0}$ (see Klimes, 1984). These are called the dynamic ray tracing equations. While the (kinematic) ray tracing equations (8) and (9) deal with the trajectory of the ray (q_i and p_i), the dynamic ray tracing equations (62) and (63) describe behavior of the neighboring rays ($Q_{ij} = \partial q_i / \partial \gamma_j$ and $P_{ij} = \partial p_i / \partial \gamma_j$). We can calculate the geometrical spreading J = det**Q** and the curvature matrix $C\mathbf{M} = C\mathbf{P}\mathbf{Q}^{-1}$ by the dynamic ray tracing equations (62) and (63) as an initial value problem at the source. Note that these are determined by **V**, the spatial second derivative of velocity. Under some approximation such as (57), we can now evaluate the wavefield at a point not only along the ray but also in the neighborhood of the ray (q_i not zero):

$$u(s, q_1, q_2, t) = \Phi\sqrt{\frac{C(s)}{\det\mathbf{Q}(s)}}\exp\left[-i\omega\left(t-\int_0^s \frac{d\zeta}{C(\zeta)}\right) + \frac{i\omega}{2}\mathbf{q}^T P Q^{-1}\mathbf{q}\right] \tag{64}$$

for scalar waves. Compared with the original form (15), the difference is a correction term in phase.

For body waves another correction is required for the polarization vector because the orientation of polarization is no longer exactly tangent or perpendicular to the reference ray when the point is not on the ray. Fig. 1 shows the example of P wave in a 2-D medium. In this case the coordinate is (s, n) instead of (s, q_1, q_2). At the point on the ray $(s, 0)$, the component of the polarization vector is only tangent to the ray **t** for P waves. However, there is an additional transverse component **n** at a point away from the ray (s, n) because of the curvature of the wavefront. Since the radius of curvature is $K = (CM)^{-1}$ as defined above, the geometrical configuration shows that the additional transverse component is $n/K = nCM = nCP/Q$. Thus, the polarization vector for P wave is corrected from **t** to $\mathbf{t} + nCM\,\mathbf{n}$. Similarly, for S waves there appears a tangential component, corrected from **n** to $\mathbf{n} + nCM\,\mathbf{t}$. It is easily extended to the 3-D case and the results are

$$\mathbf{u}(s, q_1, q_2, t) = \frac{\Phi}{\sqrt{\rho C \det\mathbf{Q}(s)}}\mathbf{s}\exp\left[-i\omega\left(t-\int_0^s \frac{d\zeta}{C(\zeta)}\right) + \frac{i\omega}{2}\mathbf{q}^T P Q^{-1}\mathbf{q}\right] \tag{65}$$

where

$$C = \alpha, \quad \mathbf{s} = \mathbf{t} + CM_{ij} q_j \mathbf{e}_i \quad \text{for P wave,}$$

$$C = \beta, \quad \mathbf{s} = \mathbf{e}_i - CM_{ij} q_j \mathbf{t} \quad \text{for S wave,}$$

Here for S waves the principal polarization vector is $\mathbf{e}_i$ $(i = 1, 2)$. This is a more general form than (24) which can be only evaluated at a point along the ray: $q_i = 0$.

Finally, we can apply the above discussions directly to the surface wave problem. For surface waves, the ray trajectories are constrained on the free surface, so the results are quite similar to 2-D body waves. The results (50) and (56) are extended to a point away from the ray (s, n) as follows:

$$\mathbf{u}(s, n, z, t) = \frac{\Phi}{[U(s) I_1(s) Q(s)]^{1/2}} \, \mathbf{s} \exp(i \psi(\mathbf{x}, t) + \frac{i\omega}{2} P Q^{-1} n^2) \tag{66}$$

where

$$\mathbf{s} = (\mathbf{n} - CMn\,\mathbf{t}) l_1(s, z) \quad \text{for Love waves,}$$

$$\mathbf{s} = (\mathbf{t} + CMn\,\mathbf{n}) r_1(s, z) + i\,\mathbf{z} r_2(s, z) \quad \text{for Rayleigh waves,}$$

Here, we used several analogies and intuitions to derive these results. The formal derivations of these results are given by the expansion of equations with respect to the ray-centered coordinate (s, q_1, q_2) in some studies: (64) for scalar waves in Červený et al. (1982), (65) for 3-D elastic body waves in Červený and Pšenčík (1983), and (66) for surface waves in Yomogida (1985). These studies showed rigorously several characteristics, for example, up to the above approximation the geometrical spreading does not need correction terms and there is no z component for Love waves in (66).

4. Superposition of Gaussian beams

The results of the paraxial ray approximation in the previous section give wavefields in the neighborhood of the ray path, so we can reduce the amount of computational time because we can merely shoot the ray close to the specific point rather than at the point itself. However, there are still some shortcomings in the above method. In some case the geometrical spreading J is close to zero and the solution becomes singular, which is called 'caustic'. To overcome such difficulties, we should correct the previous results based on formal wave theory rather than taking the high frequency limit $\omega \rightarrow \infty$ as before.

In general, the wavefield at any point (at ray coordinate (s, γ_1, γ_2)) can be expressed as the superposition of plane waves corresponding to each ray $\mathbf{u}_{\gamma_1 \gamma_2}$ such as $\mathbf{s} \exp[i\omega\tau(s, \gamma_1, \gamma_2)]$ (i.e., the Weyl integral):

$$\mathbf{u}(s, \gamma_1, \gamma_2) = \iint \Phi(\gamma_1, \gamma_2) \mathbf{u}_{\gamma_1 \gamma_2} d\gamma_1 d\gamma_2 \tag{67}$$

where $\mathbf{s}$ is the polarization vector, τ is the travel time and $\Phi(\gamma_1, \gamma_2)$ is the weighting function as given in the previous section. The integration should cover the entire region (see section 6.1. of Aki and Richards, 1980). Up to quadratic terms, we may use (57) for travel time of each ray. Even though (67) is exact at any point, it is impractical because we have to integrate the contributions from the whole region. In contrast, the geometrical ray theory in the previous section expresses wavefields by a single ray (or a small number of rays, if there are any multiple arrivals) assuming the energy concentrates along the ray as $\omega \to \infty$ or infinitesimal wavelength. In reality, with a finite wavelength, contributions far from the observed point cancel each other out by interference, and only waves close to the actual geometrical rays contribute to the result in (67). One natural way to describe such a phenomenon is to take the variables bold P and $\mathbf{Q}$, in formulations such as (64), as complex rather than real. This is a kind of correspondence principle. In this case $\mathbf{M}(s)$ has both real and imaginary parts and, for example, (64) becomes

$$u(s,q_1,q_2,t) = \Phi\sqrt{\frac{C(s)}{\det\mathbf{Q}(s)}}\exp\left[-i\omega\left(t-\int_0^s\frac{d\zeta}{C(\zeta)}\right)+\frac{i\omega}{2}\mathbf{q}^T\mathrm{Re}\mathbf{M}(s)\mathbf{q}\right]$$
$$\cdot\exp\left[-\frac{\omega}{2}\mathbf{q}^T\mathrm{Im}\mathbf{M}(s)\mathbf{q}\right]. \tag{68}$$

Thus, if the matrix $\mathrm{Im}\mathbf{M}(s)$ is positive-definite, the amplitude of (68) decays like a Gaussian function in the direction perpendicular to the ray. This agrees with the concept that the energy concentrates only close to the geometrical ray. As $\omega \to \infty$, the concentration should be stronger and the character of the wavefield becomes closer to that of the geometrical ray theory.

There are many ways to take $\mathbf{P}$ and $\mathbf{Q}$ as complex. Let us consider the 2-D case. Since the dynamic ray tracing equations are two first-order ordinary differential equations, there are two independent solutions. So the complex variables P and Q can be expressed by the linear combination of two solutions for different initial conditions with complex coefficients. In general, we choose the following two initial values: one for a plane wave and the other for a line source. See the detailed discussions of Červený et al. (1982) and Madariaga (1984). For 3-D case, see Červený (1985a, b). One important point is that J ($=\det\mathbf{Q}$) never becomes zero, that is, non-singular anywhere, if we take $\mathbf{P}$ and $\mathbf{Q}$ as complex.

For surface wave problems, we just replace the integration in (67) by the summation of Gaussian beams (66) with complex P and Q:

$$\mathbf{u}(s,n,t) = \sum_\gamma \Phi(\gamma)\left[\frac{Q(0)U(0)I_1(0)}{Q(s)U(s)I_1(s)}\right]^{1/2} \mathrm{s}\exp[i\psi(s,z)+\frac{i\omega}{2}P(s)Q^{-1}(s)n^2]\Delta\gamma. \tag{69}$$

The weighting function Φ for a point source is given by equations (58) and (59) in Yomogida and Aki (1985). Here we may remind the readers that there is still some room for studies on the criteria for choosing the appropriate complex coefficients for $\mathbf{P}$ and $\mathbf{Q}$.

One alternative way to avoid the singularity of the ray method was introduced to seismology by Chapman and Drummond (1982), and is called the Maslov method. Recently, Saastamoinen (1986) applied it to surface wave problems. It uses Liouville's theorem: the incompressibility of a canonical phase space. As shown in (33), the canonical coordinate in this case is $(\mathbf{x}, \mathbf{k})$ or $(\mathbf{x}, \mathbf{p})$, where $\mathbf{p}$ is the slowness vector (5) if ω is fixed, as in the body wave case. Liouville's theorem states that at a singular point such as a caustic, the volume of $\mathbf{x}$ space becomes zero but that of $\mathbf{p}$ space is not zero. Thus, we can evaluate the wavefields at the singular point by formulations expressed in $\mathbf{p}$ space rather than the conventional $\mathbf{x}$ space, by using Legendre transformation. The close relationships between the Gaussian beam method and the Maslov method are found in Klimes (1984), Madariaga (1984), and Thomson and Chapman (1985).

Even though the above methods are not perfect, we can calculate seismograms in laterally heterogeneous media quickly. Surface wave problems especially have some advantages over problems that involve other types of waves: 1) The summation of beams is essentially over 2-D horizontal space rather than full 3-D space, so the amount of calculations is significantly reduced. 2) Since the heterogeneities in the lateral direction are in general sufficiently smooth, the assumptions used in the above derivations are satisfied. This is the reasons why the results of examples in Yomogida and Aki (1985) are not so sensitive to the beam parameters (the choice of complex coefficients for P and Q) while for body waves the ambiguities of these parameters are very critical in some cases (see some examples of Nowack and Aki (1984)).

5. Ray tracing on the sphere

So far, we have presented formulations of surface waves for the flat Earth model. The Earth is spherical, so we need some modifications. To make the situation simpler, the effect of curvature of the Earth is assumed to be negligible compared to the wavelength. This assumption means that the undulation of the surface is smooth and that condition (27) is satisfied: $\varepsilon \ll 1$. The wavelength of surface waves at period 100 s is less than 500 km, so at periods less than 100 s, such an assumption is valid.

Then, the only correction is related to ray tracing equations. In the spherical coordinates (θ,ϕ) equations (8) and (9) may be written as

$$\frac{d\theta}{ds} = \frac{C}{R} p_\theta, \qquad \frac{d\phi}{ds} = \frac{C}{R\sin\theta} p_\phi, \tag{70}$$

$$\frac{dp_\theta}{ds} = \frac{C\cot\theta}{R} {p_\phi}^2 - \frac{1}{RC^2}\frac{\partial C}{\partial\theta}, \qquad \frac{dp_\theta}{ds} = -\frac{C\cot\theta}{R} p_\theta p_\phi - \frac{1}{RC^2\sin\theta}\frac{\partial C}{\partial\phi}.$$

where R is the radius of the Earth. Jobert and Jobert (1983) proposed the use of a Mercator projection to transform these coordinates to Cartesian coordinates. Let us introduce a new variable Θ:

$$\Theta = \ln[\tan(\theta/2)] \qquad 0 < \theta < \pi \tag{71}$$

and new phase and group velocities:

$$V(\Theta,\phi) = C(\theta,\phi)/R\sin\theta, \tag{72a}$$

$$\tilde{U}(\Theta,\phi) = U(\theta,\phi)/R\sin\theta. \tag{72b}$$

Defining the increment of distance along the ray as $d's = \tilde{U}dt = ds/R\sin\theta$, equation (70) becomes identical to (8) and (9), with $\mathbf{x} = (\Theta,\phi)$, $\mathbf{p} = (\tilde{p}_\Theta,\tilde{p}_\phi)$, $ds \to d's$ and $C \to V(\Theta,\phi)$ where the slowness vector $\mathbf{p}$ is the reciprocal of $V(\Theta,\phi)$. For an Earth with a small ellipticity, see Appendix of Yomogida and Aki (1985).

The variables P and Q are also different on the Mercator projection. The geometrical spreading $\tilde{Q}$ on the Mercator projection is related to Q on the sphere as follows:

$$\tilde{Q} = \left[\left(\frac{\partial \Theta}{\partial \gamma}\right)^2 + \left(\frac{\partial \phi}{\partial \gamma}\right)^2 \right]^{1/2} = \left[\frac{1}{\sin^2\theta}\left(\frac{\partial \theta}{\partial \gamma}\right)^2 + \left(\frac{\partial \phi}{\partial \gamma}\right) \right]^{1/2} = Q/R\sin\theta. \tag{73}$$

Similarly the variable $\tilde{P}$ is expressed as

$$\tilde{P} = P\ R\ \sin\theta. \tag{74}$$

Then, the dynamic ray tracing equations (62) and (63) on the Mercator projection are the same as those on the sphere. The ray centered coordinate n should be divided by $R\sin\theta$ on the Mercator projection, so the correction term for the phase, $\exp[(i\ \omega/2)(P/Q)n^2]$, has the identical form on both the sphere and the Mercator projection. On the other hand, the amplitude term $[Q(s)U(s)I_1(s)]^{-1/2}$ do not, so $Q(s)$ and $U(s)$ should be values on the sphere: values on the Mercator projection multipled by $R\sin\theta$ as (72a) and (73).

In summary, after transforming to the Mercator projection, we can calculate ray trajectories and the variables P and Q by formulations in the Cartesian coordinates (Θ,ϕ). Woodhouse and Wong (1986) employed another linearized formulation for surface wave propagation on the sphere.

6. Inverse problems: Born approximation

In the previous section we have completed the procedures for obtaining seismograms of surface waves in laterally heterogeneous media:

1. Construct phase velocity map on the Mercator projection (72a).
2. Trace ray trajectories in the (Θ,ϕ) plane by using ray tracing equations (8) and (9).
3. Calculate $P(s)$ and $Q(s)$ along the ray paths, with two different sets of initial conditions (e.g., a line source + a plane wave), by the dynamic ray tracing equations (62) and (63).
4. Find the point on each ray closest to the observation point and get the values of the ray centered coordinates s and n.
5. With complex variables $P(s)$ and $Q(s)$, simply sum all the beams with some weighting function Φ, as in (69).

Yomogida and Aki (1985) showed several examples of traced rays and seismograms by the above approach.

The ultimate goal is learning how to invert the data, the whole seismograms including amplitudes in this case, to determine the lateral heterogeneities. We come across one large difficulty compared to conventional travel time inversion. The travel time τ is expressed as (7). If the velocity of the media is slightly perturbed, we may have both a change in the integrand $C + \delta C$ and a change in the ray paths. Because of Fermat's principle (36), the effect of the velocity perturbation may be written as

$$\delta\tau = -\int_0^s \frac{\delta C(\zeta)}{C(\zeta)^2} d\zeta. \tag{75}$$

The important point is that the path of integration is the same as the original path up to the first order. Thus, (75) gives the linear relationship between $\delta\tau$ and δC. On the other hand, the change of the ray path affects the perturbation of amplitude with the same order as does the the change in the integrand δC. Thus, the relationship between the perturbations of amplitude and velocity is non-linear, and we may solve such problems only by iterative procedures. This is the reason why the interpretation of amplitude terms is difficult for studies of lateral heterogeneities, even though amplitude terms are very sensitive to heterogeneities. Thomson and Gubbins (1982) and Thomson (1983) inverted the amplitude data of NORSAR for three-dimensional structures under the array by using iterative procedures with formulations of the bending methods given by Julian and Gubbins (1977). Their results look encouraging, but the problem is that it required a large amount of computations. Since fast computational methods have been developed, we shall show how to use them to invert the entire seismograms to find lateral heterogeneities.

Here the inverse scheme of the scalar wave equation by Tarantolà (1984) is given. This scheme, in the frequency domain rather than in the original time domain, was used for phase velocity inversions of the Pacific Ocean basin by Yomogida and Aki (1987).

Including the force term f at $\mathbf{r}_s$, (2) may be written in the frequency domain as

$$\left[\frac{\omega^2}{C(\mathbf{r})^2} + \nabla^2\right] w(\mathbf{r}, \omega; \mathbf{r}_s) = f(\mathbf{r}_s, \omega). \tag{76}$$

Let us perturb the velocity by $\delta C(\mathbf{r})$, which produces a small change in the wavefield, $\delta w(\mathbf{r},\omega;\mathbf{r}_s)$. Neglecting higher order terms, (76) becomes

$$\left[\frac{\omega^2}{C(\mathbf{r})^2} + \nabla^2\right] \delta w(\mathbf{r}, \omega; \mathbf{r}_s) = \Delta f(\mathbf{r}, \omega; \mathbf{r}_s) \tag{77}$$

where

$$\Delta f(\mathbf{r}, \omega; \mathbf{r}_s) = \frac{2\omega^2}{C(\mathbf{r})^3} \delta C(\mathbf{r}) w(\mathbf{r}, \omega; \mathbf{r}_s). \tag{78}$$

This is the Born approximation. By introducing the Green's function G,

$$\left[\frac{\omega^2}{C(\mathbf{r})^2} + \nabla^2\right] G(\mathbf{r}, \omega; \mathbf{r}') = -\delta(\mathbf{r}-\mathbf{r}'), \tag{79}$$

δw may be expressed as

$$\begin{aligned}\delta w(\mathbf{r}_g, \omega; \mathbf{r}_s) &= -\int G(\mathbf{r}_g, \omega; \mathbf{r}')\Delta f(\mathbf{r}', \omega; \mathbf{r}_s) d\mathbf{r}' \\ &= \int \left[-\frac{2\omega^2}{C(\mathbf{r}')^3} G(\mathbf{r}_g, \omega; \mathbf{r}') w(\mathbf{r}', \omega; \mathbf{r}_s) \right] \delta C(\mathbf{r}') d\mathbf{r}'.\end{aligned} \tag{80}$$

Thus, the Fréchet derivative for inversion is

$$F(\mathbf{r}_g, \omega; \mathbf{r}_s \mid \mathbf{r}) = \frac{\partial w(\mathbf{r}_g, \omega; \mathbf{r}_s)}{\partial C(\mathbf{r})} = -\frac{2\omega^2}{C(\mathbf{r})^3} G(\mathbf{r}_g, \omega; \mathbf{r}) w(\mathbf{r}, \omega; \mathbf{r}_s). \tag{81}$$

Instead of using this form directly, it may be better to use the logarithms of field variables $\phi = \ln[w(\mathbf{r}, \omega; \mathbf{r}_s)]$ and model parameter $m = \ln C(\mathbf{r})$. This is called Rytov's method (e.g., Chernov, 1960). In this case, the Fréchet derivative is naturally normalized as

$$F(\mathbf{r}_g, \omega; \mathbf{r}_s \mid \mathbf{r}) = \frac{\partial\phi(\mathbf{r}_g, \omega; \mathbf{r}_s)}{\partial m(\mathbf{r})} = \frac{C(\mathbf{r})}{w(\mathbf{r}_g, \omega; \mathbf{r}_s)} \frac{w(\mathbf{r}_s, \omega; \mathbf{r}_s)}{\delta C(\mathbf{r})}$$

$$= -\frac{2\omega^2}{C(\mathbf{r})^2} G(\mathbf{r}_g, \omega; \mathbf{r}) w(\mathbf{r}, \omega; \mathbf{r}_s) / w(\mathbf{r}_g, \omega; \mathbf{r}_s). \tag{82}$$

In the frequency domain ϕ may be written as

$$\phi(\mathbf{r}_g, \omega; \mathbf{r}_s) = \ln[w(\mathbf{r}_g, \omega; \mathbf{r}_s)] = \ln A(\mathbf{r}_g, \omega; \mathbf{r}_s) + i\,\psi(\mathbf{r}_g, \omega; \mathbf{r}_s) \tag{83}$$

where A is the amplitude and ψ is the phase term.

Then, we obtain the final inversion formulation of Tarantola and Valette (1982)

$$m_{k+1}(\mathbf{r}) = m_k(\mathbf{r}) + (I + C_m F_k{}^+ C_\phi{}^{-1} F_k)^{-1} \cdot$$

$$\cdot\left[C_m F_k{}^+ C_\phi{}^{-1}[\phi(\mathbf{r}_g, \omega; \mathbf{r}_s) - \phi_k(\mathbf{r}_g, \omega; \mathbf{r}_s)] - [m_k(\mathbf{r}) - m_0(\mathbf{r})]\right] \tag{84}$$

so that we minimize the following quantity:

$$[\phi - \mathbf{f}(\mathbf{m}_k)]^+ \mathbf{C}_\phi{}^{-1}[\phi - \mathbf{f}(\mathbf{m}_k)] + (\mathbf{m}_{k+1} - \mathbf{m}_0)^+ \mathbf{C}_m{}^{-1}(\mathbf{m}_{k+1} - \mathbf{m}_0) \tag{85}$$

where $m_k(\mathbf{r})$ is the value at k-th iteration step, F_k are the Fréchet derivatives with $m_k(\mathbf{r})$ in (82), and $\mathbf{f}(\mathbf{m}_k)$ is the data prediction of $m_k(\mathbf{r})$ (i.e., $\mathbf{F}_k = [\partial\mathbf{f}/\partial\mathbf{m}]_{\mathbf{m}_k}$). $\phi_k(\mathbf{r}_g, \omega; \mathbf{r}_s)$ are the solutions of the forward problem for the model $m_k(\mathbf{r})$, I is the unit operator, the daggar means the adjoint operator and C_m and C_ϕ are the covariance functions of model and data, respectively. In the above formulations, we have to calculate the following wavefields at each iteration step: 1) the wavefield $w(\mathbf{r}, \omega; \mathbf{r}_g)$ at the model point $\mathbf{r}$ from the source $\mathbf{r}_s$, 2) the Green's function $G(\mathbf{r}, \omega; \mathbf{r}_g)$ from the receiver to the model point $\mathbf{r}$ and 3) the wavefield $w(\mathbf{r}_g, \omega; \mathbf{r}_s)$ at the receiver from the source to calculate the data residuals $\delta w(\mathbf{r}_g, \omega; \mathbf{r}_s)$. That is, we need wavefields at each iteration step for all the combinations of the model points $\mathbf{r}$, the sources $\mathbf{r}_s$ and the receiver $\mathbf{r}_g$. These are calculated by the procedures in the previous sections or by the other methods for laterally heterogeneous media.

The readers may refer to Yomogida and Aki (1987) for details about the application of this kind of approach to the actual data. In the examples of this paper we notice both advantages and disadvantages. The problem is that we need large amount of calculations, even though modern computers can handle many interesting problems with realistic CPU times. As one advantage, we can determine the heterogeneities with much better resolution than by the conventional approach, because the effect of finite wavelength is included, which enhances the path coverage over the model space (e.g., see Fig. 10 of Yomogida and Aki (1987)). Moreover, we can deal with not only phase terms but also amplitude anomalies, which is becoming one of the important subjects in recent seismology (e.g., Lay and Kanamori, 1985).

7. Final remarks

At the present time (1986), the efforts of theoreticians and the advances of computers enables us to achieve one of the ultimate goals in seismology: the use of whole seismograms for studies on lateral heterogeneities. Still, the conventional approaches, such as travel time inversion, have some advantages because of their simplicity and robustness. Both approaches will compete with each other in the next decade.

However, we need one more step to understand the Earth's interiors by seismic waves. So far, we have made the fundamental assumptions that the heterogeneities are weak or smooth. In many cases of seismology, the scattering of the data is much more significant than the predictions of the above approaches. Apparently, real media are much more complicated and the distortion of waveforms is much larger than expected from weak heterogeneities. In surface wave problems, severe mode-mode conversions may occur frequently in some regions such as ocean-continental boundaries.

To deal with these problems we may have two approaches. One is to establish the theory for mode-mode conversions or multiple scatterings such as Kennett (1985) and Frazer (1985). The other is a purely numerical approach such as finite-difference and finite-element methods. So far, this approach is constrained only to 2-D and forward modeling, such as the work of Frankel and Clayton (1986) and Sword et al. (1986), because it requires extremely large memories and CPU time. Numerical approaches can handle any degree and configuration of heterogeneities, but we cannot calculate many interesting problems even with the present supercomputers. On the other hand, the theoretical approach can reduce the amount of computations but the validity may become difficult to evaluate as the media become more complex. These two approaches should stimulate each other to make progress in the study of strong lateral heterogeneities in the Earth in the years to come.

Acknowledgments

The author owes much of his understandings of the present subject to discussions with K. Aki and I. Pšenčík. C. Sword gave me some comments on an early manuscript of this paper. This research was supported by the Earth Science Section, National Science Foundation Grant No. EAR 86-07694 at Caltech and EAR.86-18855 at Stanford. Contribution No. 4411, Division of Geological and Planetary Sciences, California Institute of Technology, Pasadena, California 91125.

References

Aki, K. and Richards, P.G., 1980. *Quantitative Seismology: Theory and Methods*, **1** and **2**, W.H. Freeman, San Francisco.

Babich, V.M. and Rusakova, N.Ya., 1962. The propagation of Rayleigh waves along the surface of an inhomogeneous elastic body of arbitrary shape, *J. Comp. Math. Phys. (Zhurnal vychisl. mat. i matem. fiziki)*, **2**, No.4, 652-665.

Babich, V.M., Chikhachev, B.A. and Yanovskaya, T.B., 1976. Surface waves in a vertically inhomogeneous elastic half space with weak horizontal inhomogeneity, *Izv. Earth Phys.*, **4**, 24-31.

Burridge, R. and Weinberg, H., 1976. Horizontal rays and vertical modes, *Wave Propagation and Underwater Acoustics*, Lecture Notes in Physics, **70**, 86-152.

Červený, V., 1985a. The application of ray tracing to the numerical modeling of seismic wavefields in complex structures, in *Seismic Shear Waves, Handbook of Geophysical Exploration*, Section I: Seismic Exploration, K. Helbig and S. Treitel (eds.), edited by G. Dohr, Geophysical Press, London, 1-124.

Červený, V., 1985b. Gaussian beam synthetic seismograms, *J. Geophys.*, **58**, 44-72.

Červený, V. and Hron, F., 1980. The ray series method and dynamic ray tracing system for 3-D inhomogeneous media, *Bull. Seismol. Soc. Am.*, **70**, 47-77.

Červený, V. and Pšenčík, I., 1983. Gaussian beam and Paraxial ray approximation in three-dimensional elastic inhomogeneous media, *J. Geophys.*, **53**, 1-15.

Červený, V., Popov, M.M. and Pšenčík, I., 1982. Computation of wave fields in inhomogeneous media - Gaussian beam approach, *Geophys. J. R. Astr. Soc.*, **70**, 109-128.

Chapman, C.H. and Drummond, R., 1982. Body wave seismograms in inhomogeneous media using Maslov asymptotic theory, *Bull. Seismol. Soc. Am.*, **72**, S277-S317.

Chernov, L.A., 1960. *Wave Propagation in a Random Medium*,,McGraw-Hill, New York.

Claerbout, J.F., 1985. *Imaging the Earth's Interior*,Blackwell Scientific Publications. Inc., Palo Alto.

DeNoyer, J., 1961. The effect of variations in layer thickness on Love waves, *Bull. Seismol. Soc. Am.*, **51**, 227-235.

Frankel, A. and Clayton, R.W., 1986. Finite difference simulations of seismic scattering: Implications for the propagation of short-period seismic waves in the crust and models of crustal heterogeneity, *J. Geophys. Res.*, **91**, 6465-6489.

Frazer, L.N., 1983. Feynman path integral synthetic seismograms, *Eos Trans. AGU*, **64**, 772.

Gjevik, B., 1973. A variational method for Love waves in nonhorizontally layered structures, *Bull. Seismol. Soc. Am.*, **63**, 1013-1023.

Hudson, J.A., 1981. A parabolic approximation for surface waves, *Geophys. J. R. Astr. Soc.*, **67**, 755-770.

Jobert, N. and Jobert, G., 1983. An application of ray theory to the propagation of waves along a laterally heterogeneous spherical surface, *Geophys. Res. Lett.*, **10**, 1148-1151.

Julian, B.R. and Gubbins, D., 1977. Tree-dimensional seismic ray tracing, *J. Geophys.*, **43**, 95-113.

Kennett, B.L.N., 1984. Guided wave propagation in laterally varying media - I. Theoretical development, *Geophys. J. R. Astr. Soc.*, **79**, 235-255.

Kirpichnikova, N.Y., 1969. Rayleigh waves concentrated near a ray on the surface of an inhomogeneous elastic body, *Mathematical Problems in Wave Propagation Theory, Part II. Seminar in Mathematics*, **15**, Steklov Mathematical Institute, Nauka, Leningrad, 49-62.

Klimes, L., 1984. The relation between Gaussian beams and Maslov asymptotic theory, *Studia geophys. geod.*, **28**, 237-247.

Landau, L.D. and Lifshitz, E.M., 1975. *The Classical Theory of Fields*, 4th ed., Pergamon Press, New York.

Lay, T. and Kanamori, H., 1985. Geometric effects of global lateral heterogeneity on long period surface wave propagation, *J. Geophys. Res.*, **90**, 605-621.

Madariaga, R., 1984. Gaussian beam synthetic seismograms in a vertically varying medium, *Geophys. J. R. Astr. Soc.*, **79**, 589-612.

Nowack, R.L. and Aki, K., 1984. The two-dimensional Gaussian beam synthetic method: Testing and application, *J. Geophys. Res.*, **89**, 7797-7819.

Pierce, A.D., 1965. Extension of the method of normal modes to sound propagation in an almost stratified medium, *J. Acoust. Soc. Am.*, **37**,19-27.

Popov, M.M. and Pšenčík, I., 1970. Computation of ray amplitudes in inhomogeneous media with curved interfaces, *Studia geophys. geod.*, **22**,248-258.

Saastamoinen, P.R., 1986. Maslov method and lateral continuation of surface waves in a laterally slowly and smoothly varying elastic waveguide, *Terra Cognita* (abs.), **6**, 310.

Sword, C., Claerbout, J.F. and Sleep, N.H., 1986. Finite-element propagation of acoustic waves on a spherical shell, *Open Report of Stanford Exploration Project*, **50** , 43-77.

Tanimoto, T., 1986. Surface wave ray tracing equations and Fermat's principle in an anisotropic earth, *Geophys. J. R. Astr. Soc.*, in press.

Tarantola, A., 1984. Non-linear inverse problem for a heterogeneous acoustic medium, *Geophysics*, **49**, 1259-1266.

Tarantola, A. and Valette, B., 1982. Generalized nonlinear inverse problems solved using the least-squares criterion, *Rev. Geophys. Space Phys.*, **20**, 219-232.

Thomson, C.J., 1983. Ray-theoretical amplitude inversion for laterally varying velocity structure below NORSAR, *Geophys. J. R. Astr. Soc.*, **74**, 525-558.

Thomson, C.J. and Chapman, C.H., 1985. An introduction to Maslov's asymptotic method, *Geophys. J. R. Astr. Soc.*, **83**, 143-168.

Thomson, C.J. and Gubbins, D., 1982. Three-dimensional lithospheric modelling at NORSAR: linearity of the method and amplitude variations from the anomalies, *Geophys. J. R. Astr. Soc.*, **71**, 1-36.

Woodhouse, J.H., 1974. Surface waves in a laterally varying layered structures, *Geophys. J. R. Astr. Soc.*, **37**, 461-490.

Woodhouse, J.H. and Dziewonski, A.M., 1984. Mapping the upper mantle: three-dimensional modelling of Earth structure by the inversion of seismic waveforms, *J. Geophys. Res.*, **84**, 5953-5986.

Woodhouse, J.H. and Wong, Y.K., 1986. Amplitude, phase and path anomalies of mantle waves, *Geophys. J. R. Astr. Soc.*, **87**, 753-773.

Yomogida, K., 1985. Gaussian beams for surface waves in laterally slowly-varying media, *Geophys. J. R. Astr. Soc.*, **82**, 511-533.

Yomogida, K., 1987. Gaussian beams for surface waves in transversely isotropic media, *Geophys. J. R. Astr. Soc.*, **88**, 297-304.

Yomogida, K. and Aki, K., 1985. Waveform synthesis of surface waves in a laterally heterogeneous Earth by the Gaussian beam method, *J. Geophys. Res.*, **90**, 7665-7688.

Yomogida, K. and Aki, K., 1987. Amplitude and phase data inversions for phase velocity anomalies in the Pacific Ocean basin, *Geophys. J. R. Astr. Soc.*, **88**, 161-204.

K. YOMOGIDA, Department of Geophysics, Stanford University, Stanford CA 94305, USA.

Chapter 4

On the connection between ray theory and scattering theory for surface waves

Roel Snieder

A proof is presented of the first order equivalence of ray theory for surface waves, and surface wave scattering theory, for the case of smooth lateral heterogeneity.

1. Introduction

Recently, a formalism was developed for linearized three dimensional surface wave scattering by buried heterogeneities (Snieder, 1986a), or by surface topography (Snieder, 1986b,c). In this theory the Born approximation is used, a plane geometry is assumed, and the far field limit is used throughout. These simplifications lead to a scattering formalism which is simple enough to allow mathematical manipulations, and provides an efficient method for computing synthetic seismograms for scattered surface waves.

The resulting expression for the scattered surface waves contains an integral over the inhomogeneity. By using a stationary phase approximation for this integral one selects (at least for a smooth medium) the ray geometrical solution from this scattering integral. In this way, the first order ray geometrical effects (focussing, ray bending and phase shifting) can be determined by computing simple great circle integrals. This allows for an efficient scheme for computing synthetic seismograms in a smoothly varying medium. Furthermore, the fact that ray geometrical effects are contained in the scattering integral has consequences for the way we analyze surface wave data.

N.J. Vlaar, G. Nolet, M.J.R. Wortel & S.A.P.L. Cloetingh (eds), Mathematical Geophysics, 77-84.

2. Scattering theory for surface waves

Suppose that the total displacement field in a laterally heterogeneous medium is decomposed as follows:

$$\mathbf{u} = \mathbf{u}^0 + \mathbf{u}^1 \ , \tag{2.1}$$

where $\mathbf{u}^0$ is the displacement in a laterally homogeneous reference medium, and $\mathbf{u}^1$ describes the effect of the lateral heterogeneities. Using a dyadic decomposition of the Green's function, the wavefield $\mathbf{u}^0$ excited by a point force $\mathbf{f}$ in $\mathbf{r}_s$ is given by (Snieder, 1986a):

$$\mathbf{u}^0(\mathbf{r}) = \sum_\nu \mathbf{p}^\nu(z,\phi) \frac{\exp i(k_\nu X + \frac{\pi}{4})}{(\frac{\pi}{2} k_\nu X)^{1/2}} (\mathbf{p}^{\nu *}(z_s,\phi) \cdot \mathbf{f}) \ . \tag{2.2}$$

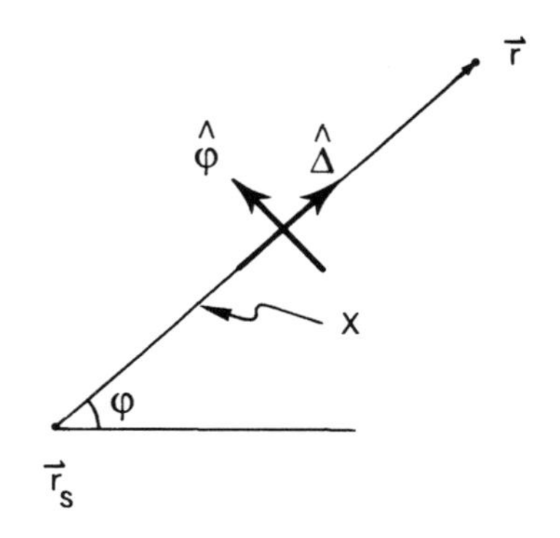

Figure 1. Definition of the geometric variables for the direct wave $\mathbf{u}^0$.

(See figure 1 for the definition of the geometric variables.) The $\mathbf{p}^\nu$ vectors are the polarization vectors, the Greek indices indicate mode numbers. These are for Love waves

$$\mathbf{p}^\nu(z,\phi) = l_1^\nu(z)\hat{\boldsymbol{\phi}} \ , \tag{2.3}$$

and for Rayleigh waves

$$\mathbf{p}^\nu(z,\phi) = r_1^\nu(z)\hat{\boldsymbol{\Delta}} + i r_2^\nu(z)\hat{\mathbf{z}} \ . \tag{2.4}$$

In these expressions l_1^ν, r_1^ν and r_2^ν are the surface wave eigenfunctions as defined in Aki and Richards (1980), normalized as in Snieder (1986a).

Reading (2.2) from right to left one follows the life history of the direct wave $\mathbf{u}^0$. At the source the wave field is excited, the excitation is given by the projection of the point force $\mathbf{f}$ on the polarization vector $\mathbf{p}^\nu(z_s,\phi)$. (In Snieder, 1986a, the extension to a moment tensor excitation is shown.) After this, the wave travels to the receiver, experiencing a phase shift and a geometrical spreading. Finally, the oscillation at the receiver is given by the polarization vector $\mathbf{p}^\nu(z,\phi)$. A summation over modes (index ν) superposes the contributions of different modes.

The distortion of the wavefield ($\mathbf{u}^1$) can be expressed in a similar way (Snieder, 1986a):

$$\mathbf{u}^1(\mathbf{r}) = \sum_{\sigma,\nu} \iint \mathbf{p}^\sigma(z,\phi_2) \frac{\exp i(k_\sigma X_2+\frac{\pi}{4})}{(\frac{\pi}{2}k_\sigma X_2)^{1/2}} V^{\sigma\nu}(x_o,y_o) \frac{\exp i(k_\nu X_1+\frac{\pi}{4})}{(\frac{\pi}{2}k_\nu X_1)^{1/2}} (\mathbf{p}^{\nu *}(z_s,\phi_1)\cdot\mathbf{f})\, dx_o\, dy_o \tag{2.5}$$

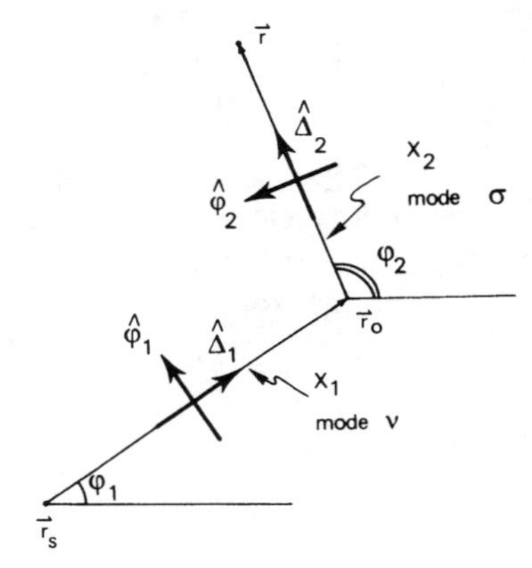

Figure 2 Definition of the geometric variables for the wave distortion $\mathbf{u}^1$.

(See figure 2 for the geometric variables.) In Snieder (1986a) a "single scattering interpretation" of this expression is presented. Note that we now have a double sum over modes (σ,ν), and that mode conversion and scattering is described by the interaction terms $V^{\sigma\nu}$. For buried scatterers the interaction terms are simple depth integrals containing the heterogeneity and the surface wave eigenfunctions (Snieder, 1986a). For perturbations in the surface topography the interaction terms contain the topography height and the surface wave eigenfunctions at the surface (Snieder, 1986b).

In Snieder (1986b) it is shown with a simple renormalization technique that the interaction terms for unconverted waves and forward scattering are closely related to the phase speed perturbations:

$$\left(\frac{\delta c}{c}\right)^\nu = \frac{-2}{k_\nu^2} V^{\nu\nu}(forward) \ . \tag{2.6}$$

3. The relation with ray theory

Now consider a smooth medium, i.e., let us assume that the horizontal scale at which the heterogeneity varies is much larger than the wavelength under consideration. In that case the surface wave modes decouple (Bretherton, 1968; Woodhouse, 1974). We shall therefore consider one mode, and drop the index ν. We take a coordinate system as shown in figure 3. The expression for the wave distortion can be written as

$$\mathbf{u}^1(s_r,n=0) = \iint \mathbf{R}(s,n)\, V(s,n)\, e^{ik(X_1+X_2)}\, dsdn \quad , \tag{3.1}$$

with

$$\mathbf{R} = i\,\mathbf{p}_{receiver}\,(\mathbf{p}^*_{source}\cdot\mathbf{f})\,/\,\frac{\pi}{2}k\,(X_1X_2)^{1/2} \ . \tag{3.2}$$

In general, the phase in (3.1) is rapidly fluctuating except near the "great circle" (n=0) where

$$X_1 + X_2 \approx s_r + \tfrac{1}{2}\left[\frac{1}{s} + \frac{1}{s_r - s}\right] n^2 \ . \tag{3.3}$$

If the heterogeneity (V) is smooth, the n-integral in (3.1) can be solved with the stationary phase approximation. Only a zone around the great circle contributes to the n-integral, the width of this zone is determined by the condition that the phase change (with n) should be less than π. This leads to the width of the Fresnel zone

$$|n| < \left[\frac{2\pi}{k}\,\frac{s\,(s_r - s)}{s_r}\right]^{1/2} \ . \tag{3.4}$$

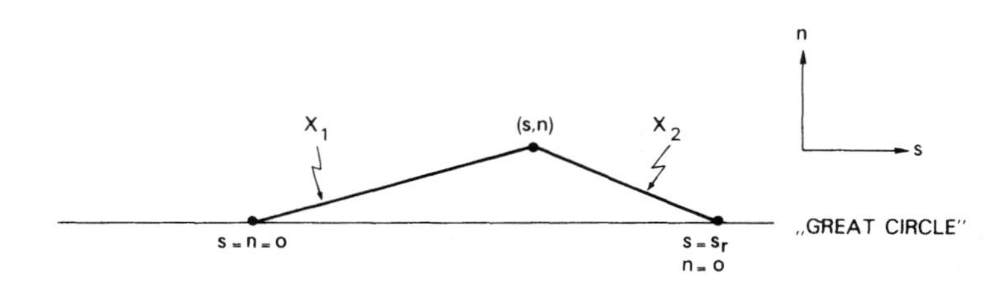

Figure 3 Definition of the geometric variables used in the stationary phase evaluation of (3.1).

This means that

$$\left[\frac{n}{s}\right]^2 < 2\pi\,\frac{s_r - s}{s_r}\,\frac{1}{ks} \ll 1 \ , \tag{3.5}$$

because of the far field assumption. A similar relation holds for $(n/(s_r - s))^2$.

The fact that only small n-values contribute to the scattering integral (3.1) allows a Taylor expansion of the terms **R** and V in the transverse coordinate n. If one just assumes the great circle values for these functions (setting n=0) one obtains the traditional great circle theorem (Jordan, 1978; Dahlen, 1979) in a flat geometry. Focussing and ray bending effects are retrieved, if second order expansions are made in the transverse coordinate n.

The interaction terms for unconverted waves have the following form (Snieder, 1986a,b):

$$V = V^{(0)} + V^{(1)} \cos\phi + V^{(2)} \cos 2\phi \ , \tag{3.6}$$

where ϕ is the scattering angle. Ignoring terms of relative order $1/ks_r$, the cosines can be replaced by 1. Using (2.6), the interaction term can then be expanded as

$$V(s,n) = -\tfrac{1}{2}k^2 \left[(\frac{\delta c}{c})(s,0) + n\partial_n(\frac{\delta c}{c})(s,0) + \tfrac{1}{2}n^2\partial_{nn}(\frac{\delta c}{c})(s,0) \right] \ . \tag{3.7}$$

Likewise, the polarization vectors at the source and the receiver are in the far field limit given by:

$$\mathbf{p}_s = \mathbf{p}_s^0 - \frac{n}{s}\,\mathbf{q}_s^0 \ , \tag{3.8.a}$$

$$\mathbf{p}_r = \mathbf{p}_r^0 + \frac{n}{s_r - s}\,\mathbf{q}_r^0 \ . \tag{3.8.b}$$

In these expressions $\mathbf{p}^0$ is the polarization vector for propagation along the great circle, while $\mathbf{q}^0$ is the rotated polarization vector:

$$\mathbf{q}^0 = \hat{\mathbf{z}} \times \mathbf{p}^0 \tag{3.10}$$

Inserting (3.7),(3.8a,b) and (3.3) in (3.1), using the stationary phase approximation for the n-integral, and adding the reference wavefield $\mathbf{u}^0$, one obtains for the total wavefield

$$\mathbf{u} = \frac{\exp i\,(ks_r + \frac{\pi}{4})}{(\frac{\pi}{2}ks_r)^{\frac{1}{2}}} \left[(1 + i\Phi - \frac{F}{2})\mathbf{p}_r^0(\mathbf{p}_s^{0*}\cdot\mathbf{f}) + D_r\mathbf{q}_r^0(\mathbf{p}_s^{0*}\cdot\mathbf{f}) + D_s\mathbf{p}_r^0(\mathbf{q}_s^{0*}\cdot\mathbf{f}) \right] \ . \tag{3.10}$$

In these expressions Φ, D_s, D_r and F are great circle integrals:

$$\Phi = -k \int_0^{s_r} (\frac{\delta c}{c})\, ds \tag{3.11.a}$$

$$D_s = -\int_0^{s_r} \frac{s_r - s}{s_r}\, \partial_n(\frac{\delta c}{c})\, ds \tag{3.11.b}$$

$$D_r = \int_0^{s_r} \frac{s}{s_r}\, \partial_n(\frac{\delta c}{c})\, ds \tag{3.11.c}$$

$$F = -\int_0^{s_r} \frac{s(s_r - s)}{s_r}\, \partial_{nn}(\frac{\delta c}{c})\, ds \ , \tag{3.11.d}$$

where the phase velocity and it's derivatives are to be evaluated at the great circle (n=0).

Up to first order in the heterogeneity, (3.10) can be rewritten as:

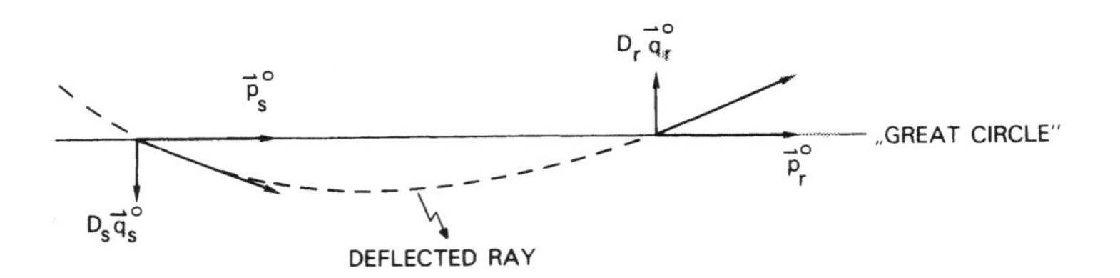

Figure 4 The relation between the ray deflection and the perturbation of the polarization vectors, for the horizontal component of the Rayleigh wave polarization vectors.

$$\mathbf{u} = (\mathbf{p}_r^0 + D_r \mathbf{q}_r^0) \frac{\exp i\,(kX + \Phi + \frac{\pi}{4})}{(\frac{\pi}{2} k s_r (1+F))^{1/2}} ((\mathbf{p}_s^{0*} + D_s \mathbf{q}_s^{0*}) \cdot \mathbf{F}) \ . \tag{3.12}$$

Note the similarity between this expression, and the expression for the direct wave in the laterally homogeneous medium (2.2). The only difference is the great circle integrals appearing in (3.12). The phase integral Φ is just the phase shift due to the phase velocity perturbation at the great circle. (In Snieder (1986b) arguments are presented why it is actually better to use $e^{i\Phi}$ instead of $1+i\,\Phi$.) The surface wave polarization vectors $\mathbf{p}_s^0$ and $\mathbf{p}_r^0$ are perturbed by the deflection integrals D_s and D_r. These integrals simply describe the ray bending effects, leading to an apparent rotation of the source and the receiver. As an example, this is shown in figure 4 for the horizontal component of the Rayleigh wave polarization vector. Finally, the focussing integral F describes the focussing due to the transverse curvature of the velocity profile.

The results (3.11) and (3.12) can also be obtained by applying perturbation theory to the ray tracing equations, as showed by Woodhouse and Wong (1986). For a straight reference ray their theory leads to the same expressions. Romanowicz (1987) derived a similar result in a spherical geometry using normal mode theory.

It is instructive to compare the order of magnitude of the terms (3.11.a-d) with the terms that have been ignored in the derivation. Suppose the velocity varies on a horizontal scale L, a simple scale analysis then shows that

$$\Phi : F : D_{rs} : N \approx 1 : \frac{1}{ks_r}\left[\frac{s_r}{L}\right]^2 : \frac{1}{ks_r}\left[\frac{s_r}{L}\right] : \frac{1}{ks_r} \ , \tag{3.13}$$

where 'N' stands for the terms that have been neglected in the calculation. Note that the phase term is larger than the other terms by a relative factor ks_r, but that for the focussing and deflection terms this factor is reduced by the cumulative effect of the transverse velocity derivatives (the s_r/L terms). As a numerical example, consider a wavelength of 100 km, an epicentral distance s_r of 2000 km, and a horizontal length scale L of the velocity

perturbation of 400 km. In that case

$$\Phi : F : D_{rs} : N \approx 1 : 0.21 : 0.04 : 0.01 \quad ,$$

which means that the focussing term, and possibly the deflection terms, can be significant.

4. Discussion

The importance of the derivation presented here, is that it shows that ray geometrical effects like ray bending and focussing are contained within the expression for surface wave scattering (2.5). This has consequences for the way we deal with surface wave data.

Up to this point, it was customary to (implicitly) divide the lateral heterogeneity in a smooth part and a rough part. The smooth part gives rise to ray geometrical effects, and formed the basis for the measurement of the dispersive properties of surface waves. The rough part of the heterogeneity generates scattered and diffracted surface waves, thus producing the surface wave coda. This part of the signal is usually regarded as noise, and is filtered out.

The separation between the smooth part of the heterogeneity and the rough part is highly artificial, and is not really necessary. The derivation in the previous section showed that both the ray geometrical effects, and the scattering effects are described by the same expression for the wave distortion (2.5), provided these effects are small. Since (2.5) constitutes a linearized relation between the inhomogeneities and the distortion of the wavefield, this expression can be used to invert for the inhomogeneities in the earth. As has been shown by Spakman (1986), extremely large noisy linear systems can be inverted sucessfully. This makes it possible to use a large data set of "surface wave residuals" for a reconstruction of the heterogeneities in the earth, incorporating ray geometrical effects and scattering effects in a unified way.

References

Aki, K. and Richards, P.G., 1980. *Quantitative Seismology* , **1** , Freeman,San Francisco.

Bretherton, F.B., 1968. Propagation in slowly varying waveguides, *Proc. Roy. Soc.*, **A302** , 555-576.

Dahlen, F.A., 1979. The spectra of unresolved split normal mode multiplets, *Geophys. J. R. Astr. Soc.*, **58** , 1-33.

Jordan, T.H., 1978. A procedure for estimating lateral variations from low frequency eigenspectra data, *Geophys. J. R. Astr. Soc.*, **52** , 441-455.

Romanowicz, B., 1987. Multiplet-multiplet coupling due to lateral heterogeneity: asymptotic effects on the amplitude and frequency of the Earth's normal modes, *Geophys. J. R. Astr. Soc.*, in press.

Snieder, R., 1986a. 3D Linearized scattering of surface waves and a formalism for surface wave holography, *Geophys. J. R. Astr. Soc.*, **84** , 581-605.

Snieder, R., 1986b. The influence of topography on the propagation and scattering of surface waves, *Phys. Earth Plan. Int.*, **44** , 226-241.

Snieder, R., 1986c. Phase speed perturbations and three-dimensional scattering effects of surface waves due to topography, *Bull. Seismol. Soc. Am.*, **76** , 1385-1392.

Spakman, W., 1986. Subduction beneath Eurasia in connection with the Mesozoic Thethys, *Geologie en Mijnbouw*, **65** , 145-154.

Woodhouse, J.H., 1974. Surface waves in a laterally varying layered structure, *Geophys. J. R. Astr. Soc.*, **37** , 461-490.

Woodhouse, J.H. and Wong, Y.K., 1986. Amplitude, phase and path anomalies of mantle waves, *Geophys. J. R. Astr. Soc.*, **87** , 753-774.

ROEL SNIEDER, Department of Theoretical Geophysics, University of Utrecht, PO Box 80.021, 3508 TA Utrecht, The Netherlands.

Chapter 5

A hybrid solution for wave propagation problems in inhomogeneous media

Arie van den Berg

This chapter deals with a method of computing wave fields in inhomogeneous media. It is based on a hybrid formulation - in the frequency domain - in terms of finite elements and a boundary integral representation. Numerical aspects of the method are discussed, and computational results for a simple model configuration are compared with analytical results. Finally the method is applied in a seismic modeling experiment. In this experiment the effect of a low velocity sediment fill at the surface of a halfspace on an incident teleseismic wave is modelled. The results clearly illustrate the strong amplification effect of the sediments on the surface displacements.

1. Introduction

For the computation of synthetic seismograms many different methods are available, each suitable for a limited class of models of the medium (Chin, Hedstrom and Thigpen, 1984). Among existing methods for the computation of the complete solution of the equations of motion we can distinguish between solutions based on an analytical approach, and purely numerical methods.
The analytical approaches reduce the equation of motion to a number of ordinary differential equations, using separation of variables (Ben-Menahem and Singh, 1981) a technique that works only in media, where the physical parameters - density and elastic constants - are separable in a suitable coordinate system (Cartesian, spherical, cylindrical). For Earth models (flat or spherical) with constant parameters in all but one (depth,radial)

N.J. Vlaar, G. Nolet, M.J.R. Wortel & S.A.P.L. Cloetingh (eds), Mathematical Geophysics, 85-116.

direction this procedure can be applied.
In purely numerical discretization methods, such as the finite difference (Alterman and Karal, 1968; Mitchel, 1969; Boore, 1972; Alford et al, 1974; Kelly et al, 1976) and finite element method (Lysmer and Drake, 1971; Smith, 1975; Zienkiewicz, 1977; Marfurt, 1984), the physical parameters can be prescribed arbitrarily. Here the governing partial differential equation is reduced, by some discretization technique, to a system of linear algebraic equations, that can be solved numerically.
The restrictions for the numerical methods are of a practical nature. Because of finite computer memory size and processor speed, the models usually have to be truncated, which creates the problem of unphysical boundary effects, that can only be solved approximately (Smith, 1974; Reynolds, 1978).
An approach, that does not belong to either one of the above classes is the boundary integral equation method (Banaugh and Goldsmith, 1963; Sharma, 1967; Schenck, 1968; Wong and Jennings, 1975; Schuster, 1984). It is based on an integral representation for the wavefield, in terms of its values on the boundary surface of the domain (Kupradze, 1956,1963; De Hoop, 1958; Tan, 1975; Aki and Richards, 1980; Colton and Kress, 1983a). An important advantage of boundary integral equation methods is, that in numerical implementations only the boundary surface has to be discretized, instead of the whole domain , as in the purely numerical techniques.
In this chapter results will be presented of a hybrid method, that combines a discretization method and a boundary integral equation technique. The method includes a finite element formalism for a finite region, with arbitrary physical parameters.
The finite region is embedded in a - possibly infinite - regular domain or background medium. For this background medium a boundary integral representation of the wavefield is applied. This combination gives a full description of the wavefield in the whole medium, thus eliminating the need for an artificial truncation of the model. Previous work on hybrid methods, combining discretization approaches with boundary integral techniques can be found in Olson and Hwang (1971), Zienkiewicz et al (1977), Berkhoff (1976) and Wilton (1978).
In the integral representation, integral kernels are used related to the Green's function of the background medium. This dependence on a Green's function restricts the applicability of the hybrid method to problems for which an algorithm for the Green's function is available.
In the following the formalism of the hybrid method will be derived for the general 3-D elastic case. Results will be presented for the 2-D scalar case.
Since we intend to use algorithms to compute Green's function in the frequency domain, the formalism will be derived in the frequency domain. In the implementation of the method, wavefield frequency spectra are computed for a range of discrete frequencies. The Fast Fourier Transform (FFT) algorithm is then applied for the inverse transformation to the time domain.

2. Finite element formulation

The equation of motion to be solved is the elastodynamic equation

$$-\partial_j \tau_{ij} - \rho\omega^2 u_i = f_i \tag{1}$$

in terms of the displacement $\mathbf{u}(\mathbf{x},\omega)$, the stress tensor $\tau(\mathbf{x},\omega)$ and the body force field $\mathbf{f}(\mathbf{x},\omega)$.

The physical parameters enter the equation in the mass density $\rho(\mathbf{x})$ and in the stress tensor through the stress-strain relation. Assuming a linear elastic medium, we have

$$\tau_{ij} = c_{ijkl} \partial_k u_l \tag{2}$$

where the elements of the fourth order stress-strain tensor are the elastic coefficients of a general anisotropic medium (Aki and Richards, 1980).

In the finite element method, a grid $\mathbf{G}$ of nodal points is defined, dividing the solution domain in a finite number of elements, consisting of connected nodal points. The unknown field can be approximated - using Lagrange interpolation - in terms of its nodal point values.

Denoting approximated field values by an overbar $\bar{\mathbf{u}}(\mathbf{x}) \approx \mathbf{u}(\mathbf{x})$ we get, for a single component of the wavefield

$$u_i(\mathbf{x}) \approx \bar{u}_i(\mathbf{x}) = \sum_{M=1}^{N} u_i(\mathbf{x}_M)\phi_M(\mathbf{x}) \quad , \mathbf{x} \in V \tag{3}$$

where the $\mathbf{x}_M$,$M=1, \cdots ,N$ are the nodal points and the scalars $\phi_M(\mathbf{x})$ are the Lagrange interpolating functions, with

$$\phi_M(\mathbf{x}_K) = \delta_{MK} \tag{4}$$

$\mathbf{x}_K$ a nodal point, δ_{MK} the Kronecker delta (Ralston, 1965).

An example of a piecewise linear interpolation function on a 2-D domain is given in fig.1. Examples of higher order approximations can be found in Zienkiewicz (1977).

Next the displacement components $u_i(\mathbf{x}_M), M=1, \cdots ,N$ are rearranged into a single vector of dN components, d the number of components of the wavefield vector $\mathbf{u}(\mathbf{x})$

$$u_i(\mathbf{x}_M) = U_K \ , K=d(M-1)+i \ , M=1, \cdots ,N \tag{5}$$

Substitution of (5) into (3) results in:

$$\bar{u}_i(\mathbf{x}) = \sum_{K=1}^{dN} U_K \Phi_{iK}(\mathbf{x}) \tag{6}$$

where the $d \times dN$ interpolation matrix $\Phi(\mathbf{x})$ is given (for d=3) by

$$\Phi(\mathbf{x}) = \begin{bmatrix} \phi_1(\mathbf{x}) & ,0 & ,0 & , \cdots & ,\phi_N(\mathbf{x}) & ,0 & ,0 \\ 0 & ,\phi_1(\mathbf{x}) & ,0 & , \cdots & ,0 & ,\phi_N(\mathbf{x}) & ,0 \\ 0 & ,0 & ,\phi_1(\mathbf{x}) & , \cdots & ,0 & ,0 & ,\phi_N(\mathbf{x}) \end{bmatrix} \tag{7}$$

On the boundary surface ∂V of the domain V we introduce the traction field as a separate unknown quantity and we express it in terms of a number of boundary nodal point values

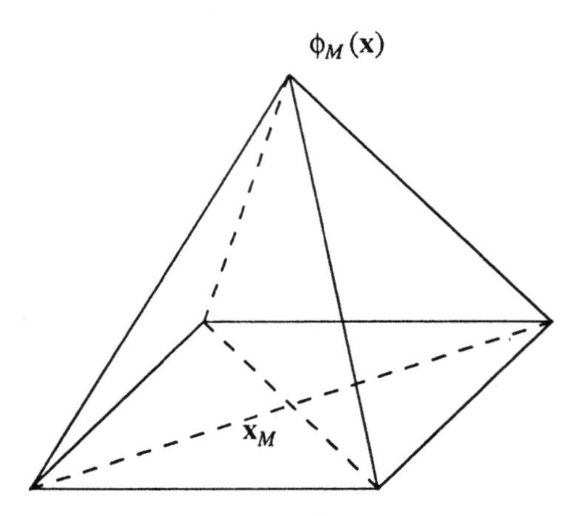

Figure 1. A piecewise first order interpolating function on a 2-D grid, with triangular elements.

$$\tau_{ij} n_j = t_i \approx \bar{t}_i = \sum_{J=1}^{dN} \Phi_{iJ}(\mathbf{x}) T_J \quad , \mathbf{x} \in \partial V \tag{8}$$

Throughout the domain volume V the elements of the stress tensor are expressed in terms of the nodal point displacements. For a linear elastic medium, where (2) holds we get

$$\tau_{ij}(\mathbf{x}) \approx \bar{\tau}_{ij}(\mathbf{x}) = \tau_{ij}[\bar{u}(\mathbf{x})] = \sum_{K=1}^{dN} U_K \tau_{ij}[\Phi_K(\mathbf{x})] \ , \mathbf{x} \in V \tag{9}$$

where $\Phi_K(\mathbf{x})$ denotes the Kth column of the matrix $\Phi(\mathbf{x})$ defined in (7).
With the expressions (6), (8) and (9) we have specified the discretized elastodynamic state $\{u_i, \tau_{ij}\}$ on the domain V in terms of a finite number of parameters; nodal point displacements throughout V and nodal point tractions on the boundary ∂V.

2.1 Algebraic equations for the discretized fields

In the foregoing we defined a discretized elastodynamic state in terms of a finite number of field parameters

$$U_K \ , K=1, \cdots ,dN \ , T_J \ , J=1, \cdots ,dN_b$$

N_b the number of nodal points on the boundary.
Next the partial differential equation (1) will be reduced into a system of algebraic equations, in the unknown field parameters. To this end we will use Galerkin's principle, which formulates the problem in a weak form (Strang and Fix, 1973; Zienkiewicz, 1977). Taking the inner product of (1) with the column vector $\Phi_L(\mathbf{x})$ and integrating over the domain V we have

$$\int_V \Phi_{iL} \{-\partial_j \tau_{ij} - \rho\omega^2 u_i - f_i \} \, dV = 0 \,, \qquad L=1,\cdots,dN \tag{10}$$

This is known as Galerkin's principle applied to the elastodynamic equation, the principle states, that the residue of the differential equation (1) must be perpendicular to the interpolating functions Φ_L in the sense of the natural functional inner product. Integrating (10) by parts we get

$$-\int_{\partial V} \Phi_{iL} \tau_{ij} n_j dA + \int_V \{\tau_{ij} \partial_j \Phi_{iL} - \rho\omega^2 u_i \Phi_{iL} - f_i \Phi_{iL} \} dV = 0 \tag{11}$$

Substitution of the discretized traction (8), the discretized stress tensor (9) and the discretized displacement field (6) we get

$$-\sum_{J=1}^{dN} T_J \int_{\partial V} \Phi_{iL} \Phi_{iJ} dA + \sum_{K=1}^{dN} U_K \int_V \{\tau_{ij} [\Phi_K] \partial_j \Phi_{iL} - \rho\omega^2 \Phi_{iK} \Phi_{iL} \} dV = \int_V f_i \Phi_{iL} dV \;, L=1,\cdots,dN \tag{12}$$

These are the finite element equations for the elastodynamic problem; a set of algebraic equations in the unknown field parameters $U_K \, , K=1,\cdots,dN$, and $T_J \, , J=1,\cdots,dN_b$. Using matrix notation we get

$$-\sum_{J=1}^{dN} D_{LJ} T_J + \sum_{K=1}^{dN} (S_{LK} - \omega^2 M_{LK}) U_K = F_L \;, L=1,\cdots,dN \tag{13}$$

where the static stiffness matrix **S** is defined by

$$S_{LK} = \int_V \tau_{ij} [\Phi_K] \partial_j \Phi_{iL} \, dV \tag{14}$$

and the mass matrix **M** is given by

$$M_{LK} = \int_V \rho \Phi_{iK} \Phi_{iL} \, dV \tag{15}$$

the boundary traction matrix **D** is defined by

$$D_{LJ} = \int_{\partial V} \Phi_{iL} \Phi_{iJ} \, dA \tag{16}$$

The frequency dependent force vector **F** is defined by

$$F_L = \int_V \Phi_{iL} f_i \, dV \tag{17}$$

The structure of the matrices **S**, **M**, **D** depends on the choice of the base functions $\phi_M \, , M=1,\cdots,N$ involved in the approximation of the fields. In general however, because the base functions have a local support - zero valued outside a limited number of neighbouring elements - it follows, that the matrices are sparse.

The following properties of the matrices **S**, **M**, **D** will prove to be useful in implementations.
From the definition (14) it follows, that the stiffness matrix **S** is symmetric.

$$S_{KL} = \int_V \tau_{ij}[\Phi_L]\partial_j \Phi_{iK}\, dV = \int_V c_{ijpq}\, \partial_p \Phi_{qL} \partial_j \Phi_{iK}\, dV$$

Using the symmetry properties of the stress strain tensor, $c_{ijpq} = c_{jiqp} = c_{qpji}$ (Aki and Richards, 1980), we have

$$S_{KL} = \int_V c_{qpji}\, \partial_j \Phi_{iK} \partial_p \Phi_{qL}\, dV = S_{LK}$$

Both **M** and **D** are symmetric and positive definite. Symmetry follows from inspection of (15) and (16).
To prove, that **M** is positive definite, define an inner product on a suitable space of vector fields on **V**.

$$(\mathbf{f}\,.\,\mathbf{g}) = \int_V \rho(\mathbf{x}) f_i(\mathbf{x}) g_i(\mathbf{x})\, dV$$

then we have for the interpolation $\overline{\mathbf{f}}(\mathbf{x}) = \Phi(\mathbf{x})\mathbf{F}$

$$(\overline{\mathbf{f}}\,.\,\overline{\mathbf{f}}) = \int_V \rho(\mathbf{x}) \overline{f}_i(\mathbf{x}) \overline{f}_i(\mathbf{x})\, dV = \int_V \rho(\mathbf{x})\, \| \overline{f}_i(\mathbf{x}) \|^2\, dV \;\; > 0 \;\; , \overline{\mathbf{f}}(\mathbf{x}) \neq \mathbf{0}$$

and

$$(\overline{\mathbf{f}}\,.\,\overline{\mathbf{f}}) = (\sum_K F_K \Phi_{iK} \,.\, \sum_L F_L \Phi_{iL}) = \sum_K \sum_L \int_V \rho \Phi_{iK} \Phi_{iL}\, dV\, F_K F_L = \sum_K \sum_L \mathbf{M}_{KL} F_K F_L > 0$$

which concludes the proof for **M**.
Considering a space of vector fields on ∂V and substitution of $\rho(\mathbf{x}) = 1$, results in a similar proof for **D**.

Characteristic for the finite element method is the fact, that the matrices (14) , · · · , (17) are evaluated in a matrix assembly procedure, that sums the contributions of the finite elements to the corresponding integrals for the matrix elements. Thus, the stiffness matrix (14) is assembled using

$$S_{LK} = \int_V \tau_{ij}[\Phi_K]\partial_j \Phi_{iL}\, dV = \sum_{e=1}^{N_e} \int_{V_e} \tau_{ij}[\Phi_K]\partial_j \Phi_{iL}\, dV \tag{18}$$

where N_e is the number of finite elements in the domain, V_e a finite element.
For the evaluation of the finite element matrices, standard software is available, see Desai and Abel (1972) and Zienkiwicz (1977) for examples and Fortran source listings. We have used a package of general purpose finite element subroutines (Segal, 1981) for the implementation of the finite element part of the present hybrid method.

2.2 Boundary conditions

In order to obtain a unique solution of the equation of motion (1) on the volume V suitable boundary conditions must be specified on the boundary ∂V. A similar situation occurs when we discretize the solution and formulate a discrete solution in terms of a system of linear equations. There are $d(N+N_b)$ unknowns - dN displacement components and dN_b traction components - and only dN equations, so the system is underdetermined and we have to specify boundary conditions to obtain a unique solution. Two easily implementable types of boundary conditions are available : zero boundary traction,

$$t_i(\mathbf{x}) = 0 \quad , i=1,2,3 \, , \mathbf{x} \in \partial V \tag{19}$$

or zero boundary displacement,

$$u_i(\mathbf{x}) = 0 \quad , i=1,2,3 \, , \mathbf{x} \in \partial V \tag{20}$$

(19) and (20) correspond to the homogeneous Neumann and Dirichlet conditions from potential theory (Courant and Hilbert, 1968). In case we specify zero traction on the boundary ∂V the traction terms in (13) vanish and we have a system of dN equations in the dN nodal point displacements, which can be solved.
When we specify zero displacements we may use an indirect argument, in that case the total number of unknown displacements in the discretized problem is $d(N-N_b)$. In this case we can construct a solution using only those $N-N_b$ base functions Φ_L that are zero on the boundary ∂V .
Applying Galerkin's principle to this reduced set of base functions, the boundary traction term will again vanish from equation (13) , this time because the matrix $\mathbf{D}$ is zero, as follows from the definition (16). We now have a system of $N-N_b$ equations in just as many unknowns, which can be solved numerically.
In many practical situations neither one of the homogeneous boundary conditions (19) or (20) corresponds with the actual physical conditions. If for instance we want to model wave propagation in a whole space we will have to limit the domain V because of computer limitations. This introduces an artificial boundary ∂V on which boundary conditions must be prescribed, which are however generally unknown. When conditions (19) or (20) are applied in this problem the artificial boundary will produce "unphysical" boundary effects in the form of reflected and diffracted waves. Several methods have been proposed to reduce these artificial waves (Smith, 1974; Reynolds, 1978), but non of these methods is entirely satisfactory.
We will follow a different approach and supply a complete description of the field outside a bounded volume V using a boundary integral equation on ∂V. Matching the integral equation to the finite element equations derived in section 1.2 we will obtain a full system of equations.

3. Integral representations for the wavefield

In the present hybrid method we will use an additional description of the wave field outside a finite volume V , an integral representation for the field in the region $V_\infty - V$, where V_∞ is the background medium, see fig.2, which can be defined as the limit case $\lim_{R \to \infty} V_R = V_\infty$.

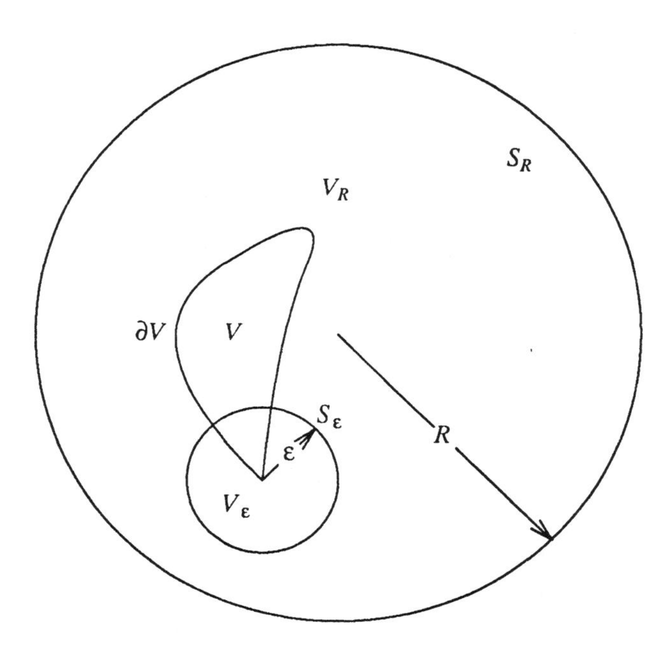

Figure 2. Domain of the integral representation with an embedded scatterer, and a small sphere surrounding a point on the boundary surface ∂V.

This integral representation accounts for the influence of the background medium on the wavefield. The integral equation derived from the representation will - after discretization - result in an underdetermined system of linear algebraic equations.
Combining this system of equations with the finite element equations derived in section 1 will result in a full system of linear algebraic equations.
However it can be shown (Schenck, 1968; Colton and Kress, 1983a; van den Berg, 1987), that the description of the field with the kind of integral equation used, breaks down for certain characteristic frequencies, the eigenfrequencies of the related internal Dirichlet boundary value problem.
Following Schenck (1968) and Tobocman (1984) so called null field equations are added to the integral equation, to overcome these problems. After adding these supplementary equations, we end up with an overdetermined system, that can be solved numerically. An efficient solution scheme for this extended set of integral equations, combined with the finite element equations is derived in section 3.
A similar hybrid approach was proposed in Zienkiewicz et al. (1977) for general field problems, and in Wilton (1978), where it was applied to problems of single frequency scattering in fluids, with an elastic wave scatterer.
Theoretical investigations of integral equations in elastodynamic problems are described in Kupradze (1956, 1963), De Hoop (1958) and Tan (1975). A rigorous treatment of integral equations for the acoustic problem is given by Colton and Kress (1983a).
Wong and Jennings (1975) used a similar integral equation in a seismological modeling

problem, for the special case of SH polarization, to study the effect of topography in the free surface of a halfspace on incident waves. Tan (1975a,b) used a set of coupled integral equations, to study wave diffraction in homogeneous media with homogeneous obstacles. Schuster (1984) discusses several types of integral equations for the acoustic case, applied to seismic modeling problems.

3.1 Introduction of the boundary integral representation

The boundary integral representation used here is based on Betti's theorem, the equivalent in elastodynamics of Green's theorem in potential theory. A derivation of the representation theorem and the resulting integral equations can be found in Kupradze (1963), Tan (1975a,b) and Aki and Richards (1980). The integral representation for the wavefield excited by a body force field $\mathbf{f}(\mathbf{x})$ is

$$\fint_{\partial V} (u_i \tau_{ij}^k - u_i^k \tau_{ij}) n_j dA + \int_{V_\infty - V} f_i u_i^k dV = \begin{cases} 0 & , \mathbf{x}_p \in V \\ c_i^k(\mathbf{x}_p) u_i(\mathbf{x}_p) & , \mathbf{x}_p \in \partial V \\ u_k(\mathbf{x}_p) & , \mathbf{x}_p \in V_\infty - V \end{cases} \qquad (1.a,b,c)$$

where $\{u_i, \tau_{ij}\}$ is the unknown wavefield and $\{u_i^k(\mathbf{x},\mathbf{x}_p), \tau_{ij}^k(\mathbf{x},\mathbf{x}_p)\}$ is the Green's state - the wavefield generated by a point force in the x_k direction, situated in $\mathbf{x}_p$.
The point $\mathbf{x}_p$ is a singular point of the Green's state and in case $\mathbf{x}_p \in \partial V$, the surface integral is understood to be the Cauchy principal value for the singular point, denoted by the bar in integral sign. The tensor $c_i^k(\mathbf{x}_p)$ is defined through the principal value, in which the boundary surface ∂V , the inner boundary of $V_\infty - V$ is replaced by the inner boundary of $V_\infty - V - V_\varepsilon$. V_ε is the intersection of a small sphere of radius ε, centered in $\mathbf{x}_p$, and the exterior medium $V_\infty - V$ (see fig.2). $c_i^k(\mathbf{x}_p)$ is the contribution to the surface integral over the boundary surface S_ε of V_ε. Kupradze has shown, that $c_i^k(\mathbf{x}_p)=1/2\delta_{ik}$ for smooth surfaces (Kupradze, 1963). For surfaces with corners a procedure to compute c_i^k is given in van den Berg (1984).
In section 2.2 the numerical evaluation of the boundary integral is discussed. In (1) the wavefield excited by a body force field is given. A useful alternative expression can be derived for the case of excitation by a incident wavefield. In that case we define the total field to be the sum of an incident field $\mathbf{u}^0$ - that would be observed in the absence of scattering obstacles - and a scattered field $\mathbf{u}^{sc}$, $\mathbf{u}=\mathbf{u}^0+\mathbf{u}^{sc}$ (see for the general case van den Berg (1987)).
Here we consider the special case where the incident field $\mathbf{u}^0$ is excited by a known body force field $\mathbf{f}$. The product integral of the body force and the Green's displacement in (1) can be identified as the wave field excited by the body force distribution $\mathbf{f}(\mathbf{x})$ in the volume $V_\infty - V$, in the absence of the scattering volume V. Therefore we define the term

$$\int_{V_\infty - V} f_i u_i^k dV = u_k^0(\mathbf{x}_p)$$

as the incident wavefield. Subtracting $\mathbf{u}^0$ from both sides of (1), we then have the integral representation

$$\oint_{\partial V} (u_i \tau_{ij}^k - u_i^k \tau_{ij}) n_j dA = \begin{cases} u_k^{sc}(\mathbf{x}_p) & , \mathbf{x}_p \in V_\infty - V \\ c_i^k(\mathbf{x}_p) u_i(\mathbf{x}_p) - u_k^0(\mathbf{x}_p) & , \mathbf{x}_p \in \partial V \\ -u_k^0(\mathbf{x}_p) & , \mathbf{x}_p \in V \end{cases} \quad (2.a,b,c)$$

(1.b) and (2.b) represent inhomogeneous integral equations for the unknown total field $\{u_i, \tau_{ij}\}$ on the boundary ∂V. (1.a) and (2.c), for interior evaluation points $\mathbf{x}_p$, are the null field equations for the elastodynamic problem.
Once the field on the boundary is known, the field off the boundary can be calculated using (1.c) or (2.a).
Two special cases of (2) can be distinguished with homogeneous boundary conditions of either Dirichlet type

$$u_i(\mathbf{x}) = 0 \quad , i=1,2,3 \quad , \mathbf{x} \in \partial V \quad (3.a)$$

representing a rigid obstacle, or of Neumann type

$$\tau_{ij}(\mathbf{x}) n_j = t_i(\mathbf{x}) = 0 \quad , i=1,2,3 \quad , \mathbf{x} \in \partial V \quad (3.b)$$

representing a traction free obstacle such as a cavity.
A half space with a negative topography can be considered as a special case of a cavity reaching the free surface. It can be shown, that the integral equation breaks down for so called characteristic frequencies, the spectrum of internal Dirichlet eigenfrequencies of the volume V, see for a rigorous treatment of the scalar case Colton and Kress (1983a), and for the elastodynamic case Kupradze (1963), van den Berg (1987).
It is shown there, that both

$$\oint_{\partial V} u_i \tau_{ij}^k n_j dA - c_i^k(\mathbf{x}_p) u_i(\mathbf{x}_p) = 0 \quad (4.a)$$

and

$$\oint_{\partial V} u_i^k \tau_{ij} n_j dA = 0 \quad (4.a)$$

have nontrivial solutions for the characteristic frequencies.
This means, that the special cases (3a,b) do not have unique solutions for the characteristic frequencies. Another consequence is, that the combination of finite element equation 1(13) and (discretized) integral equation (1.b),(2.b) cannot be solved numerically, with the solution scheme derived in section 3, for any of the characteristic frequencies. This limits the applicability of the integral equations to single frequency computations, and frequencies not close to one of the characteristic frequencies. The deficiency of integral equations of the type (1.b),(2.b), with either one of the boundary conditions (3.a,b) is well known (Hoenl et al, 1961; Copley, 1968; Schenck, 1968; Colton and Kress, 1983)
Several solutions for the non uniqueness problem have been proposed. Schenck uses the null field equations (1.a),(2.c) for the interior of the volume V. It can be shown (Colton and Kress, 1983b; van den Berg, 1987), that we can obtain a set of equations with a unique

solution if we combine the null field equations with the integral equations (1.b),(2,b), for either one of the homogeneous boundary conditions (3.a,b). Schenck (1968) applied this method to acoustic radiation problems, for a range of frequencies. In Bolomey and Tabbara (1973) and Tobocman (1984) the method is applied to acoustic scattering problems.
Several authors have used other integral equations (Burton and Miller, 1973; Colton and Kress, 1983), involving strongly singular integrals.
In the present hybrid - finite element / boundary integral method we adopt Schenck's solution of additional null field equations to overcome the non uniqueness problems of the integral operators involved.

3.2 Discretization of the integral representation

The integral expressions (1,2) in terms of the wavefield can be transformed into linear algebraic expressions in a finite number of wavefield values.
To this end we discretize the boundary surface ∂V, replacing it by a grid of connected nodal points coinciding with the boundary points of the finite element grid **G** discussed in section 1.
Next we use an approximation of the field on the boundary, using an interpolation scheme in terms of a discrete set of field values, identical to the one applied in the derivation of the finite element equations in section 1.

$$u_i(\mathbf{x}) \approx \bar{u}_i(\mathbf{x}) = \sum_J U_J \Phi_{iJ}(\mathbf{x}) \quad , \mathbf{x} \in \partial V \tag{5}$$

$$\tau_{ij}(\mathbf{x}) n_j = t_i(\mathbf{x}) \approx \bar{t}_i(\mathbf{x}) = \sum_L T_L \Psi_{iL}(\mathbf{x}) \quad , \mathbf{x} \in \partial V \tag{6}$$

where **U** is a vector of nodal point displacement component values, as defined in 1(3) and **T** is a vector of traction component values on the boundary surface. $\Psi(\mathbf{x})$ is an interpolation matrix like the matrix $\Phi(\mathbf{x})$ defined in 1(7), in our implementation we have taken Ψ=Φ.
The interpolating expressions are substituted into the integral representation (2.b,c), taking the evaluation point $\mathbf{x}_p$ first as a nodal point on the boundary, $\mathbf{x}_p = \mathbf{x}_I \in \mathbf{G}$, (2.a) and second as an evaluation point of the null field equation in the interior of V , (2.b).
The contribution from the Green's stress term to the integral equation (2.b) can be expressed directly in terms of the nodal displacement components of the boundary evaluation point $\mathbf{x}_I$.

$$c_i^k(\mathbf{x}_I) u_i(\mathbf{x}_I) = \sum_{i=1}^{3} c_i^k(\mathbf{x}_I) u_i(\mathbf{x}_I) = \sum_{i=1}^{3} c_i^k(\mathbf{x}_I) \sum_J \delta_{J\{d(I-1)+i\}} U_J$$
$$= \sum_J \sum_{i=1}^{3} c_i^k(\mathbf{x}_I)\, \delta_{J\{d(I-1)+i\}} U_J = \sum_J C_{IJ}^k U_J$$

Substituting into the integral representation (2.b,c) we get

$$\oint_{\partial V} \tau_{ij}^k \sum_J U_J \Phi_{iJ} n_j dA - \oint_{\partial V} u_i^k \sum_L T_L \Psi_{iL} dA = \sum_J C_{IJ}^k U_J - u_k^0(\mathbf{x}_I) \quad , I=1, \cdots, N_b \tag{7.a}$$

$$\int_{\partial V} \tau_{ij}^k \sum_J U_J \Phi_{iJ} n_j dA - \int_{\partial V} u_i^k \sum_L T_L \Psi_{iL} dA = -u_k^0(\mathbf{x}_I) \quad ,I=N_b+1,\cdots,N_b+N_s \tag{7.b}$$

$$\sum_J U_J \{ \oint_{\partial V} \tau_{ij}^k n_j \Phi_{iJ} dA - C_{IJ}^k \} - \sum_L T_L \oint_{\partial V} u_i^k \Psi_{iL} dA = E_I^k \quad ,I=1,\cdots,N_b \tag{8.a}$$

$$\sum_J U_J \int_{\partial V} \tau_{ij}^k n_j \Phi_{iJ} dA - \sum_L T_L \int_{\partial V} u_i^k \Psi_{iL} dA = E_I^k \quad ,I=N_b+1,\cdots,N_b+N_s \tag{8.b}$$

where $k = 1,\cdots,d$ in (7),(8). Defining :

$$\bar{B}_{IJ}^k = \oint_{\partial V} \tau_{ij}^k(\mathbf{x},\mathbf{x}_I) n_j \Phi_{iJ}(\mathbf{x}) dA(\mathbf{x}) \tag{9}$$

$$B_{IJ}^k = \bar{B}_{IJ}^k - C_{IJ}^k \quad ,I=1,\cdots,N_b \tag{10.a}$$

$$B_{IJ}^k = \bar{B}_{IJ}^k \quad ,I=N_b+1,\cdots,N_b+N_s \tag{10.b}$$

$$A_{IL}^k = \oint_{\partial V} u_i^k(\mathbf{x},\mathbf{x}_I) \Psi_{iL}(\mathbf{x},\mathbf{x}_I) dA(\mathbf{x}) \tag{11}$$

we get a set of linear algebraic equations :

$$\sum_J B_{IJ}^k - \sum_L A_{IL}^k T_L = E_I^k \quad ,I=1,\cdots,N_b+N_s \tag{12}$$

where $E_I^k = -u_k^0(\mathbf{x}_I)$,N_b the number of nodal points on the boundary, $J,L=1,\cdots,dN_b$, dN_b the number of elements in the vector of discrete boundary field values, d the number of components in the displacement field and N_s the number of evaluation points of the null field equation (2.c).
The system of equations (12) can be written schematically as,

$$\begin{bmatrix} \mathbf{B} , -\mathbf{A} \end{bmatrix} \begin{bmatrix} \mathbf{U} \\ \mathbf{T} \end{bmatrix} = \mathbf{E} \tag{13}$$

From (13) we can identify the special case of a halfspace boundary topography by substitution of the zero traction condition, $\mathbf{T} = \mathbf{0}$

$$\mathbf{BU} = \mathbf{E} \tag{14}$$

For $N_s > 0$ this is an overdetermined set of equations, that can be solved for the displacement field on the free boundary ∂V.

4. Coupled finite element / boundary integral equations

Both the finite element equation and the discretized integral equation are generally underdetermined. However, combining both sets of equations results in a full system, in the unknown field variables. Adding the null field equations will make the resultant set an overdetermined system.

The dimensions of the resulting matrix will in practice be large. Assuming a total of N nodal points in the finite element grid, with d unknown displacement components per point, and N_b nodal points on the boundary, with d unknown traction values per boundary point, we have $d(N+N_b)$ unknowns. If we use N_s null field evaluation points we have a total of $d(N+N_b+N_s)$ equations

In order to economize the numerical solution procedure of the algebraic equations, we make use of the special structure of the matrix, to derive a new set of equations, with a considerably smaller matrix.

This is done through successively eliminating the internal degrees of freedom of the system (displacements of internal nodal points) and the tractions in the boundary nodal points. The result is an equation in the unknown boundary displacements.

A similar approach, based on stepwise reduction of the degrees of freedom is followed in Berkhoff (1974), Zienkiewicz et al. (1977) and Wilton (1978).

4.1 A solution scheme for the coupled equations

The finite element equation 1(13) is

$$\sum_K (S_{LK} - \omega^2 M_{LK}) U_K - \sum_J D_{LJ} T_J = F_L \,, \qquad L=1,\cdots,dN \tag{1}$$

The algebraic equation derived from the integral representation is

$$\sum_{J=1}^{dN_b} B_{IJ}^k U_J - \sum_{L=1}^{dN_b} A_{IL}^k T_L = E_I^k \,, I=1,\cdots,N_b+N_s \,, \; k=1,\cdots,d \tag{2}$$

In the following an abbreviated notation for the dynamic stiffness matrix will be used

$$\mathbf{Z} = \mathbf{S} - \omega^2 \mathbf{M} \tag{3}$$

Next we partition the vector of unknown quantities,

$$(\mathbf{U},\mathbf{T})^T = (\mathbf{U}_I, \mathbf{U}_R, \mathbf{T})^T \tag{4}$$

where $\mathbf{U}_I$ is a vector of displacement components of the N_i internal grid points, ($N_b+N_i=N$, the total number of grid points), $\mathbf{T}$ is the vector of traction component values.

Substitution of (4) into (1) and (2) results in:

$$\begin{bmatrix} \mathbf{Z}_{II} & ,\mathbf{Z}_{IR} & ,\mathbf{0} \\ \mathbf{Z}_{RI} & ,\mathbf{Z}_{RR} & ,-\mathbf{D} \\ \mathbf{0} & ,\mathbf{B} & ,-\mathbf{A} \end{bmatrix} \begin{bmatrix} \mathbf{U}_I \\ \mathbf{U}_R \\ \mathbf{T} \end{bmatrix} = \begin{bmatrix} \mathbf{F}_I \\ \mathbf{F}_R \\ \mathbf{E} \end{bmatrix} \tag{5}$$

With $n = d(N+N_b+N_s)$ equation (5) is a set of n linear algebraic equations in $d(N+N_b)$ unknown field values. For $N_s>0$ the system is overdetermined.

A standard solution of the equations in (5), using the full complex matrix would require $2n^2$ locations of computer memory. Supposing $N_s \ll N$ and $n \approx 10^3$), it will take $\approx 2\times10^6$ memory locations to store the matrix. The number of arithmetic operations required for the solution of the full system depends even stronger on the dimension of the matrix n, ($O(n^3)$ against $O(n^2)$) (Wilkinson, 1965).

In order to reduce both the amount of memory and the computation time needed for the solution of (5) an alternative solution scheme is used, that exploits the special structure of the matrix in (5).

A and **B** are full matrices, whereas the partitions of **Z** are sparse. Solving $\mathbf{U}_I$ from the first row of (5) we get,

$$\mathbf{U}_I = \mathbf{Z}_{II}^{-1}(\mathbf{F}_I - \mathbf{Z}_{IR}\mathbf{U}_R) \tag{6}$$

Substituting (6) into (5) we can express the traction value vector **T** as

$$\mathbf{T} = -\mathbf{D}^{-1}(\mathbf{F}_R - \mathbf{Z}_{RI}\mathbf{Z}_{II}^{-1}(\mathbf{F}_I - \mathbf{Z}_{IR}U_R) - \mathbf{Z}_{RR}\mathbf{U}_R)$$

Defining

$$\mathbf{W} = \mathbf{Z}_{RR} - \mathbf{Z}_{RI}\mathbf{Z}_{II}^{-1}\mathbf{Z}_{IR}$$

we get

$$\mathbf{T} = \mathbf{D}^{-1}\mathbf{W}\mathbf{U}_R - \mathbf{D}^{-1}(\mathbf{F}_R - \mathbf{Z}_{RI}\mathbf{Z}_{II}^{-1}\mathbf{F}_I) \tag{7}$$

Substitution of (7) into (5) finally results in

$$(\mathbf{B} - \mathbf{A}\mathbf{D}^{-1}\mathbf{W})\mathbf{U}_R = \mathbf{E} + \mathbf{A}\mathbf{D}^{-1}(\mathbf{Z}_{RI}\mathbf{Z}_{II}^{-1}\mathbf{F}_I - \mathbf{F}_R) \tag{8}$$

The matrix in (8) is now $d(N_b+N_s)\times dN_b$, whereas the original matrix in (5) is $d(N+N_b+N_s)\times d(N+N_b)$. In general the dimensions of the matrix will be reduced considerably using the reduction scheme leading from (5) to (8).

The matrix **Z** is sparse and symmetric, $\mathbf{Z}_{II}$ may be stored as a band matrix (Zienkiewicz, 1977) and $\mathbf{Z}_{IR} = \mathbf{Z}_{RI}^T$.

In the computation of the matrix **W** the explicit inversion of $\mathbf{Z}_{II}$ should be avoided (this would lead to a full matrix), instead **W** can be computed column wise with the Kth column given by

$$\mathbf{W}^K = \mathbf{Z}_{RR}^K - \mathbf{Z}_{RI}\mathbf{X}^K \quad ,K=1,\cdots,dN_b \tag{9}$$

$$\mathbf{X}^K = \mathbf{Z}_{II}^{-1}\mathbf{Z}_{IR}^K \quad ,K=1,\cdots,dN_b \tag{10}$$

where $\mathbf{Z}_{RR}^K$ and $\mathbf{Z}_{IR}^K$ are the Kth column of $\mathbf{Z}_{RR}$ and $\mathbf{Z}_{IR}$ respectively. The vector $\mathbf{X}^K$ is obtained solving the equation

$$\mathbf{Z}_{II}\mathbf{X}^K = \mathbf{Z}_{IR}^K \tag{11}$$

The advantage of this scheme is, that in order to solve for the dN_b vectors $\mathbf{X}^K$ the band matrix $\mathbf{Z}_{II}$ can be decomposed or reduced to triangular form in place (without losing its band structure). Furthermore this decomposition has to be performed only once for every frequency ω.

Once the matrix decomposition is available the vectors $\mathbf{X}^K$ can be computed using the same decomposed matrix (Wilkinson, 1965; Dongarra et al, 1979).

The dynamic stiffness matrix $\mathbf{Z}_{II}$ becomes singular for a spectrum of real frequencies ω_j , $j=1,\cdots,N_i$, the eigenvalues of the generalized eigenvalue problem

$$\mathbf{Z}_{II}\mathbf{x} = (\mathbf{S} - \omega^2\mathbf{M}_{II})\mathbf{x} = \mathbf{0} \tag{12}$$

This follows from the symmetry of $\mathbf{S}_{II}$ and $\mathbf{M}_{II}$ and the fact, that $\mathbf{M}_{II}$ is positive definite (section 1) (Wilkinson, 1965).
In order to avoid numerical problems connected with the singular frequencies of $\mathbf{Z}_{II}$ we have introduced damping in the model (Aki and Richards, 1980). Using complex material parameters to model an anelastic medium, the singularities of $\mathbf{Z}_{II}^{-1}$ move from the real ω axis into the lower half complex ω plane.
Consider a slightly anelastic medium with homogeneous damping characteristics $c_{ijkl} = (1 + i\alpha)c_{ijkl}$, $|\alpha| \ll 1$.
From the definition of the stiffness matrix 1(14) it follows $S'_{KL} = (1+i\alpha)S_{KL}$. The eigenvalue equation then becomes

$$(\mathbf{S}' - \omega'^2\mathbf{M}_{II})\mathbf{x} = \mathbf{0}$$

or

$$(\mathbf{S} - \frac{\omega'^2}{1+i\alpha}\mathbf{M}_{II})\mathbf{x} = \mathbf{0}$$

the eigenvalues are

$$\omega_j'^2 = \omega_j^2(1+i\alpha)$$

defining $\alpha = -\,sign(\omega)\,|\alpha|$ we have

$$\omega_j' = \omega_j(1 - i\ sign(\omega_j)\frac{|\alpha|}{2} + O(\alpha^2))$$

for $\omega \in \mathbf{R}$, $\alpha \in \mathbf{R}$, $|\alpha| \ll 1$ we have $\omega_j' \in \mathbf{C}$. It can be shown that causality requires the elasticity parameters and the quality factor $Q = \alpha^{-1}$ to be frequency dependent (Aki and Richards, 1980). Since the present formulation is in the frequency domain, this causes no extra problems. However for small damping ($Q \approx 10^{2}$), and limited frequencies, the effect of the frequency dependence can be neglected (Aki and Richards, 1980; Kennett, 1983). In the numerical experiments we have taken $Q = 10^2$ independent of frequency. Having moved the singularities into the lower half plane, we can integrate our results over the real frequency axis, - in the inverse Fourier transform to the time domain.
Using complex elasticity parameters will result in complex matrices **S** ,**Z** For the decomposition of the complex matrix $\mathbf{Z}_{II}$ we have used a standard Gauss elimination algorithm with partial pivoting (Stoer, 1972) for complex band matrices from the LINPACK library (Dongarra et al, 1979). The matrix **D** in (8) is sparse, real, symmetric and positive definite (see section 1), which implies **D** is nonsingular and that no pivoting is required in the numerical solution (Wilkinson, 1965), so that we may economize the storage of the matrix using an equation solver for profile matrices (Zienkiewicz, 1977). As before we do not compute $\mathbf{D}^{-1}$ explicitly, instead the matrix is decomposed,

$$\mathbf{D} = \mathbf{L}\mathbf{Q}\mathbf{L}^T \tag{13}$$

with **L** a lower triangular matrix with unit diagonal, **Q** a diagonal matrix, thereby preserving the sparse nature of **D** in the matrix **L** (Wilkinson, 1965).
The decomposition (13) has to be carried out only once in a complete program run ,

because $\mathbf{D}$ is independent of frequency. Once the decomposition (13) is available, the matrix multiplication with $\mathbf{D}^{-1}$ can be computed column wise, solving the related equation

$$\begin{aligned} \mathbf{X}^K &= \mathbf{D}^{-1}\mathbf{W}^K \quad , K=1,\cdots,dN_b \\ \mathbf{D}\mathbf{X}^K &= \mathbf{W}^K \quad , K=1,\cdots,dN_b \end{aligned} \tag{14}$$

The solution of (14) is readily obtained from (13).
Having computed $\mathbf{D}^{-1}\mathbf{W}$ the result can be substituted in (8), giving a generally overdetermined system of equations in the unknown boundary displacements $\mathbf{U}_R$

$$\mathbf{C}\mathbf{U}_R = \mathbf{R} \tag{15}$$

with

$$\begin{aligned} \mathbf{C} &= \mathbf{B} - \mathbf{A}\mathbf{D}^{-1}\mathbf{W} \\ \mathbf{R} &= \mathbf{E} + \mathbf{A}\mathbf{D}^{-1}(\mathbf{Z}_{RI}\mathbf{Z}_{II}^{-1}\mathbf{F}_I - \mathbf{F}_R) \end{aligned}$$

$\mathbf{C}$ is a complex $d(N_b+N_s)\times dN_b$ matrix. To solve (15) we have used an orthogonalization procedure, based on Householder transformations (Stoer, 1972) from the LINPACK library (Dongarra et al, 1979).

4.2 Computation of field values in points off the boundary

Once the field values for the boundary grid points are obtained from (15), values for internal grid points can be computed solving the related Dirichlet boundary value problem. From (6) the equation for internal field values becomes

$$\mathbf{Z}_{II}\mathbf{U}_I = \mathbf{F}_I - \mathbf{Z}_{IR}\mathbf{U}_R \tag{16}$$

This equation is solved in a similar manner as equation (11) using Gauss elimination. The components of the resulting vector $\mathbf{U}_I$ are the multiplexed field values in internal finite element grid points. When an evaluation point does not coincide with a grid point, the corresponding field value can be obtained by interpolation, using formula 1(6).

$$\bar{u}_i(\mathbf{x}) = \sum_{k=1}^{dN} U_K \Phi_{iK}(\mathbf{x}) \tag{17}$$

Field values for, say N_x evaluation points in the exterior domain can be computed from the boundary values, using a discretization of the integral representation 2.2(2a) for the scattered field.
Discretization produces the explicit form

$$\mathbf{U}^{sc} = \mathbf{B}^{(x)}\mathbf{U}_R - \mathbf{A}^{(x)}\mathbf{T}_R \tag{18}$$

where $\mathbf{U}^{sc}$ is the vector of N_x multiplexed scattered field values for the external evaluation points, $\mathbf{U}_R$ and $\mathbf{T}$ as defined in (4). Note, that the $dN_x \times dN_b$ propagation matrices $\mathbf{B}^{(x)}$ and $\mathbf{A}^{(x)}$ differ from the matrices $\mathbf{B}$ and $\mathbf{A}$ in (5) (see also 2.2(10)).
The boundary traction value vector in (18) can also be expressed in terms of $\mathbf{U}_R$ using (7).
Total field values in exterior points can be computed from the results of (18) by adding the incident field values for the evaluation points

$$\mathbf{u}(\mathbf{x}_{ext}) = \mathbf{u}^{sc}(\mathbf{x}_{ext}) + \mathbf{u}^{0}(\mathbf{x}_{ext}) \tag{19}$$

5. Numerical results

In this section results will be presented of an implementation of the hybrid method, applied to a number of diffraction problems for SH waves in a homogeneous 2-D space with an inclusion of finite extent. The equations for the 2-D scalar SH case follow directly from the ones derived for the general elastodynamic 3-D case. Consider a medium, where both the physical parameters and the wavefield excitation do not depend on x_3.
The equations of motion in such a medium decouple into a coupled set describing the inplane P-SV motion and a scalar equation for the antiplane SH motion (Aki and Richards, 1980).
For an isotropic medium, where

$$c_{ijpq} = \lambda\delta_{ij}\delta_{pq} + \mu(\delta_{ip}\delta_{jq} + \delta_{iq}\delta_{jp})$$

(λ, μ the Lamé parameters), the scalar equation for SH motion is - denoting $u_3(\mathbf{x}) = u(\mathbf{x})$, $f_3(\mathbf{x}) = f(\mathbf{x})$ -

$$-\partial_j(\mu\partial_j u) - \rho\omega^2 u = f$$

This scalar equation will be used in the following. The algebraic equations 1(13) and 2(12) can be translated to the 2-D scalar case specifying the number of field components $d=1$.

5.1 Diffraction of a plane wave by a circular cylinder

Let a plane SH wave be incident on a circular cylinder of radius a and let the wave normal be perpendicular to the axis of the cylinder, resulting in a 2-D configuration (see figure 3). The wavefield can be found as the solution of the equation of motion specialized for a homogeneous medium.

$$-\mu\partial_j\partial_j u - \rho\omega^2 = f \tag{1}$$

In the absence of a body force field f we obtain the homogeneous Helmholtz equation:

$$\nabla^2 u + k^2 u = 0 \tag{2}$$

where ∇^2 is the Laplace operator and k is the wave number of the medium, $k = \omega/\beta$, β is the shear wave velocity, $\beta = (\mu/\rho)^{1/2}$, μ and ρ are the rigidity and mass density. The solution of (2) supplemented with suitable boundary conditions can be obtained by a series solution (Eringen and Suhubi, 1975).
The total field u in the medium surrounding the cylinder is split up in an incident field u^0 and a scattered field u^{sc} as in section 4.1 $u=u^0+u^{sc}$.
Both fields satisfy the equation (2) and the scattered field u^{sc} also satisfies a radiation condition.
Using separation of variables in the cylindrical coordinates r,ϕ we get

$$u^{sc}(r,\phi) = \sum_{n=0}^{\infty} W_n H_n(kr)\cos(n\phi) \tag{3}$$

where the nth order Hankel function of the first kind $H_n = H_n^{(1)}$, combined with the implicit

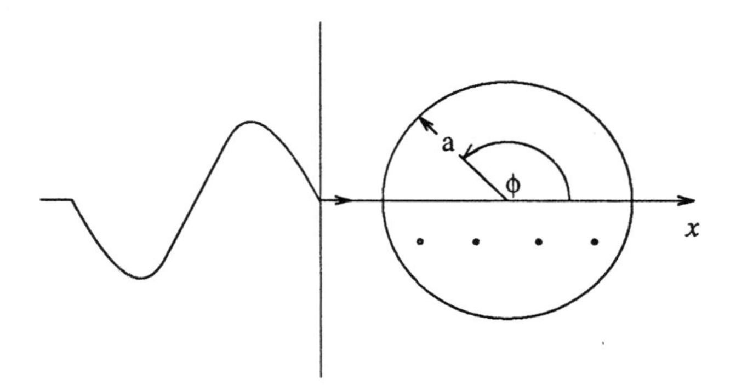

Figure 3. Plane wave traveling parallel to the x axis, incident on a cylinder with axis parallel to the z axis. Dots in lower half circle denote one of two rows of null field evaluation points, symmetric in y.

time dependence $e^{-i\omega t}$, corresponds with an outward travelling wave, as can be seen from the asymptotic expansion for large arguments (Abramowitz, 1964)

$$H_n^{(1)}(z) \sim \sqrt{2/(\pi z)}\, e^{i(z - \frac{1}{2}n\pi - \frac{1}{4}\pi)}$$

The incident field is written as a sum of standing waves.

$$u^0(r,\phi) = \sum_{n=0}^{\infty} C_n J_n(kr)\cos(n\phi) \tag{4}$$

where J_n is the Bessel function of order n.
In the following, quantities relating to the interior domain will be denoted by a superscript i. As for the incident wave, we may write for the total field in the interior:

$$u^i(r,\phi) = \sum_{n=0}^{\infty} A_n J_n(k^i r)\cos(n\phi) \tag{5}$$

Applying the boundary conditions for $r=a$,

$$u^0 + u^{sc} = u^i \tag{6.a}$$

$$\mu\frac{\partial}{\partial r}(u^0 + u^{sc}) = \mu^i \frac{\partial}{\partial r} u^i \tag{6.b}$$

to (3),(4) and (5), the excitation coefficients W_n and A_n can be expressed in the coefficients C_n of the incident field

$$\begin{bmatrix} W_n \\ A_n \end{bmatrix} = \frac{C_n}{D_n}\begin{bmatrix} \rho^i\beta^i J_n'(k^i a) & ,-\rho\beta J_n(k^i a) \\ \rho\beta H_n'(ka) & ,-\rho\beta H_n(ka) \end{bmatrix}\begin{bmatrix} J_n(ka) \\ J_n'(ka) \end{bmatrix} \tag{7}$$

where J_n' , H_n' denote the derivatives of the Bessel and Hankel functions. The determinant

D_n is given by

$$D_n = \rho\beta H_n'(ka)J_n(k^i a) - \rho^i \beta^i J_n'(k^i a)H_n(ka) \tag{8}$$

In the special case of a plane wave incident field we have $C_n = \varepsilon_n i^n$ (Eringen and Suhubi, 1975), with

$$\varepsilon_n = \begin{cases} 1 & , n=0 \\ 2 & , n>0 \end{cases}$$

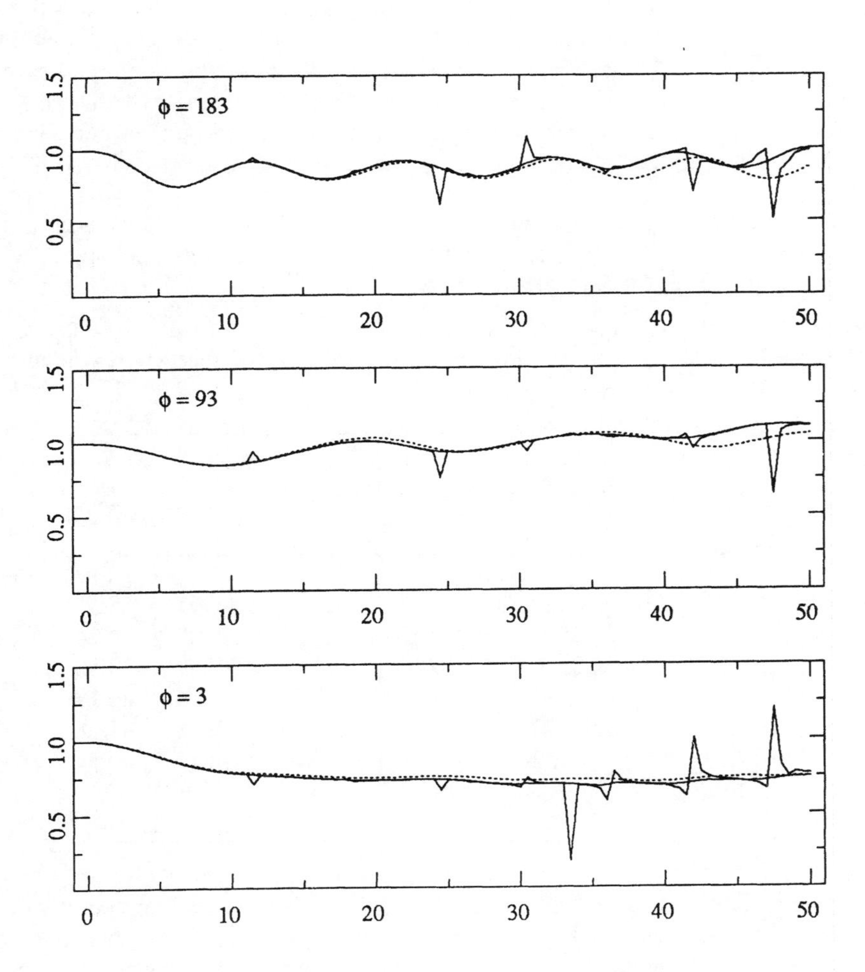

Figure 4. Overlay plots of amplitude spectra for evaluation points (ϕ=3°,93°,183°) from the hybrid method and the series solution (see text).

As a test of the accuracy of the numerical solution obtained by the hybrid method, the diffraction problem for the cylinder for the case of a transparent boundary condition has been solved by evaluation of the series solution (3).
In the computations the radius of the cylinder was taken 100 m., the shear wave velocities β^i and β for the interior respectively the exterior were 4000 and 3000 m/s . A uniform mass density of 2750 kg/m^3 was used.
The interior V was divided into 450 triangular elements, with piecewise linear interpolation of the displacement field, with 60 nodal points on the boundary and 196 in the interior. The total displacement field in the nodal points of the cylinder boundary was obtained in the frequency domain, both by the hybrid method and evaluation of the series solution. The spectra were computed for a finite number of equidistant frequencies $0<f_i \leq 50$ Hz , where $f_i = i \times \Delta f$ and $\Delta f = .2$ Hz and .5Hz for the series solution and the hybrid solution respectively. For the dimensionless wavenumber $ka = 2\pi a/\lambda$ in this experiment we have $0<ka<10.47$. Figure 4 shows an overlay plot of amplitude spectra, evaluated in three points on the cylinder boundary, $\phi_1=3°$ (forward scattering), $\phi_2=93°$ and $\phi_3=183°$ (back scattering).
The curves with the sharp indentations correspond to the result obtained without using null field evaluation points. The frequency values of these indentations correspond with the characteristic frequencies of the cylinder listed in table 1.

Table 1. Characteristic frequencies - $f_{n,s}<50$ Hz - of a cylinder of radius $a=100m$ and wave velocity $\beta=3000m$, $j_{n,s}$ is the sth root of the nth order Bessel function associated with an eigenvibration of angular order n.

(n,s)	$j_{n,s} = 2\pi f_{n,s} \frac{a}{\beta}$	$f_{n,s}$	(n,s)	$j_{n,s}$	$f_{n,s}$
0,1	2.40482	11.48	1,1	3.83171	18.30
2,1	5.13562	24.52	0,2	5.52008	26.36
3,1	6.38016	30.46	1,2	7.01559	33.50
4,1	7.58834	36.23	2,2	8.41724	40.19
0,3	8.65373	41.32	5,1	8.77148	41.88
3,2	9.76102	46.61	6,1	9.93611	47.44
1,3	10.17347	48.57			

The smooth curves closest to the previous ones result from adding eight null field evaluation points, see figure 5. Apparently the null field equations change the solution only near the characteristic frequencies. The other smooth curves are derived from the series solution. It is clear from this figure, that adding the null field equations completely eliminates the erratic behaviour of the numerical solution near the characteristic frequencies. The remaining difference between the series solution and the hybrid solution (null field equations included) is a growing function of frequency, this will be further investigated below. The frequency spectra have been transformed to the time domain using the Fast Fourier Transform algorithm (Cooley and Tukey, 1965). The spectra were multiplied with a smooth spectrum, tapering the response spectrum to zero at the limits and effectively limiting the frequency band to $0 \leq f \leq 30$Hz , where the errors in the numerical results of the hybrid method are small (see figure 4).

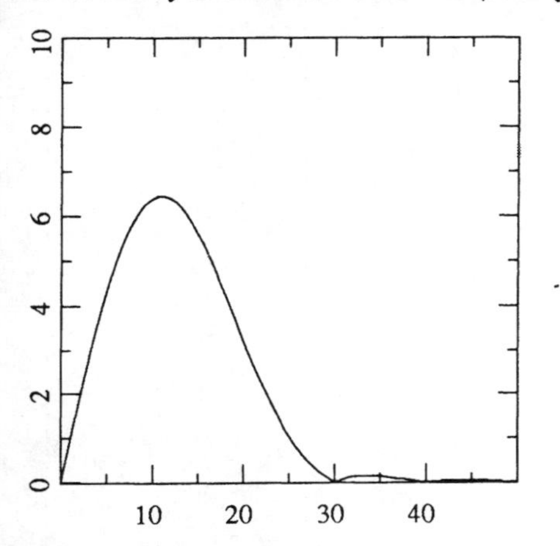

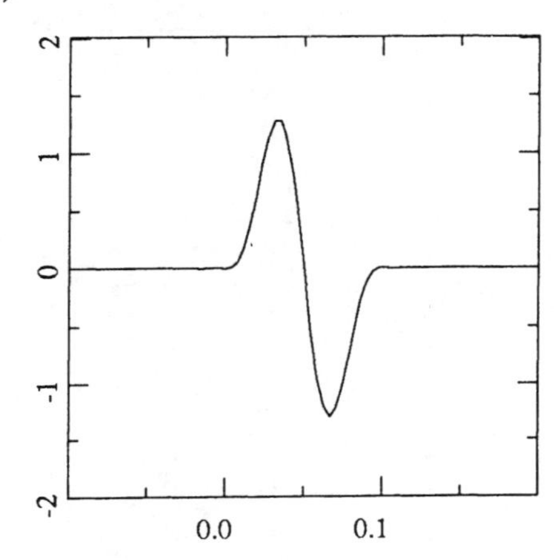

Figure 5. Amplitude spectrum, freq. in Hz. (left) and time pulse form, time in sec. (right) of the Kuepper filter $N=2$, $T=.1$ sec.

The equivalent time domain filter is one of a class of pulse forms, introduced in Kuepper, 1958), defined as

$$f(t) = \begin{cases} \sin(\delta t) - \dfrac{1}{m}\sin(m\,\delta t) & , t \in [0,T] \\ 0 & , t \notin [0,T] \end{cases}$$

where $\delta = \dfrac{N\pi}{T}$ and $m = (N+2)/N$ sec.

In the results $N = 2$, $T = .1$ sec. have been used. The pulse form and corresponding amplitude spectrum are plotted in figure 5. It appears, that after application of this lowpass Kuepper response filter to the results of the hybrid and the series solution, overlay plots of time traces, corresponding with the spectra of fig.4 agree to within a line thickness.

Time domain results for a larger set of evaluation points - including those of fig.4 - from $\phi=3°$ to $\phi=183°$ are given in figure 6.

The response is dominated by the incident pulse, the effect of the anomaly is most clearly visible on the trace with $\phi=183°$ (back scattering), where a reflected/refracted pulse with an altered pulse form arrives at about .1 second after the first motion.

An indication of the accuracy of a solution method for general problems of diffraction by transparent obstacles can be obtained by a zero contrast experiment (Berkhoff, 1976). In a zero contrast configuration, the scattering volume V is replaced by material with the the same parameters as in the exterior medium, such that the incident field u^0 satisfies the wave

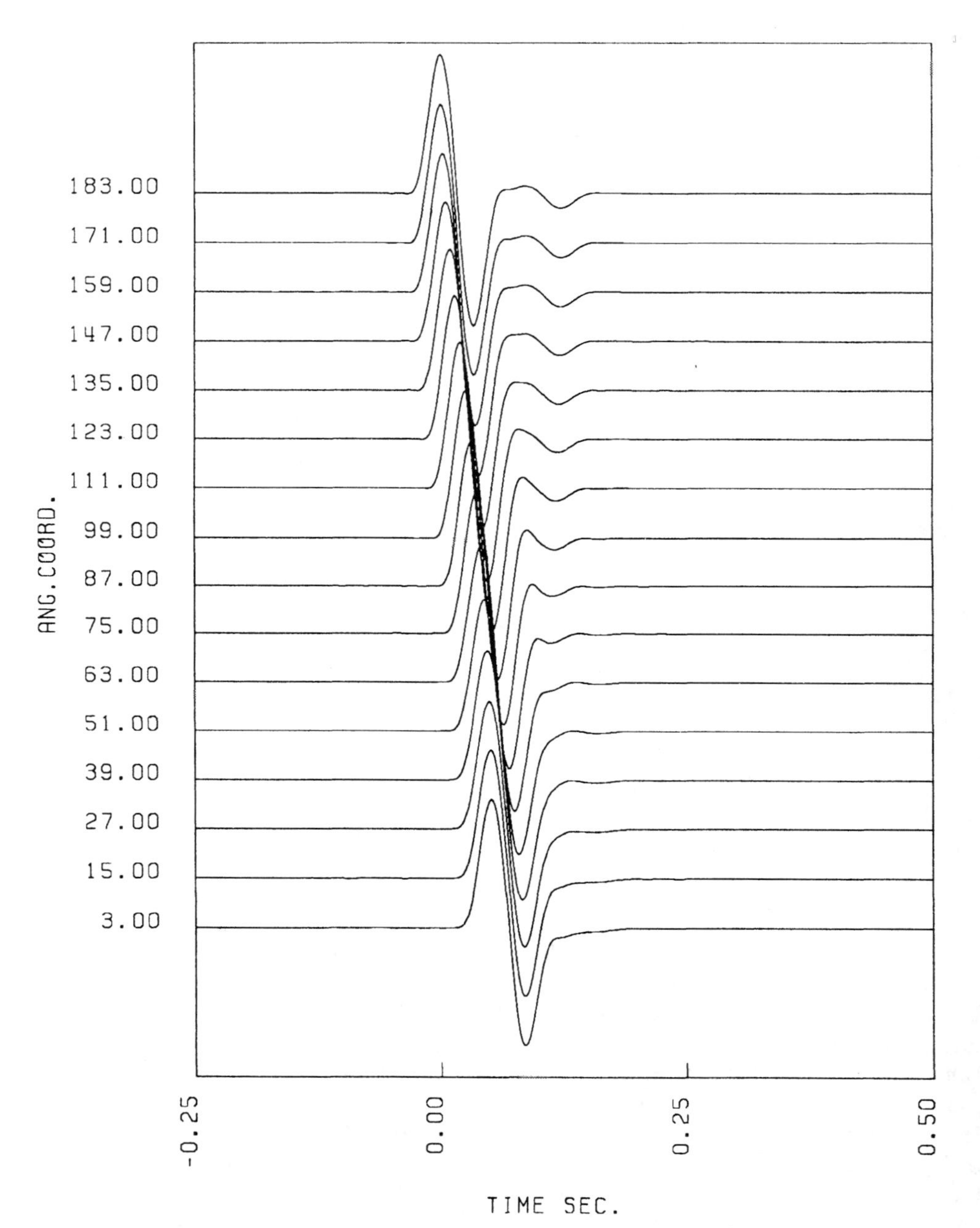

Figure 6. Time domain response for evaluation points along the boundary ∂V after applying the Kuepper filter (fig.5).

equation in the whole medium V_∞. For a homogeneous exterior medium used here, this means defining a scattering volume with homogeneous material parameters identical to the ones of the exterior domain $\rho^i=\rho$, $\beta^i=\beta$
The response to an incident plane wave for the circular scatterer with zero contrast parameters ρ=2750 kg/m^3 , β=3000 m/s was computed for the 60 nodal points on the boundary ∂V. Using the incident field u^0 as a reference a relative error norm can be defined as

$$e = \left[\frac{\sum_J |u_J - u_J^{ref}|^2}{\sum_J |u_J^{ref}|^2} \right]^{\frac{1}{2}} \tag{9}$$

where the summation is over all the boundary grid points.
The available series solution for the finite contrast diffraction problem can also be used as a reference value in (9). In figure 7 both error curves are plotted as a function of the frequency of the incident wave.
The error depends strongly on the discretization of the model, in particular on the number of elements per wavelength. This can explain the difference between the two curves, in the zero contrast test, the cylinder is filled with low velocity material, resulting in a shorter wavelength and a smaller number of elements per wave length, compared with the finite contrast case. In addition to this, in the finite contrast case only part of the wave energy penetrates into the cylinder and therefore the total response depends less strongly on the characteristics of the finite element grid.
From the error curves it can be seen, that for the frequency domain tapering used the error at the predominant frequency of 10 Hz is in the order of 1% , while the error at the 30 Hz effective upper band limit is 7% . It can also be concluded, that in order to keep the error norm below 10% , a minimum of 10 elements per wavelength is required.
The admissible error in the numerical solution depends on the type of problem to be solved. For instance in the computation of time domain results, as displayed in fig.6 the error depends mainly on the dominant frequency f_d of the incident wave, so that an error of 10% at the maximum frequency $f_{\max}$ is quite acceptable.

5.2 Effects of an anomaly at the free boundary of a halfspace

In this section some results will be presented of modeling wave propagation phenomena for an anomaly near the surface of a halfspace. In case the anomaly can be modeled as an irregularity of finite volume in an otherwise regular halfspace, the problem can be solved by the hybrid finite element / boundary integral equation method.
The problem of modeling the effects of anomalies near a halfspace boundary has been addressed in the seismological literature by several authors. Aki and Larner developed a method to compute SH wavefields based on a discrete wavenumber expansion of the field, excited by a plane incident wave (Aki and Larner, 1970; Aki and Richards, 1980). The method was extended to include P-SV problems and excitation by earthquake sources within the model (Bouchon and Aki, 1977). Bard and Bouchon further extended the method to produce time domain results (Bard and Bouchon, 1980).
Several authors have applied the finite difference method to the problem (Boore et al, 1971;

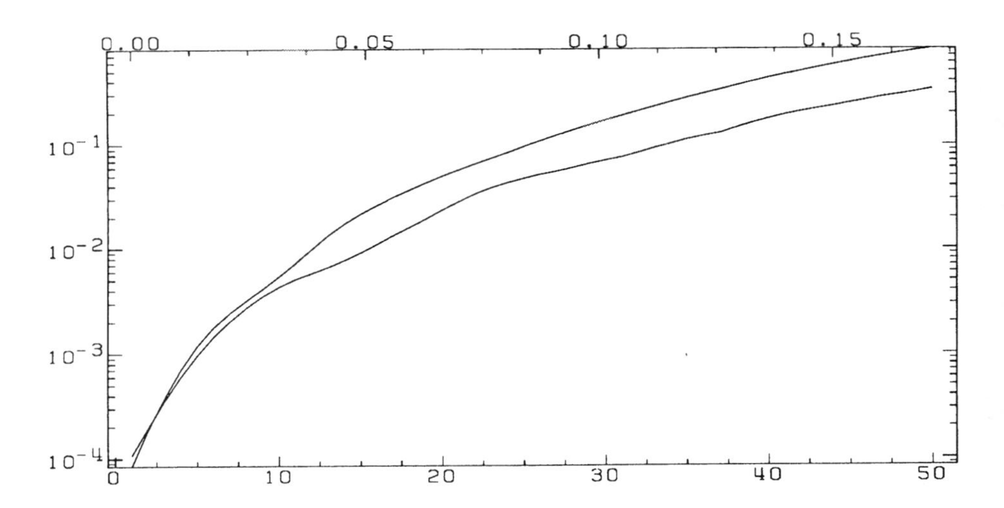

Figure 7. Relative error versus frequency in Hz (bottom) and wave lengths per element - β=3000m/s - (top) from the zero contrast test (upper curve) and comparison with the series solution (lower curve).

Kelly, 1983; Zahradnik, 1984; Levander and Hill, 1985; Hong and Bond, 1986). The finite element method was applied in Lysmer and Drake (1971), Smith (1975), Day (1977) and Crichlow (1982).
Wong and Jennings (1975) and Sills (1978) applied an integral equation technique related to the one included in the present hybrid method to the study of diffraction of plane SH waves by a surface topography in the free surface of a homogeneous 2-D halfspace.
Sanchez-Sesma (1982, 1983) used a boundary integral technique based on a plane wave expansion of the field and a least squares approximation of the boundary conditions of an anomaly.

In the following modeling experiment the effects of a low velocity sedimentary wedge on an incident wave are studied. The geometry of the model of the medium is displayed in fig.8., it consists of a triangular indentation in a half space boundary 300m wide and deep. The indentation is filled with low velocity material β=1500m/s, ρ=2225kg/m^3. The material parameters in the halfspace are ρ=2750kg/m^3, β=3000m/s. The frequency range of the computations was defined as, $f_i = i \times \Delta f$, $i=1, \cdots ,30$, $\Delta f = 1.$Hz.
The triangular fill was discretized with a regular grid of 861 nodal points - 41 points along all three sides - and 1600 triangular linear elements.
The elements were geometrically similar to the shape of the fill, sized h=7.5m in the vertical and horizontal direction. Nine null field evaluation points were used, located along the free surface of the fill, at 30 m intervals, see fig.8.
A plane wave with incident angle 45° is used as input. Three receiver arrays - AB , CD , DF in fig.8 - with respectively 40,39 and 40 sensors, spaced at 7.5m intervals, are located along the halfspace surface to register the displacement fields. Adding the two boundary edge points of the wedge, the total number of equidistant receivers is 121. Displacement amplitude spectra for the 121 receivers are plotted versus the horizontal

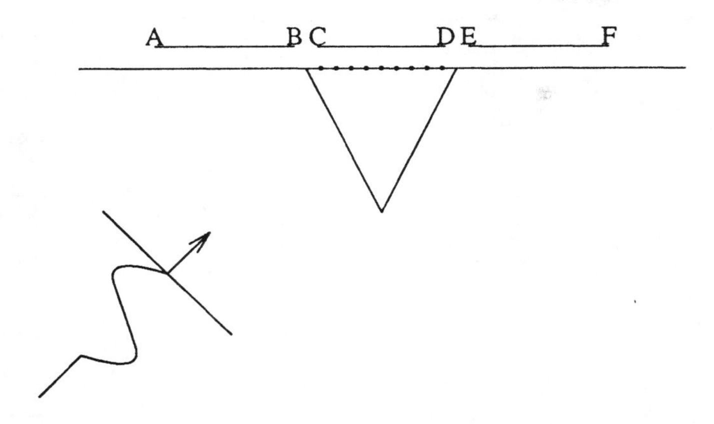

Figure 8. Model configuration - plane wave with axis parallel to the z axis incident on a low velocity fill, with 3 line arrays AB,CD,EF. dots denote 9 null field points at the surface of the fill.

coordinate x in fig.9 for a plane wave incident at 45°. A strong amplification occurs at the far side of the fill, ($0<x<100$). Note the amplification of up to 4 times the amplitude value 2 for an undisturbed halfspace.

Using the signal waveform filter displayed in fig.5, the computed frequency spectra were transformed to the time domain.
Figure 10 displays the resulting time traces for a set of receivers along the free surface (arrays *AB* and *EF*) and the walls of the fill (81 boundary nodal points). The figure shows strong primary arrivals on the left-hand side of the model, corresponding with the incident plane wave and its reflection from the left-hand side wall of the triangle (traces 1-41).
The wave diffracted from the tip along the right hand side wall into the shadow zone is followed by a wave traveling at a lower speed, inside the wedge fill. Figure 11 displays the response along the free surface of the fill (array CD). Note the primary arrivals in traces 32-39, that correspond with energy diffracted around the tip of the wedge that leaked into the fill through the right-hand side wall.
Some of the phases in fig.11 allow a ray-geometrical interpretation. Note the wave, that traverses the fill from left to right, starting at trace 1 at $t \approx 0$, then in the opposite direction arriving on trace 1 at $t \approx .35$. The second arrival on traces 1-20 (symmetric in the offset direction) corresponds with a diffraction of the incident wave at the tip of the fill. The arrival time of the incident wave in the origin of an undisturbed halfspace is $t=0$.

The results do not show any significant waveform dispersion, well known for finite element solutions of the wave equation (Alford et al, 1974; Marfurt, 1984).
This is because we have effectively limited the band width of the signals to (0,30) Hz. For the wave velocity of 1500m/s and an element size $h=7.5m$, this frequency range corresponds with a grid density range of $(6.67,\infty)$ elements per wavelength.

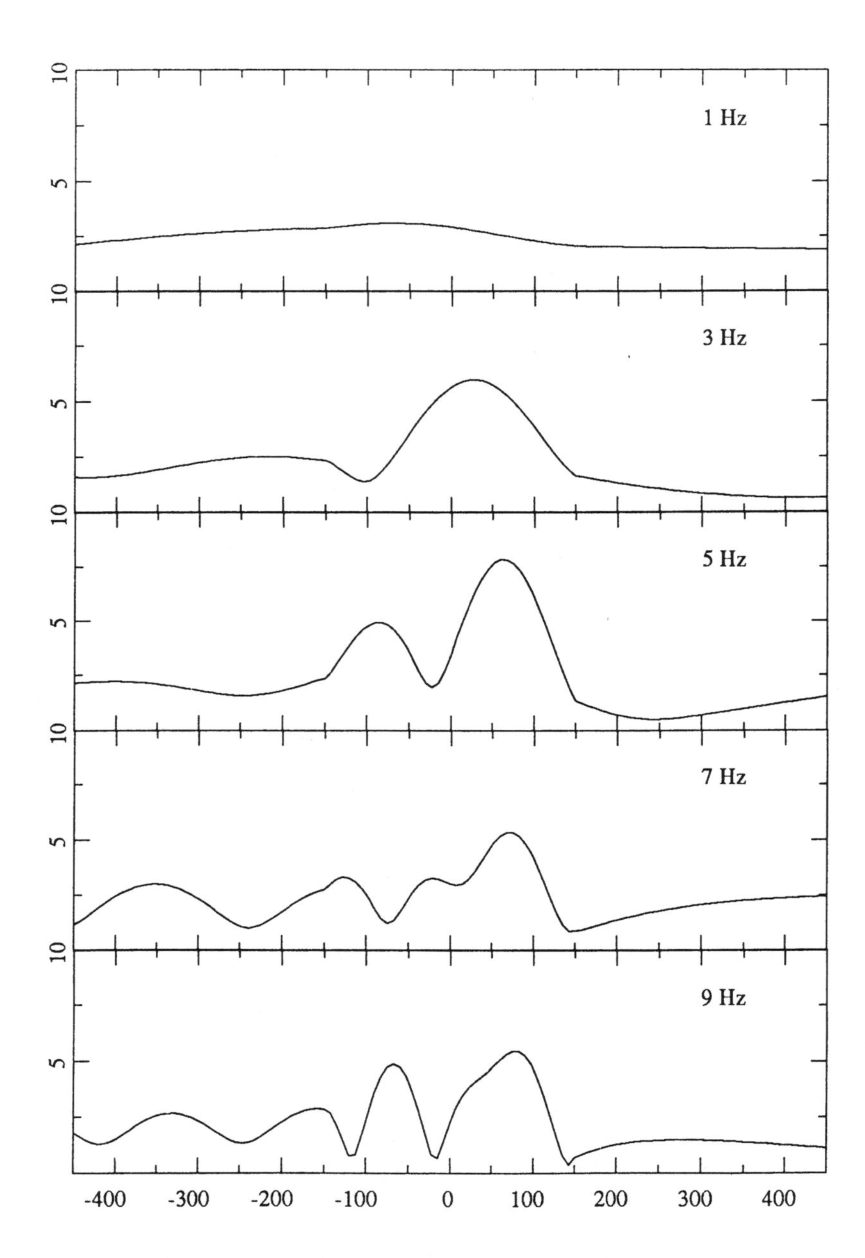

Figure 9. **Amplitude spectrum of the displacement field along the surface, including the sediment fill, for five frequencies 1,3,5,7,9 Hz, (top down).**

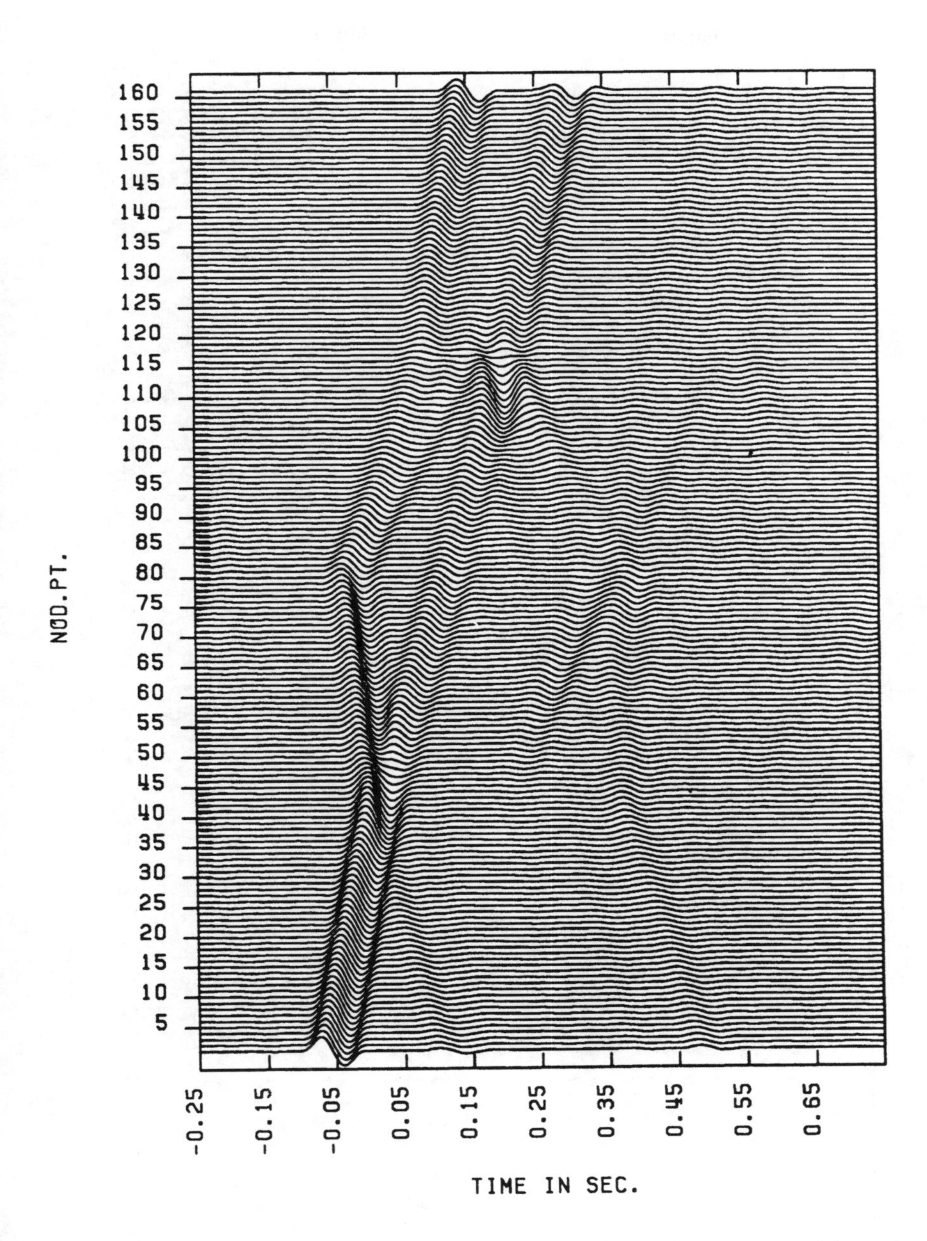

Figure 10. Time domain displacement field along the walls of the fill (traces 41-121) and the surface of the halfspace (traces 1-40,122-161)

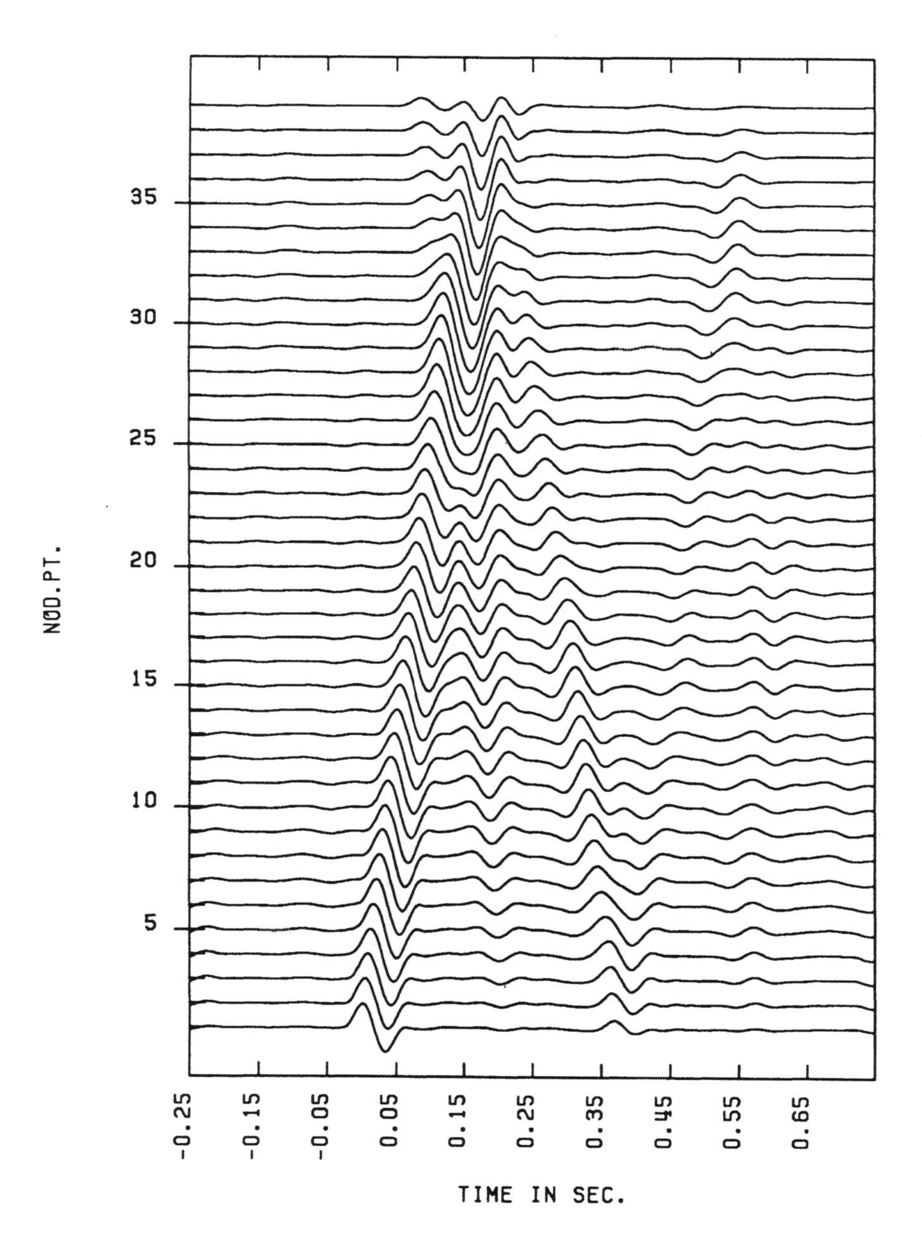

Figure 11. Time domain displacement field in the array CD along the free surface of the fill.

Dispersion curves for plane waves in a homogeneous unbounded medium, for several finite element discretizations - different element types and mass matrices - have been presented by Belytschko and Mullen (1978) and by Marfurt (1984).
For the constant strain elements in the frequency domain formulation used here, these dispersion curves show a rapid decrease in both phase and group velocity for $\nu = \lambda / h < 8$ or equivalently $f > 25$Hz, which results in the arrival of dispersive wave trains, if sufficient energy is present in this band width.

6. Concluding remarks

The present hybrid method presented has a number of advantages over pure discretization methods such as the finite element and finite difference method, namely

- Only the anomalous region has to be discretized. The background medium effects are accounted for by the integral representation. This greatly reduces the amount of computer memory required by the program. It also eliminates the effects of numerical dispersion caused by a discretization of the background medium.
- No artificial boundary conditions have to be specified as in a pure discretization method, so there are no spurious reflections to be dealt with.
- A greater variety of wavefield excitations is available, because one can either include a source in the discretized part of the model using a finite element formalism or define an incident wavefield in the background medium.
- A greater flexibility is provided in modeling experiments. The computation of the model response can be split up in several independent stages. Thus one can change part of the modeling parameters and obtain new results for the modified configuration, without repeating all the initial calculations. This makes modeling experiments with a trial and error strategy more feasible.

The implementation for 2-D scalar problems has been applied in the computation of the response of a number of models, to an incident plane wave.
The validity of the method was verified in the comparison with a series solution for a simple model geometry. It was also shown, from an analysis of the frequency dependence of the error in the solution, that - for the given type of finite elements used and a predominant frequency at 1/3 of the band width ($f_d \approx 1/3 f_{max}$) - a number of 10 elements per wavelength at the maximum frequency f_{max} gives excellent results.
In section 4.2 results for a model containing boundary edges were obtained. It was shown, that dispersion effects could be eliminated by application of a lowpass filter, that increases the number of elements per wavelength at the maximum frequency f_{max} to 6.67. This agrees with discretization criteria for full discretization methods (Alford, 1974; Smith, 1975) It can be concluded, that the discretization requirements for the finite element part of the hybrid method are the same as those for a full discretization method.
It has been shown in the modeling experiments, that Schenck's method of additional null field equations gives good results for the range of wave numbers ($ka < 10$) studied.

References

Abramowitz, M., and Stegun, I., 1964. Handbook of mathematical functions, New York, Dover.

Aki, K. and Larner, K. L., 1970. Surface Motion of a Layered Medium Having an Irregular Interface due to Incident Plane SH Waves, *J. Geophys. Res.*, **75**, 933-954.

Aki, K., and Richards, P. G., 1980. *Quantitative Seismology*, San Francisco, W.H. Freeman.

Alford, R.M., Kelly, K.R., and Boore, D.M., 1974. Accuracy of finite-difference modelling of the acoustic wave equation, *Geophysics*, **39**, 834-842.

Alterman, Z.S., and Karal, F.C., 1968. Propagation of elastic waves in layered media by finite difference methods, *Bull. Seism. Soc. Am.*, **58**, 367-398.

Banaugh, R. P., and Goldsmith, N., 1963. Diffraction of steady acoustic waves by surfaces of arbitrary shape, *J. Ac. Soc. Am.*, **35**.

Bard, P.Y. and Bouchon, M., 1980. The seismic response of sediment-filled valleys. Part 1. The case of incident SH waves, *Bull. Seism. Soc. Am.*, **vol.70**, 1263-1286.

Belytschko, T., and Mullen, R., 1978., On dispersive properties of finite element solutions. In , *Modern Problems in Elastic Wave Propagation*, , eds Julius Miklowitz, and Jan D. Achenbach. Wiley.

Ben-Menahem, A., and Singh, S. J., 1981. *Seismic waves and sources*, Springer Verlag.

Berkhoff, J.C.W., 1974. Linear wave propagation problems and the finite element method. In *Finite elements in fluids 1, Viscous flow and hydrodynamics*, proceedings "International Symp. on the f.e.m. in flow problems", Univ. College of Wales, eds. Oden, J.T., Taylor, C., Gallagher, R.H. and Zienkiewicz, O.C.

Berkhoff, J.C.W., 1976. *Mathematical models for simple harmonic linear water waves wave diffraction and refraction*, PhD Thesis TH-Delft.

Bolomey, J.-C. and Tabbara , W., 1973. Numerical aspects on coupling between complementary boundary value problems., *IEEE Transactions on Antennas and propagation*, **AP-21**, 356-363.

Boore, D. M., 1972. Finite difference methods for seismic wave propagation in heterogeneous materials. In *Seismology: Surface waves and earth oscillations*, ed B.A. Bolt, **11**, Academic press.

Boore, D. M., Larner, K., and Aki, K., 1971. Comparison of two independent methods for the solution of wave scattering problems: Respons of a sedimentary basin to vertically incident SH waves, *J. Geophys. Res.*, **76**, 558-569.

Bouchon, M. and Aki , K., 1977. Near-field of a seismic source in a layered medium with irregular interfaces, *Geophys. J. R. Astr. Soc.*, **50**, 669-684.

Burton, A.J., and Miller, G.F., 1971. The application of integral equation methods to the numerical solution of some exterior boundary value problems., *Proc. Roy. Soc. London*, Ser. A, **323**, 201-210.

Chin, R.C.Y., Hedstrom, G. and Thigpen, L., 1984. Numerical Methods in Seismology, *J. Comp. Phys.*, **54**, 18-56.

Cooley, J.W., and Tukey, S.W., 1965. An algorithm for the machine calculation of complex Fourier series, *Math. Comp.*, **19**, 297-301.

Copley, L.G., 1968. Fundamental results concerning integral representations in acoustic radiation, *J. Ac. Soc. Am.*, **44**, 28-32.

Courant, R., and Hilbert, D., 1968. *Methoden der Mathematischen Physik*, Berlin, Springer-Verlag.

Crichlow, J. M., Beckels, D., and Aspinall, W.P., 1984. Two-dimensional analysis of the effect of subsurface anomalies on the free surface respons to incident SH-waves, *Geophys. J. R. Astr. Soc.*, **79**.

Colton, D. and Kress, R., 1983. *Integral Equation Methods in Scattering Theory*, New York, John Wiley. 669-684.

Colton, D. and Kress, R. , 1983. The unique solvability of the null field equations, *Q. Jl Mech. Appl. Math.*, **36**, 87-95.

Day, S.M., 1977. *Finite element analysis of seismic scattering problems, PhD Thesis Univ. Cal., San Diego.*

Desai, C.S., and Abel, J.F., 1972. Introduction to the Finite Element Method, New York, van Nostrand Reinhold.

Dongarra, J.J., Moler, C.B., Bunch, J.R., and Stewart, G.W., 1979. *LINPACK User's Guide*, Philadelphia, SIAM.

Eringen, A. C., and Suhubi, E. S., 1975. *Elastodynamics*, New York, Academic press.

Hoenl, H., Maue, A.W., and Westfahl, K., 1961. Theorie der Beugung. In *Handbuch der Physik*, ed S. Fluegge, **25**, 218-583, Springer.

Hong, M., and Bond, L.J., 1986. Application of the finite difference method in seismic source and wave diffraction simulation, *Geophys. J. R. Astr. Soc.*, **87**.

De Hoop. A.T., 1958. *Representation theorems for the displacement in an elastic solid and their application to elastodynamic diffraction theory*, PhD Thesis Techn. Univ. Delft.

Kelly, K. R., 1983. Numerical study of Love wave propagation, *Geophysics*, **48**, 833-853.

Kelly, K. R., Ward, R. W., Treitel, S., and Alford, R. M., 1976. Synthetic seismograms: a finite difference approach, *Geophysics*, **41**, 2-27.

Kennett, B.L.N., 1983. *Seismic wave propagation in stratified media*, Cambridge university press.

Kuepper, F.J., 1958. Theoretische Untersuchungen ueber die Mehrfachaufstellung von Geophonen, *Geophys. Prosp.*, **6**, 194-256.

Kupradze, V. D., 1956. *Randwertaufgaben der Schwingungstheorie und integralgleichungen*, Berlin, VEB Deutscher Verlag der Wissenschaften.

Kupradze, V. D., 1963. Dynamical problems in elasticity. In *Progress in solid mechanics*, eds I.N. Sneddon and R. Hill, Vol. 3.

Levander, A.R., and Hill, N.R., 1985. P-SV resonances in irregular low velocity surfaces, *Bull. Seism. Soc. Am.*, **75**, 847-864.

Lysmer, J., and Drake, L. A., 1971. The propagation of Love waves across nonhorizontally layered structures, *Bull. Seism. Soc. Am.*, **61**, 1233-1251.

Marfurt, K. J., 1984. Accuracy of finite difference and finite element modeling of the scalar and elastic wave equations, *Geophysics*, **49**, 533-549.

Mitchell, A. R., 1969. *Computational methods in partial differential equations*, New York, Wiley.

Olsen, K., and Hwang, L. S., 1971. Oscillations in a bay of arbritrary shape and variable depth, *J. Geophys.Res.*, **76**, 5048-5064.

Ralston, A., 1965. *A first course in numerical analysis*, New York, McGraw-Hill.

Reynolds, A. C., 1978. Boundary conditions for the numerical solution of wave propagation problems, *Geophysics*, **43**, 1099-1111.

Sanchez-Sesma, F. J., 1983. Diffraction of elastic waves by three dimensional surface irregularities, *Bull. Seism. Soc. Am.*, **73**, 1621-1636.

Sanchez-Sesma, F. J., Herrera, I., and Aviles , J., 1982. A boundary method for elastic wave diffraction: application to scattering of SH waves by surface irregularities, *Bull. Seism. Soc. Am.*, **72**, 473-490.

Schenck, H. A., 1968. Improved integral formulation for acoustic radiation problems, *J. Ac. Soc. Am.*, **44**, 41-58.

Schuster, G.Th., 1984. *Some boundary integral equation methods and their application to seismic exploration*, PhD Thesis, Univ. Columbia.

Segal, A., 1980. *AFEP user manual*, , Dept. of Math. Techn. University Delft.

Sharma, D. L., 1967. Scattering of steady elastic waves by surfaces of arbitrary shape, *Bull. Seism. Soc. Am.*, **57**, 795-812.

Smith, W. D., 1974. A non reflecting plane boundary for wave propagation problems, *J. Comp. Phys.*, **15**, 492-503.

Smith, W. D., 1975. The application of finite element analysis to bodywave propagation problems, *Geophys. J. R. Astr. Soc.*, **42**, 747-768.

Stoer, J., 1972. *Einfuehrung in die numerische Mathematik I*, Berlin, Springer-Verlag.

Strang, G., and Fix, G.J., 1973. *An analysis of the finite element method*, Prentice-Hall.

Tan, T. H., 1975. Scattering of elastic waves by elastically transparant obstacles, *Appl. Sci. Res.*, **31**, 29-51.

Tan, T. H., 1975. *Diffraction theory for time harmonic elastic waves*, PhD Thesis, Techn. Univ. Delft.

Tobocman, W., 1984. Calculation of acoustic wave scattering by means of the Helmholtz integral equation, *J. Ac. Soc. Am.*, **76**, 599-607.

Van den Berg, A. P., 1984. A hybrid solution of wave propagation problems in regular media with bounded irregular inclusions, *Geophys. J. R. Astr. Soc.*, **79**, 3-10.

Van den Berg. A.P., 1987. *A hybrid method for the solution of seismic wave propagation problems*, PhD Thesis Univ. Utrecht.

Wilkinson, J. H., 1965. *The algebraic eigenvalue problem*, Oxford University Press.

Wilton, D.T., 1978. Acoustic radiation and scattering from elastic structures., *Int. J. Num. Meth. Eng.*, **13**, 123-138.

Wong, H. L., and Jennings, P. C., 1975. Effects of canyon topography on strong ground motion, *Bull. Seism. Soc. Am.*, **65**, 1239-1257.

Zahradnik, J., and Urban, L., 1984. Effect of a simple mountain range on underground seismic motion, *Geophys. J. R. Astr. Soc.*, **79**, 167-183.

Zienkiewicz, O. C., 1977. The finite element method , 3rd edition, McGraw-Hill.

Zienkiewicz, O. C., Kelly, D. W., and Bettes, P., 1977. The coupling of the finite element method and boundary solution procedures, *Int J. Num. Meth. Engineering*, **11**, 355-375.

ARIE van den BERG, Department of Theoretical Geophysics, University of Utrecht, PO Box 80.021, 3508 TA Utrecht, The Netherlands.

Chapter 6

Elastic wavefield inversion and the adjoint operation for the elastic wave equation

Peter Mora

Seismic data is nonlinearly related to the elastic parameters, elastic moduli and density so the analytic solution to the inverse problem of finding the earth parameters corresponding to seismic data is very difficult. In principle, the solution may be located numerically using some kind of Monte Carlo technique but this is infeasible on todays computers due to the nontrivial computer time required to solve the elastic forward problem. Therefore, a more restrictive but cheaper method such as the nonlinear least squares must be used. Crucial to any nonlinear inversion scheme based on least squares iterations is the adjoint operator. This is used to define the gradient direction on the least squares functional surface which is in turn used to evaluate the perturbations in the elastic parameters at each iteration. Tarantola (1984) has used perturbation methods to derive the adjoint operator (in terms of the elastic moduli and density) for the elastic wave equation. Mora (1986a) has rederived these equations and extended them to different model parameters including the P- and S-wave velocities and density. He also made a comparison of the inversion scheme with prestack depth migration and made some numerical computations using synthetic data to demonstrate these formulae (Mora (1986a) and Mora (1986b)). The interpretation of these equations is important because it indicates how the adjoint operation may be evaluated in terms of the forward operator. Not only does this interpretation make the computations easier but it clarifies the concepts behind the formulae. A good understanding of the concepts will hopefully lead to modifications of the current algorithm or new algorithms that will overcome the limitations (such as poor resolution of the low wavenumber model perturbations when only reflection seismic data is used).

N.J. Vlaar, G. Nolet, M.J.R. Wortel & S.A.P.L. Cloetingh (eds), Mathematical Geophysics, 117-137.

1. Introduction

The goal of seismic inversion is to obtain a quantitative estimate of the earth parameters. This is achievable thanks to our knowledge of physics which gives the relationship between the earth model and the observed seismic wavefield. Unfortunately, physics only says how to compute the observed wavefield using the knowledge of the seismic source distribution and the earth model but the inverse mapping from seismic data to earth model is not so easy. In particular, it is a nonlinear function and is non-unique (if the accuracy of the data and precision of the calculations are finite). One way to handle this nonlinearity is to use a method that only requires the solution to the forward problem in order to obtain the inverse mapping. It is possible to imagine simply using brute force to calculate the seismic response due to every earth model. The inverse solution would then be obtained by finding the earth model corresponding to the seismic response that looked most like the observed wavefield. Of course, such a method is infeasible because the number of calculations required would be infinite. An alternative is the Monte Carlo method which uses the same idea of systematically searching for the solution but it requires only a finite though large number of synthetic wavefields to be calculated. Though better than the brute force search method it is still infeasible on todays computers because of the large number of time consuming seismic wave simulations that are required. If, however, we assume that we are not totally ignorant of the earth model then we can feed in a starting guess and iteratively update this model until the computed wavefield matches the observed wavefield. At each iteration we can assume linearity (if the updates are always small enough) so the model perturbations can be obtained by a linear relationship. This iterative method requires wavefields to be calculated only a few times and so is viable. The only problem is that if the starting guess was not very good then there is a chance that the solution obtained will not be the earth model that produces the best fit wavefield. This is because of the nonlinearity resulting in a bumpy measure of fit functional. The key issues are therefore (i) how to update the earth model and (ii) what measure of fit to use. In this paper I will assume a least squares measure of fit between the two wavefields and concentrate on the first issue. The techniques outlined will be useful no matter what measure is used and hopefully, this paper will provide the reader with a good understanding of the theoretical concepts required to efficiently obtain the linear relationship between model and data from the known equations of physics (and hence the adjoint operator). The methods are generally applicable to any inverse problem and will hopefully be enlightening and informative to students who are beginning to tackle inverse problems of any kind. The adjoint calculations for the elastic wave equation are illustrated and the corresponding elastic inversion scheme is demonstrated with some synthetic examples.

2. Inversion and the adjoint operation

The aim of this section is to find expressions for the most probable set of model parameters $\mathbf{m}$ (i.e. P- and S-wave velocities and densities) given that we have made some data observations $\mathbf{d}_0$ (i.e. some seismic observations) and we know from the laws of physics (i.e. the equations for data $\mathbf{d}$ in terms of physical model parameters $\mathbf{m}$). Assuming Gaussian probability density functions for the errors in the data and model spaces leads to the least squares solution (I will assume that the implications of the choice of Gaussian distributions are well known and will not attempt give justifications here). The Gaussian probability function is given by

$$P(\mathbf{d}\,|\,\mathbf{m}) = K\exp\frac{-1}{2}\left[\Delta\mathbf{d}\mathbf{C}_{\mathbf{d}}^{-1}\Delta\mathbf{d}+\Delta\mathbf{m}\mathbf{C}_{\mathbf{m}}^{-1}\Delta\mathbf{m}\right] = K\exp\frac{-1}{2}\left[S(\mathbf{d},\mathbf{m})\right] \quad . \tag{1}$$

where $\Delta\mathbf{d} = \mathbf{d}(\mathbf{m})-\mathbf{d}_0$ is the data error (i.e. difference between the data observations $\mathbf{d}_0$ and the data computed using earth parameters $\mathbf{m}$), $\Delta\mathbf{m} = \mathbf{m}-\mathbf{m}_0$ is the model perturbation vector (i.e. difference between the model $\mathbf{m}$ and the a priori model $\mathbf{m}_0$) and the $\mathbf{C}_{\mathbf{d}}^{-1}$ and $\mathbf{C}_{\mathbf{m}}^{-1}$ are respectively the data and model covariance matrices. I have introduced this probability function to define notation and from here on I will assume the reader has a knowledge of how to solve for the maximum probability solution using nonlinear least squares inverse theory (for reviews see Menke (1984) or Tarantola (1982)). The solution for model parameters $\mathbf{m}$ using nonlinear least squares is given by the iterative formula,

$$\mathbf{m}_{n+1} = \mathbf{m}_n - \mathbf{\rho}_n \quad , \quad \mathbf{\rho}_n = \gamma_n\hat{\mathbf{H}}_n\mathbf{g}_n + \varepsilon_n\mathbf{\rho}_{n-1} \tag{2a}$$

where n is the iteration number, $\hat{\mathbf{H}}$ is the approximate inverse Hessian and $\mathbf{g}$ is the gradient vector (the direction (in the model space) of steepest ascent on the square error functional) given by

$$\mathbf{g} = \mathbf{D}^*\mathbf{C}_{\mathbf{d}}^{-1}\Delta\mathbf{d}+\mathbf{C}_{\mathbf{m}}^{-1}\Delta\mathbf{m} \quad . \tag{2b}$$

The operator $\mathbf{D} = \frac{\partial\mathbf{d}}{\partial\mathbf{m}}$ is the Frechet operator and constants γ_n and ε_n are dependent on the choice of conjugate gradient algorithm. Mora (1986a) and Mora (1986b) chose the Polak conjugate gradient algorithm because of its robustness for nonlinear functions (for a description of the conjugate gradient method see Luenberger, 1984). The approximate inverse Hessian operator $\hat{\mathbf{H}}$ used in the calculations of this paper is of the form

$$\hat{\mathbf{H}} = \frac{\mathbf{C}_{\mathbf{m}}}{\mathbf{S}^2(|\mathbf{g}|+\xi\mathbf{I})} = \begin{bmatrix} \sigma^2_\alpha\mathbf{I} & 0 & 0 \\ 0 & \sigma^2_\beta\mathbf{I} & 0 \\ 0 & 0 & \sigma^2_\rho\mathbf{I} \end{bmatrix} \frac{1}{\mathbf{S}^2(|\mathbf{g}|+\xi\mathbf{I})} \quad . \tag{2c}$$

The $\mathbf{C}_{\mathbf{m}}$ part of $\hat{\mathbf{H}}$ simply scales the different physical model parameters (P- and S- wave velocity and density respectively denoted α, β and ρ) to allow for the different physical units and variances. The $\mathbf{S}$ in the denominator is a smoothing matrix (i.e. a convolution matrix to do low pass filtering) so the denominator represents a space variable scaling to remove the slowly varying changes in magnitude in the gradient (remember that $\hat{\mathbf{H}}$ is applied to the gradient). Therefore, I have assumed that the earth has constant statistics spatially (regard the form of $\mathbf{C}_{\mathbf{m}}$) so that any gradual variation in the amplitudes in the gradient are due to non geologic effects (for instance, because of survey geometry and finite aperture etc, there will be different amounts of seismic energy passing through different parts of the model causing the resolution to vary in different parts of the model). To apply equations (2) it is clear that we need to know how to calculate the data from the model parameters (i.e. $\mathbf{d} = \mathbf{d}(\mathbf{m})$) and how to evaluate the adjoint operation (i.e. the operation of $\mathbf{D}^*$ upon some vector). The following equation to calculate the seismic data $\mathbf{d}$ from the earth properties $\mathbf{m}$ will clarify the meaning of the Frechet operator $\mathbf{D}$,

$$\mathbf{d}(\mathbf{m}+\delta\mathbf{m}) = \mathbf{d}(\mathbf{m}) + \mathbf{D}\delta\mathbf{m} + O(\delta\mathbf{m}^2) = \mathbf{d}(\mathbf{m}) + \frac{\partial \mathbf{d}}{\partial \mathbf{m}}\delta\mathbf{m} + O(\delta\mathbf{m}^2) \ . \tag{3}$$

Therefore, the operation of **D** corresponds to the linearized forward problem. Equation (2b) does not require operator **D** but rather its adjoint $\mathbf{D}^*$. Now consider the linearized forward problem,

$$\delta\mathbf{d} = \mathbf{D}\delta\mathbf{m} \ , \tag{4a}$$

or in continuous form

$$\delta\mathbf{d}(D) = \int_M dM \frac{\partial \mathbf{d}(D)}{\partial \mathbf{m}(M)} \delta\mathbf{m}(M) \ , \tag{4b}$$

where M indicates the model space and **D** the data space. This expression shows how to calculate a small perturbation in the wavefield $\delta\mathbf{d}$ resulting from a small perturbation in the model parameters $\delta\mathbf{m}$ by integrating over the model space, i.e. it is the linearized Green function representation of the forward problem or Born approximation. Once we have a representation equivalent to equation (4b) we can identify the Frechet kernel $\mathbf{D} = \dfrac{\partial \mathbf{d}}{\partial \mathbf{m}}$ and hence (using the definition of the adjoint $<\delta\mathbf{d} \,|\, \mathbf{D}\delta\mathbf{m}> = <\delta\mathbf{m} \,|\, \mathbf{D}^*\delta\mathbf{d}>$) we can compute the adjoint operation

$$\delta\hat{\mathbf{m}} = \mathbf{D}^*\delta\mathbf{d} \ , \tag{5a}$$

or in continuous form

$$\delta\hat{\mathbf{m}}(M) = \int_D dD \left[\frac{\partial \mathbf{d}(D)}{\partial \mathbf{m}(M)}\right]^* \delta\mathbf{d}(D) \ . \tag{5b}$$

Note that the ^ is used to make it clear that $\delta\mathbf{m}$ and $\delta\hat{\mathbf{m}}$ are not the same (in fact $\delta\hat{\mathbf{m}}$ does not even have the same units as $\delta\mathbf{m}$; it has units of $\left[units\,(D)\right]^2 / units\,(M)$ rather than simply $units\,(M)$). Tarantola (1984) has carried out the appropriate derivation for the case of the elastic wave equation when the model space represents the Lamé's parameters and density and Mora (1986a) obtained expressions for the adjoint in terms of other choices of model parameters such as the P- and S-wave velocities and densities. Once the adjoint is known the appropriate covariance functions may be included to obtain the gradient **g**. From Mora (1986a) the equation for the gradient in terms of the P- and S-wave velocities and density, $\mathbf{g} = \left[\delta\hat{\alpha}, \delta\hat{\beta}, \delta\hat{\rho}\right]^T$ is

$$\delta\hat{\alpha} = 2\rho\alpha\delta\hat{\lambda} + \mathbf{C}_{\alpha\alpha}^{-1}(\alpha-\alpha_0) + \cdots \ , \tag{6a}$$

$$\delta\hat{\beta} = -4\rho\beta\delta\hat{\lambda} + 2\rho\beta\delta\hat{\mu} + \mathbf{C}_{\beta\beta}^{-1}(\beta-\beta_0) + \cdots \ , \tag{6b}$$

$$\delta\hat{\rho}_\alpha = (\alpha^2-2\beta^2)\delta\hat{\lambda} + \beta^2\delta\hat{\mu} + \delta\hat{\rho} + \mathbf{C}_{\rho\rho}^{-1}(\rho-\rho_0) + \cdots \ . \tag{6c}$$

where $\delta\hat{\lambda}$, $\delta\hat{\mu}$ and $\delta\hat{\rho}$ are the $\mathbf{C}_d^{-1}$ weighted adjoint fields with respect to the Lamé parameters and density given by

$$\delta\hat{\gamma} = \sum_S \int dt\, \mathbf{C}_\mathbf{d}^{-1}(t)\left[\Omega^\gamma u_j(\mathbf{x},t)\right]\left[\Omega^\gamma \psi_j(\mathbf{x},t)\right] , \tag{7a}$$

where $\gamma = \lambda$, μ or ρ and Ω^γ is an operator to be defined shortly. Subscript α of $\delta\hat{\rho}_\alpha$ in equation (6c) indicates that this term is the gradient with respect to density when the P- and S-wave velocities are used as the other model parameters. The $+ \cdots$ at the end of each line in equations (6) is used to allow for the possibility of covariances between the different model parameters such as $\mathbf{C}_{\alpha\beta}$. In equation (7a), u_j represents the j-th component of displacement of a seismic wavefield, and ψ_j is the j-th component of the *back propagated residual wavefield* given by

$$\psi_j(\mathbf{x},t) = \sum_R \mathbf{C}_\mathbf{d}^{-1}(\mathbf{x}_R) G_{ij}(\mathbf{x},-t;\mathbf{x}_R,0) * \Delta u_i(\mathbf{x}_R,t) , \tag{7b}$$

where G_{ij} is the elastic Green's function, Δu_i is error in the displacement wavefield at the current iteration given by

$$\Delta u_i = u_i - u_{i_0} ,$$

and u_{i_0} is the observed displacement wavefield. Note that $\Omega^\gamma = \Omega^\gamma_{ijk}$ is an operator that depends on the model parameter γ and can be considered as a model parameter *unraveling* operator ($\gamma = \lambda, \mu$, or ρ). Also, note that I use an unusual convention in equation (7) to avoid complicated notation with subscripts, namely that the implied summations within $[\,]$'s must be carried out first. From Mora (1986a), the unraveling operator Ω^γ is given by

$$\Omega^\rho = i\partial_t , \quad \Omega^\lambda = i\partial_j , \quad \Omega^\mu = \frac{i}{\sqrt{2}}\left[\delta_{jk}\partial_i + \delta_{ji}\partial_k\right] . \tag{8}$$

The inversion formulae, equations (2), (6), (7) and (8), are derived and described by Mora (1986a) including a comparison with the the process of migration (see also Tarantola (1984) for the original derivation in terms of the Lamé parameters and density).

The above equations have assumed that the inverse data covariance function $\mathbf{C}_\mathbf{d}^{-1}$ can be represented as

$$\mathbf{C}_\mathbf{d}^{-1}(\mathbf{x}_R,t) = \mathbf{C}_\mathbf{d}^{-1}(\mathbf{x}_R)\mathbf{C}_\mathbf{d}^{-1}(t) . \tag{9a}$$

In the examples in this paper I specifically used diagonal forms

$$\mathbf{C}_\mathbf{d}^{-1}(t) = \frac{t^{2p}}{\sigma_\mathbf{d}^2} , \tag{9b}$$

with $p = .5$ to allow for 2D wave divergence and

$$\mathbf{C}_\mathbf{d}^{-1}(\mathbf{x}_R) = taper(\mathbf{x}_R) . \tag{9c}$$

where $taper(\mathbf{x}_R)$ gradually tapers the data at the edges of the receiver array to zero in order to decrease artifacts caused by edge effects (i.e. finite aperture artifacts).

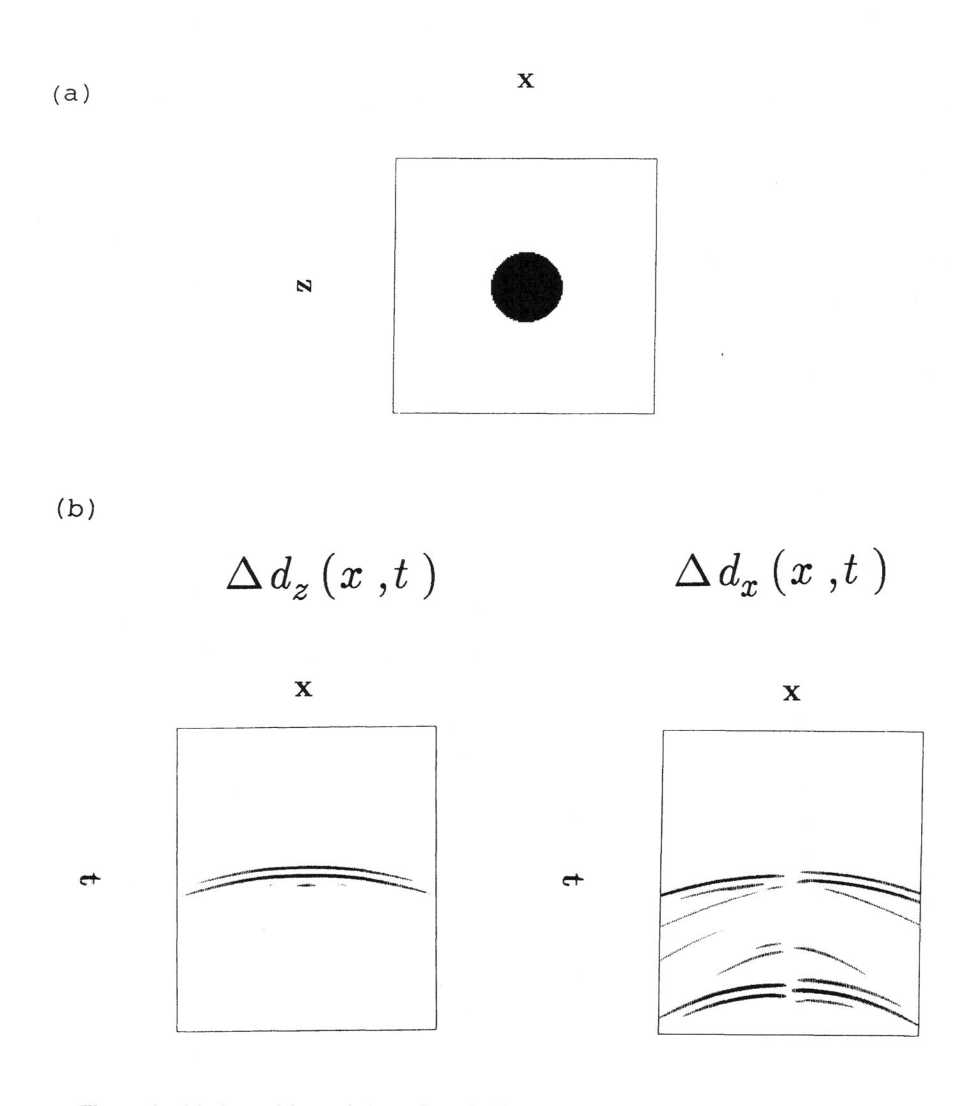

Figure 1. (a) A model consisting of a circular anomalous region embedded in a homogeneous space (Tarantola and Gauthier's camembert model). P-wave velocity anomalies are positive while S-wave velocity anomalies are negative. (b) Residual wavefield obtained by subtracting the wavefield computed by propagating waves through the homogeneous background model from the wavefield computed from the model shown in (a). The vertical component source was located at the base of the model and the two-component receivers were located at the top so the residual that can be seen is mainly in the *transmitted* waves. In the inversion, two component receivers were located along all sides though seismograms recoreded along the top of the model have the largest influence on the adjoint calculation (because these waves are most affected by the anomalous region).

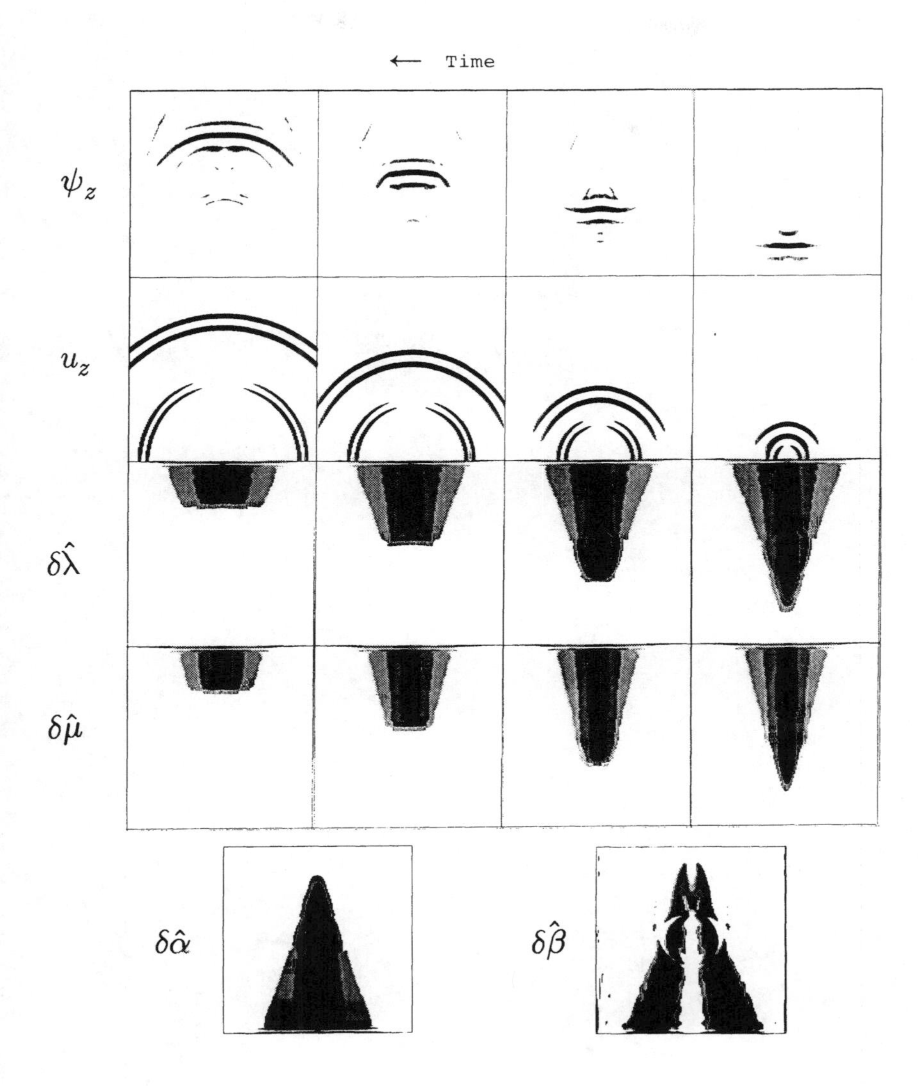

Figure 2. Pictorial illustration of the adjoint calculations corresponding to the inverse problem using the seismic transmission data of figure 1. The data residual $\Delta d_i(x,t)$ is applied as a forcing function backwards in time to generate the wavefield $\psi_i(x,z,t)$. At each instant in time the two wavefields (ψ_j and u_j, the shot wavefield) are operated on by Ω and multipled together and added into the adjoint fields $\delta\hat{\lambda}(x,z)$, $\delta\hat{\mu}(x,z)$ and $\delta\hat{\rho}(x,z)$ ($\delta\hat{\rho}$ is not shown here). Finally, the Lame's adjoint fields are converted into velocity adjoint fields $\delta\hat{\alpha}$, $\delta\hat{\beta}$ and $\delta\hat{\rho}_{\alpha}$. Amplitudes have been arbitrarily scaled for plotting. Two component data was used in the calculations though only the vertical component wavefields are shown here for brevity.

(a)

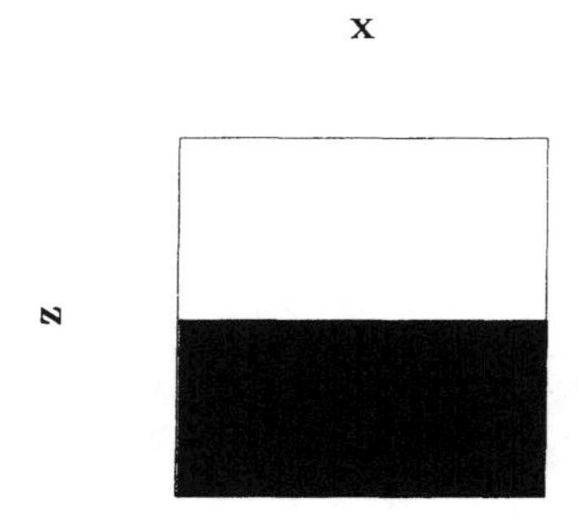

(b)

$\Delta d_z(x,t)$ $\Delta d_x(x,t)$

x x

t t

Figure 3. (a) A two layer model. The P-wave velocity anomalies of the lower layer relative to the upper layer are positive while S-wave velocity anomalies are negative. (b) Residual wavefield obtained by subtracting the wavefield computed by propagating waves through a homogeneous background model (having the properties of the upper layer) from the wavefield computed from the model shown in (a). The vertical component source and two component receivers were located on the top of the model.

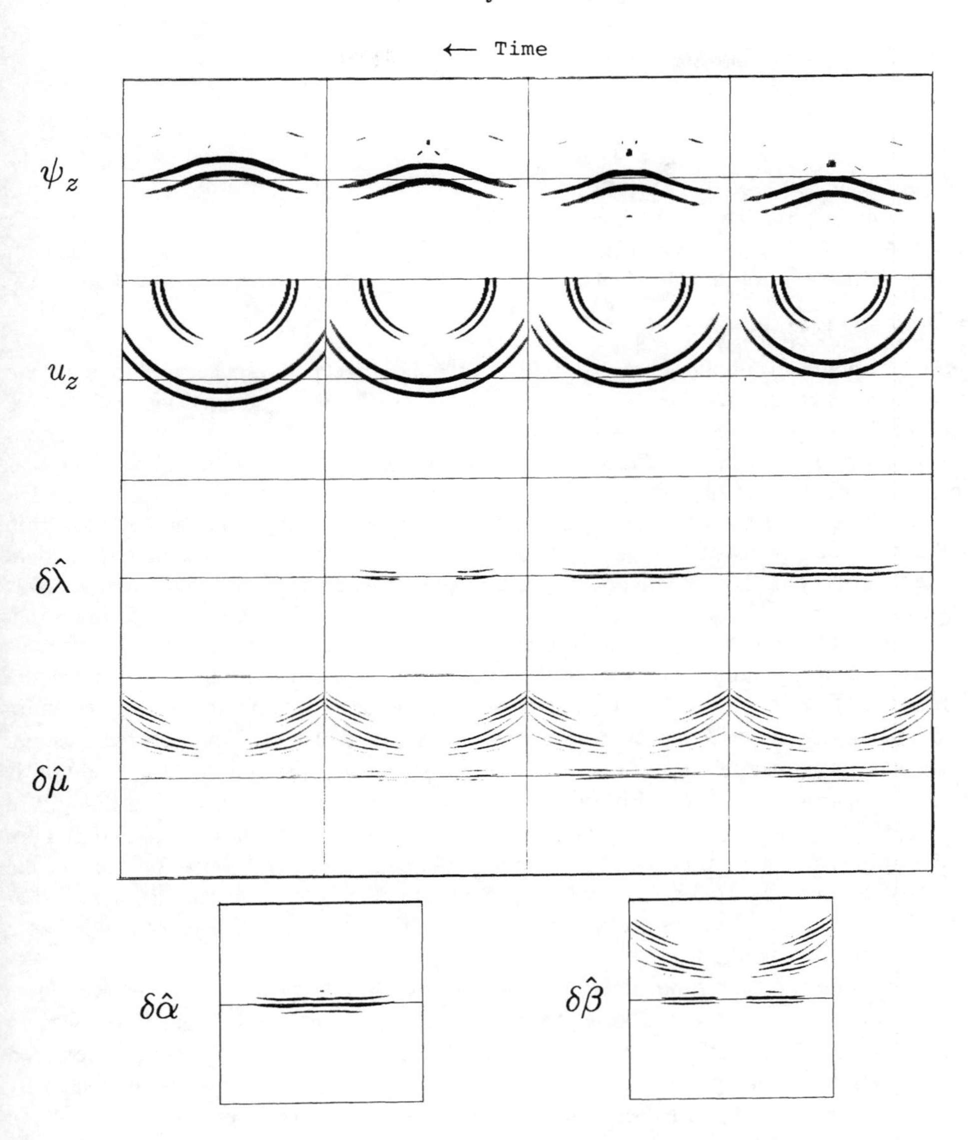

Figure 4. Pictorial illustration of the adjoint calculations corresponding to the inverse problem using the seismic reflection data of figure 3. The data residual $\Delta d_i(x,t)$ is applied as a forcing function backwards in time to generate the wavefield $\psi_i(x,z,t)$. At each instant in time the two wavefields (ψ_j and u_j, the shot wavefield) are operated on by Ω and multipled together and added into the adjoint fields $\delta\hat{\lambda}(x,z)$, $\delta\hat{\mu}(x,z)$ and $\delta\hat{\rho}(x,z)$ ($\delta\hat{\rho}$ is not shown here). Finally, the Lame's adjoint fields are converted into velocity adjoint fields $\delta\hat{\alpha}$, $\delta\hat{\beta}$ and $\delta\hat{\rho}_\alpha$. Amplitudes have been arbitrarily scaled for plotting. Two component data was used in the calculations though only the vertical component wavefields are shown here for brevity.

3. Calculating the adjoint operation

The adjoint is given by

$$\delta\hat{m} = \sum_{S}\int dt\,\Omega u_j \Omega \psi_j \quad , \tag{10}$$

(see equation (7a)). Two examples illustrate how to compute this formula and what components of the model can and cannot be resolved using reflection seismic data (i.e. shot profiles) or transmission seismic data (i.e. VSP, well-well or earthquake seismic data).

Transmission example

Figure 1a shows a circular body embedded in a homogeneous space. Relative to the homogeneous background medium, the body had a small (few percent) positive P-wave velocity anomaly and a small negative S-wave velocity anomaly. Elastic finite differences were used to do a shot simulation to generate some data. Two-component data was recorded on all sides of the model due to a vertical shot located at the base of the model. Direct waves traveling from the shot to receivers that pass through the anomalous region will be advanced (P-waves) or delayed (S-waves). This results in a significant residual (difference between the synthetic data and data generated using the homogeneous background model) as shown in figure 1b. The residual is large for transmitted waves that have passed through the anomalous region. Now, regard figure 2 which illustrates graphically how equation (10) is computed. The residuals are fed back into the model backwards in time to generate the back propagated residual wavefield ψ_j. At any instant in (backward) time, the stored background wavefield u_j and the back propagated residual wavefield ψ_j are operated on by Ω and multiplied together (the summation convention is used in the formulae so it is a little more complicated in practice). The time integral in equation (10) is evaluated by adding this contribution into the gradient fields $\delta\hat{\lambda}$, $\delta\hat{\mu}$ and $\delta\hat{\rho}$. Finally, after the Lame gradient fields are complete, equations (6) are used to convert the adjoint to be in terms of the P-wave velocity, S-wave velocity and density. Because the S-wave velocity adjoint was mainly negative, it is displayed in reverse polarity (otherwise most of the display would be white).

Observe that the back propagated wavefield ψ_j was traveling in the same direction as the background wavefield u_j (because the main data residual was in the transmitted waves) and so the adjoint is a smear running through the anomalous region. This can be compared with traveltime tomography which would also try to account for traveltime delays by distributing a velocity perturbation through the anomalous region (see Devaney (1984)). Therefore, the inversion algorithm is similar to an elastic wave equation tomography when transmitted waves are used in the inversion. Therefore, low wavenumber components of the model *can* be resolved provided transmitted waves are present in the data set. For a complete reconstruction of the circular anomaly, many shots would be required at different locations. To see this, imagine summing the adjoint field $\delta\hat{\alpha}$ with copies of itself rotated through various angles. The greater the number of shots and hence angles of waves going through the anomaly, the greater the magnitude of the central region relative to the surrounding smear. Similarly for the S-wave velocity adjoint field $\delta\hat{\beta}$. Notice that the S-wave velocity adjoint field has no significant energy in a vertical line. This is because the shot was vertical component and no significant shear wave energy passed through a vertical region above the shot and S-wave velocity can only be resolved in regions through which S-waves have propagated. However, the S-wave velocity throughout the entire region will be resolved if many shots at different locations are used to calculate the adjoint fields.

Reflection example

Figures 3 and 4 illustrate the adjoint calculations when using reflection data. The data was calculated by doing a shot simulation using elastic finite differences with the two layer model of figure 3a. The main difference between the previous adjoint calculation (using transmission data) and the current one using reflection data is that now the back propagated wavefield ψ_j is going in the opposite direction (downwards in reverse time) to the background wavefield u_j. This is essentially the same situation as a classical migration (see Claerbout (1976) where he describes migration and in particular the U/D principle which says that reflectivities can be obtained by dividing the back scattered waves with the direct wave). Therefore, just as in migration, the adjoint fields using reflection data alone are only high wavenumber approximations of the true model with about the frequency spectrum of the seismic wavelet (to see this pictorially, go through the procedure described in the the transmission section for calculating the adjoint but using figure 4).

The final P-wave velocity and S-wave velocity adjoint fields $\delta\hat{\alpha}$ and $\delta\hat{\beta}$ are shown in figure 4. The P-wave velocity result looks similar to a derivative operator located at the interface between the layers while the S-wave velocity looks like a negative derivative operator. This is the band limited solution equivalent of the positive and negative P- and S-wave velocity perturbations (see also Gathier et al. (1985) who obtained similar results for acoustic inversion, namely that transmission data resolves the complete velocity model while reflection data only resolves the high wavenumber components. Kolb et al. (1986) studied some techniques to resolve the low wavenumbers from only reflection data). Just as was the case for the transmission data example, the S-wave velocity was unresolved directly below the vertical component shot where no significant S-wave energy propagated. Notice also that there are smile-like artifacts on the S-wave velocity result because of non-zero correlations between different wave modes that did not correspond to real reflection locations (c.f. elliptical prestack migration impulse responses).

4. Results

The following results are taken from Mora (1986a) and Mora (1986b) illustrate the elastic inversion scheme summarized by equations (2), (6), (7) and (8). All forward modeling was done using elastic finite differences with a very band limited source wavelet (a fourth derivative Gaussian curve).

Impulse response

A 12 diffractor model is shown in figure 5. A two-component spilt-spread shot profile was generated with elastic finite differences and a significant amount of Gaussian noise was added. This profile (the vertical component is shown in figure 6a) was inverted and after five iterations the residual (misfit) was mainly Gaussian noise (see figure 6b). The five iteration inversion result is shown in figure 7 (it was calculated with an approximate inverse Hessian $\hat{\mathbf{H}} = \mathbf{C}_m$). Note that for plotting purposes, the results are scaled differently than

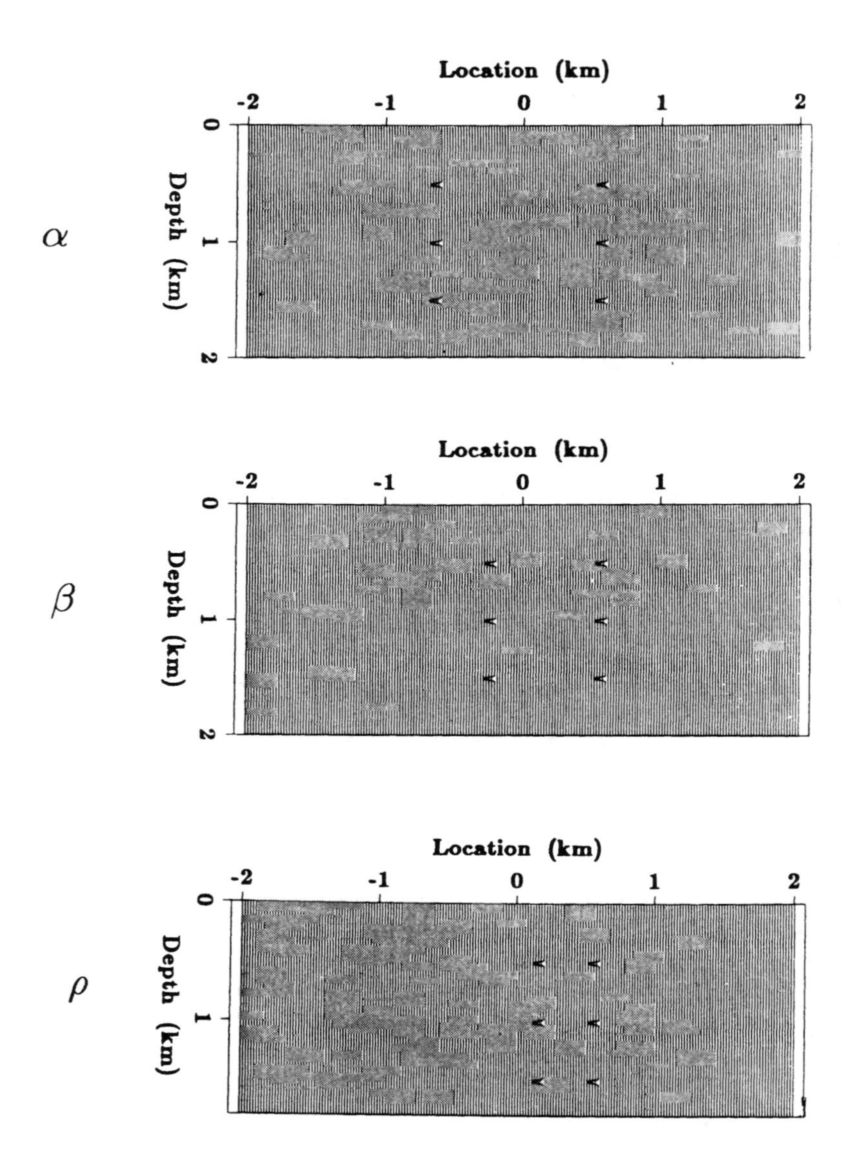

Figure 5. 12 diffractor model.

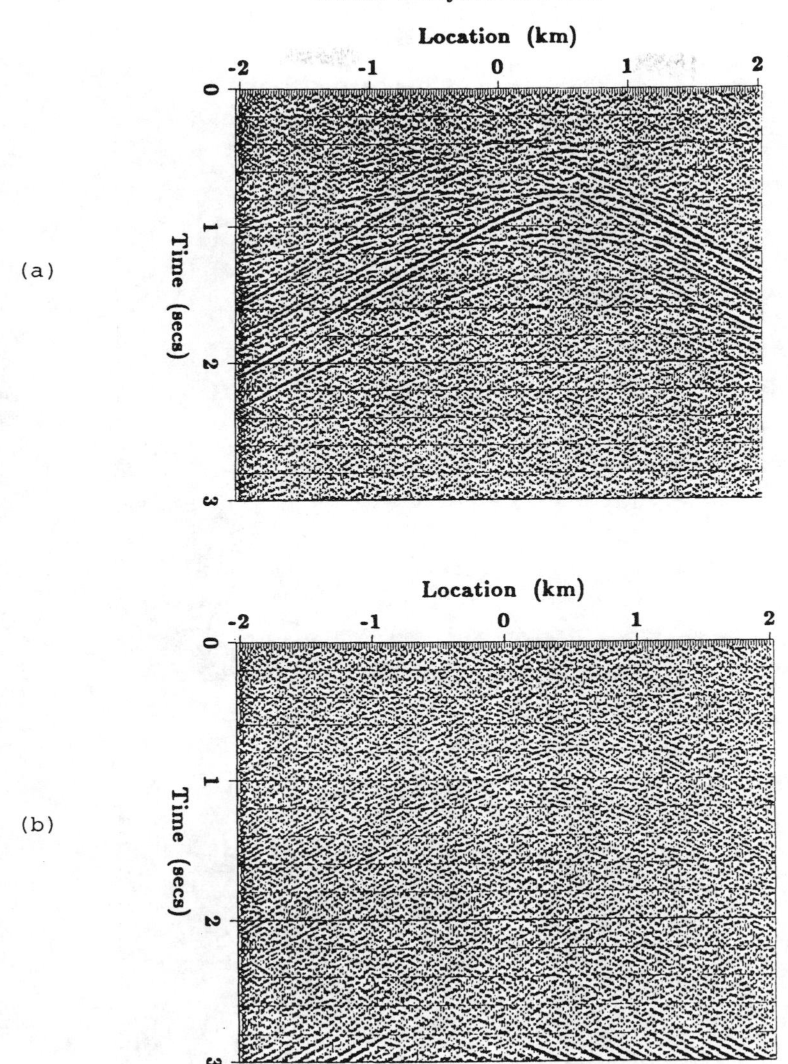

Figure 6. Noisy data generated by doing a shot simulation using the 12 diffractor model and the residual (i.e. misfit) after 5 iterations of a single shot profile inversion. The shot was vertical component and two-component receivers were used in the inversion. (a) Vertical component data (with the direct waves removed to make the diffractions more visible), and (b) vertical component residual after 5 iterations.

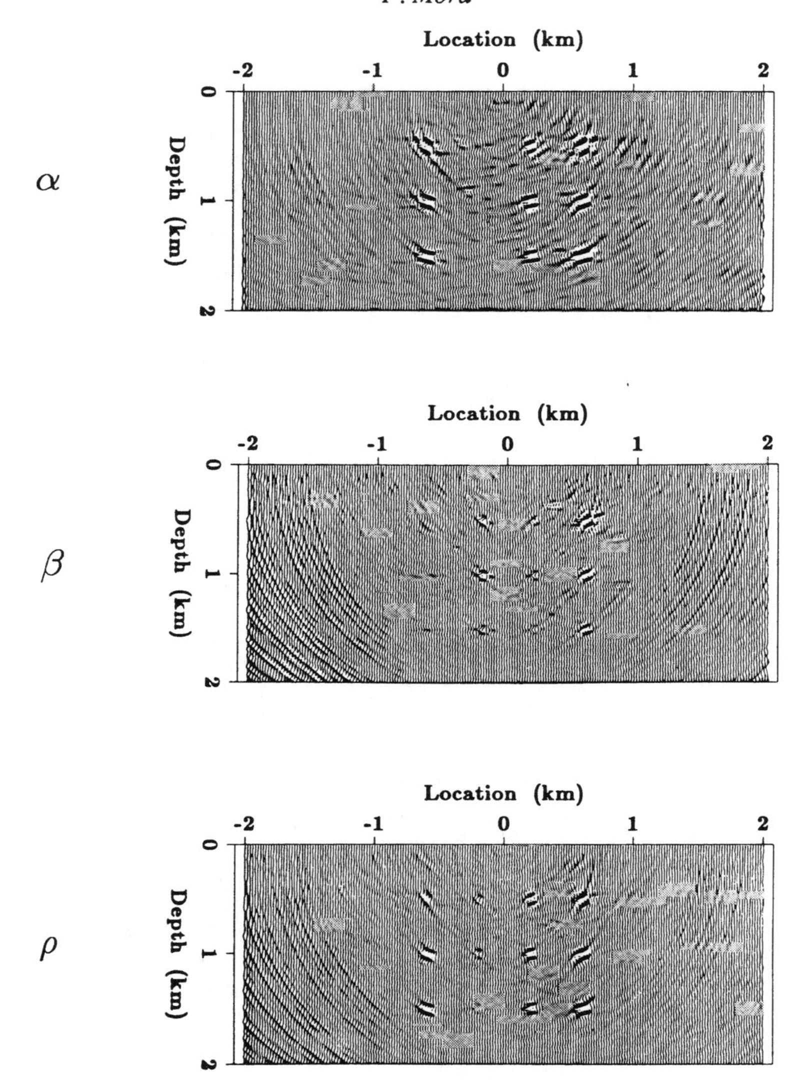

Figure 7. Inversion result after 5 iterations obtained by inverting the 12 diffractor noisy synthetic data set starting with a homogeneous initial model.

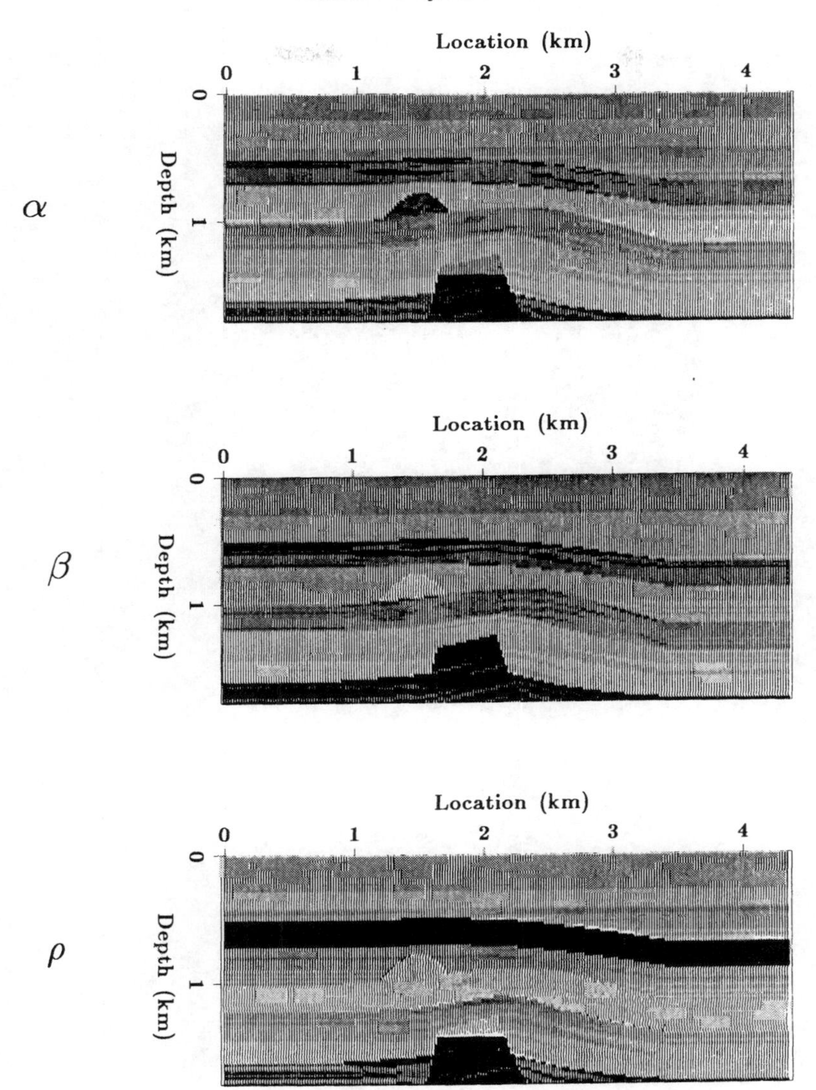

Figure 8. Horst and reef model.

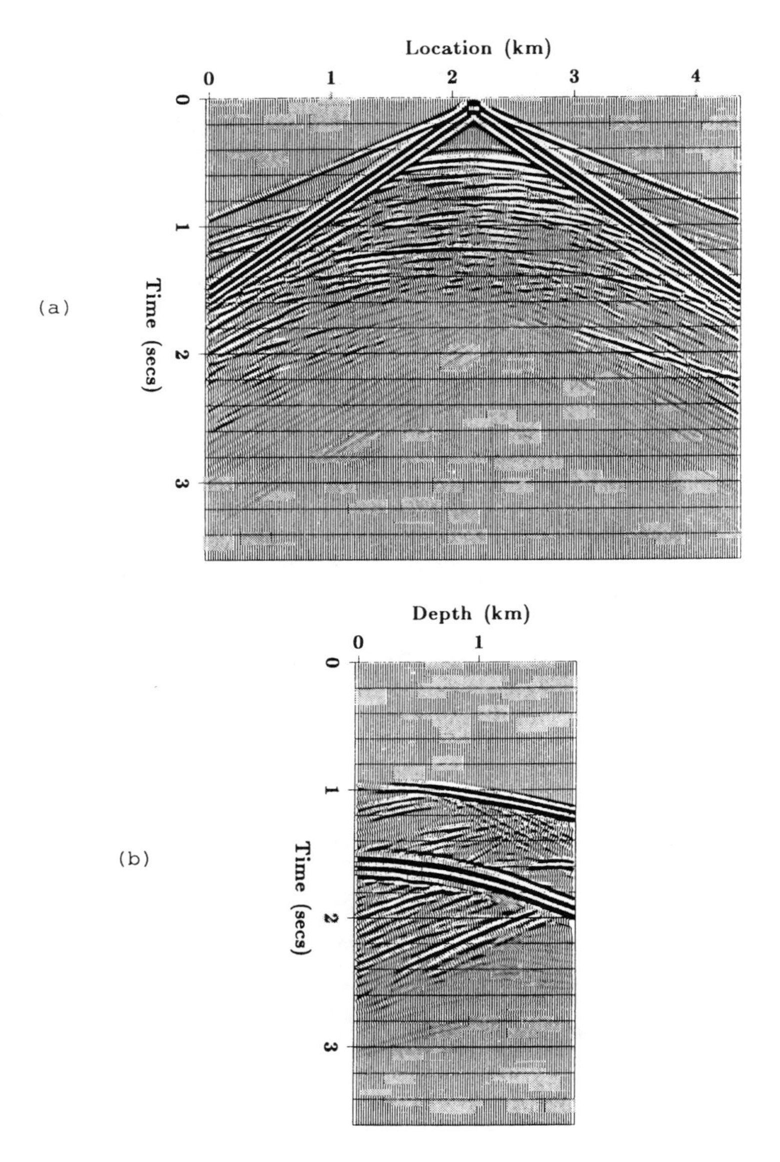

Figure 9. Typical synthetic data generated from the horst and reef model of figure 8. (a) Vertical component shot profile (from a vertical source located at $x = 2.16$ km, and (b) vertical component VSP for the well located at $x = .4$ km (from a vertical source located at $x = 2.66$ km).

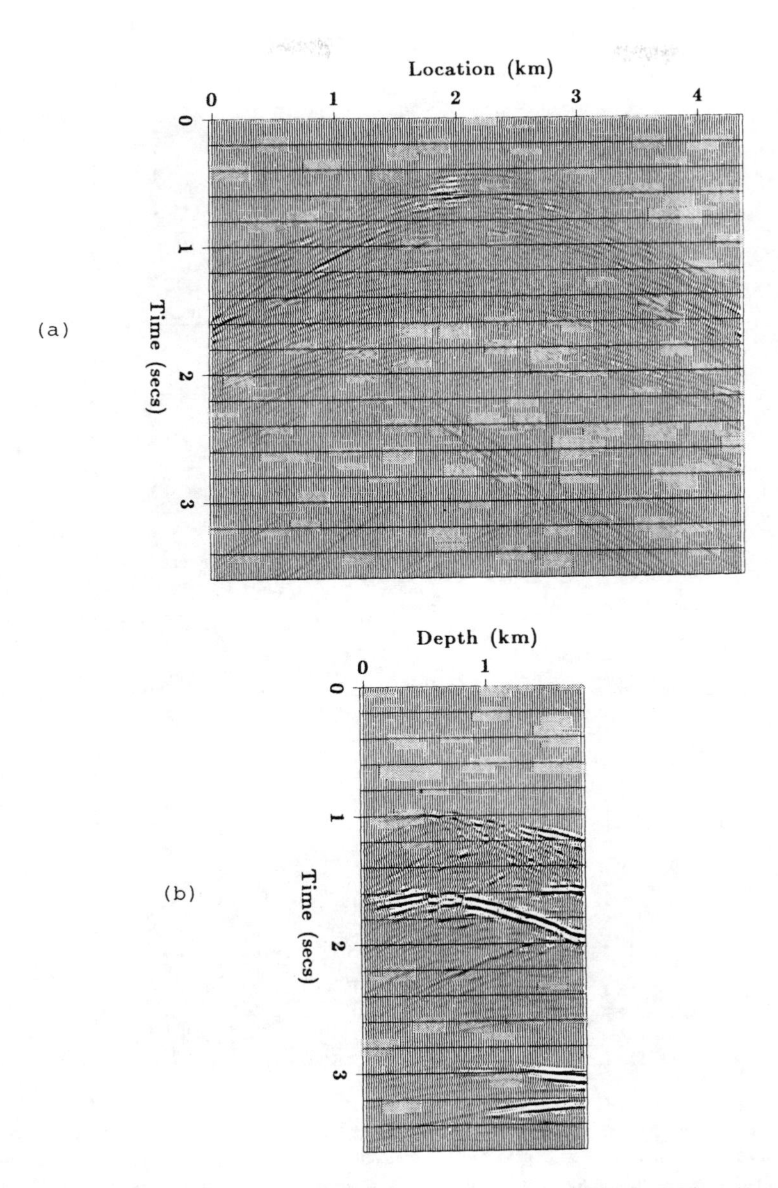

Figure 10. Typical residuals (i.e. mismatch) in shot profiles and VSP's after 10 iterations of an inversion of the horst and reef data of figure 9. The starting model for the inversion algorithm was a linear function of depth only. (a) Vertical component residual for the shot profile of figure 9a, and (b) vertical component VSP residual for the VSP of figure 9b.

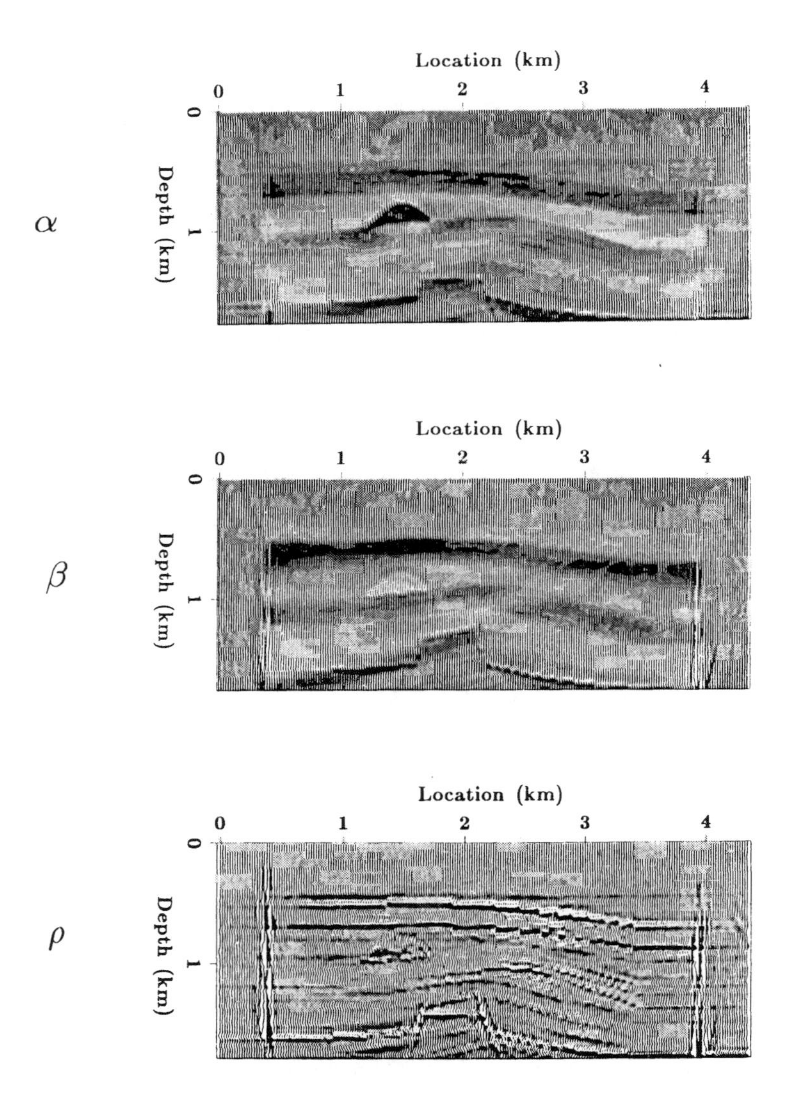

Figure 11. Inversion result after 10 iterations obtained from the horst and reef *reflection* and *transmission* data set (which consisted of 9 two-component shot profiles and 9 two-component VSP's). The 9 shots were all vertical comonent forces. The initial model used in the inversion algorithm was a linear fit of the true model that varied with depth only. Note that the P- and S-wave velocity models are almost exactly reconstructed by the inversion algorithm because the transmitted and reflected waves could respectively resolve both low and high wavenumber components of the model.

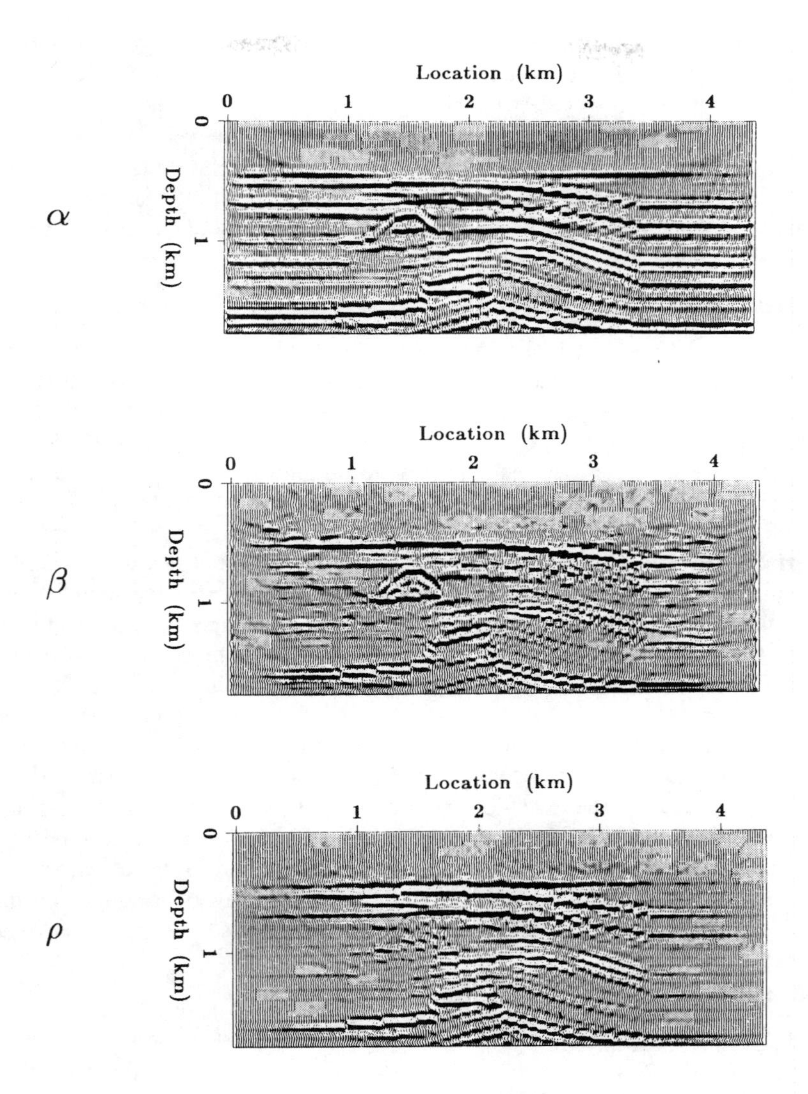

Figure 12. Inversion result for the model perturbations after 10 iterations obtained from the horst and reef *reflection* data set (which consisted of 9 two-component shot profiles). The 9 shots were all vertical component forces. The initial model used in the inversion algorithm was a linear fit of the true model that varied with depth only. Note that only the high wavenumber P- and S-wave velocity model perturbations could be resolved because no transmitted waves were recorded in the reflection data set. Transmitted waves are required by the algorithm to resolve the low wavenumbers (see figure 11).

the true model. The absolute magnitudes of the velocity and density perturbations obtained by the inversion are smaller than the true magnitudes (around 25%) because of band limitedness of the source wavelet (i.e. the null space problem). However, the diffractors are well resolved spatially and there is good resolution between the P-wave velocity diffractors and the S-wave velocity diffractors though the resolution is poor between the P-wave velocity and density. There is some noise in the solution that was caused by the Gaussian noise that was added to the data but the level is small considering that this problem is overdetermined only by a factor of about 2.75.

Complicated model inversion example

Figure 8 shows a horst and reef model that was used to generate some synthetic shot profiles and VSP's (representative data is shown in figure 9). Note that the model perturbations are small (a few %) in this example to facilitate rapid convergence. In practice, it is only necessary to know the low wavenumber velocity model adequately such that kinematics are accurate though the model perturbations may be large locally.

The VSP's correspond to wells located on either side of the model at x=.4 km and x=3.92 km. Two-component data was recoreded for 9 vertical sources located along the earths surface. An inversion was performed and after 10 iterations of the algorithm, the residual (mismatch) was small (see figure 10 where residuals for the data of figure 9 are plotted (at the same scale)). When both the VSP data (transmission data) and shot profiles (reflection data) were used in the inversion, both the high and low wavenumber components of the P-wave and S-wave velocities could be almost perfectly reconstructed (see figure 11). However, the density result is only a high wavenumber version of the true density result because low wavenumber density perturbations do not significantly affect the seismic amplitude data in comparison to the high wavenumber components. When only the shot profiles were used in the inversion, only the high wavenumber components of all three physical parameters could be resolved (see figure 12). However, the results indicate that there is always good resolution between the P-wave and S-wave velocity (whether reflection or transmission data is used) and so the elastic inversion algorithm has gained additional information not available to approaches that assume the acoustic wave equation.

5. Conclusions

Nonlinear inversion by least squares requires the adjoint operation of the forward equations. The adjoint can be defined in terms of Green's functions of the forward equations and in particular, the adjoint operation to the elastic wave equation with respect to the three (isotropic) elastic parameters can be obtained using two forward simulations. The equations have assumed nothing about the model or model complexity though the a priori model must be such that the kinematics are about correct in order to converge to the globally optimal solution. When transmitted waves are present in the data (for example, in VSP's, well-well surveys or earthquake seismic data), both the low and high wavenumber components of the P- and S-wave velocity model can be obtained by the inversion. However, when reflection data (i.e. shot profiles) is used alone it is only possible to reconstruct the high wavenumber components of the P- and S-wave velocity model. Only the high wavenumber components of density can be reconstructed in the case of both transmission and reflection data and density is not well resolved from the P-wave velocity.

However, the P- and S-wave velocities are well resolved from one another and the with some extensions to allow for the poor resolution of the low wavenumber velocity components when only reflection data is available, the iterative elastic inversion approach described should be a powerful tool to probe the earth's interior.

Acknowledgments

I acknowledge the support of the sponsors of the Stanford Exploration Project (SEP) and Jon Claerbout. Also, I acknowledge the CNRS (C_2VR) who supplied some CRAY time. Thanks to Albert Tarantola for many stimulating discussions and making possible an 8 month stay at the Institut de Physique du Globe where this work was initiated and the early stages of development carried out. Particular thanks to Alexandre Nercessian for priceless computer aid. Thanks to Odile Gauthier and Antonio Pica who are working on the acoustic equivalent of the algorithm presented and helped me to understand many aspects. Thanks to Jean Remy for computer assistance and to Jean Virieux for help in linking to the CRAY. Finally, thanks to my fellow students at Stanford, particularly, John Toldi, Dan Rothman, Kamal Al-Yahya and Jos Van Trier, for provocative and interesting discussion.

References

Devaney, A. J., 1984. Geophysical diffraction tomography, *Inst. Electr. and Electron. Eng., Trans., Geosci. and Remote Sensing*, **GE-22**, 3-13.

Gauthier, O. and Tarantola, A., 1985. Nonlinear inversion of two-dimensional seismic wavefields: Preliminary results, *Geophysics*, submitted for publication.

Kolb, P. and Canadas, G., 1986. Least-squares inversion of prestack data: simultaneous identification of density and velocity, presented at the 16-th conference on mathematical geophysics, June 22 - 28, Oosterbeek, Holland.

Luenberger, D., 1984. *Linear and nonlinear programming, second edition*, Adison-Wesley.

Menke, W., 1984. *Geophysical data analysis: discrete inverse theory*, Academic Press, Inc.

Mora, P., 1986a. Nonlinear 2D elastic inversion of multi-offset seismic data, submitted to *Geophysics*.

Mora P., 1986b. Elastic wavefield inversion of reflection and transmission data, submitted to *Geophysics*.

Tarantola, A. and Valette, B., 1982. Inverse problems = quest for information, *J. Geophys.*, **50**, 159-170.

Tarantola, A., 1984. The seismic reflection inverse problem, in: *Inverse problems of acoustic and elastic waves*, edited by: F. Santosa, Y.H. Pao, W. Symes, and Ch. Holland, SIAM, Philadelphia.

PETER MORA, Geophysics Department, Stanford University, Stanford CA 94305, USA.

Chapter 7

Subspace methods for large-scale nonlinear inversion

B.L.N. Kennett and P.R. Williamson

Many nonlinear inverse problems can be cast into the form of determining the minimum of a misfit function between observations and theoretical predictions, subject to a regularisation condition on the form of the model. For large scale problems, linearised techniques involving inversion of a Hessian matrix rapidly become difficult to handle as the size of the problem increases. It can therefore be computationally advantageous to use techniques which can achieve convergence without the inversion of large matrices. In this class are descent algorithms or modifications thereto, the simplest approach is to use a single direction of search at each step (usually related to the gradient of the misfit function). The efficiency of the search for a minimum can be improved by the introduction of a second search direction at each stage, e.g. the rate of change of the gradient, or the gradient of the regularisation term.

The success of such 2-dimensional subspace schemes encourages the use of higher dimensional subspaces, using for example the action of the Hessian on the gradient and additional direction to generate further basis vectors. With such a basis, the projection of the Hessian onto the subspace is well-conditioned and its inverse easy to calculate since the dimensionality is small. The improved representation of the behaviour means that each iterative step to the minimum is better directed and so convergence is improved.

For multi-parameter inversion simple gradient methods mix parameters of different characters. This can be particularly significant in cases where the parameters have different physical dimensions e.g. simultaneous determination of hypocentre locations and the velocity field in a region. However, by partitioning the gradient vector into parts associated with each class of parameters we can improve the situation. The analogue of introducing

N.J. Vlaar, G. Nolet, M.J.R. Wortel & S.A.P.L. Cloetingh (eds), Mathematical Geophysics, 139-154.

the rate of change of the gradient requires a similar partitioning of the Hessian to include cross-coupling between different parameters. For example, with 2 classes of parameters, 2 gradients and 4 rate of change of gradients constitute a useful 6-dimensional subspace for inversion. This scheme is illustrated for a reflection tomography problem with unknown velocities and reflector shape.

1. Introduction

In most geophysical inverse problems there is a non-linear dependence of the observable quantities on the parameters describing the model of the situation. Only a few nonlinear inverse problems have explicit solutions and then only for perfect data, e.g. the Herglotz-Wiechert formula for exact travel times in a velocity model which varies only with depth (Aki and Richards 1980 section 12.1). In general we must resort to some iterative approach to the solution of the inverse problems, with allowance for the imprecision of observed data.

The most common path is to make a local linear expansion about an assumed model and then to undertake a linear inversion for the perturbation to the model required to match the observed data (Aki and Richards 1980 section 12.3). The formulation of the model may be either in terms of continuous functions or discrete parameters, usually the smallest perturbation is sought but other criteria may be more suitable in the continuous case (Backus and Gilbert 1970). The newly derived model should now be used as the basis for a further linear inversion and the iteration continued until convergence is achieved, i.e. the model perturbations lie below a preassigned threshold. This type of inversion depends on the direct solution of a set of simultaneous linear equations which can become very computationally intensive when the number of data points and model parameters become large. Nolet (1985) has discussed a number of computation schemes which may be used for such large systems. For tomographic problems he favours the Paige-Saunders algorithm which uses an iterative solution to the linear equations via a conjugate gradient method.

Rather than use an outer iteration over the base model, with an inner iteration to solve the linearised equations, it is worthwhile to investigate direct attacks on the non-linear inversion problem (cf. Tarantola and Valette 1982). In this paper we consider iterative procedures for the non-linear case in which no large matrices need to be inverted. We introduce a single misfit function which describes the mismatch between the observational data and the corresponding theoretical predictions; the particular form of this function can be adjusted for the nature of the error statistics for the observations. The aim of the inversion is to reduce the value of the misfit function below a threshold determined by the errors in the observations. To prevent extravagant behaviour of the model parameters, we also impose some form of regularisation condition on the model. At each step in the iteration we use a locally quadratic approximation to the misfit function and try to find a path towards minimum misfit. The simplest procedure is to use a single search direction (usually related to the gradient of the misfit function). However, the efficiency of the procedure can be improved by the introduction of a limited number of further directions e.g. the direction of the rate of change of the gradient or the gradient of the regularisation term.

The set of directions form a local subspace of the model parameter space and each step works with the projection of the matrix of second derivatives of the misfit function (the Hessian matrix) onto the subspace. This projected Hessian is well-conditioned and the inverse we need can easily be calculated since the dimensionality is small. The use of the subspace gives a better local representation of the behaviour of the misfit function, thus each iterative step leads more directly to the minimum and so convergence is improved. We will illustrate these methods by application to a tomographic problem in seismic reflection: the estimation of the velocity field above a known reflector from the arrival times of reflected waves.

The use of subspace methods is particularly helpful in the situation where parameters of different types are to be determined at the same time. Such multi-parameter inversions present most difficulties when the classes of parameters differ in physical dimensions, as for example in the simultaneous estimation of earthquake hypocentres and the velocity field from the arrival times of seismic waves. The problem of dimensional dependence can be avoided by splitting up the gradient of the misfit function into a set of directions corresponding to each of the classes of parameters. This set of directions already constitutes a useable subspace but can be usefully augmented by additional directions derived from the rates of change of the partitioned gradients. This requires a similar partition of the local Hessian matrix for the misfit function. For a situation with two types of parameters, we have two directions from the gradients and a further four from their rates of change, including cross terms. Such a six dimensional space is small enough for calculation to be swift and efficient at each iteration, whilst giving relatively rapid convergence. In order to avoid purely local minima of the misfit function it is often benificial to work with a sequence of parameterisations in which the scale is reduced as the inversion proceeds (cf. Nolet et al 1986). This approach is illustrated by application to an extension of the seismic reflection tomography problem in which both the shape of the reflector and the velocity field above it are unknown.

2. The inverse problem

We suppose that we are presented with a set of observations $\mathbf{d}_o$ $\{d_{oj}, j = 1,....M\}$ and wish to use these observations to determine a discrete set of parameters $\mathbf{m}$ $\{m_k, k = 1,...N\}$ These parameters may describe the model directly or may be the coefficients in an expansion in terms of orthonormal functions (Nolet 1981)

$$m(\mathbf{x}) = \sum_{j=1}^{N} m_k h_k(\mathbf{x}), \tag{1}$$

where the basis functions $h_k(\mathbf{x})$ satisfy

$$\int_R \mathrm{d}^D x \; h_i(\mathbf{x}) h_j(\mathbf{x}) = \delta_{ij}, \tag{2}$$

over a region R of dimensionality D. In tomographic problems a suitable representation can be in terms of non-overlapping cells, in which case each of the $h_k(\mathbf{x})$ would be confined to a particular cell.

For each of the observed data values d_{oj} we must be able to calculate the predicted value $g_j(\mathbf{m})$ which will be some functional of the model parameters. We then need to establish some criterion for assessing the degree of mismatch between the observations $\mathbf{d}_o$ and the corresponding theoretical values $\mathbf{g(m)}$ Rather than work with each data point separately, which can be rather awkward when the number of data values (M) is large, we will use a single data misfit function $\Phi(\mathbf{d}_o, \mathbf{g(m)})$, to describe the match. The choice of the form of Φ depends on the character of the problem and the nature of the error statistics for the observed data. If these statistics are Gaussian then we can adopt

$$\Phi(\mathbf{d}_o, \mathbf{d}) = (\mathbf{d} - \mathbf{d}_o)^T \mathbf{C}_d^{-1} (\mathbf{d} - \mathbf{d}_o), \tag{3}$$

in terms of the data covariance matrix $\mathbf{C}_d$. In some problems it can be advantageous to "precondition" both the observations and the theoretical values using some operator R intended to reduce the non-linearity of the problem, so that

$$\Phi(\mathbf{d}_o, \mathbf{d}) = (R\,\mathbf{d} - R\,\mathbf{d}_o)^T \mathbf{C}_d^{-1} (R\,\mathbf{d} - R\,\mathbf{d}_o). \tag{4}$$

R may be needed to match up the treatment of data and theoretical values as in the direct inversion of refraction seismograms by Chapman and Orcutt (1985). When the statistics of the data errors are not Gaussian but described by a probability density $f(r)$ for data residual $r = d - d_o$, an effective choice for the misfit function in the case of independent data is

$$\Phi = -\sum_{i=1}^{M} \log_e \{f(r_i)\}, \tag{5}$$

which has been used for Jeffreys statistics in travel time studies by Sambridge and Kennett (1986), see also Menke (1984) for applications to exponential distributions.

In addition to considering the fit between data and theoretical predictions we also need to impose some restrictions on the form of possible models to prevent wild oscillations or other extravagant behaviour which may be generated in an attempt to minimise Φ. We introduce a "regularisation" function $\Psi(\mathbf{m}, \mathbf{m}_0)$ in terms of a starting model $\mathbf{m}_0$. Commonly this would represent a norm on the model e.g. the quadratic form

$$\Psi(\mathbf{m}, \mathbf{m}_0) = (\mathbf{m} - \mathbf{m}_0)^T \mathbf{C}_m^{-1} (\mathbf{m} - \mathbf{m}_0), \tag{6}$$

where $\mathbf{C}_m$ is the model covariance matrix whose properties should be chosen to fit what is known about the situation. For example, $\mathbf{C}_m$ may be designed to scale the parameters to unit a priori variance. A restriction on model norm is not the only possible choice for Ψ. For a set of non-negative parameters we can construct

$$\Psi_s = \sum_k f_k m_k \left[\log_e \left[\frac{m_k}{m_{0k}} \right] - 1 \right]. \tag{7}$$

A form suggested by configurational entropy methods used in image processing (Bryan and Skilling 1980). The f_k allow for relative weighting between parameters of different type. The form (7) is well suited to image reconstruction work, but has also been used by Harding (1984) in travel time inversion.

We now seek to reduce the misfit function Φ to below a threshold ϕ imposed by the nature of the observational errors. For independent data the quadratic form of Φ (3) is a scaled χ^2 statistic and so it is possible to generate a good statistical estimate of ϕ. In the course of the reduction of the data misfit Φ, we wish to restrain the character of the model by preventing the regularisation term Ψ from becoming large. One way of achieving this goal is to work with an objective function $F(\mathbf{d}_o, \mathbf{m})$ which is built up from a linear combination of the data misfit and regularisation terms (see e.g. Kennett 1978)

$$F(\mathbf{d}_o, \mathbf{m}) = \Phi(\mathbf{d}_o, \mathbf{d}) + \gamma \Psi(\mathbf{m}, \mathbf{m}_0), \tag{8}$$

subject to the constraint $\mathbf{d} = \mathrm{g}(\mathbf{m})$ connecting the data and model parameter spaces. The quantity γ governs the trade-off between the data fit and the model regularisation. We now seek to minimise the objective function F using an iterative procedure which will define a sequence of models in model parameter space. For perfect data we would aim for the global minimum of F, but with observed data we would terminate the iteration once the data misfit was brought below its assigned threshold. If we adopt the two quadratic forms (3), (6) we are then trying to minimise

$$F(\mathbf{d}_o, \mathbf{m}) = (\mathbf{d} - \mathbf{d}_o)^T \mathbf{C}_d^{-1} (\mathbf{d} - \mathbf{d}_o) + (\mathbf{m} - \mathbf{m}_0)^T \mathbf{C}_m^{-1} (\mathbf{m} - \mathbf{m}_0), \tag{9}$$

subject to $\mathbf{d} - \mathbf{g}(\mathbf{m}) = 0$, where we have absorbed the weighting term γ into the definition of $\mathbf{C}_m^{-1}$.

When the number of parameters is not too large, the minimum of F can be sought by a direct search in parameter space. Such an approach has been successfully used by Sambridge and Kennett (1986) to determine earthquake hypocentre locations in a known velocity model (i.e. 4 parameters) by combining a locally discrete spatial search with a bracketing of the arrival time parameter. This directed grid search has the merit that it can be applied to any choice of misfit function Φ and does not require the calculation of any derivatives of F.

For large numbers of parameters, direct search methods are impractical and we will seek a minimum by making local approximations to the behaviour of F. If F is a smooth function of the model parameters we can make a locally quadratic approximation about some current model $\mathbf{m}_c$ by truncating the Taylors series for F

$$\begin{aligned} F^Q(\mathbf{m}_c + \delta\mathbf{m}) &= F(\mathbf{m}_c) + \nabla_m F^T(\mathbf{m}_c) \cdot \delta\mathbf{m} + \tfrac{1}{2}\delta\mathbf{m}^T \cdot \nabla_m \nabla_m F(\mathbf{m}_c) \cdot \delta\mathbf{m}, \\ &= F(\mathbf{m}_c) + \boldsymbol{\theta}^T \cdot \delta\mathbf{m} + \tfrac{1}{2}\delta\mathbf{m}^T \mathbf{H} \delta\mathbf{m}, \end{aligned} \tag{10}$$

in terms of the gradient $\boldsymbol{\theta}$ and the Hessian matrix $\mathbf{H}$. The preconditioning operator R of equation (4) may be necessary to give adequate smoothness to F for the approximation (10) to be useful in a significant neighbourhood of $\mathbf{m}_c$. The minimum of the locally quadratic approximation to F lies at $\nabla_m F^Q = 0$ i.e. when $\boldsymbol{\theta} + \mathbf{H}\delta\mathbf{m} = 0$, so that the local estimate of the perturbation of the model parameters $\delta\mathbf{m}$ required to find the minimum is

$$\delta\mathbf{m} = -\mathbf{H}^{-}\boldsymbol{\theta} \tag{11}$$

where $\mathbf{H}^{-}$ is some suitable generalised inverse for the Hessian. Such "Gauss-Newton" methods are not very attractive when the number of parameters N is large since they require the construction and inversion of $N \times N$ matrices.

However, we can develop a class of very effective algorithms by restricting the local minimisation of to a relatively small n -dimensional subspace of model parameter space. We introduce n basis vectors $\{\mathbf{a}^{(j)}\}$ and a projection matrix $\mathbf{A}$ composed of the components of these vectors

$$A_{ij} = a_i^{(j)}, \quad i,j = 1,....n \tag{12}$$

The model perturbation is now to be sought in the subspace spanned by the vectors $\{\mathbf{a}^{(j)}\}$ so we take

$$\delta\mathbf{m} = \sum_{j=1}^{n} \alpha_j \mathbf{a}^{(j)}. \tag{13}$$

The coefficients $\boldsymbol{\alpha}$ are to be determined by minimising F^Q and thus we require $\partial F^Q / \partial \alpha_j = 0$. From (10) these n equations can be expressed in matrix form using the projection matrix $\mathbf{A}$ the gradient $\boldsymbol{\theta}$ and the Hessian $\mathbf{H}$ of the objective function F,

$$\mathbf{A}^T \boldsymbol{\theta} + \mathbf{A}^T \mathbf{H} \mathbf{A} \boldsymbol{\alpha} = 0 \tag{14}$$

so that

$$\boldsymbol{\alpha} = -(\mathbf{A}^T \mathbf{H} \mathbf{A})^{-1} \cdot \mathbf{A}^T \boldsymbol{\theta} \tag{15}$$

The projection onto the n-dimensional subspace by $\mathbf{A}$ gives a small $n \times n$ matrix $\mathbf{A}^T \mathbf{H} \mathbf{A}$ to be inverted. This projected Hessian is well-conditioned for sensible choice of the basis vectors $\mathbf{a}^{(j)}$

What then are appropriate choices for the directions of the basis vectors in model parameter space? We recall that we have built the objective function as a combination of a data misfit $\Phi(\mathbf{d}_o, \mathbf{g}(\mathbf{m}))$ and a regularisation term $\Psi(\mathbf{m}, \mathbf{m}_0)$.

When we are using quadratic forms for both Φ and Ψ, as in (9), the two terms appear on an equal footing and we may seek basis vectors based on the properties of the composite objective function F. An obvious choice for the first vector $\mathbf{a}^{(1)}$ is the gradient $\boldsymbol{\theta} = \nabla F$. For a second direction, we could use a conjugate gradient derived from a previous step. However, it is preferable to avoid carrying forward outdated information. We therefore look for additional information on the character of F by choosing $\mathbf{a}^{(2)}$ to lie along the direction of the rate of change of the gradient

$$\mathbf{a}^{(2)} \propto \nabla\nabla F \cdot \nabla F = \mathbf{H}\mathbf{a}^{(1)}. \tag{16}$$

Further directions can be generated by additional applications of the Hessian. For the case when all parameters lie in the same class we have found that three vectors provide a very effective basis. To avoid interdependence between the vectors we use a Gram-Schmidt orthogonalisation i.e. we construct

$$\mathbf{b}^{(2)} = \mathbf{a}^{(2)} - \frac{(\mathbf{a}^{(1)} \cdot \mathbf{a}^{(2)})}{(\mathbf{a}^{(1)} \cdot \mathbf{a}^{(1)})} \mathbf{a}^{(1)}, \tag{17}$$

and then form a normalised basis vector $\mathbf{a}^{(2)} = \mathbf{b}^{(2)} / (\mathbf{b}^{(2)} \cdot \mathbf{b}^{(2)})^{1/2}$. A consequence of this procedure is that the Hessian to be applied in (16) does not need to be found in full, a reasonable approximation will suffice.

An alternative procedure, which is particularly appropriate when the regularisation Ψ is based on an entropy measure (7), is to take the gradients of data misfits and the regularisation term as separate initial directions

$$\mathbf{a}^{(1)} \propto \nabla\Phi, \qquad \mathbf{a}^{(2)} \propto \nabla\Psi. \tag{18}$$

Further directions can then be obtained by the action of the second derivative of the data term

$$\mathbf{a}^{(3)} \propto \nabla\nabla\Phi\cdot\nabla\Phi, \qquad \mathbf{a}^{(4)} \propto \nabla\nabla\Phi\cdot\nabla\Psi. \tag{19}$$

Once again these basis vectors should be orthonormalised to ensure independence.

We can illustrate the action of the subspace schemes using the objective function (9). The gradient

$$\nabla_m F = -\mathbf{G}^T \mathbf{C}_D^{-1}(\mathbf{d}_o - \mathbf{g}(\mathbf{m})) + \mathbf{C}_m^{-1}(\mathbf{m} - \mathbf{m}_0), \tag{20}$$

where $G_{ij} = \partial g_i / \partial m_j$, and the Hessian

$$\mathbf{H} = \nabla_m \nabla_m F = \mathbf{G}^T \mathbf{C}_d^{-1}\mathbf{G} - \nabla\mathbf{G}^T(\mathbf{d}_o - \mathbf{g}(\mathbf{m})) + \mathbf{C}_m^{-1}. \tag{21}$$

In many situations, the second derivative term $\nabla G = \nabla_m \nabla_m \mathbf{g}$ is not at all easy to calculate. However, since it is modulated by the data misfit its importance diminishes as the iteration proceeds and it is often neglected. The basis vectors for the n-dimensional subspace are $\boldsymbol{\theta} = \nabla_m F$, and $\mathbf{H}\boldsymbol{\theta} = (\mathbf{H}_0 + \mathbf{C}_m^{-1})\boldsymbol{\theta}$, etc., where $\mathbf{H}_0 = \nabla\nabla\Phi$ depends only on the data misfit Φ. From (13),(15) the model perturbation

$$\delta\mathbf{m} = \mathbf{A}\boldsymbol{\alpha} = -\mathbf{A}[\mathbf{A}^T((\mathbf{H}_0 + \mathbf{C}_m^{-1})\mathbf{A}]^{-1}\mathbf{A}^T\boldsymbol{\theta}, \tag{22}$$

which resembles a projected Marquardt algorithm, but the $\mathbf{C}_m^{-1}$ need not be diagonal. Indeed, some degree of interaction between parameters can give beneficial stability in the iteration. A rather special case is provided by the 1-dimensional case with a single search direction, the gradient of F. The formula (22) then determines the step-length in this direction, which will depend on $\mathbf{C}_m^{-1}$.

In the application of the subspace methods we start with some assumed starting model m_0 for which we determine the data residual $(\mathbf{d}_o - \mathbf{g}(\mathbf{m}))$ and the Gateaux derivatives G. We can then calculate the model perturbations $\delta\mathbf{m}$ from (22) to construct

$$\mathbf{m}_1 = \mathbf{m}_0 + \delta\mathbf{m}. \tag{23}$$

We now have a choice as to the degree of emphasis we wish to place on the model $\mathbf{m}_0$. If we have good a priori reasons for retaining $\mathbf{m}_0$ as a reference level for the model parameters throughout the calculation, then we would make a quadratic approximation for the objective function F about $\mathbf{m}_1$ and find a further perturbation to the model from (22). Such an iterative development can then be carried on until the data misfit is reduced to an acceptable level, whilst trying to stay close to $\mathbf{m}_0$. Alternatively we can regard the regularisation as restraining local behaviour and update $\mathbf{m}_0$, as well as the current model, as the calculation proceeds. This latter case provides a weaker constraint on the model parameters.

Whichever approach is preferred, the subspace methods provide a powerful technique for nonlinear inverse problems with a large number of parameters. A detailed treatment of the local character of the model increment by Williamson (1986) shows that in terms of the singular value decomposition of the gradient matrix G, the dominant perturbations are associated with the large eigenvalues. There is relatively little movement for small

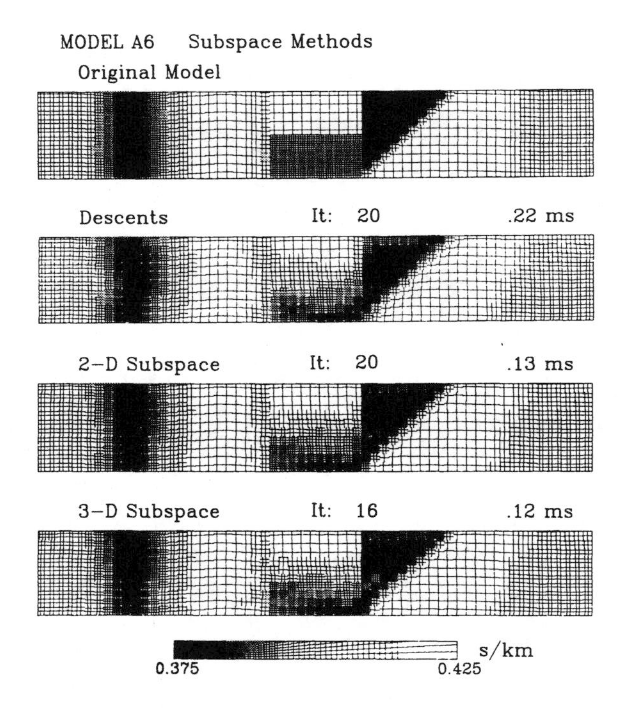

Figure 1.
Comparison of different dimensions of subspace inversion for a reflection tomography problem. The reflector is taken at the base of the model and the slowness field is represented by 600 cells within which the properties are uniform. The original model is shown in the upper panel and beneath, the results of different inversions together with their iteration count and r.m.s. data deviations

eigenvalues for which the perturbation depends on λ rather than $1/\lambda$ as in generalised inversion. The result is that the calculation schemes are stable with respect to noise and nonlinearity.

3. Reflection tomography I

As an illustration of the subspace methods described in the previous section, we consider a tomographic problem in seismic reflection: the reconstruction of a slowness field above a known reflector from the arrival times of the wave reflected from the interface. We will use a parameterisation of the slowness field in terms of a set of square cells (25 m to a side) each with uniform properties (see figure 1). This gives a simple set of derivative terms. For the ith ray and the jth cell with slowness m_j

$$\partial g_i / \partial m_j = \int_{R_{ij}} \mathrm{d}l$$

where R_{ij} is the portion of the ith raypath in the jth cell.

The model chosen for the inversion study is illustrated in the upper panel of figure 1, it contains a wide range of features and considerable variations in slowness, with a change of 10 per cent across the vertical and diagonal boundaries of the prominent dark zone of the model. The model consists of 60 cells horizontally and 10 in each vertical column, so that there are 600 slowness parameters in all. Sources and receivers are assumed to be set at the surface in the centre of each column of cells and rays traced between all sets of source/receiver pairs. This is achieved by tracing from both source and receiver to the reflector and then using Fermat's principle to select the take-off angles at source and receiver corresponding to minimum total travel time. This procedure works very well even in complex models and avoids two-point ray tracing.

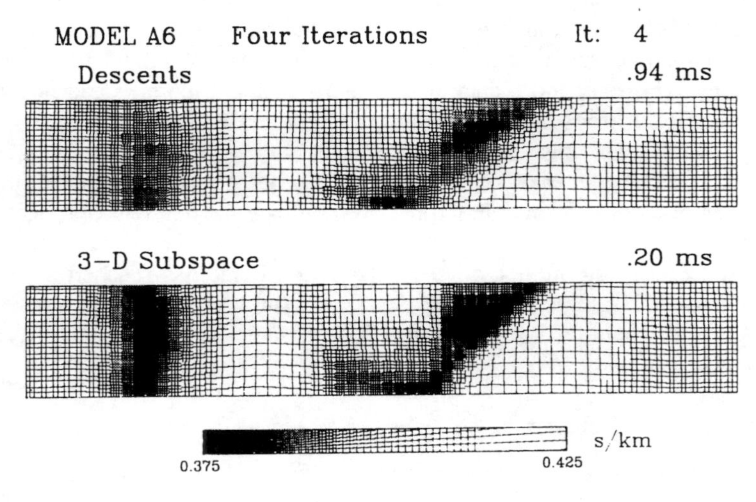

Figure 2.
Comparison of a descents method and a 3-D subspace inversion for model A6 shown in figure 1, after four iterations. The 3-D subspace reconstruction is superior to 20 iterations of the descents method.

The total number of rays is 1830 but the coverage is concentrated in the centre of the model. The derivative matrix G is therefore of dimension 1830×600 but is comparatively sparse, even so it is difficult to store the relevant elements in less than 1 Mword. Fortunately, the subspace methods do not require all this information to be available at once. In a 1-dimensional descents scheme only one row (i.e. a single ray) is required at a time. With higher order subspaces we can recalculate the terms we need from an advanced stage in the forward modelling.

In figure 1 we show a comparison of the results obtained using subspaces of increasing dimensionality, based on the calculation schemes outlined in (22),(23) above with the second derivative term excluded from the approximation of the Hessian $\mathbf{H}$. In each case the starting point was a constant slowness field of 0.4 s/km. Only local regularisation was applied to the models. The a priori model covariance was set so that the perturbations to the nearest neighbours of a particular cell are partially correlated to the perturbations of the central cell. We recognise that we are unable to recover a perturbation precisely isolated in a particular cell, and so look for a concentrated disturbance with weights 1:0.4 between the centre and its surroundings. In each of the calculations of figure 1, we have used a full nonlinear scheme and have recalculated the ray paths each iteration. Initially a constant slowness field of 0.4 s/km was assumed, with straight ray paths but the rays are far from straight for the final models shown.

For the descent scheme, the recovery of the slowness field is quite good and the r.m.s. data deviation has been reduced to quite a low level after 20 iterations. However, at this point the gradients are very low and the residual only creeps down with further iteration, since the steps in parameter space are very small. A further 10 iterations hardly changes the picture at all. With a 2-D subspace the residual is reduced to a lower level after 20 iterations and there is a noticeable improvement in the structure of the horizontal gradient zone to the left of the model and in the region of the block and staircase to the right. The use of a 3-D subspace achieves just about the same level of model reconstruction with fewer iterations. However more computation is needed at each step and in terms of total computational effort there is only a slim margin in favour of the 3-D scheme.

Even for small data misfits, the convergence of the higher dimensional subspace schemes is markedly better than a simple descents scheme, but the greatest gains occur in the early stages of the inversions. In figure 2 we compare the descents approach with a 3-D subspace after 4 iterations. The simple gradient scheme has barely begun to resolve the horizontal gradients in the model, whereas the 3-D scheme has reached a better level of reconstruction than 20 iterations of the descent method. For many purposes the lower panel of figure 2 would be adequate and this has been constructed with only modest computational effort.

4. Multiple parameter classes

We have so far concentrated on problems with a large number of parameters of similar type for which the subspace methods have worked very well. When, however, we seek to recover parameters of different types in the same inversion, direct application of the subspace methods give results which depend on the units of the parameters. One partial solution is to non-dimensionalise the parameters by dividing by their a priori variance or by some reference value, but the inversion still depends on the scaling.

This difficulty can be overcome by partitioning the problem by parameter type and establishing a sub-space system directly related to the character of the inverse problem. We will develop the approach in terms of two classes of parameter $\mathbf{s}$ and $\mathbf{b}$, which in our subsequent examples will be identified with the slowness field and the shape of the reflector, but the treatment may be readily generalised to larger numbers of parameter classes. With these two sets of parameters we can still establish an objective function F of the type (8). The structure of the data misfit Φ will be as before, but the theoretical values $\mathbf{g}(\mathbf{s},\mathbf{b})$ will depend on both parameters. The regularisation terms should include some allowance for the different parameter classes e.g. by partitioning the inverse model covariance matrix in (6)

$$\mathbf{C}_m^{-1} = \begin{bmatrix} \mathbf{C}_{ss}^{-1} & \mathbf{C}_{sb}^{-1} \\ \mathbf{C}_{bs}^{-1} & \mathbf{C}_{bb}^{-1} \end{bmatrix}. \tag{24}$$

The cross-coupling in the off-diagonal blocks in (24) may well prove difficult to estimate without a priori knowledge of the problem and should probably be ignored at first.

We will make a corresponding partition of the model parameter and gradient vectors

$$\mathbf{m} = [\mathbf{s},\mathbf{b}]^T, \qquad \mathbf{G} = [\mathbf{G}_s,\mathbf{G}_b]^T \tag{25}$$

A direct descents technique would use the gradient of the objective function F but it is preferable to partition this gradient by parameter dependence to give a 2-D subspace spanned by the vectors

$$\mathbf{a}_s = |\boldsymbol{\theta}_s|^{-1}[\boldsymbol{\theta}_s,0]^T, \qquad \mathbf{s}_b = |\boldsymbol{\theta}_b|^{-1}[0,\boldsymbol{\theta}_b]^T.$$

The model parameter perturbations

$$\delta\mathbf{m} = \begin{bmatrix} \delta\mathbf{s} \\ \delta\mathbf{b} \end{bmatrix} = -\mathbf{A}(\mathbf{A}^T\mathbf{H}\mathbf{A})^{-1}\mathbf{A}^T \begin{bmatrix} \boldsymbol{\theta}_s \\ \boldsymbol{\theta}_b \end{bmatrix}, \tag{27}$$

where $\mathbf{A}$ is the projection matrix. For the 2-D subspace

$$\mathbf{A} = \begin{bmatrix} \mathbf{a}_s & 0 \\ 0 & \mathbf{a}_b \end{bmatrix}. \tag{28}$$

If we now rescale $\mathbf{s}$ and $\mathbf{b}$ by different factors we find that the calculated model perturbations scale in the same way. This subspace scheme is not therefore affected by choices of units whereas the direction of the overall gradient is modified by the rescaling.

We have found previously that the rate of convergence of a subspace inversion scheme is improved by the inclusion of directions based on the rate of change of the gradient. We can produce comparable vectors by partitioning the Hessian

$$\mathbf{H} = \begin{bmatrix} \mathbf{H}_{ss} & \mathbf{H}_{sb} \\ \mathbf{H}_{bs} & \mathbf{H}_{bb} \end{bmatrix}. \tag{29}$$

and looking at the rate of change of each of $\boldsymbol{\theta}_s$ and $\boldsymbol{\theta}_b$ along the directions of $\boldsymbol{\theta}_s$ and $\boldsymbol{\theta}_b$. This generates four new vectors

$$[\mathbf{H}_{ss}\mathbf{a}_s,0]^T, \qquad [0,\mathbf{H}_{bb}\mathbf{a}_b]^T, \qquad [\mathbf{H}_{sb}\mathbf{a}_b,0]^T, \qquad [0,\mathbf{H}_{bs}\mathbf{a}_s]^T, \tag{30}$$

which need to be orthonormalised before we define the basis of a 6-D subspace. Six directions may appear more than is desirable, but the use of row reduction and back substitution to solve for the model perturbations (rather than direct inversion of $\mathbf{A}^T\mathbf{H}\mathbf{A}$) is straightforward and could cope with a larger subspace. Little extra work is required over that for a 2-D scheme using $\boldsymbol{\theta}$ and $\mathbf{H}\boldsymbol{\theta}$ with no allowance for the differences in parameter classes.

With this 6-D subspace we can build an iteration scheme for successive updating of the model parameters based, as before, on a locally quadratic approximation to the objective function F. The convergence of this scheme towards a minimum of F is usually good, but

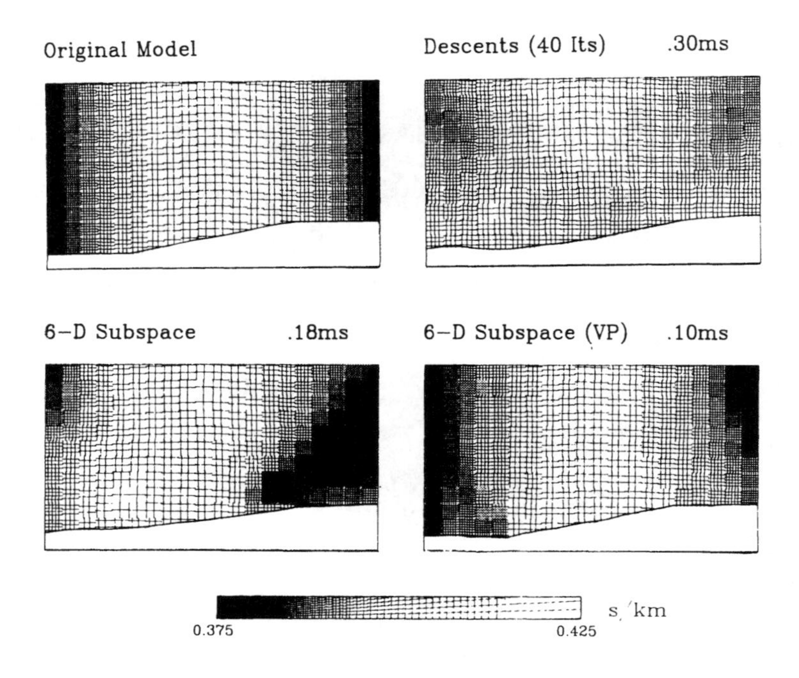

Figure 3.
Comparison of different inversions for the reflection tomography problem with unknown slowness field and reflector shape. The descents inversion uses rescaling of the model parameters by reference values and is only moderately successful. The application of the 6-D subspace method with a fine parametrisation gives a good fit to the data, but not a good recovery of the original model ("a local minimum"). The 6-D subspace method combined with variable parameterisation, indicated by VP, gives very good results.

there is no guarantee that we will actually find the global minimum we seek in the case of perfect data. Even with erroneous data we would prefer to be in the neighbourhood of the global minimum at the point we terminate the iteration.

In nonlinear problems with several classes of parameters we can get strong trade-offs between the different types of parameters. A particular feature in the data can often be matched reasonably well by a number of different styles of model. This gives strong dependence on the starting model $\mathbf{m}_0$ and a likelihood of convergence to local minima of F. One strategy that can often be successful is to work with a hierachy of scales of parameterisation (cf. Nolet et al 1986). We start with a relatively limited numbers of parameters and iteratively update the model until a current 'best' model is found. The scale length is then reduced and a further inversion is performed. This process can be continued until there is no further gain in apparent resolution, or until a prescribed parameterisation is attained. By steadily introducing detail into the model, rather than allowing it to develop ab initio, the parameters are steered away from purely local minima of F.

5. Reflection tomography II

In the previous examples of the subspace inversion method we considered the problem of recovering the slowness field above a known reflector from the arrival times of reflected waves. We now allow the reflector to vary also and try to estimate the shape at the same time as we find the slowness field.

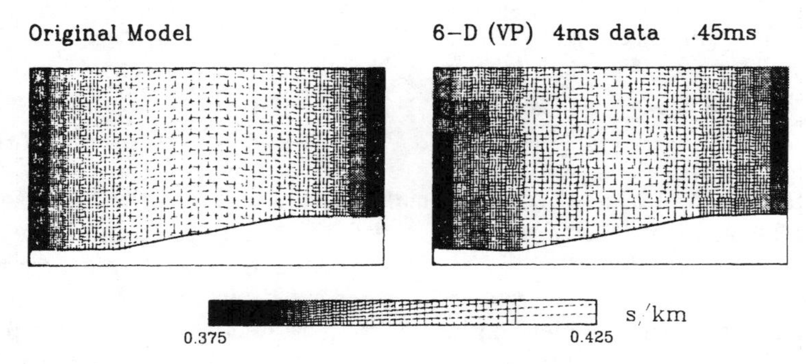

Figure 4.
Reconstruction in the presence of noise, each of the travel times were rounded to the nearest 4 ms before the inversion was commenced. The iteration was stopped when the r.m.s. data misfit matched the standard deviation of the arrival times.

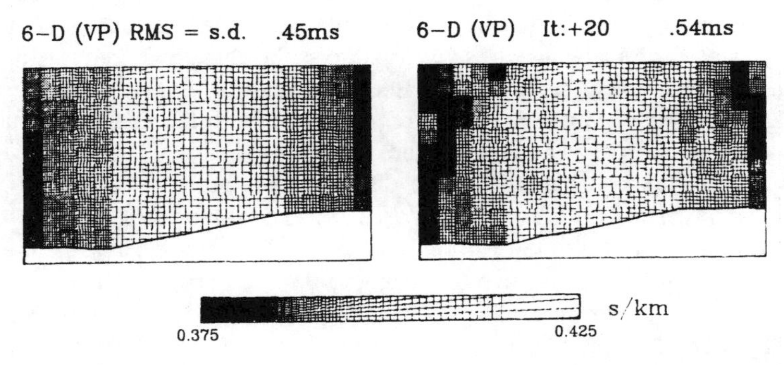

Figure 5.
Further iterations with noisy data. The left hand inversion is that shown in figure 4. On the right hand side, the calculations were not truncated at the data misfit threshold, but were continued for a further 20 iterations. The fit to the truncated data is improved, but the reconstruction of the original situation is rather poor.

Because we now have more parameters to estimate from the same data set the new problem is much more difficult than before. We will consider a model with 25 m square cells, 30 cells horizontally in 10 vertical columns, with the interface specified at each of the boundaries of the cells and linearly interpolated in between. In order to give an adequate coverage of rays in the central portion of the model we arrange the original model to have uniform zones 5 cells wide on the two sides of the 20 cell wide central zone in which the structure is placed. Figure 3 shows this central portion of the original model and a sequence of reconstructions using various inversion schemes.

The gradient method does not include any separation by parameter class. Instead, the model parameter values are divided by those for a reference (and starting) model which is uniform in the central section with a slowness 0.4 s/km and a reflector depth of 200 m. This means that we are working with relative perturbations to this model. We use strong regularisation constraints on the outermost rows of cells to damp down any tendency to change the model in this region. The results of the procedure even after 40 iterations are rather disappointing. The main features of the slowness field are only just beginning to appear and the reflector shape is relatively poorly represented. However, the r.m.s. data misfit is not too large because there have been trade-offs between the slowness and the shape of the interface.

The use of the direct 6-D subspace method achieves a considerable reduction in the r.m.s. data misfit after 16 iterations, but it is clear that although the data is well fit, we have not recovered the initial model. We have fallen into a local rather than a global minimum of the objective function. The pattern of slowness deviations form the original model correlates strongly with the inferred shape of the reflecting interface. Where the reflector is too high, the slownesses are too large above this zone and where the reflector is too low a zone of reduced slownesses is induced above it in order to compensate the travel times of the reflected waves. Since the ray paths can have significant angles to the vertical, these anomalous slowness zones tend to be wedge shaped.

The final panel in figure 3 shows the use of the subspace scheme in conjunction with a sequence of scales of parameterisation. At first, a 2-D scheme is used but once the number of parameters is large we switch to the 6-D subspace. No more than 12 iterations were used at each stage, with a change of scale as soon as the change of data misfit at an iteration is less than one per cent of its previous value. The final result gives a good representation of the original, relatively little detail has been filled in with the final 25 m stage and much of the structure was imposed at a 50 m scale.

We have so far considered perfect data, but in figure 4 we show the result of performing an inversion on a 'noisy' data set. The arrival times for the reflected waves are rounded to the nearest multiple of 4 ms to simulate the usual discrete nature of reflection data. This data set is then inverted using the 6-D subspace method with variable parameterisation and the results are surprisingly good. The final iteration on 25 m cells was terminated when the r.m.s. data misfit dropped below the standard deviation of the discretised data. Once again most of the structure is imposed on a 50 m scale.

Our final example (figure 5) shows the importance of a realistic termination criterion for the inversion of noisy data. We have used the same data set as in figure 4 and repeat the previous inversion in the left hand panel. On the right hand side we have allowed the iteration scheme to proceed for a further 20 steps. The result is a model which gives a better detailed fit to the noisy data but which is significantly worse than before at representing the original situation.

We have shown that the subspace approach can give good results in multiparameter inversion, especially when coupled with successive changes in the scales of parameterisation. We are unlikely to do much better with methods which rely on local information on the character of the objective function.

In these examples we have been able to judge the success of our inversions because we know what the original model looked like. In general, however, we will find it difficult to distinguish whether we have been trapped in the neighbourhood of a 'local' minimum sufficiently deep to invoke our termination criterion. By using a number of different starting models, we can gain some confidence that certain features or trends are common to the different reconstructions, but have to be wary of claiming the resolution of detail in our models. Only if the solution we obtain is linearly close to the true solution can we rely on the resolution analysis of Backus and Gilbert (1970), but it can often provide a good guide to the behaviour of the inversion.

6. Caveat

The results of an inversion are no better than the theoretical model $\mathbf{g}(\mathbf{m})$ used to match the observed data values. Inversion cannot reinstate portions of the problem which have been ignored in the analysis leading to $\mathbf{g}(\mathbf{m})$. For example, in the reflection tomography examples we have restricted attention to two spatial dimensions, but there may well be three-dimensional spatial variations affecting the travel times. The three-dimensional effects will be mapped into apparent two-dimensional structure.

Acknowledgements

B.L.N.K. is grateful to Holland Sea Search for making his attendance at the 16th Conference on Mathematical Geophysics possible. This work was begun at the Department of Applied Mathematics and Theoretical Physics at the University of Cambridge and P.W. was supported by a grant from Shell U.K. Exploration and Production.

The ideas in this paper have been influenced by discussions (and arguments!) with Alastair Harding, Malcolm Sambridge, Albert Tarantola and Guust Nolet.

References

Aki, K. and Richards P.G., 1980. *Quantitative Seismology*, W.H. Freeman, San Francisco.

Backus, G.E. and Gilbert, F., 1970. Uniqueness in the inversion of inaccurate gross earth data, *Phil. Trans. R. Soc. Lond.*, **266A**, 123-192.

Bryan, R. and Skilling, J., 1980. Deconvolution by maximum entropy, as illustrated by application to jet of M87, *Mon. Not. R. Astr. Soc.*, **191**, 69-76.

Chapman, C.H. and Orcutt, J., 1985. Least-squares fitting of marine seismic refraction data, *Geophys. J. R. Astr. Soc*, **82**, 339-374.

Harding, A.J., 1984. *Slowness Methods in Seismology*, Ph.D. Thesis, University of Cambridge.

Kennett, B.L.N., 1978. Some aspects of nonlinearity in inversion, *Geophys. J. R. Astr. Soc*, 55, 373-391.

Menke, W., 1984. *Geophysical Data Analysis: Discrete Inverse Theory*, Academic Press, New York.

Nolet, G., 1981. Linearised inversion of teleseismic data, in *The solution of the inverse problem in geophysical interpretation*, 9-37, ed. R. Cassinis, Plenum Press, New York.

Nolet, G., 1985. Solving or resolving inadequate and noisy tomographic systems, *J. Comp. Phys.*, 61, 463-482.

Nolet, G., van Trier, J. and Huisman, R., 1986. A formalism for nonlinear inversion of seismic surface waves, *Geophys. Res. Letters*, 13, 26-29.

Sambridge, M. and Kennett, B.L.N., 1986. A novel method of hypocentre location, *Geophys. J. R. Astr. Soc*, **87**, 679-697.

Williamson, P.R., 1986. Tomographic inversion of travel time data in reflection seismology, Ph.D. Thesis, University of Cambridge.

B.L.N. KENNETT and P.R. WILLIAMSON, Research School of Earth Sciences, Australian National University, GPO Box 4, Canberra ACT 2601, Australia.

Chapter 8

Imaging algorithms, accuracy and resolution in delay time tomography

W. Spakman and G. Nolet

In this chapter we will discuss a linearized tomographic method for the simultaneous inversion of body wave velocity, earthquake relocation parameters and station corrections using delay time data. We apply this method to investigate how well two different iterative least squares algorithms are able to solve tomographic problems in which many model parameters and a large amount of data are used. Especially we will demonstrate how estimates of accuracy and spatial resolution can be obtained and discuss the different image-distorting effects of the employed algorithms in the results of tomographic mapping.

1. Introduction

Tomographic images of the Earth's interior reflect, apart from true velocity heterogeneities, the effects of data errors, imperfect illumination by seismic waves, model parameterization, linearization and algorithm performance. These influences cannot be separated easily and often result in artificial anomalies in the final image with amplitudes comparable to those of the velocity heterogeneities we seek to resolve. If we also realize that up to now large scale tomographic results are not often supported by any attempt to estimate the spatial resolution in the inverted model, it is not hard to understand the reluctance one feels in accepting tomographic images as an accurate portrayal of actual velocity heterogeneities, or even of their sign.

Obtaining an image is not the most difficult problem in the process of a tomographic research. The major problem lies in identifying the influence of all of the above mentioned

N.J. Vlaar, G. Nolet, M.J.R. Wortel & S.A.P.L. Cloetingh (eds), Mathematical Geophysics, 155-187.

effects in the final image. In this chapter we will attempt to tackle some of this problem from a practical point of view. When using many model parameters in seismic tomography, at present no feasible method exist that allows us to estimate the spatial resolution directly from the data (see also Nolet 1985). However, we may study the image-distorting effects by applying sensitivity tests to the system of equations using synthetic models of velocity heterogeneity. We will use this technique to study the spatial resolution in a practical experiment, which is typical for large scale body wave tomography. In the same experiment we will focus our attention on how well two different iterative algorithms perform on the same tomographic problem. This will enable us to explicitely study the influence of the equation solvers on the tomographic results. The algorithms in question are the conjugate gradient solver LSQR of Paige and Saunders (1982) and a member of the SIRT family of backprojection methods. Both methods are being used in large scale body wave tomography using P-wave data. Spakman (1985) applied LSQR to solve for the upper mantle heterogeneities in Eurasian-Mediterranean area, SIRT has been used by Clayton and Comer (1984) to image the Earth's lower mantle heterogeneities.

2. The tomographic problem

The data commonly used in body wave tomography are delay times, i.e. the difference between the observed travel time of a seismic wave and the computed travel time from a (not necessarily radial symmetric) reference Earth velocity model. In the ISC earthquake location procedure the delay times are the residuals of the minimization process in which observed travel times are compared to the Jeffreys-Bullen travel time tables. Although these delays contain observational errors and their values depend on the event location procedure, in seismic tomography we also rely on their descriptive value for the Earth's lateral heterogeneity.

We consider the tomographic problem arising from a division of a *part* of the Earth's interior into a large number of cells. Our interest is to estimate the slowness heterogeneities in these cells by inversion of the delay times. The lateral heterogeneity of the Earth induces a delay time d_i (i-th ray), which we can divide into

$$d_i = d_i^M + d_i^R + d_i^O \tag{1a}$$

d_i^M is the delay time caused by the lateral heterogeneity in that part of the Earth that is covered by the cells, d_i^R is the error in origin time and event location, caused by the fact that we use a reference model instead of the real Earth to locate the earthquake, and d_i^O represents the delay time acquired by the ray segment that does not intersect the cells.

We assume that the actual Earth's heterogeneous velocity structure deviates only by a small amount from the reference model. We can linearize the ray-integral for the travel time with respect to the reference velocity model by applying Fermats principle leading to a description in terms of the reference raypath L_i and obtain an approximate expression of d_i^M in terms of the cell slowness anomalies s_l (l-th cell) as

$$d_i^M = \sum_{l=1}^{N} L_{il} s_l \tag{1b}$$

in which L_{il} is the length of the (reference) raypath segment in cell l. Although the total number of cells N is large, only $O(N^{1/3})$ L_{il} will be non-zero.

We can express d_i^R in terms of the change in total travel time T due to a small change $\mathbf{g}(\Delta t_0,\Delta \mathbf{r}_s)$ in the source vector $\boldsymbol{\chi}_s(t_0,\mathbf{r}_s)$, where t_0 is the origin time and $\mathbf{r}_s$ the source location vector. In general we shall use a local cartesian coordinate system so that we write $\boldsymbol{\chi}_s=(t_0,x_s,y_s,z_s)$. We then have for the j-th event and the i-th ray in first order approximation:

$$d_i^R = \nabla_\chi T_{ij} \cdot \mathbf{g}_j = \sum_{m=1}^{4} G_{ij}^m g_j^m \tag{1c}$$

where ∇_χ contains derivatives with respect to the components of $\boldsymbol{\chi}_s$ and where G_{ij}^m is the m-th component of $\nabla_\chi T_{ij}$ evaluated at the (reference) source location $\mathbf{r}_s^j$. Note that any combination of j and m represents one unknown, hence the j-th event adds 4 unknowns g_j^m, origin time error respectively and source location error, to equation (1a).

The last term in (1a), d_i^O, is difficult to parameterize because we do not plan to cover the remaining part of the Earth's volume with cells. Yet for an observing station outside the cell model we may at least attempt to invert for the 'near' station heterogeneity by placing a cell beneath the station, replacing d_i^O by

$$d_i^O = H_{ik} h_k \tag{1d}$$

with H_{ij} the raypath segment of ray i in the station cell k with slowness anomaly h_k. Existing estimates of station delay time corrections (e.g. Dziewonski & Anderson 1983) can be used to correct the delay time prior to inversion. In that case the station delay correction (1d) can be deleted from (1a).

Combining equations (1a-d) we arrive at the basic set of equations describing the linearized cell-tomographic problem, which for M rays reads:

$$d_i = \sum_{l=1}^{N} L_{il} s_l + \sum_{m=1}^{4} G_{ij}^m g_j^m + H_{ik} h_k \;, \quad i=1,...,M \tag{1e}$$

where j and k are uniquely determined by the ray index i.

In large scale body wave tomography the number of rays M amounts easily to $O(10^6)$ and this system of equations has to be solved for $N_T = N+4N_E+N_S$, $O(10^4)$, model parameters, where N_E is the number of events and N_S is the number of stations for which a station delay time correction is estimated.

At present only radially symmetric reference models are used in large scale delay time tomography. The determination of the delay time data, the event location and the calculation of the reference ray paths, L_i, depend on the choice of this velocity model. Locally, both laterally and in depth, the *average* Earth's velocity structure may differ too strongly from the employed model, which may induce (i) systematic errors in the event location which (ii) biases the estimation of the delay times toward the reference model and (iii) may cause the actual ray paths to differ systematically from the ones we employ in (1e).

These are highly important problems which cause hardly traceable bias and artifacts in the slowness solution. In this chapter we shall not deal further with these complications, but rather concentrate on how well we can solve (1e) assuming that it is a correct description of the effects of the Earth's lateral heterogeneity.

One characteristic of tomographic experiments is that (i) many unknowns are not resolved, despite the fact that $M \gg N_T$ and (ii) that data errors are large so that a large redundancy is needed to obtain a reliable statistical estimate. The addition of station and event corrections has an important function. It allows part of the data to be explained by other factors than lateral heterogeneity. For as far as these added unkowns to the system (1e) are independent they will certainly improve the data fit. But results to be discussed later show that they also add an ambiguity: on the one hand their inclusion improves or at least does not impair the data fit, but on the other hand our experiments show that most of the event and station corrections are among the least constrained unknowns and hence their estimation may absorb part of the delay time which may also be explained by velocity heterogeneity. In a joint inversion, as we consider here, this will have a damping effect on the estimation of the heterogeneities. Since the corrections have a firm physical basis we do not think this is too conservative but rather welcome the damping effect.

As for the event corrections, we expect these to be correlated with the lateral heterogeneity, and thus with the event location. In large scale tomography the number of events can be large, $O(10^4)$, and since an event adds 4 unknowns to (1e), we would probably overdo the damping if we would attach corrections to every event separately. This can be prevented by assigning the same corrections to all events within a cell, and hence instead of relocating every single event, we estimate the 4 corrections for event 'clusters'.

As an alternative for a simultanuous inversion for heterogeneity, event and station corrections one may iterate between relocating events and estimating the slowness heterogeneity, using the updated slowness model in every next relocation step (Dziewonski, 1984). Convergence of this process will lead to a quasi simultaneous inversion of heterogeneity and event relocation parameters, but there is poor control on how much of the delay time that is actually due to heterogeneity is finally absorbed in the relocation of events or vice versa.

Generally speaking (1e) can be applied to a limited part of the Earth's interior as well as to the entire Earth. However applications of (1e) to very large volumes would at present necessarily require a parameterization in large cells. This may be sufficient for lower mantle and core tomography, but it is certainly unwanted for upper mantle imaging, since short wavelength heterogeneities certainly exist in the upper mantle. And then, even if we *could* parameterize the entire upper mantle in cells of 100×100×100 km, an exercise which would involve 356,000 unknowns for merely the slowness anomalies, this would result in a model that combines regions with a dense ray coverage (Europe, N. America) with regions that are covered very badly (oceans). In upper mantle tomography one might as well apply (1e) to a relative small region (i.e. compared to the entire upper mantle) taking advantage of the possibility of using 'small' cells, and try to correct for the delay time contributions of those raypath segments outside the cell model.

If we have events located *inside* the cell model and an observing station at teleseismic distance that 'sees' the cell model with a relatively small spatial angle, most rays will acquire delays from the same mantle region outside the cell model. We anticipate that the station correction will absorb most of the contribution to the delay time which is caused by the intermediate mantle region. The same holds for an event at teleseismic distances which is observed by stations at the surface of the cell model. In this case the estimation of event mislocation and origin time may be biased, but conveniently absorb the influence of the mantle heterogeneities outside the cell model.

3. Least squares inversion

We will set up a system of equations for least squares inversion of (1e). In matrix notation (1e) can be written as:

$$\hat{\mathbf{A}}\,\mathbf{x} = (\,\mathbf{L}\mid\mathbf{G}\mid\mathbf{H}\,)\,\mathbf{x} = \mathbf{d} \quad , \mathbf{x} = \begin{bmatrix}\mathbf{s}\\ \mathbf{g}\\ \mathbf{h}\end{bmatrix} \tag{2}$$

with **L** the matrix of ray cell pathlengths, **G** the matrix of relocation coefficients, **H** the matrix of station correction coefficients, $(\,\mathbf{s}\;\mathbf{g}\;\mathbf{h}\,)^{\mathrm{T}}$ the corresponding parts of the solution vector **x** and **d** the delay time vector.

Eq. (2) will be solved by least squares. It is well known that the least squares solution is an unbiased estimate if $(\hat{\mathbf{A}}^{\mathrm{T}}\hat{\mathbf{A}})^{-1}$ exists (e.g. Draper and Smith, 1982, page 87). In tomography, however $\hat{\mathbf{A}}^{\mathrm{T}}\hat{\mathbf{A}}$ is usually singular. We may obtain a Bayesian estimate of the solution by adding our a priori expectations **Ix=0** as 'phoney' data to equation (2). In addition, if we scale (2) with a priori uncertainties we find:

$$\mathbf{z} = \mathbf{F}_x^{-1/2}\mathbf{x} \;, \quad \begin{bmatrix}\mathbf{F}_d^{-1/2}\hat{\mathbf{A}}\mathbf{F}_x^{+1/2}\\ \eta\mathbf{I}\end{bmatrix}\mathbf{z} = \begin{bmatrix}\mathbf{F}_d^{-1/2}\mathbf{d}\\ \mathbf{0}\end{bmatrix} \tag{3a}$$

where $\mathbf{F}_d$ and $\mathbf{F}_x$ are the covariance matrices of the data and model parameters, respectively and η is a constant governing the amount of solution damping that is applied by the inclusion of **Iz=0**. Note that we do not need to compute the inverse $\mathbf{F}_x^{-1/2}$ explicitly. After solving the second equation of (3a) for **z**, we obtain the solution **x** from $\mathbf{x} = \mathbf{F}_x^{+1/2}\mathbf{z}$. Equation (3a) has a unique solution, which can be shown to be the maximum likelihood estimate if both types of 'data' have a Gaussian distribution. This estimate is biased, however, towards **0**.

We will solve our least squares problems with two different *iterative* row action methods, i.e. SIRT and LSQR, of which details will be discussed in the next section. Both methods possess *intrinsic* damping properties of which the effect on the solution decreases with increasing iteration number (Van der Sluis and van der Vorst 1987). Since we anticipate to stop the iterative least squares processes after a relative small number of iterations, we may take advantage of the intrinsic damping properties to obtain a solution of (2) that is not biased by explicit damping, by selecting among all possible solutions the **x** that minimizes $\mathbf{x}^{\mathrm{T}}\mathbf{F}_x^{-1}\mathbf{x}$. Again with $\mathbf{z} = \mathbf{F}_x^{-1/2}\mathbf{x}$ we then solve:

$$\mathbf{F}_d^{-1/2}\hat{\mathbf{A}}\mathbf{F}_x^{1/2}\mathbf{z} = \mathbf{F}_d^{-1/2}\mathbf{d} \tag{3b}$$

subject to the condition $Min\,(\mathbf{z}^{\mathrm{T}}\mathbf{z})$. This equation is the same as (3a) but with η=0. The bias introduced by implicit damping is one of the algorithm characteristics that we shall study in this paper.

Equations (3a) and (3b) are subjective choices for setting up a least squares system of equations. Alternatives exist which favour a more elaborate minimum model criterion (for a brief review see Nolet 1987).

We will now discuss our a priori estimates of $\mathbf{F}_x$. Let us write

$$\mathbf{F}_x = \begin{bmatrix} \mathbf{F}_x^{(s)} & & 0 \\ & \mathbf{F}_x^{(g)} & \\ 0 & & \mathbf{F}_x^{(h)} \end{bmatrix} \tag{4}$$

with $\mathbf{F}_x^{(s)}=diag(f_l^{(s)})$, $\mathbf{F}_x^{(g)}=diag(f_{jm}^{(g)})$ and $\mathbf{F}_x^{(h)}=diag(f_k^{(h)})$ in correspondence with the subdivision of $\mathbf{x}$ in (2). The notation jm denotes the m-th relocation parameter of the j-th event.

We would like to scale the model parameters to a priori unit variance using for $\mathbf{F}_x$ the covariance matrix of $\mathbf{x}$, but, since we do not have the reliable uncertainty estimates nor information on the correlation between the individual x_j, we assume $\mathbf{F}_x$ diagonal and apply a simple, though sensible, uniform length scaling using the subjectively expected slowness amplitude of s_l of 0.002 km/s. To this purpose, we scaled the amplitudes of the elements of $\mathbf{g}$, using a priori estimates of 1 second for the standard error in the origin time and a 10 km mislocation of the source coordinates. Let the scaling factors be denoted by $\gamma_m^{1/2}$, m=1,..,4 then $f_{jm}^{(g)}=\gamma_m$.

We also wish to control the expected trade off between cell slowness heterogeneity $\mathbf{s}$, the relocation parameters (now $\mathbf{F}_x^{(g)-1/2}\mathbf{g}$) and station corrections $\mathbf{h}$ if these are not independently constrained. A scaling of the latter two, both with one constant, say μ resp. κ, will accomplish this. Now $f_{jm}^{(g)}=\mu^{-1}\gamma_m$ and $f_k^{(h)}=\kappa^{-1}$. The starting value for μ and κ in the present scaling is 1. But this can (subjectively) be changed, after some experience has developed with test inversions, on the basis of the amplitude behaviour of the different parts of $\mathbf{x}$.

In the case that cells of different sizes are involved, the different ray sampling, combined with the model minimization tends to attribute larger slowness anomalies to large cells with respect to the anomalies in the smaller ones. Adjacent cells of different sizes that are not spatially resolved may hence develop a heterogeneity contrast which is unwanted. We may partially prevent this by scaling the cell heterogeneities s_l of $\mathbf{z}$ by some typical cell size dependent number δ_l. One choice is taking $\delta_l = v_l^{-1/2}$, v_l being the cell volume of the l-th cell. This scaling conforms to projecting the cell part of the solution on the space of all possible Earth models spanned by the orthonormal 'cell-functions' $C_l(\mathbf{r})=v_l^{-1/2}$ if $\mathbf{r}$ is inside cell l and $C_l(\mathbf{r})=0$ elsewhere (Nolet, 1987). We will use the cell volume scaling only in the case that we solve our tomographic problem with the LSQR method, since , as we shall discuss later, the SIRT algorithm employs *implicitly* its own cell-size dependent scaling.

We adopt $f_l^{(s)}=v_l^{-1}$ (and 1 in case of the SIRT method), i.e. $x_l=z_l f_l^{+1/2}$. Note that we now destroyed the property that the components of $\mathbf{z}$ are of approximate equal absolute magnitude, since we now have expected slowness heterogeneity amplitudes on the order of $0.002\times v_l^{+1/2}$, but the expected amplitudes of $z_{jm}(=\gamma^{-1/2}g_{jm})$ are still on the order of 0.002. This is easily amended by an additional scaling of $f_{jm}^{(g)}$ with the reciproke of the average cell volume $\hat{v}^{-1}$.

With $\mathbf{F}_x^{(s)}=diag(v_l^{-1})$, $\mathbf{F}_x^{(g)}=diag(\mu^{-1}\hat{v}^{-1}\gamma_m)$ and $\mathbf{F}_x^{(h)}=diag(\kappa^{-1}v_k^{-1})$, $\mathbf{F}_x$ as in (4) and $\mathbf{F}_d=\mathbf{I}$ (assuming unit variance in the data), the system (2) is transformed to

$$\begin{aligned} \mathbf{x} &= \mathbf{F}_x^{1/2}\mathbf{z} \\ \hat{\mathbf{A}}\mathbf{F}_x^{1/2}\mathbf{z} &= \mathbf{d} \end{aligned} \tag{5}$$

In solving (5) we wish to minimize $\mathbf{z}^T\mathbf{z}$ $(=\mathbf{x}^T\mathbf{F}_x^{-1}\mathbf{x})$.

Due to large errors in the delay time data and often poor station-event distribution we may not expect to resolve more from (3) than rather smoothly varying anomalies. Using our physical intuition, we also expect, apart from regions where large slowness heterogeneity contrasts may be expected (e.g. slab regions), that adjacent cells will have correllating velocity heterogeneities. On the other hand errors and resolution artifacts may lead to poorly constrained solutions in cells, more than once resulting in large heterogeneity contrast between those cells and neighbouring cells with good ray coverage.

We can combine expectations and physical intuition and at the same time constrain (i.e. stabilize) the inversion by applying a smoothing procedure to the solution $\mathbf{z}$. There are two possible ways to perform this. A posteriori smoothing of $\mathbf{z}$ will result in a smooth picture but also destroy the least squares character of the fit. The alternative is to include smoothing in (3) and obtain a smoothed $\mathbf{z}$ which is also in accordance with our least squares procedure. This approach is easily incorporated in (5).
Let $\mathbf{S}$ be a smoothing operator (e.g. a matrix with row sums 1) that maps $\hat{\mathbf{z}}$ on $\mathbf{z}$. Then we have

$$\begin{aligned} \mathbf{x} &= \mathbf{F}_x^{1/2}\mathbf{z} \\ \mathbf{z} &= \mathbf{S}\hat{\mathbf{z}} \\ \hat{\mathbf{A}}\mathbf{F}_x^{1/2}\mathbf{S}\,\hat{\mathbf{z}} &= \mathbf{d} \end{aligned} \tag{6}$$

We solve the last equation of (6) for $\hat{\mathbf{z}}$ subject to $Min(\mathbf{z}^T\mathbf{z})$ and obtain the smoothed solution from $\mathbf{z}=\mathbf{S}\hat{\mathbf{z}}$ and finally the 'physical' solution from $\mathbf{x}=\mathbf{F}_x^{+1/2}\mathbf{z}$. Notice that the approach of this smoothing procedure is an another example of scaling the columns of $\hat{\mathbf{A}}$ (cf. 3), in this case fixing correlations between the unknowns. Indeed one may look upon $\mathbf{S}^2$, or its generalized equivalent, as an (our subjective) approximation of the covariance matrix of $\mathbf{z}$. We remark that we take the smoothing operation $\mathbf{S}$ confined to the cell heterogeneities, thus if we write $\hat{\mathbf{z}}$ as $\hat{\mathbf{z}}=(\hat{\mathbf{z}}^{(s)},\hat{\mathbf{z}}^{(g)},\hat{\mathbf{z}}^{(h)})$, then $\mathbf{S}\hat{\mathbf{z}}=(\mathbf{z}^{(s)},\hat{\mathbf{z}}^{(g)},\hat{\mathbf{z}}^{(h)})$.

The last adaptation we will perform of the tomographic system of equations (6) concerns a row averaging procedure. Since in most seismic regions events are closely clustered, we may sum rays from predetermined event clusters arriving in one station into one row of the last equation of (6). In the same way we combine the corresponding delay times into one delay time belonging to what one may call a *composite* ray. Any desired row scaling can easily be incorporated (i.e. we may take a weighted combination of rays and corresponding delay times to construct the composite ray). This procedure serves the purposes of averaging the data errors and smoothing the solution, and it also reduces the number of rows involved in the last equation of (6). Mathematically this row condensing operation can be expressed as a left multiplication of the third equation of (6) with a $M'\times M$ matrix $\mathbf{D}$.

Then we finally arrive at the tomographic system of equations, which we will further employ in this chapter

$$\begin{aligned} \mathbf{x} &= \mathbf{F}_x^{1/2}\mathbf{z} \\ \mathbf{z} &= \mathbf{S}\hat{\mathbf{z}} \\ \mathbf{A} &= \mathbf{D}\hat{\mathbf{A}}\mathbf{F}_x^{1/2}\mathbf{S} \\ \mathbf{A}\hat{\mathbf{z}} &= \mathbf{D}\mathbf{d} = \hat{\mathbf{d}} \end{aligned} \tag{7}$$

from which we want obtain **x** by least squares inversion of $\mathbf{A}\hat{\mathbf{z}}=\hat{\mathbf{d}}$ subject to minimizing $\hat{\mathbf{z}}^T\hat{\mathbf{z}}$.

4. Algorithms

We present a brief introduction of the LSQR and SIRT algorithms which we will investigate.

In large scale tomography, using many parameters and data, **A** is sparse and very large ($O(10^{+10})$ elements is common). Finding a solution of (5) using techniques that require explicit memory storage of **A** will therefore not be possible. Row action methods like LSQR and SIRT require only access to one row of **A** at a time and consequently **A** may reside on secondary storage. Row action methods are iterative methods. At each iterative step an approximate solution of (7) is obtained, which is used as a starting point for the solution of the next iteration. More than once the convergence of the iterative solution to the least squares solution is measured from the decrease in data residual (i.e. the difference between the actual data and those predicted by the approximate model solution) as a function of the iteration number. Our results will show that this is a misleading quantity for observing the solution convergence, especially when comparing the convergence that different algorithms can achieve in the same number of iterations.

Two different classes of iterative algorithms have been employed so far in linearized seismic tomography. The most widely used methods are the stationary iterative backprojection methods called SIRT (Simultanuous Iterative Reconstruction Technique), such as the method discussed by Ivansson (1983) or (the different) method used by the Caltech group (as described in Hager et al. 1985). The second class are the conjugent gradient methods (CG). Paige and Saunders (1982) proposed a CG method, which they called LSQR, for the inversion of large sparse matrix systems. Nolet (1985) introduced this method to seismic tomography, applying it in a numerical experiment in which he compared the performance of LSQR to a SIRT method in solving sparse linear systems and Spakman (1985, 1986a,b) has used LSQR in large scale upper mantle tomography.

In the following sections we will discuss the way these methods solve tomographic system of equations of the general form:

$$\mathbf{A}\mathbf{x} = \mathbf{t} \tag{8}$$

with **x** the model vector and **t** the data vector.

5. LSQR

The LSQR variant of the CG method is somewhat similar to inversion methods based on singular value decomposition (SVD). The SVD solution is constructed in a p-dimensional subspace of the model space, spanned by the eigenvectors of $\mathbf{A}^T\mathbf{A}$ belonging to the largest eigenvalues $\lambda_1,..,\lambda_p$. Since the variance σ_j^2 of the solution x_j is inversely proportional to the magnitude of the smallest eigenvalue λ_p (e.g., Nolet, 1981), this limits the propagation of data errors into the model, at the expense of a strong limitation in the degrees of freedom of the model - which practically means that the model is heavily smoothed and that those parts of the model that are not illuminated by seismic rays have slowness deviations equal or close to zero.

With systems of large size, it is not possible to calculate the eigenvectors of $\mathbf{A}^T\mathbf{A}$ in any reasonable amount of CPU time. But instead of diagonalizing $\mathbf{A}^T\mathbf{A}$ we may tri-diagonalize it with a simple scheme, originally due to Lanczos, in which the columns of the transformation matrix $\mathbf{V}$ turn out to be orthogonal. Briefly the method works as follows. First normalize $\mathbf{A}^T\mathbf{t}$ with $\beta_1=|\mathbf{A}^T\mathbf{t}|$ and choose this as the first column $\mathbf{v}^{(1)}$ of $\mathbf{V}$, or, in other words, as the first 'basis' vector of a subspace to be constructed in the model space. The next basis vectors are now essentially determined by repeated multiplications with $\mathbf{A}^T\mathbf{A}$ and subsequent orthogonalization and normalization. Thus, to find the next vector we first construct $\mathbf{w}^{(1)} = \mathbf{A}^T\mathbf{A}\mathbf{v}^{(1)} - \alpha_1\mathbf{v}^{(1)}$ and choose $\alpha_1=\mathbf{v}^{(1)T}\mathbf{A}^T\mathbf{A}\mathbf{v}^{(1)}$ so that $\mathbf{w}^{(1)T}\mathbf{v}^{(1)}=0$, and then set $\mathbf{v}^{(2)}=\mathbf{w}^{(1)}/|\mathbf{w}^{(1)}|$. To construct $\mathbf{v}^{(3)}$ we must orthogonalize it to the *two* previous vectors :

$$\begin{aligned}\mathbf{w}^{(2)} &= \mathbf{A}^T\mathbf{A}\mathbf{v}^{(2)} - \alpha_2\mathbf{v}^{(2)} - \beta_2\mathbf{v}^{(1)}\\ \mathbf{v}^{(3)} &= \mathbf{w}^{(2)}/|\mathbf{w}^{(2)}|\end{aligned} \tag{9}$$

with $\alpha_3=\mathbf{v}^{(3)}\mathbf{A}^T\mathbf{A}\mathbf{v}^{(3)}$ and $\beta_3=\mathbf{v}^{(2)T}\mathbf{A}^T\mathbf{A}\mathbf{v}^{(2)}$. One can show that a 3- term recursion of this kind suffices to orthogonalize the whole set of vectors $\mathbf{v}^{(1)}, \ldots, \mathbf{v}^{(p)}$, constructed in this way:

$$\gamma_{j+1}\mathbf{v}^{(j+1)} = \mathbf{A}^T\mathbf{A}\mathbf{v}^{(j)} - \alpha_j\mathbf{v}^{(j)} - \beta_j\mathbf{v}^{(j-1)} \tag{10}$$

Multiplying (10) with $\mathbf{v}^{(j+1)T}$ easily shows that the normalizing factor $\gamma_{j+1}=\beta_{j+1}$. Reorganizing (10), and assembling $\mathbf{v}^{(1)}, \ldots, \mathbf{v}^{(p)}$ as columns of a matrix $\mathbf{V}_p$ we obtain:

$$\mathbf{A}^T\mathbf{A}\mathbf{V}_p = \mathbf{V}_{p+1}\mathbf{T}_p \tag{11}$$

where $\mathbf{T}_p$ is a tri-diagonal $(p+1)\times p$ matrix with upper subdiagonal $(\beta_2, \ldots, \beta_{p+1})$, diagonal $(\alpha_1, \ldots, \alpha_p)$ and lower sub-diagonal $(\beta_2, \ldots, \beta_p)$.

In analogy with the SVD method, the LSQR solution is found by expanding the p-th approximation $\mathbf{x}^{(p)}$ to $\mathbf{x}$ in terms of the basis vectors $\mathbf{v}^{(1)}, \ldots, \mathbf{v}^{(p)}$

$$\mathbf{x}^{(p)} = \mathbf{V}_p\mathbf{y}_p \tag{12}$$

so that the least squares system $\mathbf{A}^T\mathbf{A}\mathbf{x} = \mathbf{A}^T\mathbf{t} = \beta_1\mathbf{v}^{(1)}$ is reduced to a system $\mathbf{A}^T\mathbf{A}\mathbf{V}_p\mathbf{y}_p = \mathbf{V}_{p+1}\mathbf{T}_p\mathbf{y}_p = \beta_1\mathbf{v}^{(1)}$, or, after multiplication with $\mathbf{V}_{p+1}^T$:

$$\mathbf{T}_p\mathbf{y}_p = \beta_1\hat{\mathbf{e}}_1 \tag{13}$$

which is a tridiagonal system of $(p+1)$ equations with p unknowns, that can be solved by least squares or damped least squares, with very little extra computational effort.

In the LSQR method, Paige and Saunders (1982) avoid explicit use of $\mathbf{A}^T\mathbf{A}$ and further reduce $\mathbf{T}_p$ to a bidiagonal matrix. A simple Ratfor version of their algorithm is given by Nolet (1987).

The LSQR method seeks the $\mathbf{x}$ that minimizes $|\mathbf{Ax} - \mathbf{t}|$ subject to minimizing $\mathbf{x}^T\mathbf{x}$. Since it uses conjugate gradients, the residual vectors after each iteration are mutually orthogonal. For a detailed description of the properties of the LSQR algorithm see Paige and Saunders (1982) or Van der Sluis and Van der Vorst (1987).

Simple solution damping (regularization) can formally be achieved by adding N rows $\eta\mathbf{I}=\mathbf{0}$ to (8) (section 3). In practice, for the LSQR algorithm, we only need to add η to the α_i and damping is applied as if the damping equations are specified explicitly (Paige and Saunders, 1982).

Since LSQR constructs the solution from a set of orthogonal vectors, the algorithm should theoretically converge in N steps or less (N being the dimension of the model space). In practice CPU time limitations force us to stop the calculation after a few iterations ($p \ll N$), but nevertheless we expect that the orthogonalization will result in favourable convergence properties of LSQR with respect to iterative methods that do not employ orthogonality properties. Moreover van der Sluis and van der Vorst (1987) show that the LSQR algorithm starts the construction of the solution by neglecting those components belonging to the very smallest eigenvalues of $\mathbf{A}^T\mathbf{A}$. The contributions of the smallest eigenvalues eventually enter the solution more and more as iteration proceeds. This behaviour attributes to the *intrinsic* damping properties of the algorithm, which are similar to the SVD method with sharp eigenvalue cut-off.

6. SIRT

SIRT methods operate in a way quite different from CG methods. The SIRT method we will employ can be described by the following iterative process:

$$
\begin{aligned}
x_j^{(0)} &= 0 \text{ for } j=1,N \\
\text{for } p &= 0,1,2,.. \\
r_i^{(p)} &= t_i - \sum_j^N A_{ij} x_j^{(p)} \ , \ i=1,M \\
\Delta x_j^{(p)} &= \omega \frac{1}{c_j} \sum_i^M \frac{A_{ij} r_i^{(p)}}{\rho_i} \\
x_j^{(p+1)} &= x_j^{(p)} + \Delta x_j^{(p)}
\end{aligned}
\tag{14}
$$

with $c_j = \sum_i^M |A_{ij}|$ and $\rho_i = \sum_j^N |A_{ij}|$.

If we use the matrix notation of van der Sluis and van der Vorst (1987), equation (14) reads with $\mathbf{C} = diag\,(c_j)$ and $\mathbf{R} = diag\,(\rho_i)$:

$$
\begin{aligned}
\mathbf{x}^{(0)} &= 0 \\
\text{for } p &= 1,2,.. \\
\mathbf{r}^{(p)} &= \mathbf{t}-\mathbf{A}\mathbf{x}^{(p)} \\
\Delta\mathbf{x}^{(p)} &= \omega\mathbf{C}^{-1}\mathbf{A}^{\mathrm{T}}\mathbf{R}^{-1}\mathbf{r}^{(p)} \\
\mathbf{x}^{(p+1)} &= \mathbf{x}^{(p)}+\Delta\mathbf{x}^{(p)}
\end{aligned}
\qquad (15)
$$

In which, after p iterations $\mathbf{r}^{(p)}$ is the data residual ($\mathbf{r}^{(0)}$=$\mathbf{d}$), and $\Delta\mathbf{x}^{(p)}$ is the update of the approximate solution $\mathbf{x}^{(p)}$. ω is a relaxation parameter that can be set to a value between 0 and 2 (e.g. Van der Sluis and Van der Vorst 1987). We adopted a value of ω=1. (15) is the SIRT process as is described in Hager et al. (1985), although starting vector and ω are not specified in that paper and could be different.

We may clarify the proces (15) as follows (for the sake of simplicity we restrict the discussion to the cell part of the solution and for the moment identify $\mathbf{x}$ with $\mathbf{s}$). The data residual $r_i^{(p)}$ belonging to the i-the ray is backprojected along the raypath in a weighted manner as the time correction $A_{ij}r_i^{(p)}/\rho_i$ (ρ_i being the ray length), which is to be explained by $x_j^{(p+1)}$. The weight A_{ij}/ρ_i permits the slowness anomaly in those cells with larger ray intersections to explain most of the data residual. To allow $\mathbf{A}$ to be accessed by rows, the single-row contributions $A_{ij}r_i^{(p)}/\rho_i$ are accumulated in the computer memory as $\bar{r}_j^{(p)}=\sum A_{ij}r_i^{(p)}/\rho_i$. At the end, when all rows have been processed and $\bar{r}_j^{(p)}$ contains the total time correction. A reasonable estimate for the slowness correction $\Delta x_j^{(p)}$ is obtained by dividing $\bar{r}_j^{(p)}$ by the total length of all ray path segments intersecting cell j, i.e.

$$
\Delta x_j^{(p)}=\frac{\bar{r}_j^{(p)}}{c_j} \qquad (16)
$$

Rewriting $\Delta x_j^{(p)}$ as a simple average over all row contributions we obtain

$$
\Delta x_j^{(p)}=\frac{1}{N_j}\sum_i^M\frac{A_{ij}}{L_j}\frac{r_i^{(p)}}{\rho_i} \qquad (17)
$$

with N_j the number rays visiting cell j, i.e the cell hitcount and L_j the average length of ray segments in cell j, we may identify the i-th row update of the cell slowness in cell j, say $\Delta x_{ij}^{(p)}$, as

$$
\Delta x_{ij}^{(p)}=\frac{A_{ij}}{L_j}\frac{r_i^{(p)}}{\rho_i} \qquad (18)
$$

Since $A_{ij}/L_j\approx 1$, we can observe from (18) that, for a given residual $r_i^{(p)}$, the single row contributions to $\Delta x_j^{(p)}$ are of the same order for both well and poorly visited cells. However, the single cell solution update (17) is inversily weighted by the cell hitcount N_j. Especially in tomographic problems with strongly varying cell-hitcount, we expect that the weight $1/N_j$ will suppress the amplitudes $x_j^{(p)}$ in the well visited cells with respect to those in the poorly visited regions of the cell model. This will prove to be a useful observation later on. We may also observe from (18) that the single-row contribution $\Delta x_{ij}^{(p)}$ to (17) will, on the average, be systematically larger for shorter rays, due to the weight factor ρ_i

which biases $\Delta x_j^{(p)}$, and thus $x_j^{(p+1)}$, in the direction of short-ray contributions. Since there is no physical reason why this should be the case, this is an unwanted property of SIRT.

Indeed, van der Sluis and van der Vorst (1987) prove that the SIRT process (15) does not solve (8) but instead the reweighted least squares problem

$$\mathbf{R}^{-1/2}\mathbf{A}\mathbf{x}=\mathbf{R}^{-1/2}\mathbf{t} \tag{19}$$

They also prove that the least squares solution is minimal in the norm $\mathbf{x}^T\mathbf{C}\mathbf{x}$, i.e there is an additional scaling involved of the columns of $\mathbf{A}$ with $\mathbf{C}^{-1/2}$. Note that the LSQR algorithm seeks a solution of $\mathbf{Ax=t}$, subject to minimizing $\mathbf{x}^T\mathbf{x}$. The $1/c_j$ scaling serves the same purpose as the column scaling we apply in case of the LSQR algorithm (section 3), where we *explicitly* divide the columns of $\mathbf{A}$ with the square root of the cell volume in order to account for the differences in total expected ray path length in a cell. Note however, that in general the values of c_j may vary much more over the model than the cell volumes. Our results will show that the column scaling influences the solution. Since the x_j are expected to be of the same order of magnitude, the minimization of $\mathbf{x}^T\mathbf{C}\mathbf{x}$ implies that the SIRT process allows larger amplitudes to the solution in poorly visited cells (which are expected to be the least constrained) with respect to the solution in well visited cells. We already noted the same effect in our discussion of equation (17).

Solution damping can be achieved by adding a positive constant η to the column sum c_j. Physically this means that we try to explain the same time correction $\bar{r}_j$ in cell j by a larger total ray length $c_j+\eta$ (cf. 16), which serves the same objective as adding damping equations to (8). Mathematically, however, this way of regularization is quit different from the one discussed in section 3. This is most easily expressed by the fact that the SIRT-type damping affects every column of $\mathbf{A}^T\mathbf{A}$, whereas the damping discussed in section 3 only affects its diagonal elements.

Backprojection methods like the Algebraic Reconstruction Technique update the solution after *one* single row has been processed, with updates similar to (18). This causes the solution to depend on the order in which the equations are supplied to the algorithm. SIRT methods do not have this deficiency. The summation in (17), which is used to obtain the solution update, is independent of the order of the equations.

7. The experiment

Nolet (1985) compared LSQR to the SIRT variant, which is discussed by Ivansson (1983), in solving sparse tomographic systems. For a rather small numerical model (the matrix was 200×400) he found superior convergence properties for the LSQR method. However it remains to corroborate these results for matrices of more realistic size ($10^4\times10^5$) resulting from actual seismic tomography and different SIRT algorithms as well, in particular the method described in Hager et al. (1985) that is reported to have fast convergence as well (Clayton, personal communication 1985).

In this section we will do so by applying the LSQR and SIRT methods to the same data as used by Spakman (1985) in estimating the P-wave velocity heterogeneities of the upper mantle beneath Central Europe, the Mediterranean and the Middle East. We will first give an outline of the structure of the tomographic equations he used.

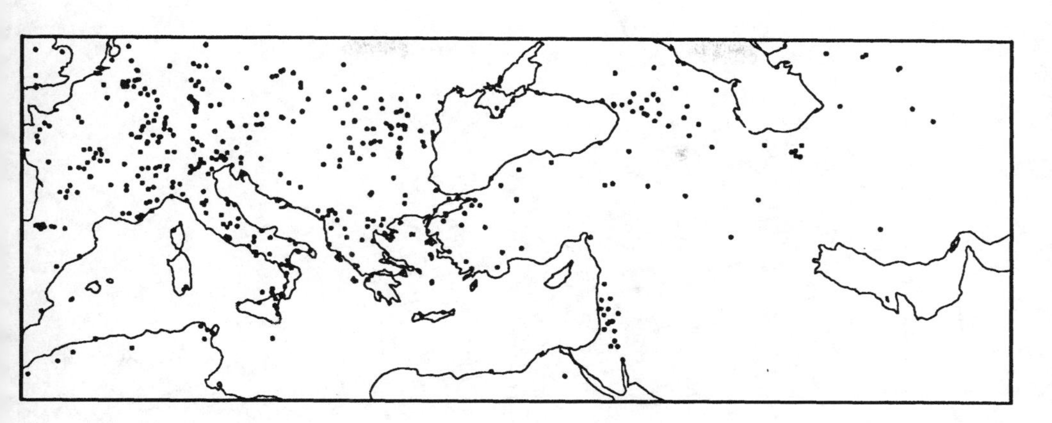

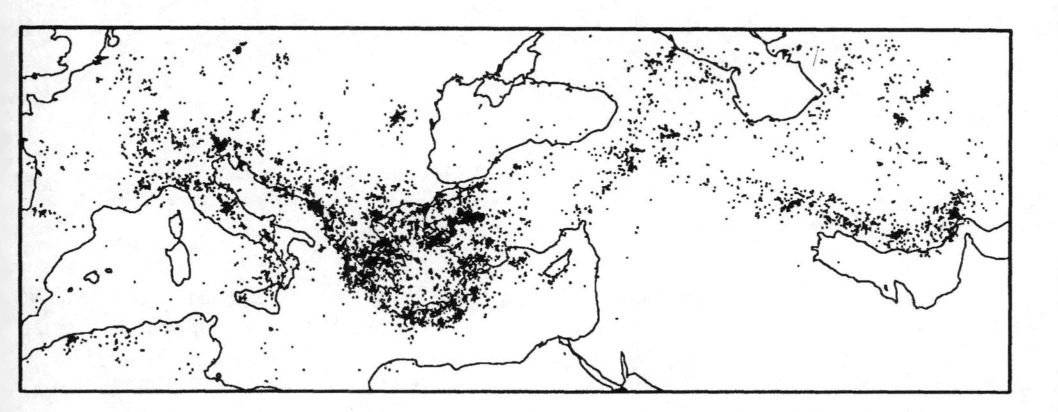

Figure 1. The geographical region beneath which the upper mantle is tomographically scanned. Top panel shows the regional stations and the bottom panel displays a subset (M > 3) of seismicity.

A cursory look at figure 1, displaying the locations of stations and events in this area, is enough to envisage the possibility of studying its upper mantle by seismic tomography. The region is well covered, although not ideally, with a large number of stations and exhibits a well spread and high seismicity. A selection of the ISC-bulletin tapes (years 1964-1982) for *regional* events that are observed by at least 10 stations, results in about 30,000 mainly weak and poorly determined events. Taking an absolute delay time threshold of 5 seconds, the ISC reports for these events 500,550 delay times, observed by 937 regional and teleseismic stations up to 90° epicentral distance.

The upper mantle cell model which is used to parameterize the slowness heterogeneties is shown in figure 2. It consists of 9 layers, which are subdivided in 52×20 approximately 1°×1° cells. The layer thickness increases with depth from 33 km for the first layer to 130 km for the last one, which reaches a depth of 670 km. The inclusion in (7) of 30,000 events would lead to the estimation of 120,000 unknown hypocentre coordinates, which are not expected to be well resolved. For reasons discussed in section 2, events were grouped in

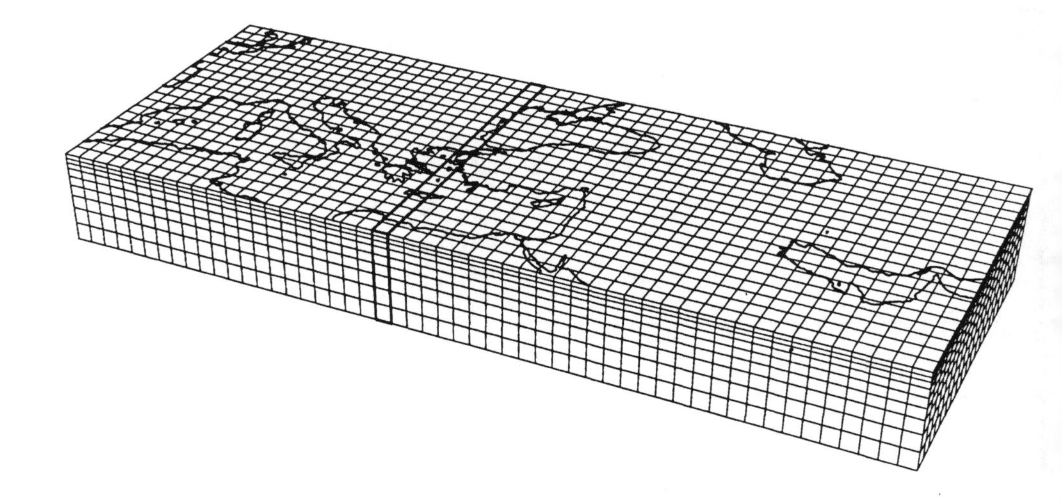

Figure 2. The cell parameterization of the upper mantle into 9 layers of increasing thickness. Each layer consists of 1040 $1^0\times1^0$ cells. The thick lines near the centre of the model indicate the upper mantle cross section which is subject of our discussion.

50×50×33 km cells and instead of relocating each single event, a cluster of events is relocated. This decreased the dimension of g to 10,484. Station corrections are estimated only for those stations outside the cell model for which no station delay time is reported by Dziewonski and Anderson (1983), which is the case for 226 stations. A 80 km thick cell is placed beneath these stations to account for the station slowness anomaly. In this way the total number of parameters amounted to 20,070.

A smoothing matrix **S** is chosen to apply only smoothing within a horizontal layer (retaining optimal depth resolution), in a way that the slowness heterogeneity in a cell is required to correlate by 20% of its amplitude with the solution in every adjacent cell. The raypaths are computed from the Jeffreys-Bullen reference model for P-waves, using the ISC hypocenter locations. Finally a row condensing operation is applied by composing every row of $\mathbf{D\hat{A}}$ in (7) by at most 4 rays coming from a cluster of events and arriving at the same station. This reduced the number of rows in (7) to 292,451.

The solution to this system of tomographic equations has been approximated by Spakman (1985) in 16 iterations using the LSQR algorithm (see Spakman, 1986a, 1986b for some of the upper mantle results). He also used sensitivity tests with synthetic velocity models to estimate the resolving power of the inversion procedure. We will use this application of (7) in our experiments to study the algorithm performance and spatial resolution. First we have to make some introductory remarks.

In section 2 we briefly discussed the problems arising from reference model differences with respect to the actual 'average' Earth. The most important error introduced by an inadequate model is spatial bias (i.e. displacements with respect to the actual location) of the inferred heterogeneities. We use the JB model as background model. The model is characterized by the absence of a low velocity layer and first order discontinuities in the upper mantle. The model bias will prove to be relevant to our results only in the case that we are discussing the spatial resolution obtained when inverting the *real* data.

In solving systems like (7), it is important to realize that *how* an algorithm performs is quite independent of whether the system of equations is an adequate description of the physical reality or not. All shortcomings mentioned in the previous sections will result in a tomographic image that differs from actual slowness heterogeneity, but this does not affect the algorithm performance. On the other hand, each numerical method is afflicted with its own shortcomings, bringing errors to the solution which we wish to investigate.

Merely applying LSQR or SIRT to invert real data, we do not obtain a measure of the spatial resolution. Sensitivity tests with synthetic velocity models and data can partly help to overcome this problem. With a known model $\hat{\mathbf{x}}$ we can compute artificial data $\mathbf{t}$ by computing the forward problem $\mathbf{A}\hat{\mathbf{x}}=\mathbf{t}$. We may also add an error vector $\boldsymbol{\varepsilon}$ to the data and then solve the inverse problem $\mathbf{Ax}=\mathbf{t}+\boldsymbol{\varepsilon}$ for $\mathbf{x}$. By performing the same sensitivity test on both algorithms we obtain, what we will call, inversion responses $\mathbf{x}_{lsqr}$ and $\mathbf{x}_{sirt}$, which can be compared to the synthetic model $\hat{\mathbf{x}}$ for spatial resolution analysis, and it also permits us to draw conclusions about the performance of the algorithms.

For reliable resolution estimates we need to devise sensitivity tests that mimic the actual data case. A restriction we can impose is that we obtain the same relative variance reduction as in the *real data* case, i.e. $|\mathbf{r}^{(p)}|/|\mathbf{t}|$. The maximum allowed relative drop of the variance is reached when $|\mathbf{r}^{(p)}|$ is roughly equal to the expected length of the data noise vector, since the number of degrees of freedom that we allow in the solution is small with respect to the number of data. A better fit is an indication that the algorithm is matching the model to the data errors. For the construction of our sensitivity tests this implies that we can mimic the real data variance reduction best (in a conservative sense) by adding as much errors $\boldsymbol{\varepsilon}$ to the synthetic data $\mathbf{t}$ as are needed to create the same drop in variance reduction as in the real data case, i.e. $|\boldsymbol{\varepsilon}|/|\mathbf{t}+\boldsymbol{\varepsilon}| \approx |\mathbf{r}^{(p)}|/|\mathbf{t}|$. A second requirement is that $\hat{\mathbf{x}}$ must be chosen in such a way that it enables us to derive conclusions which are also applicable to the real data case. We used cell-spike and (quasi) harmonic sensisitivity models, which, as we will discuss, serve this purpose.

We include the relocation and station parameters in the tests since otherwise the inversions cannot be representative for the actual inversion response. We set these variables to zero values. In this way we can study how much of the synthetic data (strictly caused by the cell-anomalies) is explained by the event relocation and station delays after inversion of the synthetic data.

In most cases we will show inversion results obtained after 16 iterations. The primary reason for stopping after 16 iterations is that the tests are expensive in computations with large systems. With the matrix elements stored on disc, $O(100$ Mbytes$)$, the time needed for I/O operations is at least as important as the CPU time. We performed our computations on a CDC 855 mainframe computer. Per iteration the required amount of computing time is 120 CPU seconds and 500 I/O seconds for LSQR and for SIRT: 100 CPU seconds and 250 I/O seconds. Our implementation of LSQR needs a double amount of I/O because two separate matrix-vector multiplications, one with $\mathbf{A}$ and one with $\mathbf{A}^{\mathrm{T}}$, are performed per iteration, but this excess I/O can be avoided by more efficient programming. For SIRT the

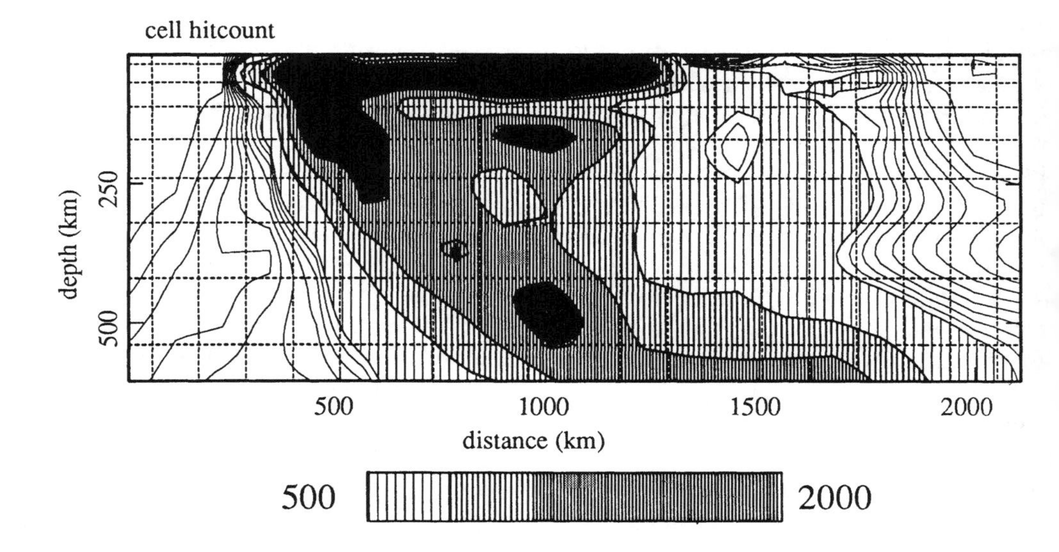

Figure 3. Cell hitcount. The line contouring (in increments of 50) indicates the regions of poor illumination. The shading denotes the well visited cells and the black areas denote those cells with hitcounts between 2000 and 6500. Dash-dotted lines represent the cell division in the cross section. Notice that the outline of the cross section runs trough the centers of the edge cells.

multiplications with **A** and $\mathbf{A}^T$ is performed simultaneously with only one disk access per row.

The second reason is that in the first 10 iterations both algorithms have already reached about 90% of the total variance reduction that is achieved after 16 iterations, which indicates that the residual after 16 iterations is near the noise level of the data. The third reason is that in the absence of damping, especially for SIRT, solutions in some cells are already developing large heterogeneity contrast with adjacent cell solutions after 7 iterations, which is indicative for an unstable solution. Most of our results have been obtained without explicit solution damping, because only in that case artifacts of the inversion can be clearly revealed.

Convergence of an iterative least squares process is commonly measured by the reduction of the data residual (variance reduction), $|\mathbf{t}-\mathbf{Ax}|$ as a function of the iteration number. This is the quantity which is minimized by LSQR. The SIRT method minimizes the residual $|\mathbf{R}^{-1/2}\mathbf{t} - \mathbf{R}^{-1/2}\mathbf{Ax}|$ (section 3). Yet the *model fit* to the actual (univariant and otherwise not reweighted) data in terms of the variance reduction is best expressed by $|\mathbf{t}-\mathbf{Ax}|$ and therefore we use this quantity to denote the data residual.

Since the number of unknowns is very large we will restrict our discussion of the inversion results to the cross section of the cell model which is indicated in figure 2. As we will see this cross section exhibits all the characteristics which one may encounter in seismic tomography. Note that all the results are obtained from the 3D inversion.

One of the first things to consider in seismic tomography is how adequate the cells are visited by seismic rays, because that gives qualitative information on how well constrained the cell solutions are. We can get an impression of the cell illumination by seismic rays by comparing the distribution of station locations with respect to epicentres in figure 1. We observe that the ray illumination is not ideal since in most of the event regions the density of the station distribution is poor. The best visited cells are located beneath Central and South-East Europe, where we may expect to obtain the highest spatial resolution in the model. On the other hand, the edges of the model are hardly visited by rays and consequently the cell slowness anomalies in these regions are poorly constrained.

The illumination of cell j can be described in terms of the cell hitcount N_j, defined as the number of rays intersecting cell j. For the cells in the cross section the hitcount is shown in figure 3. Both poorly visited (line contoured) as very well visited regions (hatched) are present, in fact the cross section exhibits a strongly inhomogeneous cell hitcount. We can also recognize preferred directions in ray illumination, even in this region which is relativily well covered by seismic rays, with respect to other areas. In this figure, as in all that follow, we use a bilinear interpolation of the function displayed, that connects the function at grid point values defined at the center of the cells.

8. Real data inversion

Figure 4 displays the estimates of the velocity anomalies inferred from the inversion of the actual P-delay times with both algorithms. The shading denotes anomalies in percentages of the corresponding reference mantle velocity. Positive values (cross-hatched) indicate relative high velocity regions. White areas represent either values between -0.1% and 0.1% or unresolved cells. The black shading indicates values that are beyond the limits of the contour scale.

Both images posses the same main features of which the most prominent one, the strongly dipping high velocity anomaly, is interpreted as the Aegean slab (Spakman 1986a). The differences are largest where the hitcount is low. The LSQR solution shows small or zero amplitude anomalies corresponding to the poorly visited cells (to the left and right in figure 4, except for the anomaly in the middle right). whereas in the large hitcount areas of the cross section the heterogeneities exhibit modestly varying anomalies between -3% to +3% In the upper left and to the right in the cross section, the SIRT image exhibits a correlation between small hitcount and large heterogeneity contrast, which are on the order of 7-10% over distances as small as 100 km. Although this may not be unrealistic from a geodynamical point of view, we have to explore the possibility that these contrasts indicate an unstable solution in the corresponding cells. The anomaly amplitudes in the well visited cells are systematically smaller, not only with respect to those obtained in the ill-illuminated part but also with respect to corresponding anomalies in the LSQR solution.

Some characteristics of the inversions are listed in table 1, in which, where relevant, units are in seconds and σ denotes standard deviation, e.g. σ_r is the standard deviation in the residual $\mathbf{r}^{(p)}$. The other quantaties have already been defined in section 7. From table 1 we see that after 16 iterations the algorithms achieved about the same variance reduction of about 13%. The differences between the LSQR and SIRT image lead to the observation that evidently variance reduction alone is not a useful measure for judging the reliability of a tomographic image.

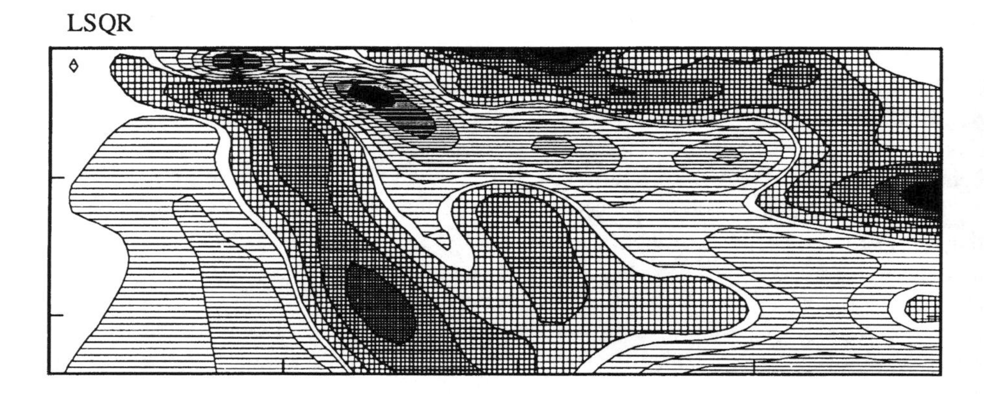

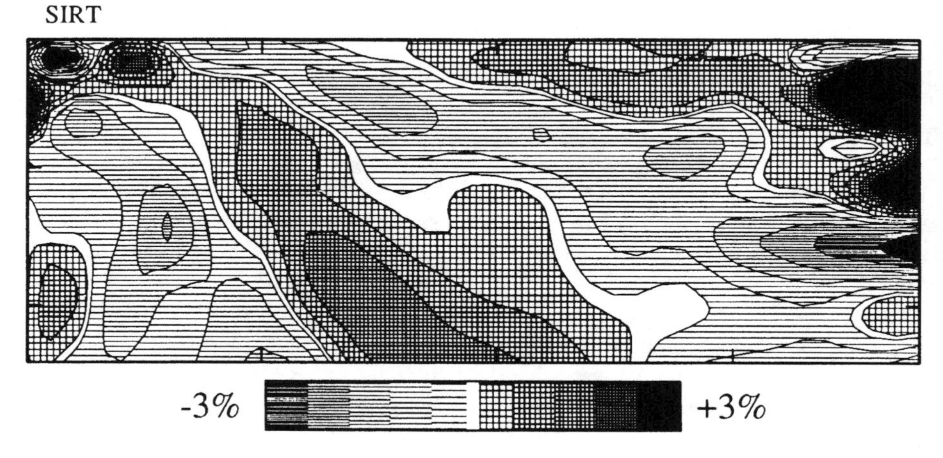

Figure 4. The velocity anomalies in the cross section obtained from the inversion of the P-delay time data after 16 iterations for LSQR (top) and SIRT (bottom). The shading denotes velocity anomalies in percentages relative to the Jeffreys-Bullen mantle velocity. Low (high) velocity regions are (cross) hatched. Black regions indicate anomaly amplitudes beyond the scale limits. The geodynamocal significance of the anomalies is discussed in Spakman 1986a.

A conclusion we can draw at this point is that LSQR treats (at least in the first 16 iterations) the well visited cells with a higher priority, or, to state this differently, implicitly damps the solution in the ill-illuminated cells. In the SIRT solution, the amplitudes are all of are of about the same size, i.e on the order of 0.5-1.5%, apart from the possible instability of the solution in the upper left corner and to the right side of the cross section. More conclusions are not warranted because we do not know the *actual* velocity heterogeneity, hence we cannot know if the LSQR or the SIRT solution is closest to reality and we have so far not been able to establish the reliability of the inferred anomalies.

Table 1

test/data	p	σ_{noise}	$\lvert\varepsilon\rvert$	$\lvert t+\varepsilon\rvert$	$\sigma_{t+\varepsilon}$	LSQR / SIRT $\lvert \mathbf{r}^{(p)}\rvert/\lvert t+\varepsilon\rvert$	σ_r
real data	16	-	-	821	1.52	0.871 / 0.869	1.32 / 1.32
spike 1	16	0	0	210	0.388	0.580 / 0.570	0.225 / 0.221
spike 2	16	0.5	270	342	0.632	0.854 / 0.850	0.540 / 0.538
harmo 1	16	0	0	396	0.732	0.270 / 0.280	0.197 / 0.205
harmo 2	16	1.3	702	806	1.49	0.870 / 0.866	1.30 / 1.29
perm. data	10	-	-	821	1.52	0.979 / 0.973	1.49 / 1.48
harmo 3	16	1.3	702	806	1.49	0.868 / 0.871	1.29 / 1.30
harmo 4	16	1.3	702	806	1.49	0.861 / -	1.28 / -
harmo 4	7	1.3	702	806	1.49	0.868 / -	1.29 / -

Especially the strong correlation between the slab geometry and the preference in ray directions raises the question if the slab actually exists or that the image is merely due to lack of spatial resolution. We have to seek refuge to sensitivity tests in order to explain what is observed.

9. Cell-spike test

Cell-spike models are characterized by zero slowness except for a few cells, which all have the same deviation in slowness value (apart from a possible difference in sign). If $\hat{\mathbf{x}}_{spike}$ consists of only one cell with non-zero slowness anomaly, the solution $\mathbf{x}_{spike}$ is called the cell response function, which is in fact a column or a row of the resolution matrix $\mathbf{\Pi} = (\mathbf{A}^T\mathbf{A})^{-G}\mathbf{A}^T\mathbf{A}$, where '$G$' denotes generalized inverse. In a problem using only a few cells and few data it would be worthwhile computing $\mathbf{\Pi}$, but in our problem this is impractical. However as we shall see later on, cell response functions are quite sharply peaked around the anomalous cell, and have much smaller amplitudes in the rest of the model. By distributing a number of well separated cell-spikes over the model it is possible to approximate the central part of many of the response functions in one inversion run. Cell response functions then give a proper estimate of the image blurring caused by the combined effects of model, data and algorithm imperfections. In practice they indicate the maximum detail that can be resolved in large contrast heterogeneity models.

The employed cell-spike model consists in every layer of the cell model of 40 equidistantly spaced cell-spikes. In adjacent layers the pattern is shifted in both horizontal directions by half of the spacing between the spikes. In this way the cell-spikes are always separated by at least 2 empty cells within a layer and by at least one in vertical direction. A 5% velocity contrast with respect to the surrounding upper mantle velocity is attributed to the spikes. The amplitude alternates in sign from one spike-cell to another. Inversions have been performed of the exact synthetic data and also of the same data with the addition of a realistic amount of normal distributed random noise. The results for the cross section are shown in figures 5a-d and table 1 (i.e. tests spike 1 and spike 2, resp. noise-free and noise

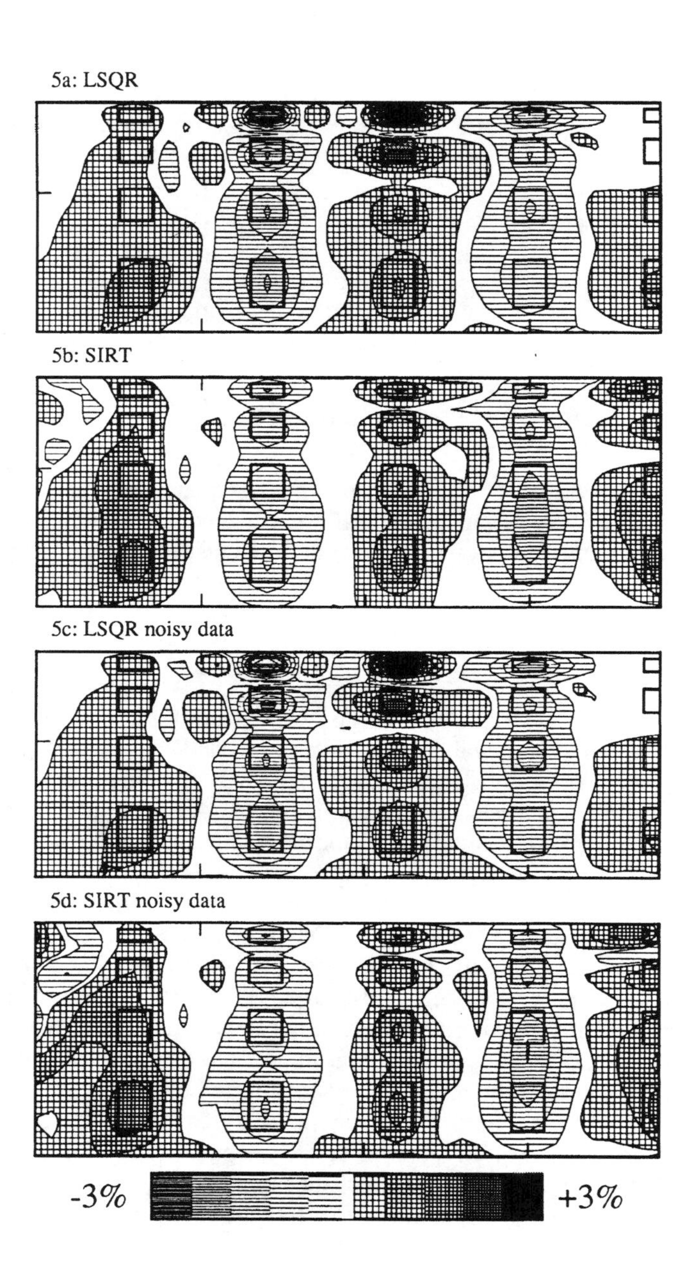

Figure 5. The inversion response to the spike sensitivity test. Rectangles denote the location of the cell-spikes to which a 5% velocity anomaly contrast is attributed with respect to the surrounding upper mantle velocity. Shading as in figure 4, but with a different scale. 5a-b; inversion result using exact synthetic data for LSQR (a) and SIRT (b). 5c-d; same as 5a-b but with noise added. See table 1 for details (tests spike 1 and 2).

added). The rectangles in figure 5 indicate the true position and spatial size of the cell-spikes.

In the noise-free case LSQR is able to resolve 30-40% of the input spike amplitude in the lower central part of the cross sections and up to 70% in the upper central area. If we take the half-width of the cell response function as a measure of spatial resolution then single cell anomalies are reasonably well resolved. Notice that a small bias exists in the location of the central peaks with respect to the actual location. This is a consequence of the preference in ray direction, as are the elongated shapes some of the cell response functions. The blurring of the image is primarily due to the imperfect cell illumination, the applied smoothing and the finite number of iterations. The ill-illuminated cells are either hardly resolved or not resolved at all, but the sign of the input anomalies is always recovered (except for the upper right corner, where LSQR has damps the solution to zero). At the top in figure 5a, between the first and second 'column' of spikes, small anomalies exist that are side lobes of responses belonging to spikes located outside the cross section.

The SIRT solution in the perfect data case (fig. 5b) is only able resolve 15-30% of the amplitude contrasts and achieves a poorer spatial resolution than LSQR. To the lower right a clear example exists of two spikes that cannot be separated after 16 iterations, and to the upper right we can observe a slight dislocation of the cell anomaly with respect to the actual cell-spike location. Note that $\mathbf{x}_{sirt}$ has a better amplitude response than $\mathbf{x}_{lsqr}$ in the poorly visited regions of the cross section.

The addition of Gaussian noise with σ=0.5 s to the synthetic data (table 1) results for both methods in a slight decrease in the matching of amplitudes in the well visited cells, but does not dramatically influence the spatial resolution (figures 5c-d). We can amend this by applying a proper amount of explicit damping to the solution, this will be shown when we discuss the results of the harmonic sensitivity test.

10. Harmonic model test

The harmonic sensitivity model consists in every layer of a superposition of two sine functions mimicking a 2D harmonic slowness anomaly pattern with wavelengths of 6 cells in both lateral directions and maximum amplitudes of -3% and +3% relative to the surrounding upper mantle velocities. In the adjacent layers the pattern is shifted by half a wavelength in order to obtain a 3% anomaly discontinuity at layer interfaces. The wavelength of 6 cells (660 km) has been choosen in order to obtain resolution estimates for heterogeneities of a length scale typical for complicated upper mantle regions (in a geodynamical sense).

The inversion results for the noisy data case are presented in figures 6a-b and table 1 (i.e. test harmo 2; test harmo 1 gives the numbers on the noise-free case). The contour lines indicate the true location of the -1.5% and +1.5% velocity anomaly amplitude levels of the harmonic model.

The same amplitude and resolution characteristics are observed as in the spike test. In addition both solutions show a larger convergence to the actual harmonic model, i.e. amplitudes are better resolved than in the spike test, but again the resolving power of the LSQR method is much higher in the well illuminated area. In fact in the upper part the spatial resolution is to nearly single cell precision. In the *poorly* visited cells SIRT has the

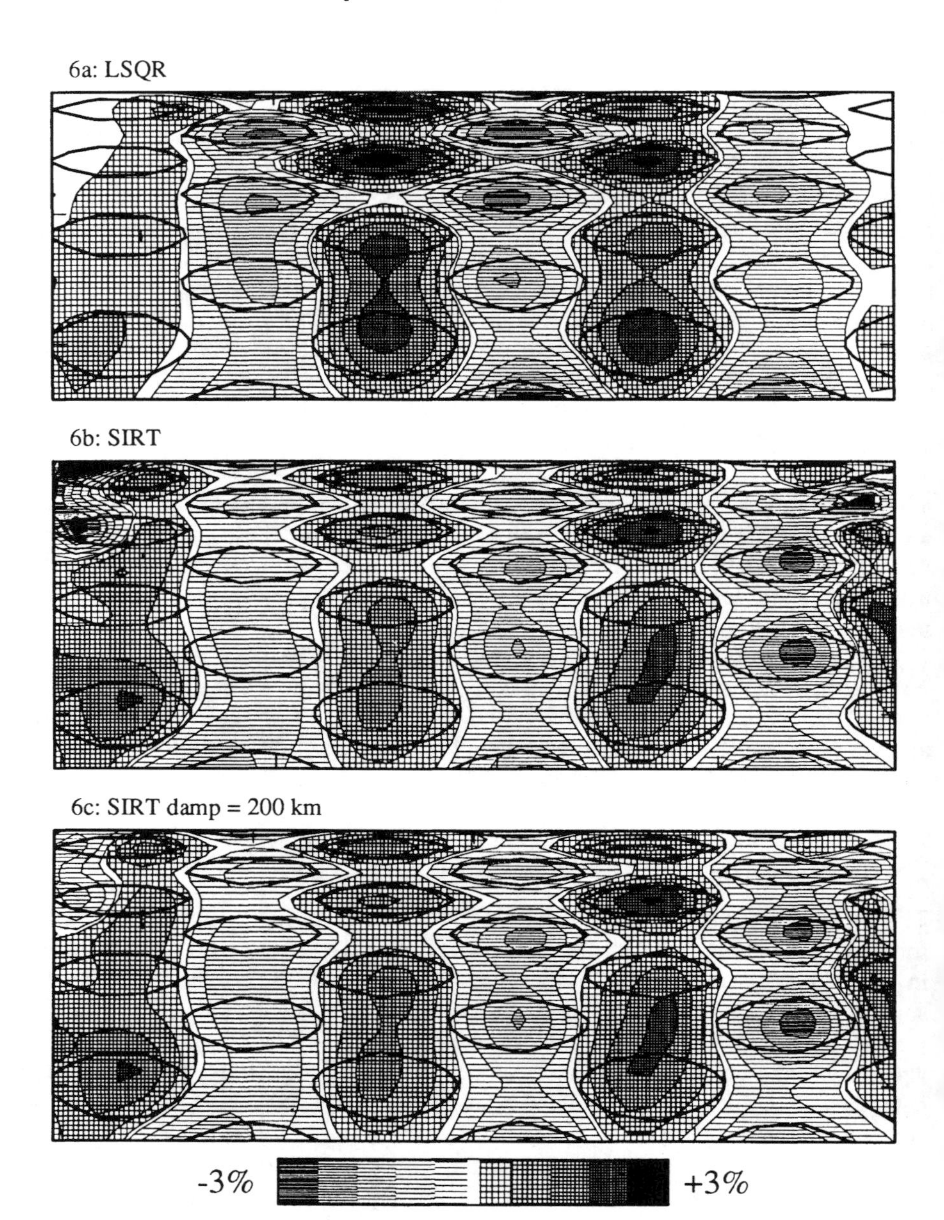

Figure 6. The inversion response to the harmonic sensitivity test using noisy data. Contour lines indicate the -1.5% and +1.5% amplitude level of the harmonic model, which has -3% and +3% maximum amplitudes and a wavelength of 6 cells. The pattern is shifted by half a wavelength in adjacent layers. Shading as in figure 4, but with a different scale. 6a-b inversion results for LSQR (a) and SIRT (b). See table 1 for details (harmo 1 and 2). 6c SIRT solution with explicit damping; 200 km ray path length is added to the c_j.

better average amplitude response. Notice however the irregular large amplitude anomalies in the upper corners of the cross section. A comparison with the noise-free results (not shown here) shows that the size of these amplitudes is caused by the added noise. Also in the spike test we observed the sensitivity of the SIRT solution to noise in the poorly illuminated cells. In the harmonic test this effect proves to be much more dramatic. Figure 6c shows the SIRT inversion response when a small amount of damping is applied. We added 200 km of ray length to the total ray path length in every cell. The large amplitudes in the upper corners have decreased to an acceptable value, but the error in the location of the anomaly in the upper right corner still remains. Note that the damping only affects the anomalies in the poorly illuminated cells.

Although we scaled the x_j to approximate uniform variance prior to inversion, evidently LSQR accepts a bias toward zero anomalies for poorly visited cells, constructing a solution primarily in the well visited cells (corresponding to the colums of **A** with large column sums), i.e. in the first 16 iterations. SIRT however attempts to compensate for the column bias, by scaling the columns of **A** with $\mathbf{C}^{-1/2}$, but at the expense of an increase in the variance of the solution, especially visible in the upper right and left corners of the cross section.

In order to obtain a convenient image of the spatial resolution and amplitude response we computed a measure of the fit f_j in the vicinity of cell j between the harmonic sensitivity model $\hat{\mathbf{x}}$ and the inversion response $\mathbf{x}$ defined as

$$f_j = 1 - \frac{\left((\hat{\mathbf{x}}-\mathbf{x})^T \mathbf{W}_j \, (\hat{\mathbf{x}}-\mathbf{x}) \right)^{1/2}}{(\hat{\mathbf{x}}^T \mathbf{W}_j \, \hat{\mathbf{x}})^{1/2}} \tag{18}$$

in which $\mathbf{W}_j$ is a *three*-dimensional smoothing window centered at cell j, that rapidly falls off to the sides as 0.5^k, with k the number of cells separating a cell from cell j. $\mathbf{W}_j$ serves as a penalty fuction that weights spatial errors (i.e errors in the location of the anomaly). Note that the fit function f_j equals 1 if $\mathbf{x}=\hat{\mathbf{x}}$ and equals 0 for a 100% amplitude misfit. Figure 7 displays the results. Clearly LSQR exhibits superior convergence to the actual solution. Only in the lower left and middle right the SIRT response is slightly closer to the actual synthetic model. The model fit of the damped SIRT solution is somewhat improved in the upper corners, but is still poor. The misfit (i.e. $1-f_j$) is the combined effect of amplitude errors and spatial errors and hence it does not represent a purely spatial error estimate. Notice that the reason for large model misfit is quite different for LSQR or SIRT. LSQR exhibits small or, zero amplitudes in the poorly visited regions, wheras the SIRT solution has very large amplitudes in these regions, both leading to misfits on the order of 100%. By comparing figures 6a and 7a we may tentativily attribute a spatial error of half the cell size to a fit of 60-70%. In the region of poor fit, say 10-20%, the fact that the sign of the anomalies is always resolved (fig. 6a) indicates a spatial error on the order of twice to three times the cell size. The central part of the cross section is expected to be resolved with a spatial error varying from 50 to 120 km.

11. Permuted data test

How do we assess a correlation between the delay times and the variable ray coverage? If we lack independent evidence on the existence of the inferred anomalies obtained from

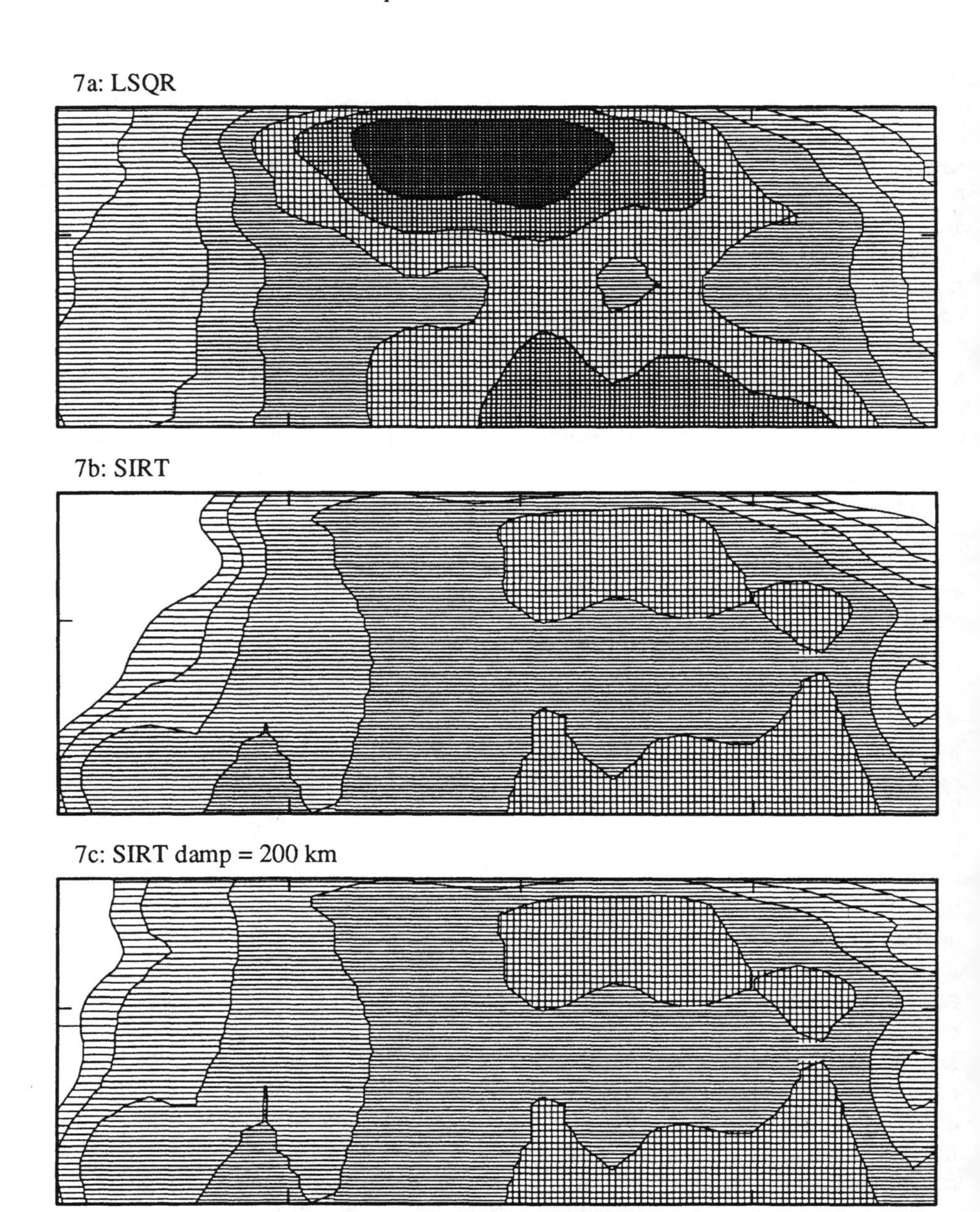

Figure 7. Model fit for harmonic sensitivity test using noisy data. a- LSQR, b- SIRT, c- SIRT with damping. The shading is proportional to the fit. White areas indicate *misfits* of more than 100%.

seismic tomography, the following test can be helpful. We may destroy the correlation of $\mathbf{t}$ with the physical structure, while retaining all statistical properties of the data t_i, if we randomly permute the data, obtaining $\mathbf{t}_{pm}$, but keeping the order of the rows of $\mathbf{A}$ fixed. The next step is to invert $\mathbf{A}\mathbf{x}_{pm}=\mathbf{t}_{pm}$. If $\mathbf{x}_{pm}$ exhibits strong correlations with $\mathbf{x}$, (the solution obtained from the actual data case) we cannot be sure of anything, since the problem now depends on the type of structure we invert for (in the case $\mathbf{x}$ is a very smooth or constant velocity model possibly $|\mathbf{x}|\approx|\mathbf{x}_{pm}|$). However if we obtain, with respect to the original problem, only a small variance reduction and $|\mathbf{x}_{pm}|\ll|\mathbf{x}|$ we may conclude that there exists a significant correlation between cell illumination and the data and hence the solution $\mathbf{x}$ can contain significant information on the Earth.

We will briefly discuss the results of the permuted data inversion. These are displayed in figure 8 and table 1. Only 10 iterations are performed on these very 'noisy' data. The LSQR algorithm behaves very well, showing only modest amplitudes after a 2.1% reduction of the residual. The undamped SIRT solution seems to be very sensitive to the 'noise' after a 2.7% variance reduction, especially in the poorly visited cells. We remark that dominance of positive velocity anomalies indicates that the algorithms try to explain the mean delay time (-0.034 seconds) of the data set. This also attributes to the bulk of the observed variance reduction. The lack of correlation between figure 8 and figure 4 and the small decrease in residual indicate that the actual delay times correlate well with the cell illumination and hence there is significant 'signal' in the data.

12. The Aegean slab

After performing all of the above tests we can answer the question: does the slab in figure 4 actually exist? First of all it follows from the permuted data test that the anomalies in figure 4 result from existing heterogeneities and not from pure noise. Secondly it follows from the sensitivity tests that existing anomalies are resolved in the well visited cells of the cross section. Since the LSQR results demonstrate superior convergence to the sensitivity models in the well illuminated region we only will consider figures 5c and 6b and 4 to obtain the answer. To make the comparison easier we plotted the outlines of the high velocity anomalies of figure 4 on top of the harmonic sensitivity test results. This is shown in figure 9. Clearly the sign of the slab anomaly is well resolved. As for the precise geometry we cannot be certain. For example, the upper left part of the slab is not well resolved. The entire outline of the slab may by biased due to the preferent ray directions, this we can conclude from the elongated shapes of some of the harmonic responses. But considering the heterogeneity that *can* be resolved (e.g. harmonic, spike model) we can conclude that the slab is present in the Aegean upper mantle and that the image may be accurate up to say a 50-100 km spatial error.

The above argument holds if the JB-model introduces negligible model bias with respect to the actual 'average' Upper Mantle model in the Mediterranean-Central European region, and if the actual amplitude of the slab anomaly is not much larger than what is resolved (1-2 %), since that would deflect the actual ray geometry away from the reference raypaths. Both problems are difficult to assess and if we realise that the 'best' reference model (in view of the entire 3D cell heterogeneity solution) is a model that can only be

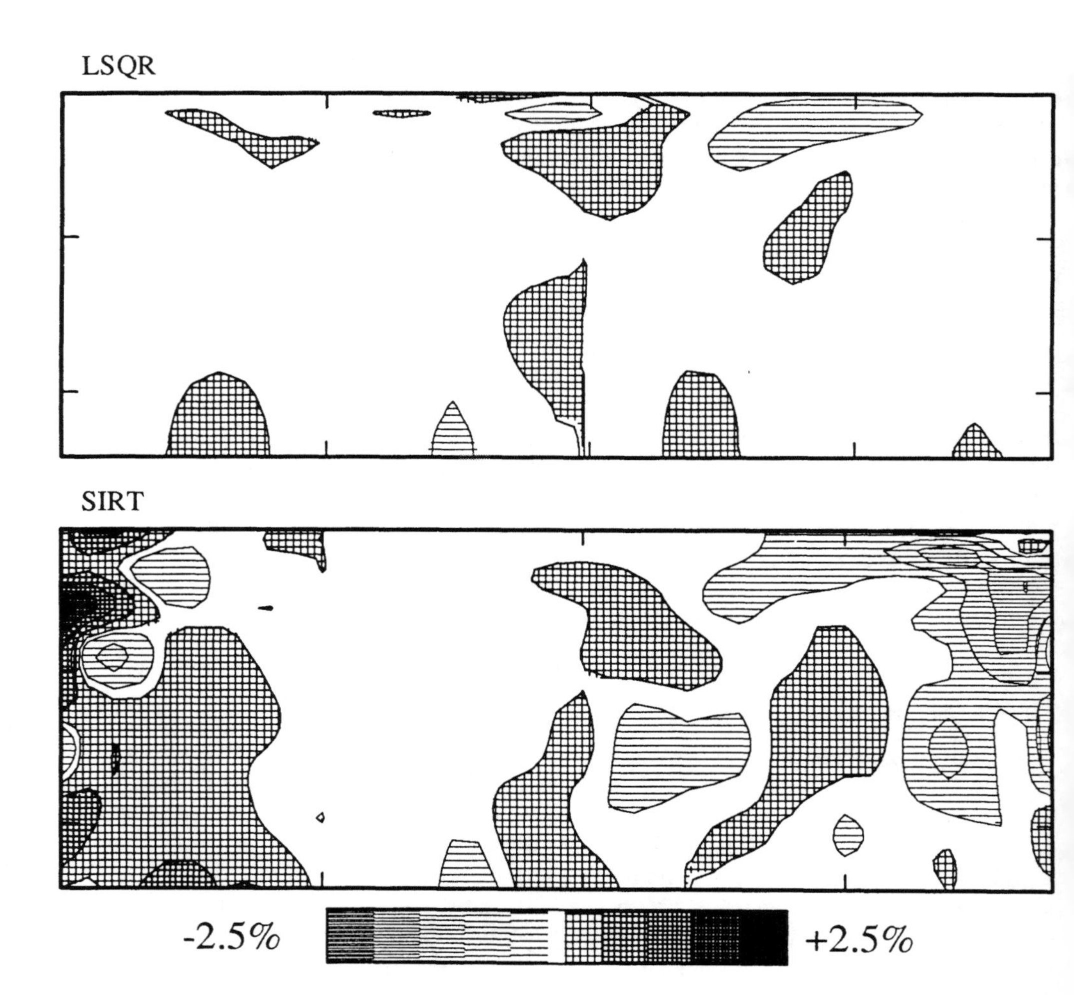

Figure 8. Permuted data inversion results after 10 iterations. LSQR (top) and SIRT (bottom). See table 1 for details. Shading as in figure 4, but with a different scale.

significant locally, solving these problems thoroughly calls for fast and efficient 3D ray tracing algorithms, that can deal with $O(10^6)$ rays in a reasonable amount of computation time.

13. Event and station corrections

In all inversions the values for the trade off controlling parameters μ and κ (section 3) were fixed to 0.5 and 2 respectively. In inverting the *real* data we obtained station corrections on the order of 0.5 seconds and event (cluster) corrections on the order of 5 km, using these values. Recall that we set the amplitudes of **g** and **h** to zero values in the sensitivity tests. The inversion responses of these parameters in the performed tests showed that their

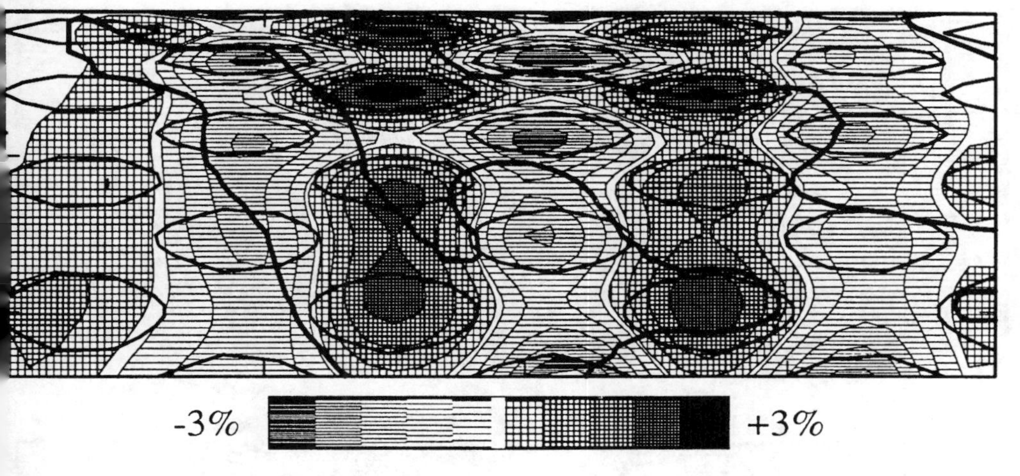

Figure 9. Figure 6a with superimposed the outlines of the high velocity anomalies of figure 4.

estimation is not independent of estimating the cell heterogeneities. The station corrections achieved amplitudes of ~ 0.05 seconds, in the harmonic test (with the addition of noise), which is a small response, but the relocation parameters obtained amplitudes of several kilometers. The amplitudes show a similar behaviour with respect to the the number of rays contributing to the estimation of one event parameter (i.e. event hitcount), as we observed for the cell heterogeneities. For SIRT the poorest constrained solutions exhibit the largest amplitudes. For LSQR this is just the other way around. On the average SIRT attributes somewhat larger amplitudes to the event corrections.

Suppressing **g** by taking a larger value of μ results in a small increase of the magnitudes of the cell slowness anomalies, but the influence of the data errors, especially for SIRT, becomes more visible, since the errors are less allowed to be absorbed by the (on the average ill-constrained) event relocations. For LSQR and SIRT we observed that this scaling will primarily influence the overall absolute magnitude of the heterogeneities, but hardly the shape of the image itself.

We also performed harmonic sensitivity tests *without* inverting for the relocation and station corrections. Note that we use the same synthetic data as in the other harmonic tests, since in the forward calculation of the synthetic delay times **g** and **h** are zero vectors. Figure 10 shows the model fit that can be obtained when the estimation of **g** is omitted from the inversion scheme. There is a general increase in accuracy and spatial resolution in the results of both algorithms when compared to figure 7. In spite of this increase we do *not* favour the neglection of **g** and **h**, since their inclusion is physically required. The expansion of the linear system by including **g** and **h** has two effects: (i) for as far as the added columns of the matrix are independent, we observe that the expansion results in a decrease of the amplitude contrasts between adjacent cells leading to smoother anomalies and (ii) we observe an average decrease in slowness amplitudes. This decrease is caused by the

10a: LSQR

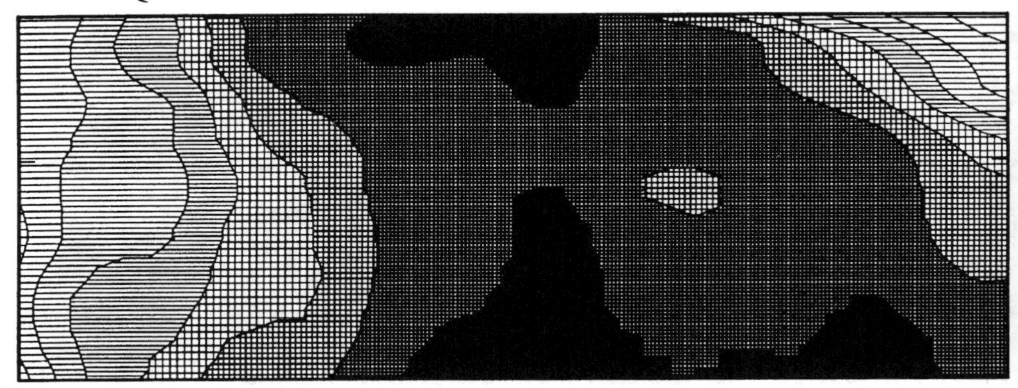

10b: SIRT

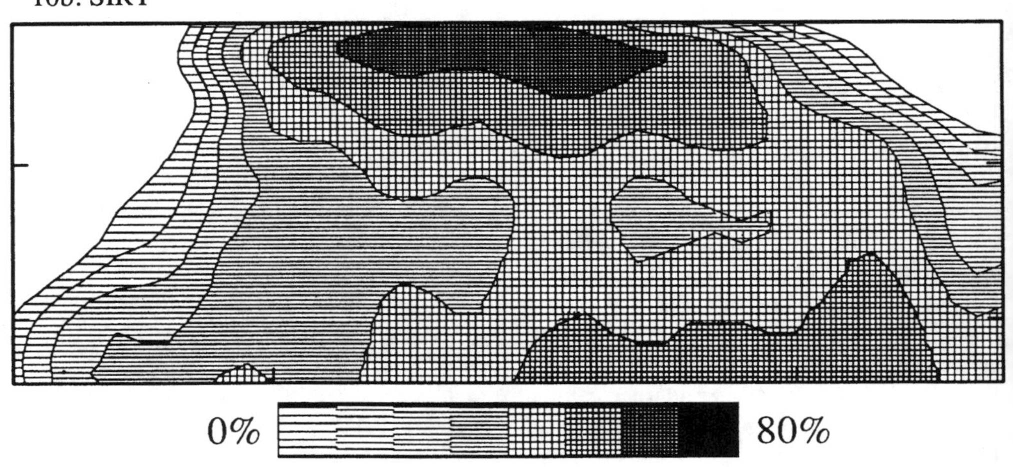

Figure 10. The model fit of harmonic sensitivity tests in which the estimation of relocation and station corrections are deleted from the inversion scheme. Synthetic noise has been added to the data. 10a: LSQR response , 10b: SIRT response, both after 16 iterations. See also table 1 (test harmo 3).

removal of 'signal' from the delay times for the estimation of **g** and **h** and by the lack of independency of the columns of **A** belonging to **g**, **h** and and the slowness vector **s**, respectivily. We remark that the corrections for one event are determined from only $O(20)$ rows of the matrix equation. Because of this we expect that inconsistencies in the entire data set are easily absorbed by the estimation of the components of **g**, i.e. in a way many of the relocation parameters are used as 'waste baskets' for data errors, which has a stabilizing effect on the estimation of the better constrained slowness anomalies.

14. Discussion

Our experiments demonstrate that the LSQR method exhibits superior convergence toward the actual solution in large parts of the cell model in a realistic tomographic problem. Only in some poorly illuminated areas the SIRT solution tends to have a somewhat better amplitude response. LSQR proves, at least in the first 16 iterations, to be much more stable with respect to large data errors. An advantage of the SIRT method is that we also obtain a solution in the poorly visited regions of the cell model, where LSQR accepts a bias toward zero amplitudes. These differences in algorithm performance are not reflected in the achieved variance reductions after 16 iterations, which, in the tomographic problems that we considered, are of comparable size for both methods.

We can explain the performance of LSQR by the way it constructs the solution. In the first iterations the algorithm starts with estimating only those contributions to the solution that belong to larger eigenvalues λ_n of $\mathbf{A}^T\mathbf{A}$ (van der Sluis and van der Vorst 1987). While iterating, eventually more and more smaller eigenvalues enter the solution. This is also reflected by the condition number of $\mathbf{T}_p$, equation (13). The condition number gives for every iteration the ratio between the largest and the smallest λ_n that entered the problem. In the harmonic test, with the inclusion of noise, this number increases from 1 in the first iteration to 45 when p=16. The same increase is observed when inverting the real data. Although the condition number is only a relative measure, it indicates that the smallest eigenvalues did not enter the problem in the first 16 iterations when solving *our* problem (8). Hence the variance in the solution components x_j is only due to a small eigenvalue range and is not expected to vary dramatically over the model. Note that in general this does not imply that the errors cannot be large, since this depends largely on the smallest eigenvalue that entered the problem and the noise in the data.

Data errors combined with the ill-conditioning of $\mathbf{A}^T\mathbf{A}$, which is caused by the (nearly) linear dependency of the columns of this matrix, add small eigenvalues to the problem. Since LSQR neglects the smaller eigenvalues in the first iterations the ill-conditioning of $\mathbf{A}^T\mathbf{A}$ will have small effect. LSQR, however, starts to give large amplitude anomalies in well visited cells in some regions not shown inthis paper when p=16. In this respect we like to remark that where the LSQR solution shows signs of instabilities, the SIRT solution is already clearly unstable. Note, however, that this depends on the scaling of the problem. We will clarify this below.

Both LSQR and SIRT seek least squares solutions to a given problem. One of the fundamental differences in the way they achieve the solution is the additional column scaling with $\mathbf{C}^{-1/2}$ which is *implicitly* employed by the SIRT method, but not by LSQR (section 6). In order to obtain more insight in the effect of this scaling, we can apply the LSQR method to solve $\mathbf{x}$ from the *SIRT−scaled* problem:

$$\mathbf{x} = \mathbf{C}^{-1/2}\mathbf{u} \ , \quad \mathbf{A}\mathbf{C}^{-1/2}\mathbf{u} = \mathbf{t} \tag{20}$$

subject to $Min(\mathbf{u}^T\mathbf{u})$. This is equivalent to solving the problem: $\mathbf{Ax} = \mathbf{t}$ subject to $Min(\mathbf{x}^T\mathbf{Cx})$. Note that we neglect in (20) the SIRT-*rowscaling* with $\mathbf{R}^{-1/2}$ (cf. (19)).

To this purpose we only performed a harmonic sensitivity test without the estimation of relocation and station corrections, i.e. the same tomographic problem as discussed in section 13. We can compare the model fit results of this test, which are shown in figure 11

11a: LSQR 'sirt-scaling' p=16

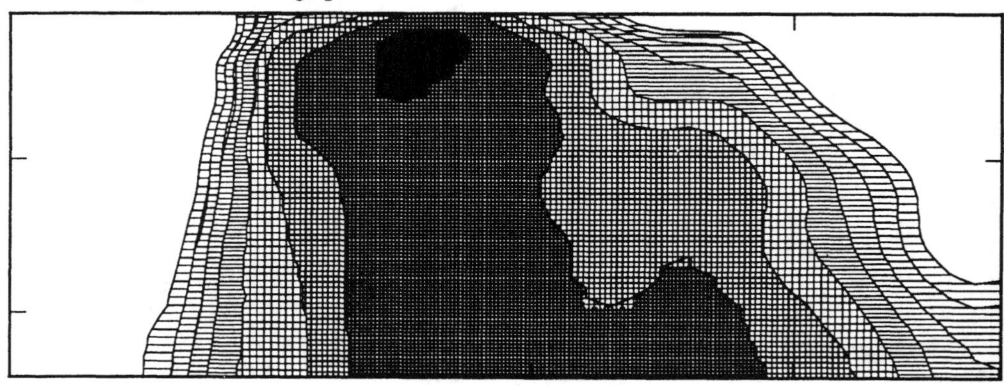

11b: LSQR 'sirt-scaling' p=7

Figure 11. The model fit of the response of the *SIRT–scaled* harmonic sensitivity test. 11a: p=16 , 11b: p=7. See also table 1 (test harmo 4)

and table 1 (i.e. test harmo 3), with those displayed in figure 10. In figure 11, the regions of worst fit are caused by irregular large amplitude anomalies, similar to what we observed for the SIRT solution (fig. 10b) and not by zero-biased anomalies, which we obtained for the *unscaled* LSQR solution (fig. 10a). The result for p=16 (fig. 11a) shows a much better fit in the well visited cells than the comparable SIRT solution (fig. 10b). Already after 7 iterations the SIRT-scaled LSQR solution obtained a variance reduction up to the noise level of the data (table 1). Since it is expected that in the next 9 iterations parts of the model tend to match the synthetic errors, we also give the p=7 result (fig. 11b). The model fit of this solution is much better than the p=16 result obtained with SIRT (fig. 10b), however the fit is less when compared to the *unscaled* LSQR fit (fig 10a).

Evidently the column scaling with $\mathbf{C}^{-1/2}$, allows the LSQR algorithm to estimate larger amplitude anomalies in the poorly visited regions of the model. When solving the *unscaled* problem, these solution components are hardly in range of the algorithm in the first 16 iterations (fig. 10a). We note that in the absence of synthetic noise (results which are not shown here), anomalies in the poorly visited regions remain to exist, with reduced amplitudes, but still much larger than what LSQR would achieve in the *unscaled* problem. From this comparison it appears that the 'SIRT-scaling' makes the solution very sensitive to noise in the ill-illuminated cells. We have also observed this for the comparable SIRT solutions.

This test leads to the conclusion that we can obtain SIRT-type solutions by applying the LSQR algorithm to the problem (20), with the great advantages that (i) we need much less iterations to obtain comparable results and moreover (ii) we do not need to employ the rowscaling with $\mathbf{R}^{-1/2}$, which we regard as unphysical. Thus, it is also possible to obtain solutions in the poorly illuminated regions of the model using the LSQR algorithm. In general, the reliability of these poorly determined components of $\mathbf{x}$ is hard to assess, especially when one is inverting real data. Although regularization improves the image, it also introduces heavily biased estimates in the poorly visited cells and hence it does not improve the reliability of the image.

Another observation, not principally connected to the column scaling, is that the model fit in the central area of the cross section (fig. 11a) benifits from the additional 9 iterations, although the data variance only drops by 0.7 % (table 1). This is of course a small reduction considering the amount of iterations. At the same time, the fit to both sides of the central area is destroyed, where indeed errors have large impact on the solution.

In view of the results of this study and the variance reductions listed in table 1 we must conclude that the variance reduction is difficult to use as a stopping criterium for the iteration.

It is possible to speed up the convergence rate of SIRT methods by using a large relaxation parameter ω (section 6). We investigated this by performing sensitivity test using ω=2. The results show, when compared to the SIRT solutions obtained with ω=1, a better fit in the central region of the cross section, but larger instabilities in the poorly illuminated cells. The model fit is still less than in the comparable LSQR solutions.

On the whole, we favour the LSQR algorithm for its superior convergence and the possibility to apply a priori solution scaling, i.e. column scaling, for example to mimic SIRT-type solutions. We like to remark that our results depend of course on the employed ray geometry, which was the same in all tests. However, an independent study using a different ray-illumination (Nolet 1985) and the theoretical work of Van der Sluis and Van der Vorst (1987) are both in favour of the superior convergence of the LSQR algorithm with respect to SIRT type methods.

An observation, very important for the interpretation of tomographic images of the Earth, concerns the correlation between cell hitcount and anomaly amplitudes. In 'SIRT-type' solutions dominant anomalies may only express poor cell hitcount in relation to large data errors. In the 'conventional' LSQR solutions, i.e. where we solve $\mathbf{Ax}=\mathbf{t}$ subject to $Min\,(\mathbf{x}^T\mathbf{x})$, zero-biased anomalies may correlate with poorly illuminated regions. In the first case non-existing anomalies can be imaged or existing anomalies exaggerated, whereas in the latter case existing anomalies can be overlooked.

Sensitivity tests have proven to be very useful in scrutinizing these effects and also in estimating upper limits of attainable accuracy and spatial resolution in tomographic images. As for the accuracy we note that both algorithms tend to underestimate amplitudes systematically in large parts of the model in the first 16 iterations.

We extensivily discussed our method of large scale cell tomography to image a part of the Earth's interior and illustrated it with a practical example. Our main objective was to reveal the main characteristics of two imaging algorithms that are used in seismic tomography. We hope that the results of this research enable a clarifying view on existing and future tomographic results stemming from either method.

Acknowledgement

We have greatly benefitted from a number of discussions with Prof. A. van der Sluis.

References

Clayton, R.W. and P. Comer, 1984. A tomographic analysis of mantle heterogeneities, *Terra Cognita*, **4**, 282-283.

Draper, N and H. Smith, 1981. *Applied regression analysis*, 2nd ed., Wiley, New York, 709 pp.

Dziewonski, A.M. 1984. Mapping the lower mantle: determination of lateral heterogeneity in P-velocity up to degree and order 6. *J. Geophys. Res.*, **89**, 5929-5952.

Dziewonski, A.M. and D.L. Anderson, 1983. Travel times and station corrections for P waves at teleseismic distances, *J. Geophys. Res.*, **88**, 3295-3314.

Hager, B.H., R.W. Clayton, M.A. Richards, R.P. Comer and A.M. Dziewonski, 1985. Lower mantle heterogeneity, dynamic topography and the geoid, *Nature*, **313**, 541-545.

Ivansson, S., 1983. Remark on an earlier proposed iterative tomograpic algorithm, *Geophys. J. R. Astr. Soc.*, **75**, 855-860.

Nolet, G., 1985. Solving or resolving inadequate and noisy tomographic systems, *J.Comp.Phys.*, **61**, 463-482.

Nolet, G. 1981. Linearized inversion of (teleseismic) data, In: *The solution of the inverse problem in geophysical interpretation*, R. Cassines (ed.), Plenum Press, New York, 9-37.

Nolet, G. 1987. Seismic wave propagation and seismic tomography, In: *Seismic tomography*, G. Nolet (ed.), Reidel, Dordrecht, 1987, 7-29.

Paige , C.C. and M.A. Saunders, 1982. LSQR: an algorithm for sparse linear equations and sparse least squares, *ACM Trans.Math.Soft.*, **8**, 43-71 and 195-209.

Spakman, W. 1985. A Tomographic Image of the Upper Mantle in the Eurasian-African-Arabian Collision Zone, *EOS*, Vol. **66**, No. 46, 975.

Spakman, W. 1986a. Subduction beneath Eurasia in connection with the Mesozoic Tethys, *Geol.Mijnbouw*, **65**, 145-153.

Spakman, W. 1986b. The upper mantle structure in the Central European-Mediterranean region, In: *European Geotraverse (EGT) Project, the central segment*, R. Freeman, St. Mueller and P. Giese (eds.), European Science Foundation, Strassbourg, 215-222.

Van der Sluis, A. and van de Vorst, H.A., 1987. Numerical solution of large, sparse linear systems arising from tomographic problems, In: *Seismic tomography*, G. Nolet (ed.), Reidel, Dordrecht, 1987, 53-87.

W. SPAKMAN and G. NOLET, Department of Theoretical Geophysics, University of Utrecht, PO Box 80.021, 3508 TA Utrecht, The Netherlands.

Chapter 9

The determination of fluid flow at the core surface from geomagnetic observations

J. Bloxham

We examine the problem of using observations of the Earth's magnetic field to study the kinematics of fluid flow at the surface of the core. In particular, we study two nonlinear inverse problems: the problem of inferring maps of the magnetic field at the core-mantle boundary from observations of the field made at the earth's surface; and the problem of using maps of the field at the core-mantle boundary to infer the fluid flow in the core immediately beneath the boundary. Both inverse problems are continuous and unstable: adequate regularization conditions must be employed if meaningful solutions are to be obtained. We highlight the inadequacy of the traditional approach to these problems, that of arbitrarily truncating the resulting expansions in terms of spherical harmonic basis functions, in the framework of stochastic inversion, to regularize the solutions based on the requirement that the solutions should be spatially smoothly-varying. We present models of the field at the core-mantle boundary for 1842, 1905, 1969 and 1980. In order to invert these models of the magnetic field for fluid motions the field must satisfy certain necessary conditions. We find some weak evidence that the field violates these conditions, resulting from the effects of electrical diffusion in the core, over the period 1969—1980. We find very simple, but necessarily nonunique, fluid flow solutions for the periods 1842—1905, 1905—1980 and 1966—1980: the effects of diffusion are clearly apparent in the 1905—1980 solution. We investigate a simple explanation for the rapid diffusion observed during the period 1905—1980, and discuss how diffusive effects may be incorporated into the solutions.

N.J. Vlaar, G. Nolet, M.J.R. Wortel & S.A.P.L. Cloetingh (eds), Mathematical Geophysics, 189-208.

1. Introduction

It is now widely accepted that Earth's magnetic field, which is known to have existed over almost the entire geological time span, is the result of a self-regenerating dynamo process operating in Earth's fluid outer core. The dynamo process involves complicated interactions between the velocity and magnetic fields in the core, resulting in the regeneration of magnetic field, balancing the losses from Ohmic decay. Although considerable advances have been made in understanding this dynamo process, theoreticians are still far from a concensus on even some of the most general aspects of the type of dynamo process at work in the core.

The problem is not only open to theoretical attack: observations of the field have been made for several centuries and are potentially an extremely valuable data set for constraining dynamo theories. In particular, we seek to use these observations to produce maps of the magnetic field at the core-mantle boundary (CMB). From these maps we seek to infer the fluid motions in the core responsible for the secular variation (the gradual changes in the field observed over the course of several years), from which we hope, in turn, to be able to deduce the pattern of convection in the core and so provide input to the dynamo problem.

In Section 2 of this chapter we examine the problem of using surface observations to infer the magnetic field at the CMB. We highlight the shortcomings of the traditional approach to this problem and discuss two new methods of field analysis which have been developed in response. In particular, we examine the method of stochastic inversion, following closely the approach of Gubbins and Bloxham (1985).

In Section 3 we consider a second inverse problem, that of inverting models of the field and its secular variation for fluid motions at the core surface. We adopt the Bullard-Gellman formalism (Bullard and Gellman 1954) and formulate the inverse problem using the approach of Whaler (1986). Certain approximations must be introduced, most importantly the frozen-flux approximation (Roberts and Scott 1965; Backus 1968). We discuss this approximation in detail, and present arguments that it may not be adequate for our purpose.

In the final section we discuss the consequences of these results and indicate where further developments in the interpretation of our observations are required.

2. The magnetic field at the core-mantle boundary

2.1 The geomagnetic field modelling inverse problem

In this section we consider the problem of modelling the magnetic field at the core-mantle boundary (CMB) using observations of the magnetic field made, of necessity, on, or just above, Earth's surface. Unfortunately, formidable difficulties are encountered in attempting to solve this problem. To highlight the nature of these difficulties we will first consider the traditional approach; for a review of the many variations on this approach see Barraclough (1978).

We consider the representation of the geomagnetic field **B** as the gradient of a scalar Φ expanded in surface spherical harmonics. We have:

$$\mathbf{B} = -\nabla\Phi \tag{1}$$

with

$$\Phi = a\sum_{l=1}^{\infty}\sum_{m=0}^{l}\left[\frac{a}{r}\right]^{l+1}(g_l^m\cos m\phi + h_l^m\sin m\phi)P_l^m(\cos\theta) \tag{2}$$

where (r, θ, ϕ) are spherical polar coordinates, with $r = a$ Earth's surface. The $P_l^m(\cos\theta)$ are Schmidt quasi-normalized associated Legendre functions. The coefficients $\{g_l^m, h_l^m\}$ are known as the geomagnetic coefficients.

The problem is one of estimating the set of geomagnetic coefficients. Difficulties are immediately apparent: first, to describe the continuous field at the CMB the set of geomagnetic coefficients is infinite; the set of available data is necessarily finite. Another problem is instability: downward continuation of the potential Φ to the CMB amounts to setting $r = c$ in the expansion, where c is the radius of the CMB. (We should note that we have implicitly assumed that the mantle is an electrical insulator: this assumption has been justified by Benton and Whaler (1983)). The factor $\left[\frac{a}{c}\right] \approx 1.8$, so that terms of increasing spherical harmonic degree l are multiplied by successive factors of ≈ 1.8. As a result, arbitrarily small changes in the data can result in arbitrarily large changes in the field at the CMB. A possible third problem is non-uniqueness: the scalar Φ is not measured directly. Instead, we measure elements of the magnetic field which are linear or nonlinear functionals of spatial derivatives of Φ. Even perfect knowledge of these functionals everywhere at Earth's surface may not be sufficient to determine Φ uniquely (see, for example, Backus 1970a).

The naive solution to these problems is to truncate the expansion at some degree $l = Lmax$. The resulting finite set of geomagnetic coefficients is then estimated using a least squares procedure. With a suitably small value of *Lmax* the problems described above disappear: the solution is well-determined by the available data, and stable solutions are obtained. Truncating the spherical harmonic expansion at, say, degree 8 (the preferred truncation level of Benton et al. (1982)) is tantamount to assuming the magnetic field originating in Earth's core consists entirely of harmonics of degree 8 and below — a wholly unwarranted assumption. The inadequacy of such methods has been highlighted recently by several authors (Whaler and Gubbins 1981; Shure, Parker and Backus 1982; Gubbins 1983; Gubbins and Bloxham 1985). However, because the geomagnetic inverse problem is an unstable continuous inverse problem, unless additional information is introduced into the problem, nothing, in principle, can be deduced about the solution (Backus 1970b). Two new methods of field analysis — stochastic inversion (Gubbins 1983; Gubbins and Bloxham 1985) and harmonic splines (Shure, Parker and Backus 1982; Parker and Shure 1982) — seek stable solutions to the inverse problem by incorporating more reasonable additional assumptions. With appropriately specified *a priori* information the methods yield identical models of the magnetic field at the CMB. The validity of each method depends upon the validity of the *a priori* information. Stochastic inversion is algebraically and computationally more straightforward than harmonic splines: we consider stochastic inversion below.

2.2 Stochastic inversion

In order to obtain stable solutions to the inverse problem it is necessary to make some assumption regarding the smoothness of the field at the CMB. The generation of field in the core is described by the induction equation (see, for example, Moffatt 1978)

$$\frac{\partial \mathbf{B}}{\partial \mathbf{t}} = \nabla \times (\mathbf{u} \times \mathbf{B}) + \eta \nabla^2 \mathbf{B} \tag{3}$$

where $\mathbf{u}$ is the fluid velocity, and $\eta = 1/(\mu_0 \sigma)$ is the magnetic diffusivity, μ_0 being the magnetic permeability and σ the electrical conductivity. The tendency of fluid motions in the core to produce ever smaller scale features in the field (described by the first term on the right hand side (RHS) of the induction equation) is limited by the effects of diffusion (described by the second term). We write the radial component of the induction equation in the form

$$\frac{\partial B_r}{\partial t} + \nabla_h \cdot (\mathbf{u}_h B_r) = \eta \left[\frac{\partial^2 B_r}{\partial r^2} + \frac{4}{r} \frac{\partial B_r}{\partial r} + \frac{2B_r}{r^2} + \frac{\Delta B_r}{r^2} \right] \tag{4}$$

where ∇_h denotes the horizontal gradient and Δ is the angular part of $r^2 \nabla^2$. The term $\nabla_h \cdot (\mathbf{u}_h B_r)$ represents the effect of the fluid motions, the terms on the RHS the effects of diffusion. The first two terms on the RHS cannot be evaluated from surface data, since they involve radial derivatives which are possibly discontinuous across the CMB; the other two represent the 'observable' part of the diffusion term. Multiplying by B_r and integrating over the CMB gives

$$\frac{d}{dt} \oint \frac{1}{2} B_r^2 \, dS + \oint B_r \nabla_h \cdot (\mathbf{u}_h B_r) \, dS = \tag{5}$$

$$\eta \oint B_r \left[\frac{\partial^2 B_r}{\partial r^2} + \frac{4}{r} \frac{\partial B_r}{\partial r} \right] dS + \eta \oint B_r \left[\frac{2B_r}{r^2} + \frac{\Delta B_r}{r^2} \right] dS$$

A natural choice of regularization condition, or additional information, is to seek solutions minimizing the quantity

$$\psi = \oint B_r \left[\frac{\Delta B_r}{r^2} \right] dS \quad . \tag{6}$$

Such solutions, for a given misfit to the data, have the least small scale structure measured by the quantity ψ. We have, in effect, prejudiced the inversion against solutions with unnecessary small scale structure, structure which would rapidly decay diffusively.

An alternative way to view this additional information is from a probability theory viewpoint. Minimizing the above term is equivalent to imposing an *a priori* probability distribution on the coefficients in the spherical harmonic expansion. (6) is equivalent to requiring

$$\mathbf{m} \sim N_P(0, \mathbf{C}_m) \tag{7}$$

meaning $\mathbf{m} = (g_1^0, g_1^1, h_1^1, g_2^0, \ldots)$ is P-dimensional Normally distributed with zero mean and covariance matrix $\mathbf{C}_m$. P is, for the time being, infinite. From a Bayesian perspective, we have expressed our *a priori* knowledge in terms of a probability distribution: we have, in effect, said that, in the absence of any other information, we believe the geomagnetic coefficients to be zero within the bounds expressed by the covariance matrix C_m. We expect our *a priori* beliefs to be modified when we consider the data. The *a priori* covariance matrix C_m corresponding to (6) is given by

$$\left[C_m^{-1}\right]_{ij} = 4\pi\left[\frac{a}{c}\right]^{2l} \frac{(l+1)^2}{2l+1} l^4 \delta_{ij} = \nu(l) \tag{8}$$

We assume that the data γ are related to $\mathbf{m}$ by a model equation of the form

$$\gamma = \mathbf{f}(\mathbf{m}) + \mathbf{e} \tag{9}$$

where $\mathbf{f}$ is a (possibly nonlinear) functional of the coefficients $\mathbf{m}$, and $\mathbf{e}$ is a vector of data-errors with

$$\mathbf{e} \sim N_d(0, C_e) \tag{10}$$

where D is the number of data.

The solution which, for a given misfit to the data γ, minimizes (6) is the mode of the *a posteriori* distribution. The solution must, for a nonlinear functional $\mathbf{f}$, be sought iteratively. We use the algorithm

$$\mathbf{m}_{i+1} = \mathbf{m}_i + \left[A_i^T C_e^{-1} A_i + \lambda C_m^{-1}\right]^{-1}\left[A_i^T C_e^{-1}(\gamma - \mathbf{f}(\mathbf{m})) - \lambda C_m^{-1}\mathbf{m}_i\right] \tag{11}$$

where

$$A_i = \frac{\partial \mathbf{f}}{\partial \mathbf{m}}\bigg|_{\mathbf{m}=\mathbf{m}_i} \tag{12}$$

λ is a damping parameter: larger values of λ result in smaller values of the dissipation integral ψ, i.e. increased reliance is placed on the additional information. In practice, we truncate the expansion at some degree *Lmax*, chosen to ensure the expansion has converged. The covariance of the solution (the *a posteriori* covariance) is given by

$$C = \hat{\sigma}^2\left[A^T C_e^{-1} A + \lambda C_m^{-1}\right]^{-1} \tag{13}$$

where

$$\hat{\sigma}^2 = \frac{(\gamma - \mathbf{f}(\mathbf{m}))^T C_e^{-1}(\gamma - \mathbf{f}(\mathbf{m}))}{D - \mathrm{tr}\left[(A^T C_e^{-1} A)(A^T C_e^{-1} A + \lambda C_m^{-1})^{-1}\right]} \tag{14}$$

and where A is evaluated at the solution $\mathbf{m}$. An important concept in the analysis of an inverse problem is resolution: the resolution matrix R is given by

$$R = \left[A^T C_e^{-1} A + \lambda C_m^{-1}\right]^{-1}\left[A^T C_e^{-1} A\right] \tag{15}$$

The trace of the resolution matrix is the effective number of degrees of freedom of the model.

The variance of an estimate of the radial field at a point on the CMB is given by

$$\text{var}\ \mathbf{a}^T\mathbf{C}\mathbf{a} = \sum_{l>Lmax} v_l \left[\frac{a}{c}\right]^{2l+4} (l+1)^2 \tag{16}$$

where the components of **a** have the form

$$(l+1)^2\left[\frac{a}{c}\right]^{l+2} P_l^m(\cos\theta)\{\cos m\phi, \sin m\phi\} \tag{17}$$

and v_l is given by the *a priori* form (8).

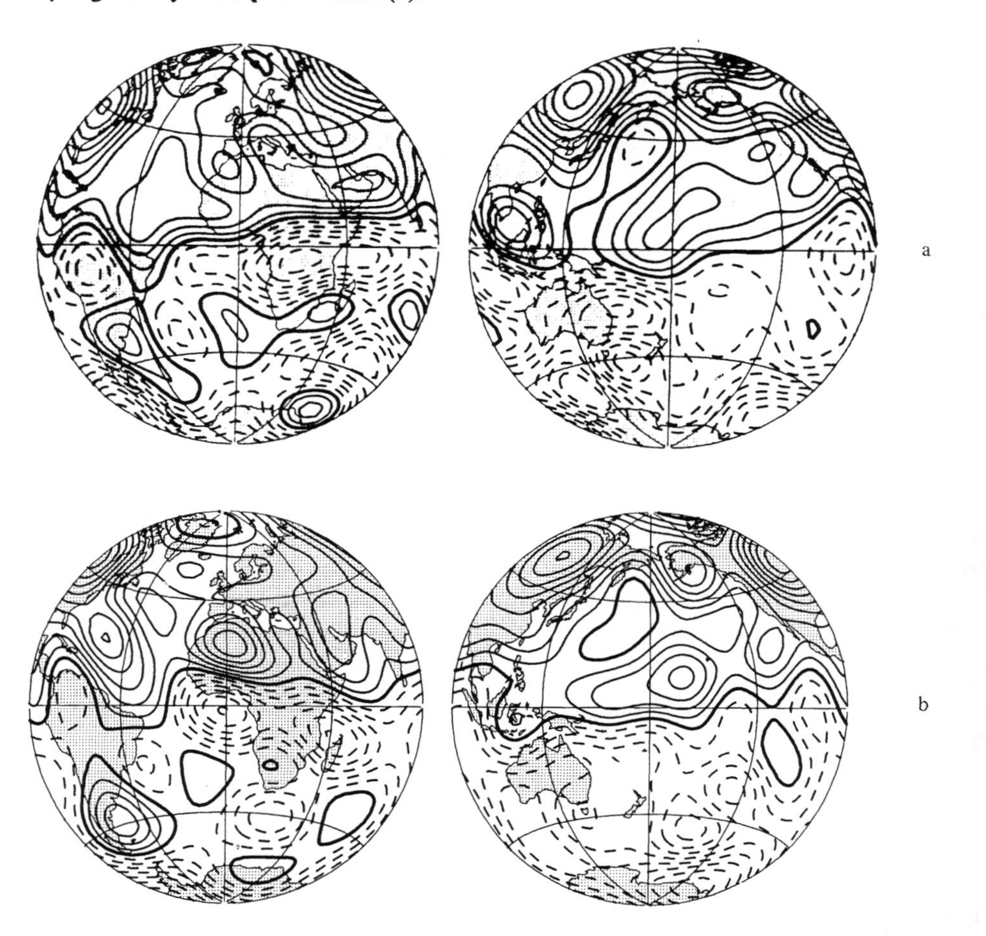

Figure 1. Maps of the radial component of the magnetic field at the core-mantle boundary. Contour interval 100μT. Continuous contours represent flux into the core, broken contours flux out of the core; bold contours represent zero radial field. 1*a* 1842, 1*b* 1905, 1*c* 1969, and 1*d* 1980.

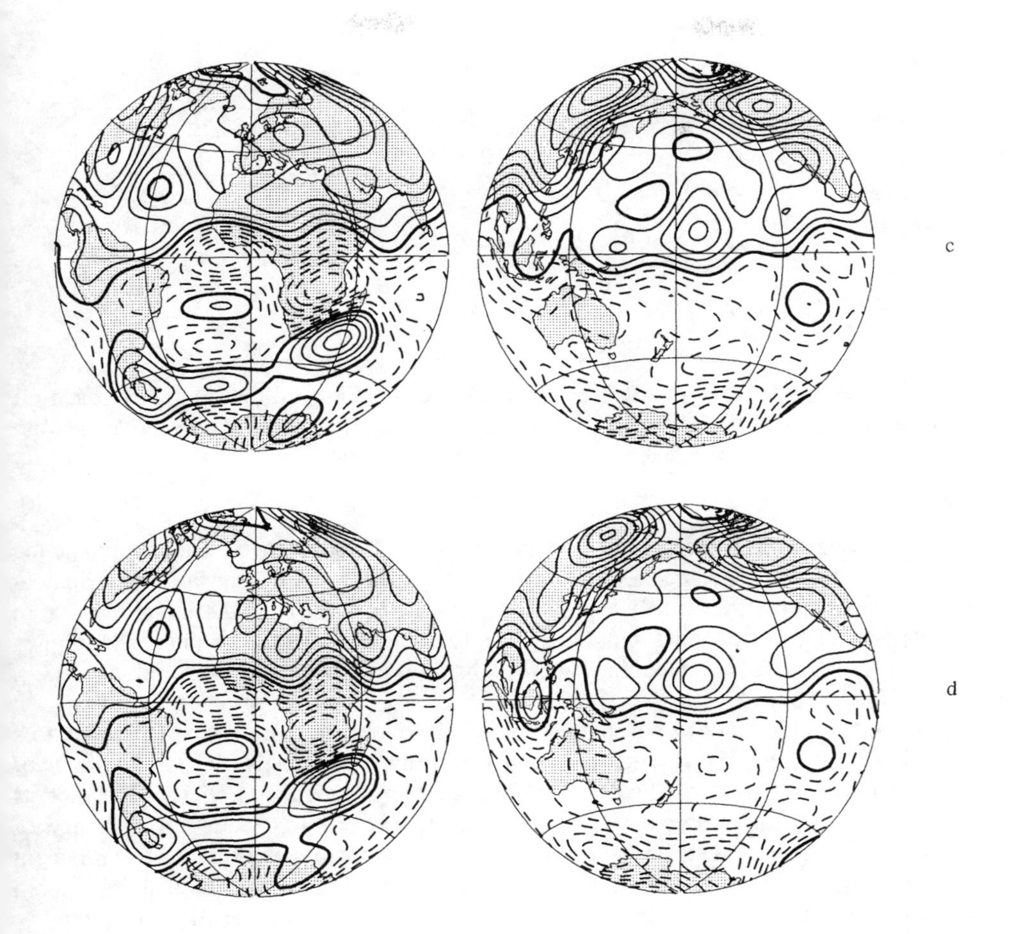

Figure 1. (continued).

Error estimates obtained this way may be too small: the additional information is strictly only valid asymptotically. The variance of the true field at the CMB at some degree l may be larger than that implied by (8). With adequate data the field is well-resolved out to about degree 12, and the *a priori* information has little effect. At very high degrees, say above degree 30, the asymptotic form (8) is likely, on energetic grounds, to be valid. At intermediate wavenumbers the variance of the field at the CMB may be larger than that given by (8), but the core signal observed at the surface is contaminated by crustal fields: unfortunately we have no means available to estimate the power in these degrees. In this state of ignorance we choose to apply the asymptotic form (8). As we must choose λ to ensure that the models are not unduly contaminated by crustal fields, it is likely that our error estimates are too small in this range. The possibility that the resultant models are nevertheless contaminated by crustal fields has been discussed by Bloxham and Gubbins

Table 1. Model Statistics.

Model	Number of Data	Damping Parameter	Norm $\times 10^{12}$	Misfit	Tr(R)	Error (μT)
1842	12398	1.3×10^{-12}	132	0.92	102	70-100
1905	26705	3.0×10^{-12}	130	1.00	112	50-70
1969	4977	1.0×10^{-11}	137	1.25	155	36
1980	4178	1.0×10^{-11}	136	1.09	150	28

(1985) and Bloxham (1986a). However, it should be remembered that methods based on the truncation at, say, degree 8 ascribe zero variance to all degrees above the truncation level and will necessarily produce much smaller error estimates.

2.3 Field models

Stochastic inversion has been applied to the analysis of geomagnetic data spanning the period 1695—1980, producing field models for 1715, 1777, 1842, and at approximately decade intervals from 1905 to 1980. Models for 1842, 1905, 1969, and 1980 are shown in Figure 1; pertinent statistics are summarised in Table 1. Full details of the production of these models may be found in Gubbins and Bloxham (1985), Bloxham and Gubbins (1985), and Bloxham (1986a).

Such models should not be considered unique: the choice of damping parameter is, to a large extent, subjective. We should really consider a range of models for various values of the damping parameter (see, for example, Gubbins and Bloxham (1985)). The models presented here are our preferred models, chosen after examining models for a range of values of the damping parameter. The choice of damping parameter is precisely equivalent to choosing the misfit to the data, the approach of the method of harmonic splines: both harmonic splines and stochastic inversion are subject to this same degree of subjectivity.

3. Fluid motions at the core-mantle boundary

3.1 Kinematics of the secular variation

As we have seen, the kinematics of the secular variation is described by the induction equation

$$\frac{\partial \mathbf{B}}{\partial t} = \nabla \times (\mathbf{u} \times \mathbf{B}) + \eta \, \nabla^2 \mathbf{B} \tag{18}$$

The first term on the RHS represents the convection of magnetic field lines, the second term the diffusion of field lines. We can associate characteristic timescales with these terms: we have a convective timescale τ_C and a diffusive timescale τ_D.

Consider a region in which the velocity and magnetic field varies on a lengthscale L, and in which the fluid has characteristic velocity U. Then

$$\tau_C \sim L/U \tag{19}$$

and

$$\tau_D \sim L^2/\eta \tag{20}$$

The ratio of these two timescales defines the magnetic Reynolds number

$$R_m = \tau_D/\tau_C = U\,L/\eta = \frac{|\nabla \times (\mathbf{u} \times \mathbf{B})|}{|\eta\,\nabla^2\mathbf{B}|} \tag{21}$$

The magnetic Reynolds number measures the relative importance of the convective and diffusive contributions to the secular variation: large values indicate that the effects of diffusion may be negligible. For Earth's core, $L \approx 10^6$m, $U \approx 5 \times 10^{-4}$ms^{-1}, and $\eta \approx 1$m^{2}s^{-1} giving $R_m \approx 500$, suggesting that we may, to a good approximation, be able to neglect the effects of diffusion. This argument may be misleading: if the magnetic field varies over a shorter lengthscale L_B than the velocity field then

$$\frac{|\nabla \times (\mathbf{u} \times \mathbf{B})|}{|\eta\,\nabla^2\mathbf{B}|} \sim R_m L_B/L \text{ or } R_m(L_B/L)^2 \tag{22}$$

depending on the exact geometry of the magnetic and velocity fields (see Moffatt 1978). In such circumstances the effects of diffusion may be significant even when the magnetic Reynolds number is large; we shall return to this possibility later.

We proceed with our consideration of the kinematics of the secular variation by assuming that the effects of diffusion are indeed negligible: the so-called frozen-flux hypothesis (Roberts and Scott 1965). Following Backus (1968), we write the radial component of the frozen-flux induction equation at the CMB in the form

$$\partial_t B_r + \mathbf{u} \,.\, \nabla_h B_r + B_r \nabla_h \,.\, \mathbf{u} = 0 \tag{23}$$

where $B_r = \mathbf{B} \,.\, \hat{\mathbf{r}}$, $\nabla_h = \nabla - \hat{\mathbf{r}}(\hat{\mathbf{r}} \,.\, \nabla)$, and $\partial_t = \partial/\partial t$ giving

$$\partial_t B_r + \nabla_h \,.\, (B_r \mathbf{u}) = 0 \tag{24}$$

Integrating over a patch S on the CMB we have

$$\int_S \partial_t B_r dS + \int_S \nabla_h \,.\, (B_r \mathbf{u}) dS = 0 \tag{25}$$

If ∂S is a contour of constant B_r, we have

$$\int_S \partial_t B_r dS = B_r \int_s \nabla_h \,.\, \mathbf{u}\, dS \tag{26}$$

Now, if we choose S such that ∂S is a contour of $B_r = 0$ (a null-flux curve) then we have

$$\int_S \partial_t B_r dS = 0 \tag{27}$$

providing conditions which must be satisfied by $\partial_t B_r$ under the frozen-flux hypothesis. From Alfven's theorem the null-flux curves are material lines yielding the equivalent conditons

$$\frac{d}{dt}\int_S B_r dS = 0 \tag{28}$$

Provided these conditions are satisfied, and the radial component of the magnetic field B_r and its time derivative $\partial_t B_r$ are known, then the radial component of the frozen-flux induction equation can be inverted for the fluid velocities $\mathbf{u}$ at the CMB. Unfortunately such inversions are subject to a fundamental nonuniqueness. Suppose we have found an eligible velocity field $\mathbf{u}_0$: then any velocity $\mathbf{u}$ where

$$B_r \mathbf{u} = B_r \mathbf{u}_0 - \hat{\mathbf{r}} \times \nabla_h \psi \tag{29}$$

is also a solution, where ψ is constant on each null-flux curve and is twice continuously differentiable on the CMB (see Backus 1968). The form of this nonuniqueness is easily understood: from maps of the radial field at the CMB, or indeed of all three components of the field, it is not possible to determine footpoints where magnetic field lines enter and leave the core; if such points could be determined then fluid motions could be traced easily. As a result we are unable, for example, to determine the component of flow around contours of B_r. Additional assumptions are required to resolve this nonuniqueness; even with sufficiently strong assumptions to resolve the formal nonuniqueness we should still be alert to the possibility that flow around contours of B_r may be poorly determined.

3.2 Testing the frozen-flux hypothesis

Flux integrals over patches bounded by null-flux curves for 1980.0 and 1969.5 are shown in Table 2, together with estimates of their error, from Bloxham and Gubbins (1986). By comparing the changes in these integrals with their error we are able to reject the hypothesis that the effects of diffusion are negligible with greater than 99.5 % confidence (details can be found in Bloxham and Gubbins 1986). This result indicates that we can resolve, above the noise, the effects of diffusion: it does not necessarily indicate that the frozen-flux approximation should be abandoned, but it does indicate that the approximation is inadequate if we wish to explain all the resolved secular variation. We should bear in mind, however, that this result is strongly dependent on our error estimates (which, as we have discussed, are to some extent subjective): doubling the estimated error on the flux integrals would be sufficient to bury the effects of diffusion in the noise. To double the error on these integrals would, however, require a much larger increase in the error in the radial field. The flux integrals are dominated by quite low degree harmonics which are well-determined and for which the error estimates are reasonable; error estimates for higher degree harmonics (say above degree 12) are liable to be optimistic but these harmonics contribute less to the flux integrals, so the result is more robust that it might at first appear to be.

It is worth noting that most of the evidence for flux diffusion comes from the region beneath the south Atlantic Ocean. Comparing flux integrals in this region for 1905 and 1969 we see much stronger evidence of diffusion (see Figure 1, and Bloxham and Gubbins 1985). The large and rapid increase in flux through the patch beneath southern Africa is well-resolved and demands an explanation: we will consider this flux intensification in more detail later. For the time being we will remark that it suggests that we should exercise caution in the interpretation of core fluid motion models obtained using the frozen-flux

Table 2. Flux Integrals MWb

Patch	1980.0	1969.5	Change	Error
Northern Hemisphere	-17547	-17469	-78	45
Southern Hemisphere	18850	18691	160	51
South Atlantic	-1274	-1250	-25	32
St Helena	-88	-50	-38	15
Easter Island	-20	-41	21	14
North Atlantic	3	5	-2	7
North-West Pacific	2	10	-8	9
North-East Pacific	33	35	-2	14
North Pole	39	66	-27	15

approximation.

3.3 The core fluid motions inverse problem

We represent the CMB velocity field in terms of a toroidal-poloidal decomposition

$$\mathbf{u} = \mathbf{u}_{\mathrm{T}} + \mathbf{u}_{\mathrm{P}} \tag{30}$$

where

$$\mathbf{u}_{\mathrm{T}} = \nabla \times (\mathbf{r}T) = \left[0, \frac{1}{\sin\theta}\frac{\partial T}{\partial \phi}, -\frac{\partial T}{\partial \theta}\right] \tag{31}$$

and

$$\mathbf{u}_{\mathrm{P}} = \nabla(rS) = \left[0, \frac{\partial S}{\partial \theta}, \frac{1}{\sin\theta}\frac{\partial S}{\partial \phi}\right] \tag{32}$$

We expand the scalar potentials S and T in surface spherical harmonics $Y_l^m(\theta,\phi)$

$$\left.\begin{aligned} T(\theta,\phi) &= \sum_{l,m} t_l^m Y_l^m(\theta,\phi) \\ S(\theta,\phi) &= \sum_{l,m} s_l^m Y_l^m(\theta,\phi) \end{aligned}\right\} \tag{33}$$

For ease of notation the coefficients t_l^m and s_l^m are complex: with this more compact notation the geomagnetic coefficients g_l^m and h_l^m are given by the mapping $g_l^m + ih_l^m \to g_l^m$. Then, substituting into the radial component of the frozen-flux induction equation, and following closely Whaler (1986), we have

$$\sum_{l_1,m_1}\left[\frac{a}{c}\right]^{l_1+2}(l_1+1)\dot{g}_{l_1}^{m_1}Y_{l_1}^{m_1}=\frac{1}{c}\sum_{l_2,m_2}\sum_{l_3,m_3}\left[\frac{a}{c}\right]^{l_2+2}(l_2+1)g_{l_2}^{m_2}$$

$$\times\left\{\frac{t_{l_3}^{m_3}}{\sin\theta}\left[\frac{\partial Y_{l_2}^{m_2}}{\partial\phi}\frac{\partial Y_{l_3}^{m_3}}{\partial\theta}-\frac{\partial Y_{l_2}^{m_2}}{\partial\theta}\frac{\partial Y_{l_3}^{m_3}}{\partial\phi}\right]\right. \tag{34}$$

$$\left.-\,s_{l_3}^{m_3}\left[\frac{\partial Y_{l_2}^{m_2}}{\partial\theta}\frac{\partial Y_{l_3}^{m_3}}{\partial\theta}+\frac{1}{\sin^2\theta}\frac{\partial Y_{l_2}^{m_2}}{\partial\phi}\frac{\partial Y_{l_2}^{m_2}}{\partial\phi}-l_3(l_3+1)Y_{l_2}^{m_2}Y_{l_3}^{m_3}\right]\right\}$$

Multiplying through by $Y_{l_1}^{m_1*}$ (the complex-conjugate of $Y_{l_1}^{m_1}$) and integrating over the CMB reduces this equation to

$$\dot{\mathsf{m}}=\mathsf{Et}+\mathsf{Gs}=\left[\mathsf{E}:\mathsf{G}\right]\left[\begin{array}{c}\mathbf{t}\\ \cdots\\ \mathbf{s}\end{array}\right] \tag{35}$$

where, by exploiting standard results from the theory of spherical harmonics, we can show that the matrices E, G have elements of the form

$$E_{l_1\,l_3}^{m_1m_3}=\frac{1}{4\pi c}\left[\frac{c}{a}\right]^{l_1+2}\frac{2l_1+1}{l_1+1}\sum_{l_2,m_2}\left[\frac{a}{c}\right]^{l_2+2}(l_2+1)g_{l_2}^{m_2}$$

$$\times\oint\left[\frac{\partial Y_{l_2}^{m_2}}{\partial\phi}\frac{\partial Y_{l_3}^{m_3}}{\partial\theta}-\frac{\partial Y_{l_2}^{m_2}}{\partial\theta}\frac{\partial Y_{l_3}^{m_3}}{\partial\phi}\right]Y_{l_1}^{m_1*}\,d\Omega \tag{36}$$

and

$$G_{l_1\,l_3}^{m_1m_3}=\frac{1}{8\pi c}\left[\frac{c}{a}\right]^{l_1+2}\frac{2l_1+1}{l_1+1}\sum_{l_2,m_2}\left[\frac{a}{c}\right]^{l_2+2}(l_2+1)$$

$$[l_1(l_1+1)+l_3(l_3+1)-l_2(l_2+1)]g_{l_2}^{m_2}\times\oint Y_{l_1}^{m_1*}Y_{l_2}^{m_2}Y_{l_3}^{m_3}\,d\Omega \tag{37}$$

Two methods are available to evaluate these matrix elements. The integrals occurring in the above expressions are the Elsasser and Gaunt integrals which are expressible in closed form as combinations of Wigner 3-j coefficients (see, for example, Winch 1974). Alternatively we can exploit Orszag's transform method (Orszag 1970), yielding a computationally more efficient means of evaluation (D. Gubbins and D. Lloyd, personal communication).

The integrals are only non-zero for certain choices of (l_1,m_1), (l_2,m_2) and (l_3,m_3): the selection rules for these choices are given by Bullard and Gellman (1954). These rules limit the degree to which the expansions (33) must be continued to the sum of the maximum degree of the magnetic field and the maximum degree of the time rate of change of the magnetic field, making the problem appear to be discrete because these two expansions are truncated at a finite degree. The resulting inversion is under-determined so additional information is required; and, most importantly, for self-consistency the velocity model should also converge.

Various additional assumptions, or regularization conditions, are suggested. The most simple is to seek the velocity field with the minimum kinetic energy, i.e. seek to minimize

$$\oint \mathbf{u}^2 d\Omega = 4\pi \sum_l \frac{l(l+1)}{2l+1} \sum_{m=0}^{l} [(t_l^m)^2 + (s_l^m)^2] \tag{38}$$

A more stringent condition is to seek solutions which are spatially smoothly varying. Regularization conditions based on minimizing the second derivative of the model are common in geophysical inverse problems, for example in determining the radial dependence of electrical conductivity in the mantle. With a vector field, such as **u**, it is less straightforward. A suitable norm is given by

$$\oint [(\nabla_h^2 u_\theta)^2 + (\nabla_h^2 u_\phi)^2] d\Omega \tag{39}$$

To leading order this adds an additional factor of l^4 to the denominator of (38).

We consider two inverse problems: a linear problem for which we have available both $\mathbf{m}(t_0)$ and $\dot{\mathbf{m}}(t_0)$, and a nonlinear problem where we have $\mathbf{m}(t_1)$ and $\mathbf{m}(t_2)$. The linear problem is described by equation (24). The solution of this continuous inverse problem proceeds along similar lines to the geomagnetic field modelling inverse problem.

To solve the nonlinear problem we must first assume that **u** is steady over the time interval $t_1 \rightarrow t_2$. The problem must also be linearised. It is instructive to examine the exact solution to (24)

$$B_r(t) = \exp\,[-(t - t_0)\nabla_h \,.\, \mathbf{u}] \circ B_r(t_0) \tag{40}$$

Expanding the exponential differential operator

$$B_r(t) = -(t - t_0)[\nabla_h \,.\, (\mathbf{u} B_r(t))] + \frac{1}{2}(t - t_0)^2 \Big[(\nabla_h \,.\, \mathbf{u})^2 \circ B_r(t_0) \Big] + \cdots \tag{41}$$

If L is the smallest lengthscale over which **u** and B_r vary then this equation can be linearised provided

$$\Delta t = t - t_0 \ll L / |\mathbf{u}| \tag{42}$$

the Courant-Levy-Friedrichs condition encountered in other advection problems. Accordingly, we linearise the problem with a time step Δt satisfying the above condition. With

$$\mathsf{H} = (\mathsf{E} : \mathsf{G}) \tag{43}$$

and

$$\mathbf{u} = \begin{bmatrix} \mathbf{t} \\ \cdots \\ \mathbf{s} \end{bmatrix} \tag{44}$$

we have

$$\mathbf{m}(t_2) = \mathbf{m}(t_1) + \sum_{n=1}^{N} \mathsf{H}(t_1 + (n-1)\Delta t)\mathbf{u} \tag{45}$$

where $\Delta t = (t_2 - t_1)/N$, and H is recomputed at each $t = t_1 + (n-1)\Delta t$.

3.4 Models of the core fluid flow

We will consider two recent solutions of the linear problem, by Voorhies (1986) and by LeMouël, Gire and Madden (1985), and solutions of the nonlinear problem (see Figure 2).

Voorhies seeks to invert for the steady part of the core surface velocity field using main field and secular variation models. Typical solutions span the period 1960—1980. These solutions, unfortunately, are based on severly truncated expansions of the main field and secular variation; Voorhies prefers a truncation level of degree 8 despite evidence that harmonics up to at least degree 10 are well-resolved at Earth's surface with recent data. The implications of such a truncation where discussed in Section 2. Voorhies also does not employ a regularization condition; instead the velocity expansions are truncated (degree 12 being typical). These solutions, as stressed by Whaler (1986), to be self-consistent should include harmonics up to degree 16: it is inadequate to deal with this problem of convergence by arbitrarily truncating.

By contrast, LeMouël, Gire and Madden (1985) employ a more reasonable truncation level, truncating (the magnetic field) at degree 10; degrees 9 and 10 contain considerable power and have a substantial effect on maps of the field at the CMB. They also employ a regularization condition to seek spatially smoothly varying solutions using norms similar to those discussed above, although it is not clear that their solutions are adequately regularized. Nonuniqueness is almost fully-resolved by assuming that the motions are geostrophic. The geostrophic constraint is discussed by LeMouël (1984), LeMouël, Gire and Madden (1985), and Backus and LeMouël (1986). Order of magnitude calculations of the terms in the momentum equation suggest that the predominant force balance for fluid particles near the CMB, but outside the boundary layer, is between Coriolis forces and pressure gradients: this imposes strong constraints on the permissible motions.

Bloxham has obtained solutions of the nonlinear core motions problem seeking the steady part of the motion between various main field models: solutions are presented for the periods 1966—1980, 1905—1980, and 1842—1905, based on the field models discussed in section 2. The solutions for 1905—1980 and 1842—1905 are only constrained by the field models at the two end-points of the timespan: the solutions are free to advect the field by any desired route between the end-points. Consequently, the nonuniqueness in the determination of the flow is not fully-resolved. Ideally, one would wish to have field models available at the end of each timestep to constrain the solution further. Of course, in the limit of small timesteps the problem would reduce to the linear steady core motions problem.

Possibly the most striking feature of the fluid velocity models produced by these three groups is their dissimilarity: the reasons for some of the differences are, however, fairly apparent. For example, although Voorhies formally solves the uniqueness problem by assuming steady motions (details can be found in Voorhies and Backus 1985) his models have considerable flow beneath the Pacific where the secular variation is low. This flow, which does not seem to be required is largely aligned with contours of radial field: the

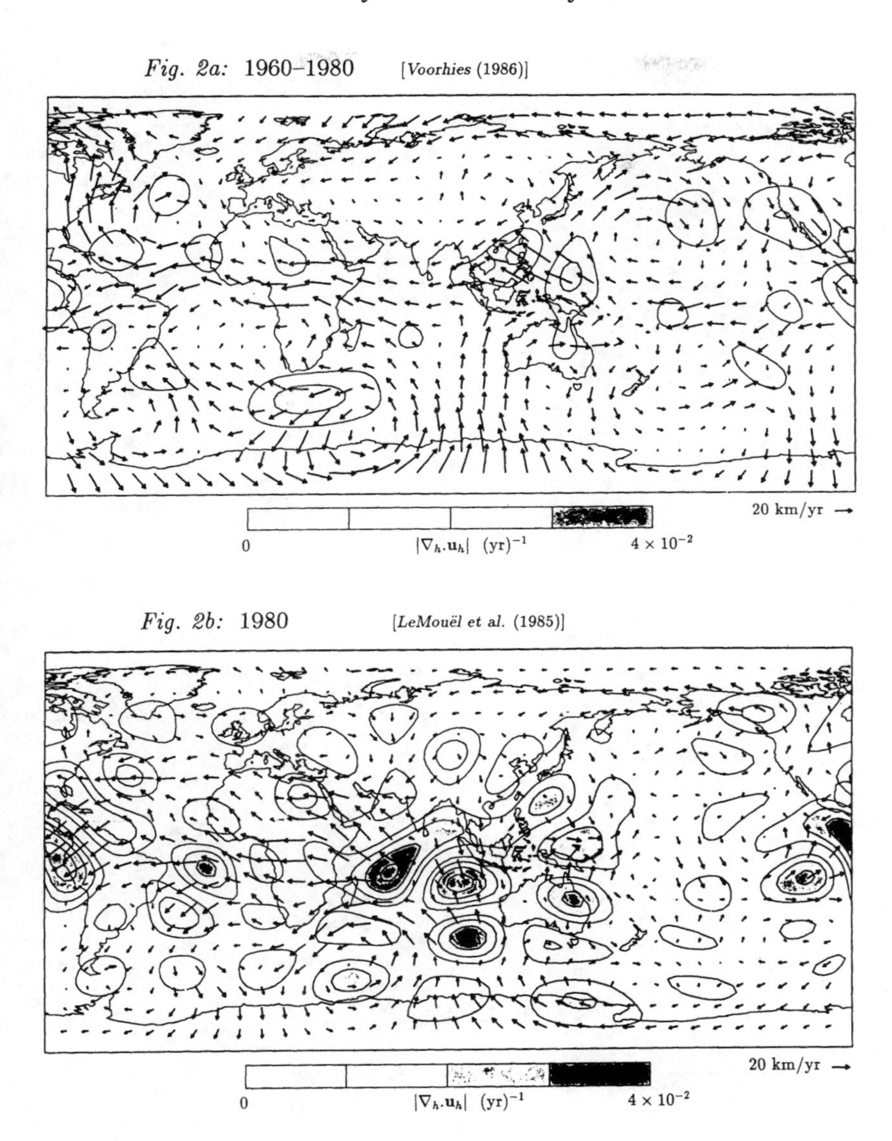

Figure 2. Maps of fluid motions at the core surface. The vectors represent the direction and strength of the surface motion; the shading represents the intensity of up and downwelling flow.

problem of ill-determined flow may still remain. Alternatively, this flow may be a result of aliasing of secular variation beneath Europe and Africa into the Pacific region as a result of the truncation level. The geostrophically-constrained flows of LeMouël, Gire and Madden (1985) show a complicated pattern of intense up and downwelling. The geostrophic

Fig. 2c: 1966–1980

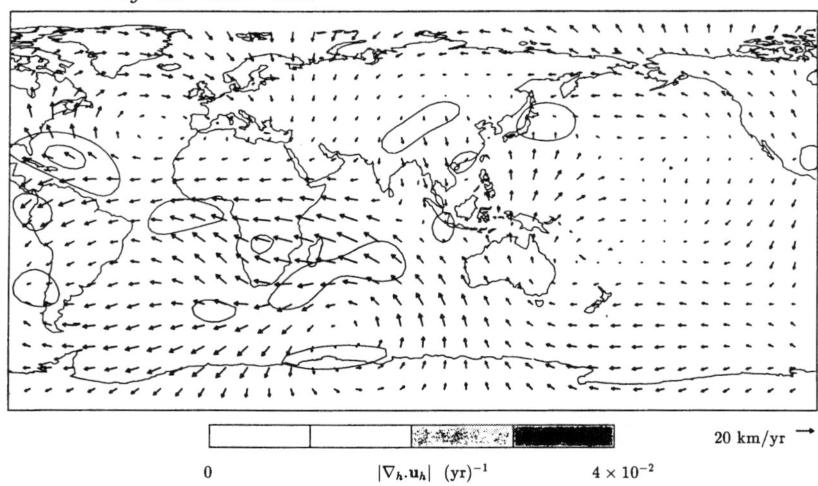

Fig. 2d: 1905–1980

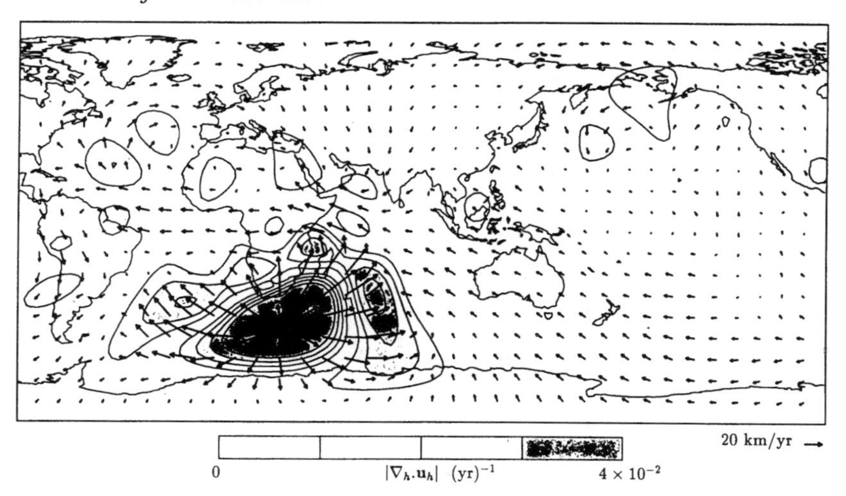

Figure 2. (continued).

condition places strong constraints on up and downwelling flow away from the equator: the consequent stretching of vortex lines must be balanced by north-south motion.

On the other hand, this author's solutions are very much simpler. This simplicity may be because comparitively rapidly-varying aspects of the flow (possibly resolved by LeMouël et al.) have been averaged out: however, given the error estimates for the field

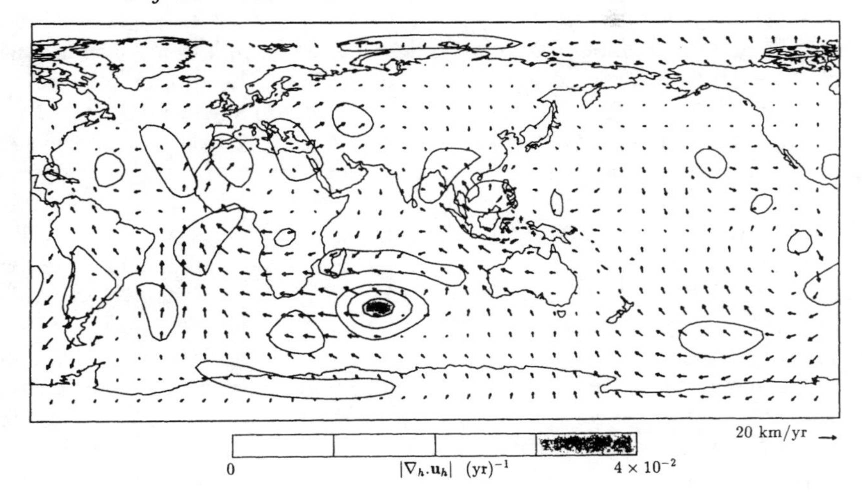

Figure 2. (continued).

models presented here (errors which are, if anything, too small) we are unable to resolve such rapid small-scale secular variation.

Certainly we should be wary of interpreting differences between flow models at different points in time as representing evidence of unsteadiness of the flow when models of the flow at the same point in time differ so widely. In fact, the series of models shown in Figure 2c-e indicate that the steady motions approximation may be reasonable even over quite long timespans. The models bear a reasonable similarity to each other, at least away from the region beneath southern Africa. These models tend to reinforce the conjecture that diffusive effects should not be ignored. Vigorous fluid motions are largely confined to the region between 90° E and 90° W: the source of these motions is a strong upwelling jet beneath southern Africa, precisely where the strongest evidence was found for flux diffusion. This jet is especially strong in the solution for the period 1905—1980: the emergence of the flux spot pair beneath southern Africa (see next section) occurred largely during the first few decades of this century. This suggests that the frozen-flux approximation may, as we suspected, be inadequate for obtaining models of the core flow: indeed, there is some indication that the frozen-flux approximation may be a worse approximation than the steady motions approximation for these models. In the next section we consider, with a simple forward model, a possible explanation for the intense secular variation observed under southern Africa.

3.5 Diffusive effects

As an example of the effects of diffusion, we consider the effect of an upwelling motion in the core on the toroidal field. It has been speculated that the toroidal field in the core (only the poloidal ingredient is observed at Earth's surface) is considerably stronger than the poloidal field. Upwelling motion in the core will tend to advect this field towards the CMB where it will become concentrated leading to diffusion of field through the CMB. This mechanism, which is illustrated in Figure 3, will result in two adjacent regions of intense flux of opposite sign: such a field configuration is indeed observed beneath southern Africa (see Figure 1d). Note that this upwelling is, to some extent, resolved by the frozen-flux solutions: the upwelling motion results in an net motion of poloidal field lines out of the region. However, we can place little confidence in the resolved intensity of the upwelling.

This problem has been investigated in the low magnetic Reynolds number regime by Allan and Bullard (1958, 1966; Allan 1961) and by Nagata and Rikitake (1961; Rikitake 1967), and recently by Bloxham (1986b) at higher values of the magnetic Reynolds number appropriate to Earth's core. The results indicate that flux can be expelled readily from the core even when the magnetic Reynolds number is high: indeed large values of the magnetic Reynolds number are certainly not sufficient for the frozen-flux approximation to be reasonable and are, in fact, likely to lead to failure of the approximation. The creation of short lengthscales in the magnetic field is characteristic of high magnetic Reynolds number flow: the importance of diffusion, as pointed out in Section 3.1, cannot necessarily be assessed on the basis of the magnetic Reynolds number.

4. Discussion

Although we now have available reliable models of the magnetic field at the CMB (although, necessarily, not obtained by completely objective arguments) and models produced by different groups are extremely similar (compare, for example, the 1980 field models of Gubbins and Bloxham (1985) and Shure, Parker and Langel (1985)) the problem of using these models to learn more about the dynamo process operating in Earth's core is far from complete. The results so-far obtained by different groups differ widely.

If the effects of diffusion are indeed, as we are now at the very least beginning to suspect, sufficiently important to render inversions for the fluid motions at the core surface unreliable in the frozen-flux approximation then we are faced with a dilemma: either we continue to attempt such inversions in the knowledge that any results obtained are possibly unreliable, or we attempt to include the effects of diffusion. But including diffusive effects, especially of the kind considered above, plunges us back into the depths of dynamo theory. The great advantage of the frozen-flux approximation is that it allows us to see the fluid motions which, presumably, are some manifestation of flow within the body of the core contributing to the dynamo, while avoiding the need to actually consider dynamo theory. We should only abandon the frozen-flux approximation very reluctantly: our current aim should be to explore more fully its domain of validity, and to seek to understand more completely to what extent we can trust the solutions which we obtain. If we are forced ultimately to abandon the frozen-flux approximation our loss will be very great, and it is unclear as to how to proceed.

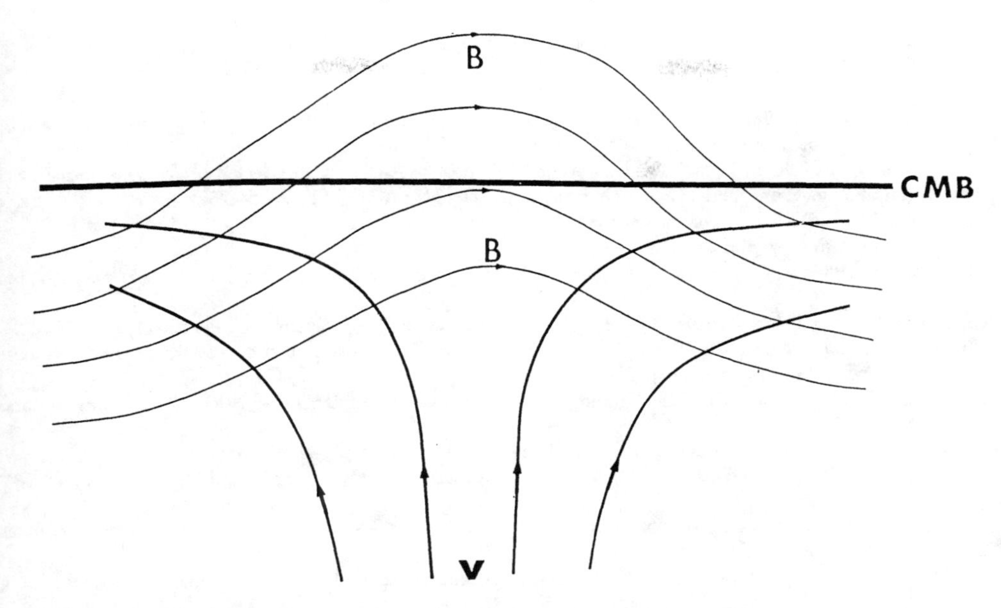

Figure 3. Schematic illustration of expulsion of toroidal field by an upwelling motion in the core.

One possibility is to use simple models of the dynamo (such as that of Busse (1975)) in conjunction with inversions in which we retain the diffusion term: certainly, it begins to appear that we need a more sophisticated theory of the secular variation to explain our observations.

Acknowledgments

I am very grateful to David Gubbins for innumerable helpful discussions throughout this study, and to Kathy Whaler for allowing me to use her FORTRAN subroutines for calculating the Gaunt and Elsasser matrices. This work was partially supported by NSF Grant EAR-8608890.

References

Allan, D.W., 1961. Secular variation of the earth's magnetic field, *Ph.D. Thesis*, University of Cambridge.

Allan, D.W. and Bullard, E.C., 1958. Distortion of a toroidal field by convection, *Rev. Mod. Phys.*, **30**, 1087-1088.

Allan, D.W. and Bullard, E.C., 1966. The secular variation of the earth's magnetic field, *Proc. Camb. Phil. Soc.*, **62**, 783-809.

Backus, G.E., 1968. Kinematics of the secular variation in a perfectly conducting core, *Phil. Trans. R. Soc.*, **A263**, 239-266.

Backus, G.E., 1970a. Non-uniqueness of the external geomagnetic field determined by surface intensity measurements, *J. Geophys. Res.*, **75**, 6339-6341.

Backus, G.E., 1970b. Inference from inadequate and inaccurate data I, *Proc. Nat. Acad. Sci.*, **65**, 1-7.

Backus, G.E. and LeMouël, J.-L., 1986. The region on the core-mantle boundary where a geostrophic velocity field can be determined from frozen-flux geomagnetic data, *Geophys. J. R. Astr. Soc.*, **85**, 617-628.

Barraclough, D.R., 1978. Spherical harmonic models of the geomagnetic field, *Geomagn. Bull. Inst. Geol. Sci.*, **8**.

Benton, E.R. and Whaler, K.A., 1983. Rapid diffusion of the poloidal geomagnetic field through the weakly-conducting mantle: A perturbation solution, *Geophys. J. R. Astr. Soc.*, **75**, 77-100.

Benton, E.R., Estes, R.H., Langel, R.A. and Muth, L.A., 1982. Sensitivity of selected geomagnetic properties to truncation level of spherical harmonic expansions, *Geophys. Res. Lett.*, **9**, 254-257.

Bloxham, J., 1986a. Models of the magnetic field at the core-mantle boundary for 1715, 1777 and 1842, *J. Geophys. Res.*, in press.

Bloxham, J., 1986b. The expulsion of magnetic flux from the Earth's core, *Geophys. J. R. Astr. Soc.*, **87**, 669-678.

Bloxham, J. and Gubbins, D., 1985. The secular variation of Earth's magnetic field, *Nature*, **317**, 777-781.

Bloxham, J. and Gubbins, D., 1986. Geomagnetic field analysis — IV: Testing the frozen-flux hypothesis, *Geophys. J. R. Astr. Soc.*, **84**, 139-152.

Bullard, E.C. and Gellman, H., 1954. Homogeneous dynamos and terrestrial magnetism, *Phil. Trans. R. Soc.*, **A247**, 213-278.

Busse, F.H., 1975. A model of the geodynamo, *Geophys. J. R.. Astr. Soc.*, **42**, 437-459.

Gubbins, D., 1983. Geomagnetic field analysis — I: Stochastic inversion, *Geophys. J. R. Astr. Soc.*, **73**, 641-652.

Gubbins, D. and Bloxham, J., 1985. Geomagnetic field analysis — III: Magnetic fields on the core-mantle boundary, *Geophys. J. R. Astr. Soc.*, **80**, 695-713.

LeMouël, J.-L., 1984. Outer core geostrophic flow and secular variation of earth's geomagnetic field, *Nature*, **311**, 734-735.

LeMouël, J.-L., Gire, C. and Madden, T., 1985. Motions at core surface in the geostrophic approximation, *Phys. Earth Planet. Inter.*, **39**, 270-287.

Moffatt, H.K., 1978. *Magnetic field generation in electrically conducting fluids*, Cambridge University Press.

Nagata, T. and Rikitake, T., 1961. Geomagnetic secular variation and poloidal magnetic fields produced by convectional motions in the earth's core, *J. Geomagn. Geoelectr.*, **13**, 42-53.

Orszag, S.A., 1970. Transform method for the calculation of vector-coupled sums: Application to the spectral form of the vorticity equation, *J. Atmos. Sci.*, **27**, 890-895.

Parker, R.L. and Shure, L., 1982. Efficient modelling of the earth's magnetic field with harmonic splines, *Geophys. Res. Lett.*, **9**, 812-815.

Rikitake, T., 1967. Non-dipole field and fluid motion in earth's core, *J. Geomagn. Geoelectr.*, **19**, 129-142.

Roberts, P.H. and Scott, S., 1965. On analysis of secular variation 1. A hydromagnetic constraint: theory, *J. Geomagn. Geoelectr.*, **17**,137-151.

Shure, L., Parker, R.L. and Backus, G.E., 1982. Harmonic splines for geomagnetic field modelling, *Phys. Earth Planet. Inter*, **32**, 114-131.

Shure, L., Parker, R.L. and Langel, R.A., 1985. A preliminary harmonic spline model from Magsat data, *J. Geophys. Res.*, **90**, 11505-11512.

Voorhies, C.V., 1986. Steady flows at the top of earth's core derived from geomagnetic field models, *J. Geophys. Res.*, **91**, 12444-12466.

Voorhies, C.V. and Backus, G.E., 1985. Steady flows at the top of the core from geomagnetic field models: the steady motions theorem, *Geophys. Astrophys. Fluid Dyn.*, **32**, 162-173.

Whaler, K.A., 1986. Geomagnetic evidence for fluid upwelling at the core-mantle boundary, *Geophys. J. R. Astr. Soc.*, **86**, 563-588.

Whaler, K.A. and Gubbins, D., 1981. Spherical harmonic analysis of the geomagnetic field: An example of a linear inverse problem, *Geophys. J. R. Astr. Soc.*, **65**, 645-693.

Winch, D.E., 1974. Evaluation of geomagnetic dynamo integrals, *J. Geomagn. Geoelectr.*, **26**, 87-94.

J. BLOXHAM, Department of Geological Sciences, Harvard University, 24 Oxford Street, Cambridge MA 02138, USA.

Chapter 10

Long wavelength features of mantle convection

G.T. Jarvis and W.R. Peltier

Two dimensional numerical solutions of the temperature field in idealized models of mantle convection are Fourier analyzed to reveal their long wavelength features. On the basis of such analyses Jarvis and Peltier (1986) previously suggested that evidence for thermal boundary layers in the mantle may be contained in the published spectral amplitudes of lateral heterogeneity, and Jarvis (1985) concluded that with current levels of resolution tomographic techniques have a better chance of detecting horizontal boundary layers than vertical plumes. This possibility is examined further here together with the nature of lateral variations of heat flow, gravity and topography at the upper surface of a convecting layer - all influenced primarily by near surface temperatures. The effects of internal heating, rigidly moving surface plates, and poor vertical resolution are also studied with respect to their impact on the spectral signatures of the model temperature solutions. Although these complications all tend to reduce the distinction between the shape of the boundary layer and interior spectra of lateral heterogeneity, a second criterion based upon the amount of power in the spectra at different depths may help to interpret the signature of boundary layers in the Earth's mantle.

1. Introduction

Geophysical interest in the thermal structure of the Earth's mantle has traditionally been focussed on the radial variation of temperature. Only recently have lateral variations become of interest; the reason being that until recently global seismology was only capable of determining the mean radial variation of seismic velocities. Consequently our view of the planetary interior was an azimuthally averaged one. With no reliable information on

N.J. Vlaar, G. Nolet, M.J.R. Wortel & S.A.P.L. Cloetingh (eds), Mathematical Geophysics, 209-226.

lateral variations of mantle structure at depth, model implications for these lateral variations were unconstrained and, therefore, of little value.

However, recent studies involving seismic tomographic imaging of the Earth's interior have begun to provide a measure of lateral heterogeneity throughout the Earth's mantle (e.g. Dziewonski et al., 1977; Masters et al. 1982; Dziewonski, 1984; Woodhouse and Dziewonski, 1984; Clayton and Comer, 1984, 1986; Hager et al., 1985; Giardini et al., 1986). The inferred lateral variations of seismic wave velocities have almost universally been attributed to variations in mantle density and, in turn, to temperature within the mantle (e.g. Anderson and Dziewonski, 1984), although crystal anisotropy and variable mineralogy may also contribute to variable seismic velocities at shallow depths. In addition the recent growth of global data sets for such fields as heat flow, gravity and chemistry of igneous rocks has led to a heightened interest in the implications and origins of the implied heterogeneity in the Earth's mantle. Since these new data may ultimately help to constrain models of mantle flow (on the basis of the lateral heterogeneity predicted by the various flow models) geophysical motivation now exists for examining the characteristic lateral heterogeneity found in models of mantle convection.

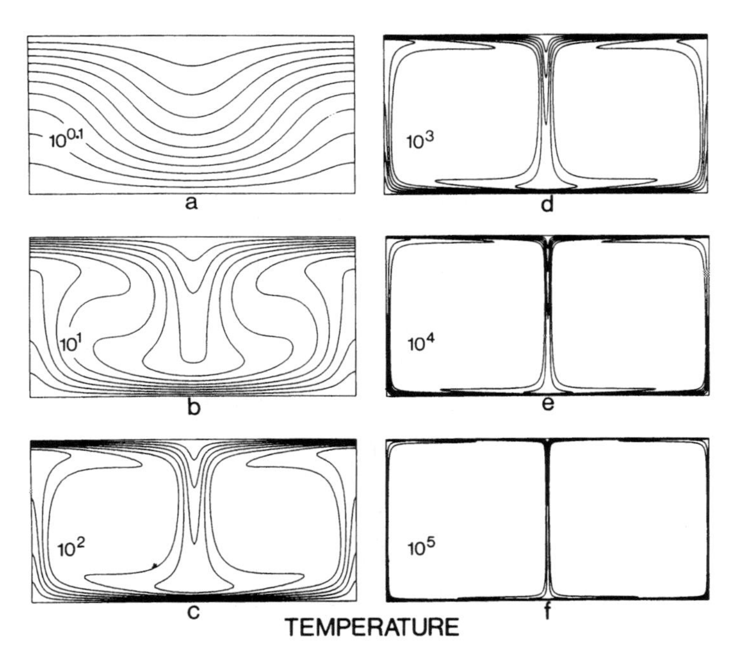

Figure 1. Steady two dimensional temperature fields for pairs of counterrotating convection rolls. Contours of temperature are plotted with a constant interval of $\Delta T/11$ in each frame. Each frame is labelled with the corresponding value of R_B/R_c. Solutions were obtained on the following numerical meshes: (a) and (b) 24 x 24; (c) 80 x 80; (d), (e) and (f) 200 x 200.

In tnis paper we discuss the signatures of lateral temperature variations predicted by simple convection models and examine possible implications for the interpretation of the seismic tomographic inversions. A major difficulty in comparing models and observations arises from the fact that due to the uneven distribution and limited amount of data currently available, only the lowest degree and order harmonic representation of the tomographic image is possible. Accordingly, published analyses to date do not attempt to extract the local variations of density or temperature. Rather they determine the first few coefficients in the spherical harmonic representation of lateral variations. Dziewonski (1984), for example, extracts the expansion coefficients from his data up to the sixth angular order (implying a lateral resolution of 3,000 km). Thus only a very smooth image, consisting of the longest wavelength harmonics, is recovered by tomographic inversion of mantle heterogeneity. In contrast, the thermal structure predicted by numerical models of whole mantle convection (see Fig. 1) is characterized by narrow plumes and boundary layers containing extreme temperature variations - that is, it contains both long and short wavelength harmonics. To enable useful comparisons of model predictions and tomographic inversions we subject the model predictions to harmonic analysis and compare the long wavelength components of the model solutions with the observations.

2. The convection model

A two-dimensional time-dependent finite difference numerical scheme is employed to solve the hydrodynamic equations governing the conservation of mass, momentum and thermal energy in a plane layer of a fluid with infinite Prandtl number. It is assumed that the fluid is incompressible with a constant viscosity and that the temperature and velocity fields are horizontally periodic and symmetric about the vertical nodal planes of the horizontal component of velocity. The wavelength of periodicity, λ, is set to $\lambda = 2d$ where d is the depth of the layer. Thus the solution domain is rectangular with an aspect ratio of one (i.e., $0 \leq x \leq \lambda/2$). For a whole mantle model $d = 2900$ km and the unit aspect ratio solutions contain features with $\lambda \leq 5{,}800$ km which is comparable to wavelengths of angular order $l = 5$ in the spherical Earth. This paper represents an extension of our earlier study (Jarvis and Peltier, 1986). The mathematical formulation and difference scheme, employed to generate the solutions which are examined in this paper, have been described previously by Jarvis (1984) and Jarvis and Peltier (1982).

3. Lateral temperature fluctuations

The nature of the lateral temperature variations found in steady convection models is illustrated by the contours of one full wavelength of the temperature fields shown in Fig. 1 for unit aspect ratio models at six different Rayleigh numbers. It is clear from Fig. 1 that lateral temperature variations are primarily due to the hot upwelling and cold downwelling streams of fluid. As the Rayleigh number increases the hot and cold plumes thin so that temperature fluctuations are confined to ever narrower regions.

Of greater geophysical interest than their variation with Rayleigh number is the manner in which lateral variations change with height for a given Rayleigh number. These changes with height are illustrated in Fig. 2(a) for a model appropriate for whole mantle convection in which the Rayleigh number, R_B, exceeds the critical value for the onset of convection, R_c, by a factor of 10^4. The uppermost profile in Fig. 2(a) is taken at a level which is one grid level Δz ($= d/200$) below the upper surface of the numerical mesh and lies well within the upper thermal boundary layer. This boundary layer profile differs significantly from the

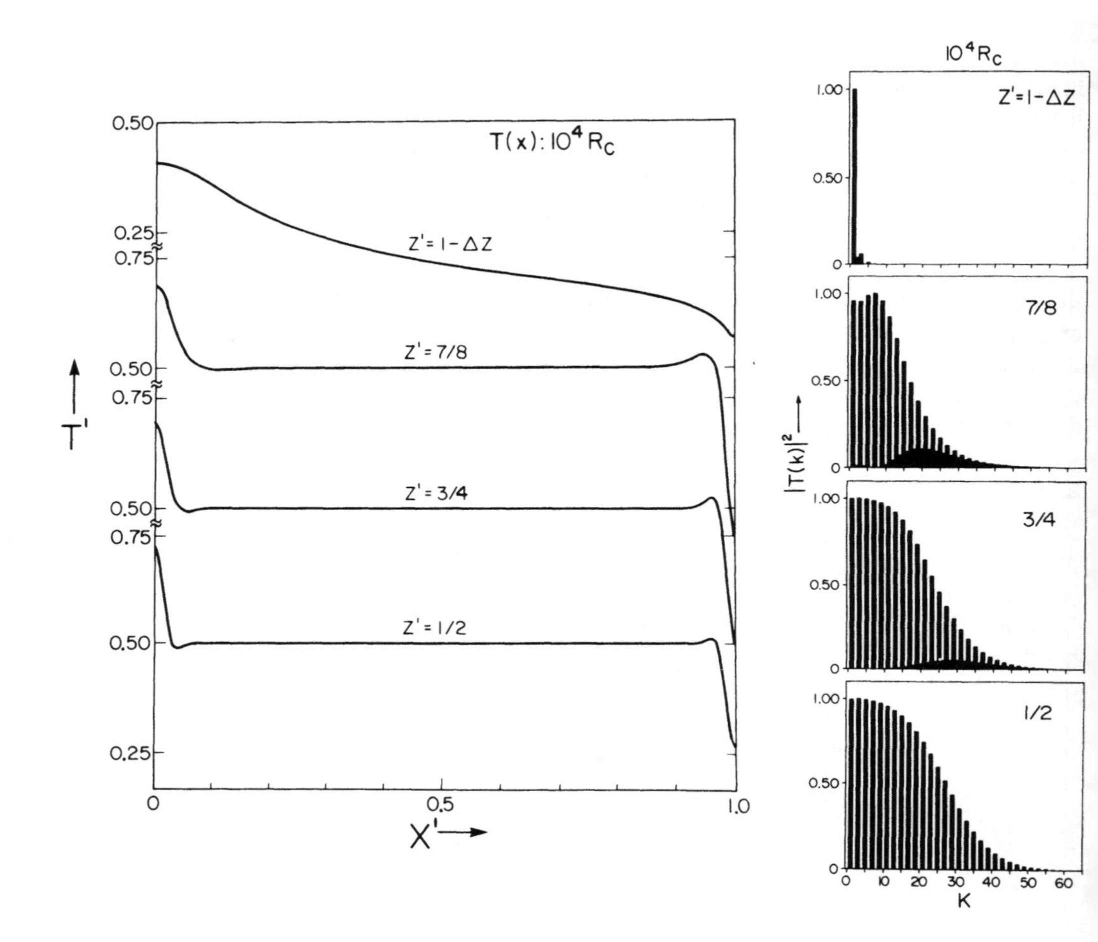

Figure 2. (a) Horizontal profiles of temperature, at various heights, across a steady convection roll for which $R_B = 10^4 R_c$. Each profile is labelled with the dimensionless height z' ($=z/d$) at which it is taken. (b) Power spectra of the lateral variation of temperature as a function of dimensionless height, corresponding to the spatial variations shown in 2(a).

deeper interior profiles in that there are no sharp fluctuations of temperature associated with hot and cold plumes. Rather, temperature varies smoothly across the profile.

The distinction between lateral variations in boundary layers and the interior of a convecting layer is equally evident in the spectral domain. To Fourier analyse the profiles of Fig. 2(a) each profile was extended to a full wavelength, $x = \lambda$, with its mirror image about the end point $x = \lambda/2$, and then subjected to the discrete Fourier transform

$$\bar{T}(k_x',z) = \sum_{n=o}^{N-1} T(n\Delta x, z) \exp\left[i\,2\pi n k_x'/N \right] \tag{1}$$

where $\bar{T}$ denotes the Fourier transform of T, k_x' is the dimensionless horizontal wavenumber, N is the number of grid points in the horizontal direction in the extended profile and Δx is the horizontal grid interval (i.e. $\Delta x = 2ad/n$ where a is the aspect ratio of the solution domain). k_x' expresses the wavenumber in integral multiples of the minimum wavenumber $k_{x(min)} = \pi/ad$. The power at each wavenumber is given by the product of $\bar{T}\,\bar{T}^*$ where $\bar{T}^*$ is the complex conjugate of $\bar{T}$. For our studies that extended temperature field has even symmetry about $x = 0$ so that $\bar{T}^* = \bar{T}$ and power is given by $|\bar{T}(k)|^2$.

In Fig. 2(b) power spectra, obtained by applying Equation (1), are juxtaposed to the corresponding temperature profiles of Fig. 2(a). The most interesting feature of these spectra is the sharp distinction between those in the interior of a steadily convecting layer, with significant power distributed over several harmonics, and that within the thermal boundary layer which is dominated by a single harmonic. (The boundary layer harmonic has a wavelength corresponding to that of the large scale circulation.) Since this distinction would be apparent even if only the first few harmonics of each spectrum were available, Jarvis and Peltier (1986) suggested it may be possible to locate thermal boundary layers within the mantle by examining the spectra of lateral heterogeneity as obtained from seismic tomography at different depths in the mantle. However, it is not as yet known whether spectra obtained for mantle heterogeneity could be expected to exhibit such a clear distinction between the thermal boundary layers and the interior of the convection cells; there are many complicating effects which may rule out this distinction. The robustness of this clear distinction will be examined, with respect to various complications, in this and future studies.

Closely related to the individual spectra of lateral heterogeneity at selected levels in the mantle is the overall image obtained tomographically. Because these images only contain long wavelength components they provide a considerably "out of focus" or blurred impression of the actual heterogeneity in the mantle. To examine the extent of this effect Jarvis (1985) performed a two-dimensional Fourier analysis of model temperature fields and extracted only the longest wavelength components from which the model fields were then re-synthesized.

To use discrete Fourier transform techniques in two dimensions the range of each temperature field was extended by generating mirror images of the field about the two planes $x = \lambda/2$ and $z = d$, such that $T(x,z) = T(\lambda-x, 2d-z)$. The horizontal extension simply completes one full wavelength of the horizontally periodic solution. The vertical extension, however, has no physical meaning; it is strictly a mathematical device used to satisfy the requirement of a periodic variation for the Fourier transform. The two dimensional transform on the extended grid (consisting of $N \times N$ mesh intervals) was obtained by first applying Equation (1) horizontally across each row of the numerical grid, and then applying the same algorithm to the vertical variation of each spectral amplitude of the horizontal transforms. The wavenumbers in the horizontal and vertical directions are denoted k_x and k_z respectively, and each have a minimum value of $k_{min}=\pi/d$ and a maximum value of $k_{max} = N\pi/2d$, where N is number of grid elements in each direction of the extended numerical mesh. Dimensionless wavenumbers are defined relative to the minimum wavenumber as $k_x' = k_x/k_{min}$ and $k_z' = k_z/k_{min}$. Because the extended temperature field has even symmetry in both x- and z-directions the spectral amplitudes $\bar{T}(k_x', k_z')$ are real, and the inverse transform is therefore obtained by re-applying Equation (1) to $\bar{T}/N$ (Kanasewich, 1981) as

$$T(nx,z) = (1/N) \sum_{k_x'=0}^{K} T(k_x',z) \exp\left[i2\pi nk_x'/N\right] \quad (2)$$

To recover the complete field $T(x,z)$, the summation limit K in Equation (2) must be $K = N-1$. However, if only a subset of low wavenumber components is to be included in the re-synthesized field the value of K is set to a lower value such that $1 \leq K \leq N-1$. Such incomplete fields will be labelled $T^*_{kxmax,kzmax}$ where the pair of subscripts indicates the maximum wavenumbers in the x- and z-directions.

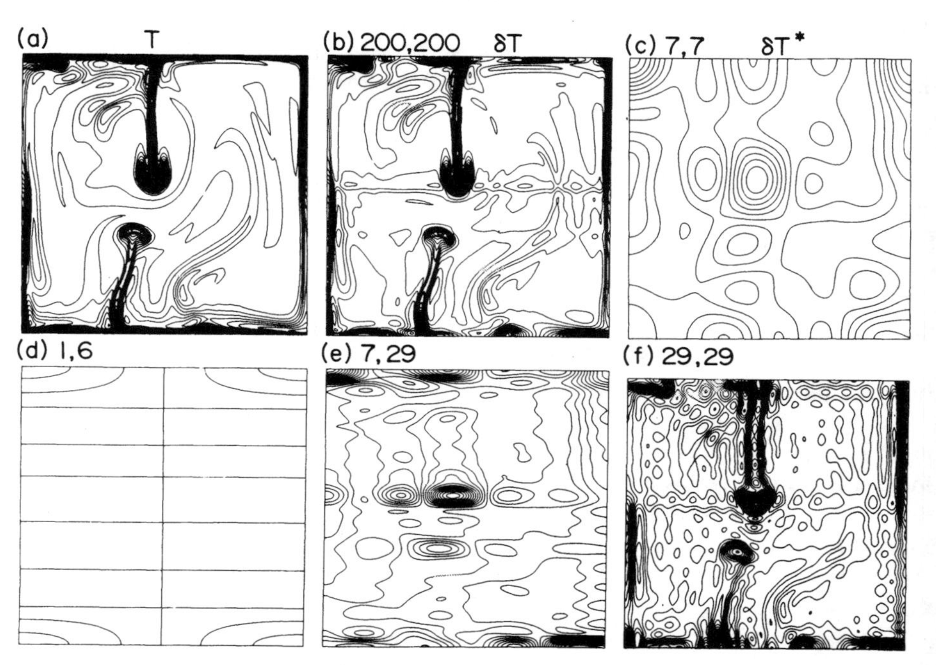

Figure 3. Isotherms for a model solution of time-dependent whole-mantle convection. This model is heated 20% from within and 80% from below. It has a constant heat flux lower boundary condition and a Rayleigh number R_R defined in terms of heat flux at the upper surface (e.g. McKenzie et al., 1974) equal to 1.5 x 10^9. Contour interval is 45°C in each frame. The complete field is shown in frame (a). Frames (d) and (e) show the truncated fields $\delta T^*_{1,6}$ and $\delta T^*_{7,29}$ which are explicitly discussed in the text, where the subscripts indicate the number of harmonics retained in the x- and z-directions in the series expansion.

Truncating the Fourier series expansion has the effect of defocussing the model fields and thus casts them into a form similar to that of the observations although it should be noted that such abrupt truncation in the wave number domain has the undersireable effect of inducing Gibb's oscillations in the reconstructed space domain fields, an effect against which the tomographic inversions have been at least partly protected. One additional major distinction, however, is that tomographic results are normally presented as lateral fluctuations about an azimuthal average which is not in general known at each level (e.g. Masters et al., 1982; Dziewonski, 1984; Hager et al., 1985). Such a presentation omits the major signature of horizontal thermal boundary layers: viz. the mean vertical variation.

The implication here is that the tomographic image of a horizontal boundary layer would appear less pronounced than would that of a vertical plume at the same level of resolution. In practice, however, resolution is usually higher in the vertical direction than it is azimuthally. This tends to mitigate the inherent bias against the detection of horizontal boundary layers.

To render the truncated model solutions closer to the format of the observations in at least this one respect, Jarvis (1985) also eliminated the mean vertical variation of temperature in the resynthesized fields by setting the first term in the summation in Equation (1) to zero [i.e., $\bar{T}(k_x'=0,z) = 0$] at each level in the numerical grid. This procedure removes the DC component from the resynthesized temperature field so that the horizontal average at every level is equal to zero.

An example of a model temperature field which has been Fourier analyzed and partially resynthesized in this way is shown in Fig. 3. The contours of Fig. 3(a) are isotherms of the temperature field at one instant in a highly time dependent flow. Figs. 3(d) and (e) show isotherms for two of several different truncated expansions of the complete field shown in Fig. 3(a). These are denoted $\delta T^*_{1,6}$ and $\delta T^*_{7,29}$ where the symbol δ denotes variations about the horizontal mean, the asterisk indicates truncation and the subscripts indicate the maximum wavenumbers in the x- and z-directions (i.e. $\delta T^*_{7,29}$ includes all harmonics with $1 \leq k_x' \leq 7$, $1 \leq k_z' \leq 29$). Frames (d) and (e) show expansions with truncation at the levels thought to approximate the resolution of Dziewonski's (1984) and Clayton and Comer's (1984) tomographic images. Both plumes and thermal boundary layers are poorly represented in Fig. 3(d) while horizontal boundary layers, but not the vertical plumes, are evident in Fig. 3(e) in the form of rows of alternating high- and low-temperature concentrations adjacent to the horizontal bounding surfaces. In addition, the relatively broad heads of the transient plumes are recovered in Fig. 3(e) but appear to be detached from the upper and lower surfaces. Thus if tomographic images were reliable at this level of resolution we might expect to see the major horizontal boundary layers but should not look for the continuity of even the major vertical structures. Unfortunately, Hager et al. (1985) have argued that the tomographic inversions are quite inaccurate at the level of resolution of Fig. 3(e) and even the low resolution of Fig. 3(d) is probably better than can be reliably recovered at present. Thus Fig. 3(d) indicates that, at best, tomography may reveal the presence of horizontal thermal boundary layers but not vertical plumes. The boundary layers would appear as isolated patches of anomalously hot and cold regions. Similar patches of anomalously slow and fast seismic velocity appear in Dziewonski's (1984) cross sectional tomographic images all along the core-mantle boundary - a probable site of a major thermal boundary layer and the spectra of his model have been interpreted in this way by Forte and Peltier (1987). If the lateral variation of seismic velocity is taken as proxy for the lateral variation of temperature, we can understand this "observed" image of the variation of the temperature field, in terms of Fig. 3, as the long wavelength component of the variations of the full field, again as discussed in Forte and Peltier (1987) and further elaborated by them elsewhere in this volume.

4. Surface features

The fact that the truncated expansions of the model fields contain more information concerning horizontal boundary layers than of other regions in the convecting layer is consistent with the variation with depth of the spectra of lateral heterogeneity shown in Fig. 2(b). Fig. 2(b) indicates that within the boundary layers most of the power is contained in the longest wavelength harmonics whereas away from boundary layers, since power is distributed over many harmonics, a smaller fraction of the total power in the spectra occurs at the longest wavelengths. This suggests that not only the lateral variations of temperature within the boundary layers but also those features which depend most strongly upon lateral temperature variations within the boundary layers will be most reliably recovered by tomographic inversions at long wavelengths.

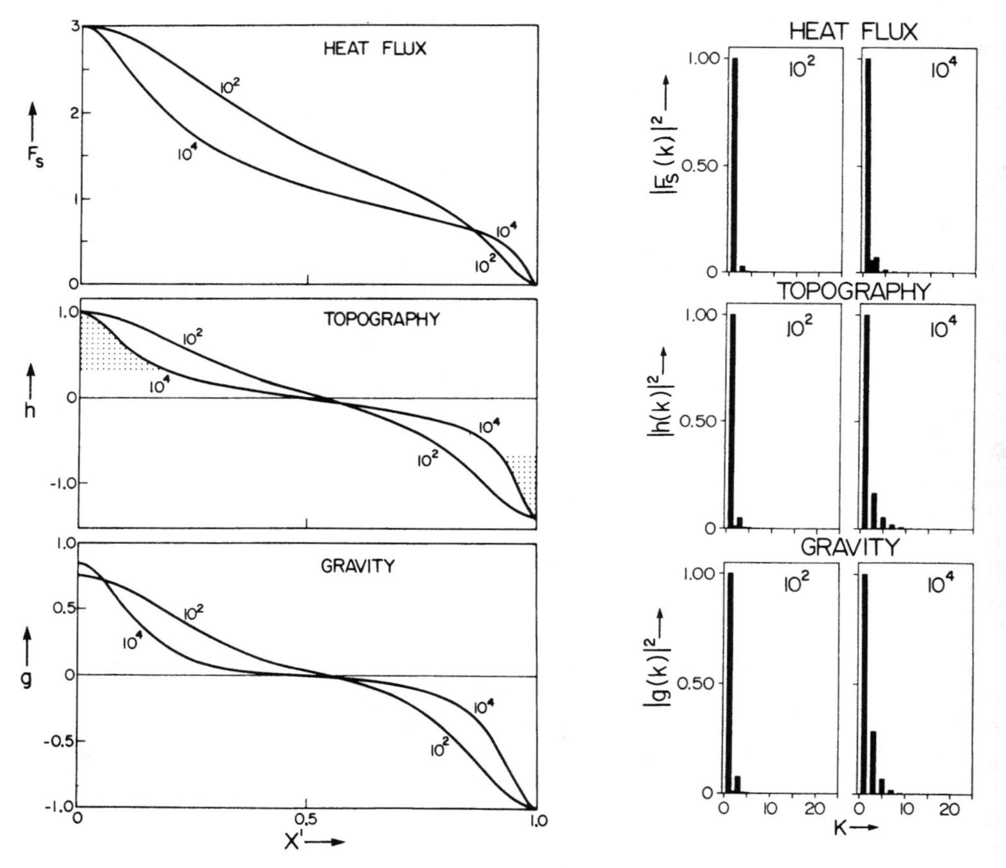

Figure 4. (a) Shapes of the surface profiles of heat flux F_s, topography h, and gravity g ,above two model convection cells, one with (Benard) Rayleigh number $R_B = 10^2 R_c$ and one with $R_B = 10^4 R_c$. Stipling on the topography curve is to highlight the asymmetry. (b) Power spectra of the profiles shown in 4(a), normalized with respect to the maximum power in each spectrum.

For example, Hager et al. (1985) estimated the magnitude of the lateral variations both of temperature within the mantle and of topographic variations at the core-mantle boundary and at the Earth's surface based upon the tomographic images of lateral heterogeneity within the mantle. However, Jarvis (1985) subsequently demonstrated that while the longest wavelength component of the lateral temperature variation contains only approximately 10% of the total variation in a numerical model of mantle convection, the topographic variation induced by this component contains approximately 50% of the total topographic variation. The more complete recovery of topography than temperature, even though the topography results from lateral variations in temperature, is due to the fact that topography depends primarily upon the near surface (i.e. boundary layer) temperatures (e.g., Parsons and Daly, 1983). The same is true for surface heat flow and the observed gravity field above a convecting layer, both of which arise as a consequence of the lateral heterogeneity of temperature. This is illustrated in Fig. 4(a) where profiles are shown of the surface heat flux, F_s, topography, h, and gravity, g, as predicted by two model solutions of Bénard convection between free boundaries with Rayleigh numbers $R_B = 10^2 R_c$ and $10^4 R_c$, where $R_c = 779.273$ is the critical value of R_B for the onset of convection. Each profile has been normalized with respect to its maximum departure from zero in order to facilitate the comparison of the *shapes* of the profiles at different Rayleigh numbers.

For each of the quantities plotted in Fig. 4(a) the lateral variations are relatively smooth and uniform at $R_B = 10^2 R_c$ but relatively sharp and concentrated toward the cell edges at $R_B = 10^4 R_c$. At both Rayleigh numbers all of the profiles of surface features plotted in Fig. 4(a) exhibit an asymmetry about their mid-points with a narrower spatial variation above the sinking plume (at right) than that above the rising plume (at left). This asymmetry is most noticeable in the profiles of surface topography and is in the same sense as the observed asymmetry between broad ocean ridges and narrow trenches. Although this asymmetry would be enhanced in models with internal heating, even without internal heating the topographic and gravitational signatures of a sinking plume are stronger and more concentrated than those of a rising plume. This is clearly due to the fact that sinking plumes originate in the upper boundary layer and their lateral temperature variation is highly concentrated, whereas rising plumes have traversed the entire depth of the convecting layer before reaching the upper boundary layer and their near-surface lateral temperature variation is therefore more diffuse.

The heat flow at the upper surface of the convecting layer was computed using the energetically complete formulation described by Jarvis and Peltier (1982), while the surface topography and gravity were computed following the approach of McKenzie et al. (1974) and McKenzie (1977). For each of these three quantities the power spectra of the profiles at both Rayleigh numbers are shown in Fig. 4(b). In general most of the power is concentrated at the lowest wavenumber, particulary for the low Rayleigh number case for which the surface profiles are relatively smooth. The spectra for the heat flux are very similar to that shown in Fig. 2(b) for the temperature just below the upper surface at $z = 1 - \Delta z$. This follows from the fact that the heat flux at this level is predominantly due to conduction and hence the lateral variation of heat flux is controlled primarily by the lateral variation of temperature immediately below the constant temperature upper surface.

The spectra for topography are similar to those for vertical velocity (Jarvis and Peltier, 1986). This similarity results from the fact that the most rapid changes in surface topography are produced by the velocity gradient $\partial w/\partial z$ (Jarvis and Peltier, 1982). Since $w = 0$ at the upper surface the lateral variation of $\partial w/\partial z$ is controlled by the lateral variation of w immediately below the surface.

The gravity spectra are similar to those of topography but have more power at $k_x' > 1$ than do the spectra for topography. The similarity stems from the fact that in constant viscosity fluids the surface topography is the major contributor to the gravity field (McKenzie et al., 1974; McKenzie, 1977). The larger power at $k_x' > 1$ is due to the fact that gravity is also sensitive to temperature variations below the upper boundary. As shown in Fig. 2(b) the temperature spectra at depths below the upper boundary layer are rich in power at $k_x' > 1$. The increase in power at $k_x' > 1$ (relative to topography) is only slight, however, because the contributions of temperature fluctuations to the gravity field decrease rapidly with depth. An important point to note from Fig. 4 is that the sign of the gravity anomaly over upwellings is opposite to that observed on the surface of the Earth where positive anomalies are associated with trenches and negative anomalies are associated with ridges (Forte and Peltier, 1987).

The concentration of most of the power in the lowest order harmonic has strong geophysical implications. If the long wavelength components are filtered out of the global gravity field, for example, (e.g. McKenzie et al., 1980) most of the power associated with the large scale flow will be eliminated and major features such as mid-ocean ridges would not be apparent in the composite (but filtered) field. If, for example, the mean plate size of approximately 6,000 km represents a typical half-wavelength of the convective flow in the whole mantle, the lowest harmonic has a wavelength of $\lambda = 12{,}000$ km and the third harmonic has $\lambda = 4{,}000$ km. In their analysis of short wavelength features of the Earth's gravity field and oceanic bathymetry, McKenzie et al. (1980) filtered out wavelengths greater than 4,000 km (i.e. $k_x' \leq 3$). Fig. 4(b) suggests that this would eliminate from the data more than 90% of the power associated with the large scale convective overturning. Accordingly, mid-ocean ridges were not apparent in their synthesized data. On the other hand, Forte and Peltier (1987) found that when only the relatively low order components of the gravity field are retained, the signature of oceanic ridges is quite apparent in the gravity field.

5. Influence of internal heating and moving plates

Because of the simplicity of the numerical models they are deficient in modelling convection in the Earth's mantle in many respects. Two important aspects of mantle convection which have not been included are: (1) the presence of internal heating by spontaneous decay of naturally occuring radioisotopes in the mantle and (2) the strong dependence of mantle viscosity on temperature (e.g. Weertman, 1970; Kolstedt and Goetze, 1974), which presumably accounts for the rigidity of the moving lithospheric plates. Rather than conduct an exhaustive survey we shall simply illustrate the nature of the influence of these factors by comparing two pairs of typical models.

The effects of internal heating on lateral variations of surface features is illustrated in Fig. 5, where spatial profiles and power spectra are compared for two models with the same

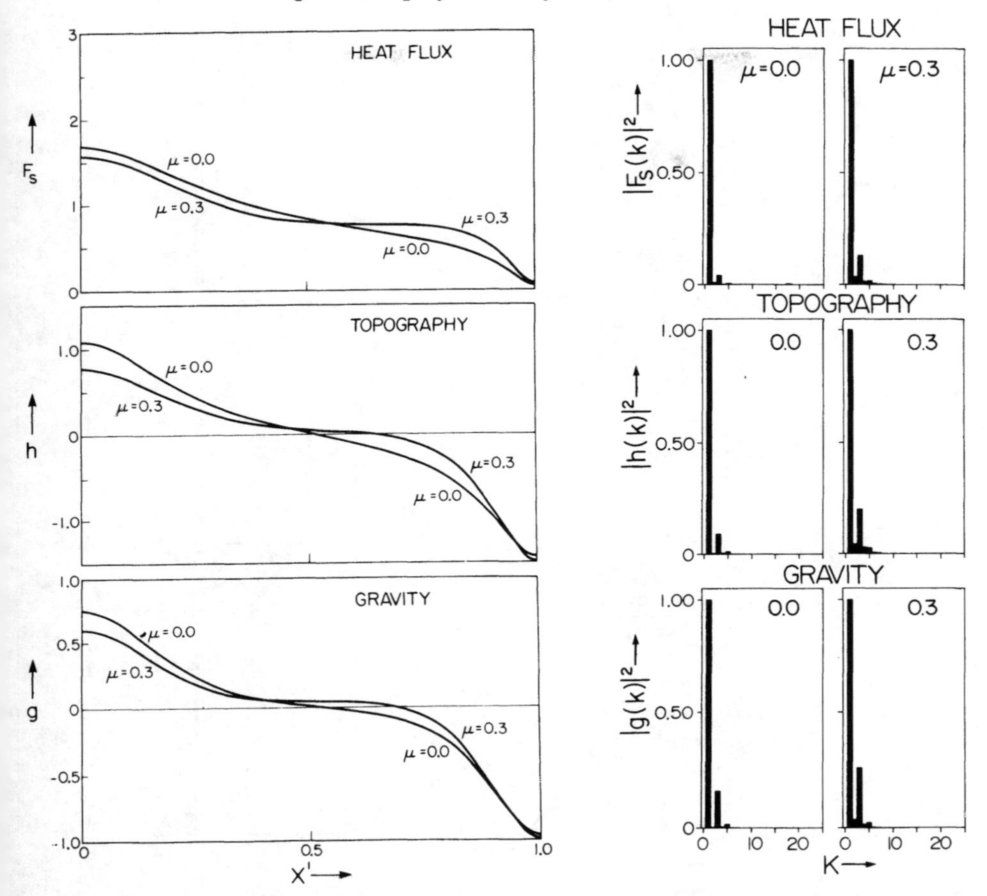

Figure 5. (a) Shapes of the profiles of heat flux Fs, topography h, and gravity g, at the upper surface of convection cells with partial internal heating. $\mu = 0.0$ indicates no internal heating; $\mu = 0.3$ indicates 30% internal heating. (b) Power spectra of the profiles shown in 5(a).

total surface heat flux but different fractional amounts of internal heating, denoted μ, equal to 0% and 30% of the surface heat flux. These plots indicate that the main effect of including partial internal heating is to produce a flattening of the profiles of each surface feature. The heat flux and gravity profiles are, respectively, the least and most sensitive to the inclusion of internal heating. One consequence of this flattening is that the asymmetry between features above upwelling and downwelling flow is accentuated in the case of partial internal heating. The influence of internal heating is registered on the corresponding power spectra (Fig. 5(b)) as small but detectable power in the even harmonics and increased power in the higher odd harmonics.

With regard to the rigidity of the surface plates, Jarvis and Peltier (1981, 1982) have argued that the role of the rigid lithosphere in integrating surface stresses to produce a uniform surface velocity can be mimiced by imposing an appropriate constant surface velocity as a boundary condition at the upper surface of the cell. By choosing the surface

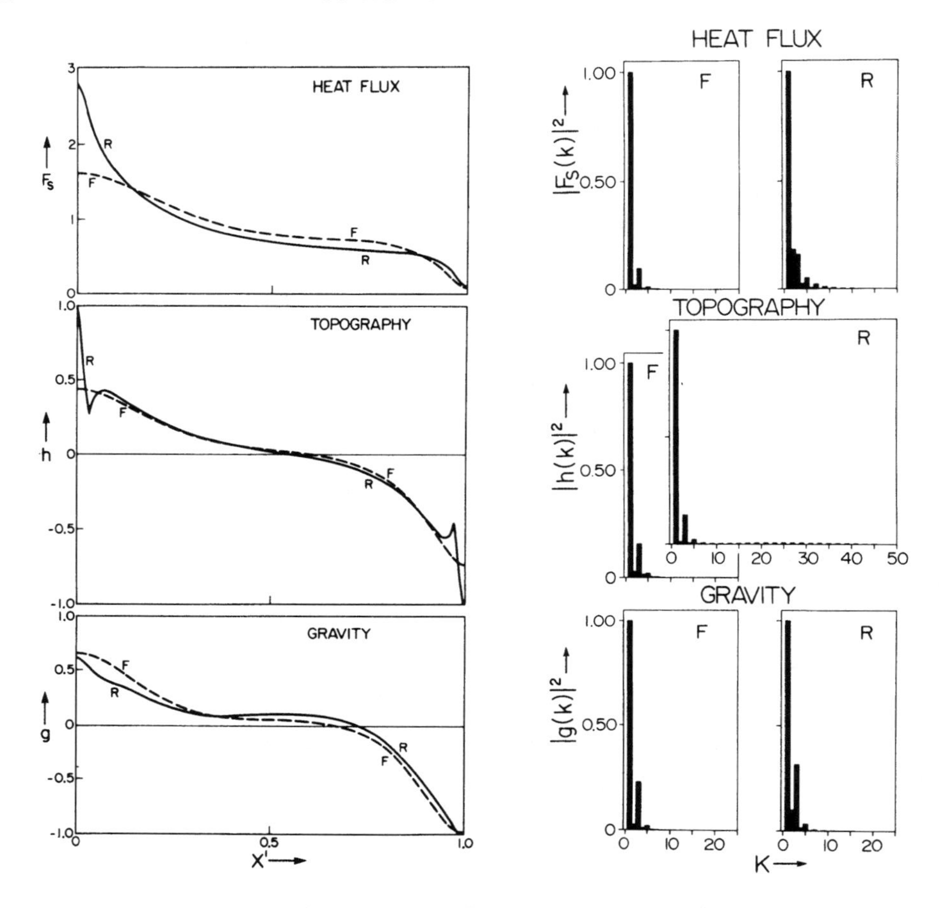

Figure 6. (a) Shapes of the profiles of heat flux Fs, topography h, and gravity g, at the upper surface of convection cells with either free boundaries (F) or a rigidly moving upper surface (R). (b) Power spectra of the profiles shown in 6(a).

velocity equal to the mean surface velocity which would occur for the same model with a free upper surface, the imposition of this boundary condition results in no net forcing of the circulation, although both positive and negative shear stresses are generated at different points along the surface. Fig. 6 presents a comparison of the surface variations of heat flux, topography and gravity in the spatial and wavenumber domains for two models with the same surface heat flux, both heated 20% from within and 80% from below (i.e. $\mu = 0.2$). The models differ in the single respect that one has a free upper boundary (indicated by the label F in Fig. 6) and the other has a rigidly moving upper surface (indicated by the label R). With the exception of a small edge effect, there is very little difference between either the spatial or wavenumber signatures of the two models with regard to topography or gravity. However, the heat flow variation is significantly altered by the change of mechanical boundary condition. In the case of a rigidly moving upper surface the heat flow profile approximates much more closely that predicted by kinematic plate models (e.g.

Parsons and Sclater, 1977). This change in the heat flow signature is expressed in the power spectrum for the heat flow as increased power at all wavenumbers greater than one.

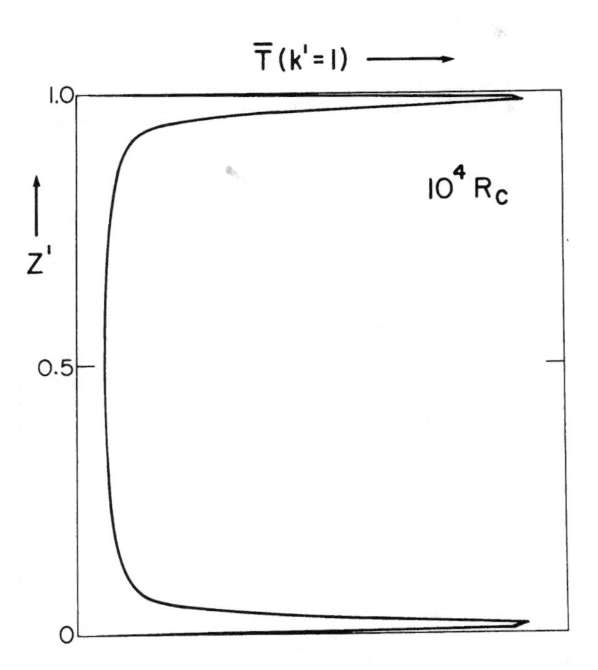

Figure 7. Vertical variation of the magnitude of the maximum spectral amplitude at each level in a steady model solution for Benard convection with a Rayleigh number of $R_B = 10^4 R_c$.

Fig.s 5 and 6 indicate that the basic conclusion drawn with regard to Fig. 4 is essentially unaffected by the additional factors of partial internal heating and rigidly moving plates at the upper surface. Thus surface features affected most strongly by near surface temperatures continue to be dominated by their long wavelength components and are, therefore, the best candidates for tomographic imaging.

6. The bottom boundary layer

A different type of complication arises when the model flows are heated entirely from within (i.e. $\mu = 1.0$) . Steady models under this condition have no thermal boundary layer at the lower surface. Yet the power spectrum at the bottom boundary is dominated by a single long wavelength harmonic (Jarvis and Peltier, 1986) and thus has the same *shape* as for a model with a thermal boundary layer at the bottom. The difference between models with and without bottom boundary layers lies in the relative amounts of power in the spectra at different levels. All of the spectra shown in this paper have been normalized by the maximum power at each level. But this maximum power is typically much larger in the boundary layers than in the interior. Fig. 7, for example, shows the variation of the magnitude of the maximum spectral amplitude with depth for a Bénard convection model at large Rayleigh number (and no internal heating). The upper and lower boundary layers both stand out in this plot as regions of extremely large amplitude relative to the interior regions.

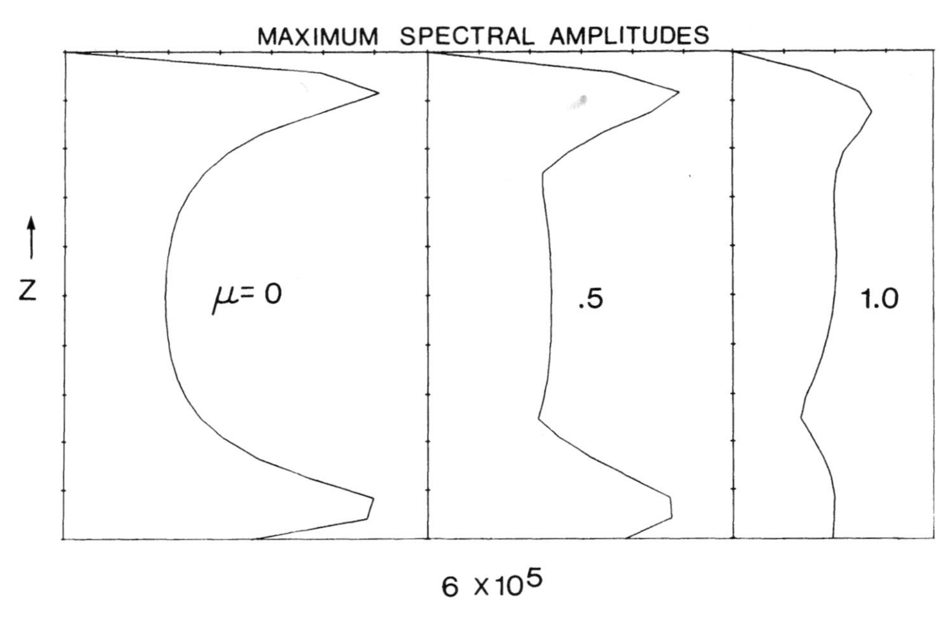

Figure 8. Vertical variations of the magnitude of the maximum spectral amplitude at each level in three steady model solutions. These models have constant heat flux bottom boundary conditions, the same surface heat flux but different ratios, μ, of internal rate of heat generation to surface heat flux. The Rayleigh number for each model is $R_R = 6x\,10^5$.

Fig. 8 shows similar plots for three steady models which all have the same surface heat flux but different degrees of internal heating, varying from 0% at the left to 100% at the right. The power spectra along the bottom row of each of these models is dominated by the fundamental harmonic of the circulation. However, when there is significant heating from below (i.e. $\mu = 0.0$ or 0.5 in Fig. 8) the power in the single harmonic at the base of the layer is greater than the power in any of the harmonics in the interior of the convection cell. In contrast, when heating is entirely from within (and accordingly, there is no bottom boundary layer) the power in the single harmonic at the base of the layer is no greater than harmonics in the spectra at higher levels. Thus a second criterion which appears to be required in order to detect a thermal boundary layer is an increased power in the spectrum relative to spectra at other levels in the fluid layer.

This criterion also appears to be applicable in time dependent flows. Fig. 9 displays plots similar to those of Fig. 8 for a Rayleigh number approximately 10 times that of the models in Fig. 8. At this Rayleigh number the flow remains steady for the model with no internal heating (i.e. $\mu = 0$) but is time dependent for the model with 60% internal heating

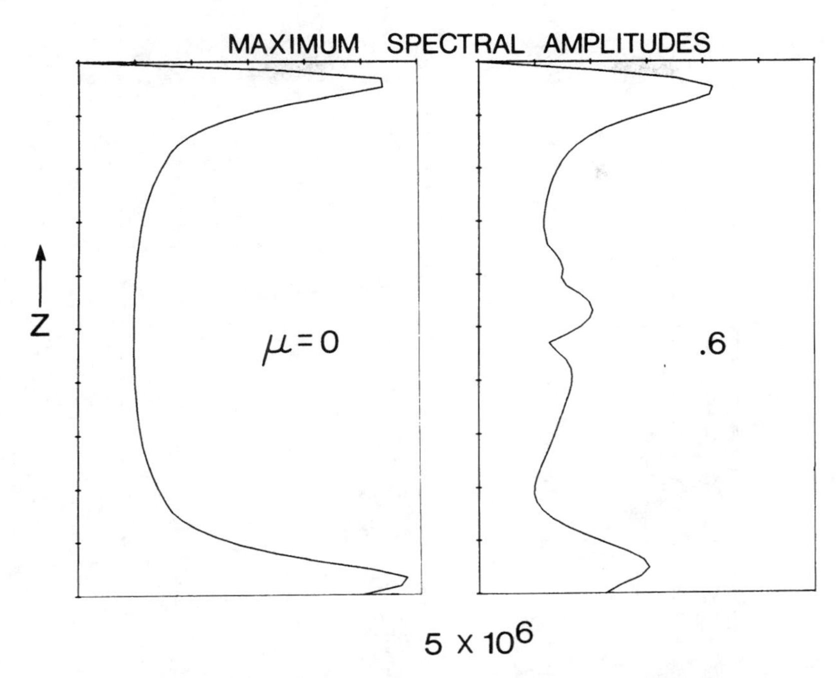

Figure 9. Vertical variations of the magnitude of the maximum spectral amplitude, as in Fig. 8, but for $R_R = 5x\,10^6$ and for $\mu = 0.0$ and 0.6. The model with 60% internal heating is time dependent at this larger Rayleigh number.

($\mu = 0.6$). Although the vertical variations are less uniform in the time dependent case, a rise in power at the lower boundary is still evident.

7. Reduced vertical resolution

Having a second criterion for detecting thermal boundary layers in the long wavelength tomographic data is especially important because of the low vertical resolution of the tomographic data. The discussions above concerning the manner in which the model spectra vary with depth have implicitly adopted the vertical resolution of the model solutions. For 200 x 200 grids this is equivalent to a 15 km vertical resolution, whereas Dziewonski's (1984) tomographic inversions of lower mantle heterogeneity, for example, have a vertical resolution of approximately 500 km. Since thermal boundary layers in the mantle are expected to be about 100 km thick (Jarvis and Peltier, 1982), the sharp distinction seen in Fig. 2 between the spectra inside and outside of thermal boundary layers may be blurred somewhat in the observations. To examine this effect we have averaged the model temperature fields over finite vertical intervals centred on the depth of interest and then computed the power spectra of the "blurred" model fields at various depths.

The result of applying this procedure to the steady whole mantle model of Fig.s 1(e) and 2 is shown in Fig. 10. At the left are spectra from individual rows of the model solution, while at the right are the spectra which result after the temperature field is

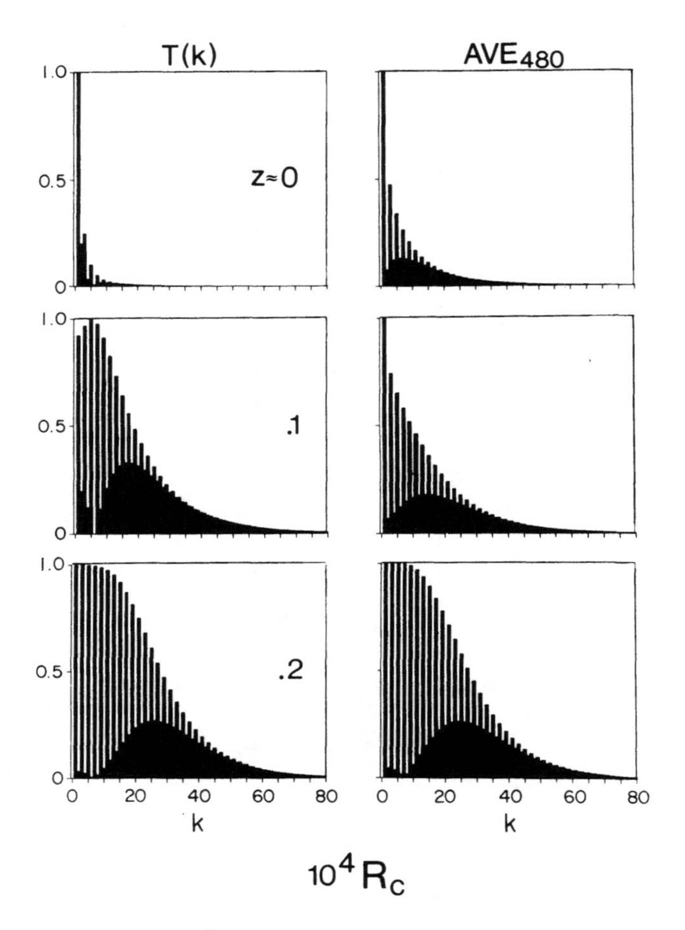

Figure 10. Amplitude spectra $|\bar{T}(k)|$ at three dimensionless heights $z' = \Delta z$, 0.1 and 0.2, in the vicinity of the lower boundary of the steady state unit aspect ratio model of Bénard convection with $R_B = 10^4 R_c$. Frames at the left are spectra for temperature variations along individual rows of grid elements in the model solution. Frames at the right are spectra for the horizontal variation of temperature which has been averaged over 33 rows of the 201 row grid (centered on the dephts $z' = 0.1$ and 0.2 in the upper two frames and rows 0 to 32 for the lower frame) which for a 2,900 km deep mantle corresponds to about 480 km.

averaged over a vertical interval of 480 km (or 33 rows of the 201 row numerical mesh). The spectra are shown for dimensionless heights of $z/d = \Delta z$, 0.1 and 0.2, with zero being the bottom of the layer and z being the vertical grid spacing (1/200 in this case). Away from the thermal boundary layer (e.g. at $z/d = 0.2$) the averaging has little effect on the shape of the spectrum. However, in the thermal boundary layer ($z/d = \Delta z$) the vertical averaging adds power at high harmonics and therefore weakens the distinction between the boundary layer and interior spectra. This weakening occurs because the averaging distance is considerably greater than the thickness of the boundary layer and the blurred boundary layer temperature profile is thus contaminated by the temperature field in the cell interior.

Given that contamination of boundary layer spectra is to be expected in the tomographic data, the second criterion discussed in Section 5 may be especially helpful in identifying thermal boundary layers in the mantle from the tomographic data.

8. Conclusions

We have modelled various complications which may affect our ability to identify thermal boundary layers in the Earth's mantle from the spectral amplitudes of lateral density variations as obtained by seismic tomography. The presence of significant amounts of internal heating (from radioactivity) and the poor vertical resolution of seismic data may seriously inhibit the occurence in the tomographic data of the sharp distinction found in simple models between the shape of the spectral envelopes within thermal boundary layers as opposed to those in the interior of the convecting layer. However, the increased power in the long wavelength components of the spectra observed within the thermal boundary layers provides a second diagnostic criterion. In this regard it is noteworthy that Dziewonski (1984) found a sharp increase in spectral amplitudes at the core-mantle boundary where a thermal boundary layer might be expected. Height dependent spectra of his model of the lateral heterogeneity have been constructed by Forte and Peltier (1987) and shown to be interpretable using the analysis reviewed here and originally presented in Jarvis and Peltier (1986). We suggest that future model studies concentrate on the signature of lateral heterogeneity within the thermal boundary layers of more complicated models and attempt to apply the findings to regions such as the core-mantle boundary where a thermal boundary layer is expected to exist in the form of the seismically observed D'' region.

Acknowledgements

Financial support in the form of operating grant funds from the Natural Science and Engineering Research Council of Canada is gratefully acknowledged. We are also indebted to the Program of Vector Computer Access operated by NSERC for scientific use of the CRAY X-MP facility at Dorval Quebec on which many of the computations discussed here were performed.

References

Anderson, D.L., and Dziewonski, A.M., 1984. Seismic tomography, *Sci. Am.*, **251**, 60-68.

Clayton, R.W., and Comer, R.P., 1984. A tomographic analysis of mantle heterogeneities, *Terra Cognita*, **4**, 282.

Dziewonski, A.M., 1984. Mapping the lower mantle: determination of lateral heterogeneity in P velocity up to degree and order 6, *J. Geophys. Res.*, **89**, 5929-5952.

Dziewonski, A.M., Hager, B.H., and O'Connell, R.J., 1977. Large scale heterogeneities in the lower mantle, *J. Geophys. Res.*, **82**, 239-235.

Forte, A.M., and Peltier, W.R., 1987. Plate tectonics and aspherical Earth structure: the importance of poloidal-toroidal coupling, *J. Geophys. Res.*, **92**, 3645-3679.

Giardini, D., Li, X., and Woodhouse, J.H., 1986. Heterogeneous models of the Earth from modal splitting functions of low order multiplets, *Terra Cognita*, **6**, 291.

Hager, B.H., Clayton, R.W., Richards, M.A., Comer, R.P., and Dziewonski, A.M., 1985. Lower mantle heterogeneity, dynamic topography and the geoid, *Nature*, **313**, 541-545.

Jarvis, G.T., 1984. Time dependent convection in the Earth's mantle, *Phys. Earth Plan. Int.*, **36**, 305-327.

Jarvis, G.T., 1985. The long wavelength component of mantle convection, *Phys. Earth Plan. Int.*, **40**, 24-42.

Jarvis, G.T., and Peltier, W.R., 1981. Effects of lithospheric rigidity on ocean floor bathymetry and heat flow, *Geophys. Res. Lett.*, **8**, 857-860.

Jarvis, G.T., and Peltier, W.R., 1982. Mantle convection as a boundary layer phenomenon, *Geophys. J. R. Astr. Soc.*, **68**, 389-427.

Jarvis, G.T., and Peltier, W.R., 1986. Lateral heterogeneity in the convecting mantle, *J. Geophys. Res.*, **91**, 435-451.

Kanasewich, E.R., 1981. *Time Sequence Analysis in Geophysics*, The University of Alberta Press, Edmonton, p 480.

Kolstedt, D.L., and Goetze, C., 1974. Low stress and high temperature creep in olivine single crystals, *J. Geophys. Res.*, **79**, 2045-2051.

Masters, G., Jordan, T.H., Silver, P.G., and Gilbert, F., 1982. Aspherical Earth structure from fundamental spheroidal-mode data, *Nature*, **298**, 609-613.

McKenzie, D.P., 1977. Surface deformation, gravity anomalies and convection, *Geophys. J. R. Astr. Soc.*, **48**, 211-238.

McKenzie, D.P., Roberts, J.M., and Weiss, N.O., 1974. Convection in the Earth's mantle: towards a numerical simulation, *J. Fluid Mech.*, **62**, 465-538.

McKenzie, D.P., Watts, A., Parsons, B., and Roufosse, M., 1980. Planform of mantle convection beneath the Pacific Ocean, *Nature*, **288**, 442-446.

Parsons, B., and Daly, S., 1983. The relationship between surface topography, gravity anomalies and the temperature structure of convection, *J. Geophys. Res.*, **88**, 1129-1144.

Parsons, B., and Sclater, J.G., 1977. An analysis of the variation of ocean floor bathymetry and heat flow with age, *J. Geophys. Res.*, **82**, 803-827.

Weertman, J., 1970. The creep strength of the Earth's mantle, *Rev. Geophys. Space Phys.*, **8**, 145-168.

Woodhouse, J.H., and Dziewonski, A.M., 1984. Mapping the upper mantle: three dimensional modeling of Earth structure by inversion of seismic waveforms. *J. Geophys. Res.*, **89**, 5953-5986.

G.T. JARVIS, Department of Earth and Atmospheric Science, York University, North York ON M3J 1P3, Canada.

W.R. PELTIER, Department of Physics, University of Toronto, Toronto ON M5S 1A7, Canada.

Chapter 11

Mean-field methods in mantle convection

Francesca Quareni and David A. Yuen

The use of the single-mode, mean field method as a simple tool in understanding the physics of mantle convection is investigated. We present here a summary of steady-state results derived from both the incompressible (Boussinesq) and the compressible (anelastic) set of equations for cartesian geometry. We have employed various types of Newtonian rheology: constant, depth dependent, temperature dependent, and both temperature and pressure dependent viscosities. Mean-field methods can be quite useful in predicting heat transfer characteristics and the averaged interior temperature of the convecting medium for a wide variety of material properties and boundary conditions. Several new findings have resulted from the mean-field method. Of foremost geophysical interest are (1.) the strong non-linear coupling between the adiabatic heating parameters and the creep parameters, particularly the activation volume, (2.) the constraint on the amount of internal heating in the mantle from consideration of melting temperatures in variable viscosity convection and (3.) the important role played by dissipative heating in compressible convection with temperature and pressure dependent viscosity. In order to satisfy the constraints of mantle adiabaticity, the activation volume in the lower mantle can not be larger than about 4 cm^3/mole. Enhanced dissipation takes place when the compressibility is sufficiently high, corresponding to Grueneisen parameters less than about 1.5. Thus there seems to exist a self-regulating mechanism in the mantle for controlling the interior temperature, owing to the non-linear interactions among internal heating, variable viscosity and mantle compressibility.

N.J. Vlaar, G. Nolet, M.J.R. Wortel & S.A.P.L. Cloetingh (eds), Mathematical Geophysics, 227-264.

1. Introduction

Studies of the thermal evolution and the interior thermal structure of the Earth's mantle require the solutions of non-linear partial differential equations which typically demand the resources of super-computers. In the last few years there has been a renaissance of the truncated modal equations (e.g. Gough, Spiegel and Toomre, 1975) as a means of describing convective processes in the geological sciences (Olson, 1981; Fleitout and Yuen, 1984a , b; Quareni and Yuen, 1984; Quareni et al., 1985; Yuen and Fleitout, 1985; Quareni et al., 1986a; Spera et al., 1986; Yuen et al., 1987; Clark et al., 1987; Olson, 1987).

These non-linear modal equations are formed by expanding the fluctuating velocity and temperature fields in a complete set of planforms, and then truncating the expansion. The most severe truncation retains all but one and is called the single-mode, mean-field equations. Within this description convection has a simple cellular structure with a prescribed horizontal planform and an associated horizontal wavelength. Mean-field also saw great usage in geodynamo problems (e.g. Krause and Raedler, 1980) and other areas in physics.

One great advantage of the single-mode, mean-field theory is that with modest computational efforts high Rayleigh number convection for various types of rheologies (Quareni et al., 1985) can be attained under an interactive environment using workstations. Another objective of the mean-field method is to serve as a reconaissance tool for exploring new physics, as in double- diffusive convection in magma chambers (Spera et al., 1986; Clark et al., 1987) and adiabatic heating in mantle convection (Quareni et al., 1986a). Having established scaling relationships on parameterized heating and mass transfer laws, one may then proceed to focus better on particular portions of the parameter space with better guidance for solving the full set of equations. Such is the philosophy which we will adopt for studying the combined effects of equation of state and variable viscosity in mantle convection (Yuen et al., 1987). A clear understanding of the steady-state solutions and the bifurcation points of the control variables proves very useful in delineating regions in the parameter space where changes in the character of the solution may likely occur (Yamaguchi et al., 1984; Golubitsky and Schaeffer, 1985).

In this paper our goal is to present an overview of the research carried out thus far in the application of the mean-field theory to mantle convection. We review the results for the Boussinesq (incompressible) level of approximation in section 2. We then introduce the anelastic (compressible) set of equations and the validity of the different levels of approximations in section 3. In section 4 we focus on the geophysically interesting phenomena associated with the non-linear coupling of rheology and compressibility of mantle materials. In the final section we will mention the limitations and the future prospects of this approximate method for geodynamicists.

2. Mean-field methods: Boussinesq (incompressible) limit

This section will be concerned with the incompressible or Boussinesq limit of the mean-field equations. This widely used approximation, useful for describing convection in thin layers, has been the mainstay in mantle convection for the past two decades. The Boussinesq system is obtained by making the following approximations: (a) the density changes come into the buoyancy term only, (b) the flow has zero velocity divergence and

(c) diffusion and advection are the only mechanisms for causing temperature changes. The reader is referred to Mihahjan (1972) and Turcotte et al. (1973) for a rigorous exposition of the expansion procedures used in the derivation of the Boussinesq convection equations.

2.1 Mean-field approximation

The mean-field equations for thermal convection was first developed by Herring (1963) and Roberts (1965) for finite Prandtl number fluids. Asymptotic treatments of mean-field equations can be found for infinite (Chan, 1971) and finite (Gough et al., 1975) Prandtl numbers. Comparison of solutions between mean-field and two-dimensional equations was conducted by Quareni et al (1985) for variable viscosity, infinite Prandtl number fluids.

When viewed in wavenumber-space, the formulation of the mean-field equations can be elucidated much clearer as a subset of the full spectral-transform equations for thermal convection (McLaughlin and Orszag, 1982; Vincent, 1985; Kimura et al., 1986). The key feature in the mean-field approximation is the assumption that the temperature field may be decomposed into a mean T and a fluctuating θ component, which is then expanded in terms of a complete set of functions specifying the horizontal planform. The velocity field can be likewise expanded with these same horizontal basis functions along with the requirement that the continuity equation is satisfied. The single-mode, mean-field approximation then consists in keeping the first term of this expansion, in contrast to the retention of additional modes (Toomre et al., 1982). The weak-coupling approximation (Gough et al., 1975) is made in which the divergence of $\mathbf{u} - \langle\mathbf{u}\rangle$ is set to zero. In this work we will assume rolls along the y-direction to be the given planforms in the single-mode framework. The temperature field may then be expressed as

$$T(x,z,t) = \bar{T}(z,t) + \theta(z,t)\cos(kx) \tag{1}$$

where k is the dimensionless wavenumber, x and z are respectively the dimensionless horizontal and vertical coordinates and t represents time. The aspect-ratio of the flow is given by $a=\pi/k$. The velocity field is given in terms of the vertical velocity w, which is a fluctuating variable with the form $W(z,t)\cos(kx)$. The horizontal velocity u, from the continuity equation, is given by

$$u = \frac{-1}{k}\frac{dW}{dz}\sin(kx) \tag{2}$$

By this simple Fourier decomposition the dimensionality of the governing non-linear partial differential equations is reduced by one. In the steady-state limit, the mean-field, Boussinesq equations with variable viscosity take the form (Quareni et al., 1985) where thermal diffusion has been used in the scaling of the equations:

$$\frac{dY}{dz} + \frac{d\eta}{dz}\left[Y - 3k^2\frac{dW}{dz}\right] + 2k^2\frac{d^2W}{dz^2} + k^4W + Ra_\infty\frac{k^2\theta}{\eta} = 0 \tag{3}$$

$$Y = \frac{d^3W}{dz^3} + \frac{1}{\eta}\frac{d\eta}{dz}\left[\frac{d^2W}{dz^2} + k^2W\right] \tag{4}$$

$$\frac{d^2\bar{T}}{dz^2} - \frac{d}{dz}(W\theta) + r = 0 \tag{5}$$

$$\frac{d^2\theta}{dz^2} - k^2\theta - W\frac{d\bar{T}}{dz} = 0 \tag{6}$$

where the Rayleigh number Ra_∞ is based on the interior viscosity η_∞. The strength of internal heating is denoted by r, which is specified according to the bottom thermal condition. For a base-heated (constant-temperature) configuration, r represents the ratio of two Rayleigh numbers, $r=Ra_H/Ra_T$, where Ra_T is the Rayleigh number appearing in the momentum equation (3), and Ra_H is the Rayleigh number based in the magnitude of internal heating H. The two Rayleigh numbers are:

$$Ra_T = \frac{g\,\alpha\Delta T d^3}{\kappa\eta_\infty}\rho_0 = Ra_\infty \tag{7a}$$

$$Ra_H = \frac{g\,\alpha H d^5}{k_0\kappa\eta_\infty}\rho_0 \tag{7b}$$

where g is the gravity acceleration, α is the thermal expansivity, ρ_0 is a reference density, k_0 is the thermal conductivity, κ is the thermal diffusivity, d is the depth of the fluid layer and ΔT is the temperature difference imposed across the layer.

In the single-mode, mean-field approximation there is no non-linear coupling between temperature and velocity fields with different wavenumbers. The driving thermal buoyancy comes from the fluctuating component of the temperature field θ, since it has a non-vanishing horizontal dependence.

For a bottom thermal condition with a given flux F, the partitioning ratio r becomes μ (Jarvis and Peltier, 1982). It now reads:

$$\mu = \frac{\rho_0 H d}{\rho_0 H d + F} \tag{8}$$

Within the present scheme of approximation, viscosity only varies with depth in so far as the mean temperature $T(z)$ and depth dependences are concerned. This then accounts for the appearance of $\eta(z)$ in eqns. (3) and (4). Eqns. (3) through (6) constitute an eighth order, non-linear ordinary differential system. We must specify eight boundary conditions at the top (z=0) and the bottom (z=1) in order to complete a well-posed two-point, boundary value problem. We demand that W and its second vertical derivative (stress- free condition) vanish both at top and bottom. The thermal boundary conditions at the top require that both T and θ are zero. For a fluid heated from below the the temperature is constant there, so that at z=1, T=1 and θ=0. For internally heated configurations, where the bottom flux is specified, the boundary conditions at z=1 are

$$\frac{d\bar{T}}{dz} = 1 - \mu \quad \text{and} \quad \frac{d\theta}{dz} = 0 \tag{9}$$

We have attained steady-state solutions to this eighth-order system by both initial-value techniques (Quareni and Yuen, 1984; Quareni et al., 1985) and by treating it as a two-point boundary value problem (Quareni et al., 1985). Solving it as a two-point boundary-value problem is about an order of magnitude faster than integrating the equations in time. In order to focus on the general physics, we have decided to study solely the steady-state behavior. The different types of steady-states will be reflected in the bifurcations of the solutions, as the critical parameters are varied. Therefore , analyzing a whole suite of steady-states will be a good strategy for exploring later time dependent solutions (Hansen, 1987). We regard the mean-field method as an extremely useful reconnaissance tool to be employed before committing any major efforts by using the full set of equations.

For solving the mean-field equations, we found it very useful to introduce a new variable Y, eqn (4), which eliminates the second derivative of the viscosity in the momentum equation (Quareni et al., 1985). Further transformation can lead to the elimination of the first derivative of the viscosity. Each successive reduction leads to greater numerical stability and computational speed.

The eighth order system is solved as a sequence of two fourth-order problems, with the momentum equation leading at each iteration. Typically we find that twenty to thirty iterations are required for the boundary conditions to be satisfied in one part out of 10^4. Successive underrelaxation methods of the temperature and velocity fields are found to be necessary for the more pressing cases (Quareni et al., 1985). The numerical methods can be found described in Quareni et al. (1985). Up to 200 unevenly spaced points have been used for the difficult situations at high Rayleigh numbers.

2.2 Rheological laws

One major reason why mean-field methods are used is to study in an economical and rapid fashion the influences of variable viscosity, in so far as the Newtonian (linear) relationship between stress and strain-rate are concerned. The deviatoric stress tensor τ_{ij} is related to the velocity field through the dynamic viscosity η as

$$\tau_{ij} = \eta\left[\frac{\partial u_i}{\partial x_j} + \frac{\partial u_j}{\partial x_i} - \frac{2}{3}\delta_{ij}\frac{\partial u_i}{\partial x_i}\right] \tag{10}$$

For incompressible fluids this takes the simpler form

$$\tau_{ij} = \eta\left[\frac{\partial u_i}{\partial x_j} + \frac{\partial u_j}{\partial x_i}\right] \tag{11}$$

Temperature and pressure dependent viscosity is one of the unique physical features in mantle convection, as a consequence of the thermally activated process in mantle creep (e.g. Sammis et al., 1977). In the past two ways of describing variable viscosity have been used for Boussinesq convection. The first would be to express η as being proportional to $\exp(H^*/RT)$, where H^* is the activation enthalpy, R the gas constant and T the temperature in Kelvin. This particular form follows the Arrhenius formula of kinetics and is correct from a physical viewpoint. A second way is based on a Taylor series expansion of the argument of the exponential and a retention of the first term of this series in the exponential. It is called the Frank-Kamenetzky approximation in the combustion literature (e.g. Buckmaster and Ludford, 1984). This form with η being proportional to $\exp(-AT)$ has been used in laboratory (Richter et al., 1983), analytical (Fowler,1985; Morris and Canright, 1984) and numerical (Christensen, 1984, 1985) investigations. This viscosity, when extended to include depth dependence, is given by

$$\eta_M(T,z) = \eta_0 \exp(-AT + cz) \tag{12}$$

where η_0 is the viscosity at the surface. The parameters A and c govern respectively the thermal and depth dependences of this viscosity. The particular form of the viscosity, given by eqn. (12), has the attribute that is becomes softer in the interior by increasing A. For this reason we have called this rheology "molle", meaning soft in Italian. We will use the subscript M for the purpose of nomenclature.

The quantity $\Delta\eta_T = \exp(A)$ measures the viscosity contrast due to the temperature dependence, while $\Delta\eta_p = \exp(c)$ is the viscosity contrast due to the depth dependence of the rheology.

A rheology, based on the Arrhenius relationship, places constraints on the interior viscosity to lie at a given value η_∞ for a given reference temperature T_∞. This parametrization is more useful than the "molle" rheology, eqn. (12) for describing deep mantle convection. It takes the form

$$\eta_D(T,z) = \eta_\infty \exp\left(\frac{H^*}{RT} - \frac{Q^*}{RT_\infty}\right) \tag{13}$$

where Q^* is the activation energy and is related to H^* by

$$H^* = Q^* + \rho(z) g V^* \tag{14}$$

For the Boussinesq approximation the density in the activation enthalpy assumes an averaged value, ρ_0. The activation volume is given by V^*. The potential dependences of Q^* and V^* with depth have been discussed in Sammis et al. (1977). The rheological parameters of candidate lower mantle substances are still uncertain, in spite of preliminary estimates (Knittle and Jeanloz, 1987).

We have called the viscosity law, given by eqn. (13), the"duro" rheology, with subscript D. It is designated "duro", which means hard in Italian, because increasing Q^* would lead to a more viscous upper boundary layer, or a harder lithosphere. "Duro" rheology is also to be distinguished from the "molle" rheology in that the temperature appears beneath the depth- dependence in the argument of the exponential function in the "duro" rheology, but not in the "molle", where the depth dependence consists of a single factor. This difference would lead to different heat transfer predictions for deep mantle convection (Quareni and Yuen, 1986). Thus, "molle" and "duro" rheology produce different dynamical effects in mantle convection. Since the "duro" rheology is more valid from physical considerations, it should be used in studying deep mantle convection, where the effects of this difference are most pronounced. A third type of flow law has also been employed. It is a depth- dependent viscosity which is intended to simulate a viscosity jump between the upper-lower mantle boundary at z=0.25. We may write this depth dependent viscosity as

$$\eta(z) = \eta_0 \left[1 + \frac{(\Delta\eta - 1)}{2} \left[1 + \tanh\left(\frac{z_1 - z}{\delta_1}\right) \right] \right] \tag{15}$$

where $z_1 = 0.25$, $\delta_1 = 0.005$ and $\Delta\eta$ is the viscosity jump. This rheology $\eta(z)$ has been employed by Koch and Yuen (1985) for studying geoid anomalies over layered convective systems.

In scaling the Rayleigh number used in the momentum equation we have used the surface viscosity η_0 in the case of the "molle" and depth- dependent rheologies, and the asymptotic viscosity η_∞ in the case of "duro" flow law.

2.3 Boussinesq results for constant and depth dependent viscosities

We commence by going over some mean-field results for constant and depth- dependent viscosities, all within the framework of the Boussinesq approximation.

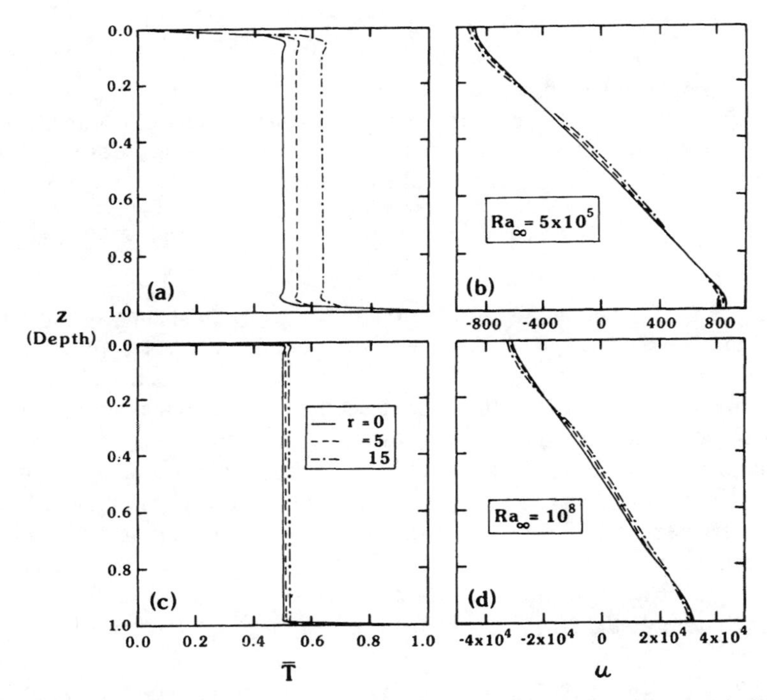

Figure 1. Mean-temperature and horizontal velocity fields associated with base-heated (r=0) and bi-modally heated ($r \neq 0$) convection. Ra_∞ is associated with constant viscosity in this case and is given in panels (b) and (d).

Figure 1 shows the mean temperature and horizontal velocity fields for base-heated, constant viscosity convection with aspect-ratio one. Two different Rayleigh numbers characterizing upper and whole mantle convection have been considered. They are

Ra_∞=5.10^5 and 10^8 respectively. The amount of internal heating has been varied from zero (r=0) to r=15. At lower Rayleigh numbers the interior is hotter for larger amounts of internal heating. The Nusselt number at the bottom can, in fact, become negative for sufficiently large r for low Rayleigh numbers. For aspect-ratio three and Ra_∞=10^4 the temperature gradient turns negative for r greater than about 8. However, this phenomenon disappears at higher Rayleigh numbers for the range of internal heating considered, r less then 20. We observe that the efficient convective transfer at Ra_∞=10^8 removes the differences in the isotherms observed at Ra_∞=5.10^5 , which can amount to a 30 % difference between r=0 and r=15. This is due to the velocity fields being about 50 times higher (cf panels (b) and (d)).

In steady-state convection the Nusselt number Nu provides a measure of the enhancement of heat transfer from convection. The definition of the Nusselt number depends sensitively on the bottom thermal boundary condition. For situations where the bottom is maintained at a constant temperature it is given by

$$Nu = \left\{ \left[\frac{dT}{dz} \right] / q_c \right\}_{z=0} \tag{16}$$

where q_c is the conductive heat flux at the surface.

When the heat flux is specified at the bottom boundary, we may define the Nusselt number in terms of the changes in the horizontally averaged temperature at the bottom as a consequence of convection. It is given by (Roberts, 1967)

$$Nu = \left[\frac{T_c}{T} \right]_{z=0} \tag{17}$$

We show in fig. 2 the Nusselt number for bimodal heating, calculated according to (16), as a function of the wavenumber k. It is to be noted that q_c increases greatly with r, the amount of internal heating. This accounts for the lowering of Nu as a function of r. Thus the efficiency in removing heat from the interior decreases with greater amounts of internal heating. It is observed that the maximum heat transfer occurs for k_{max}≈1.5π for Ra_∞=10^8. This wavenumber shifts downward to π for Ra_∞=10^5 by appealing to the highly asymmetric character of the neutral curve for the onset of convection with stress-free boundaries (Chandrasekhar, 1961). Due to the skewed nature of the neutral curve, modes with high wavenumbers would be preferentially included inside the convectively unstable envelope for higher Rayleigh numbers. It would be of interest to pursue this question concerning the nature of the $Nu(k)$ curve for high Rayleigh numbers with the fully non-linear equations.

The effects of a depth-dependent viscosity on heat transfer are illustrated in fig. 3, where a viscosity law according to eqn. (15) has been used. The maximum Nusselt number at k around π is reduced from about 30 in the isoviscous case (Quareni et al., 1985) to 13 upon a viscosity increase of 30 imposed at the upper-lower mantle interface. It is further reduced to 9 when Δη is increased to 100. Such a dramatic decrease of Nu on the viscosity stratification in the mantle would mean that many of the arguments put forward to explain two differently depleted mantle reservoirs (e.g. Jacobsen and Wasserburg, 1979) and the need for layered convection can be reconciled by an increase of viscosity with depth, which

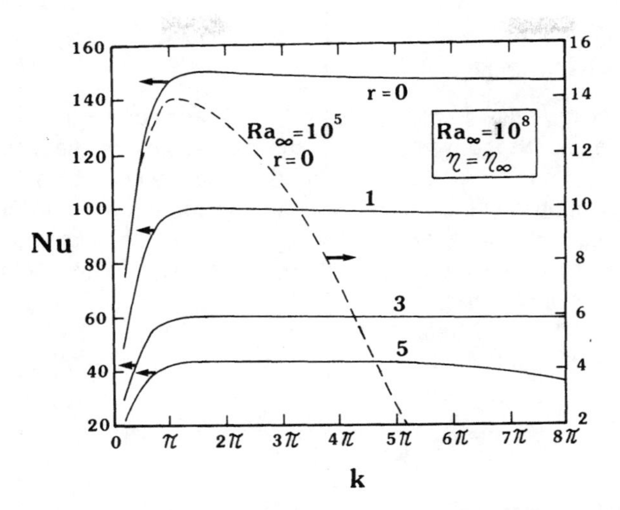

Figure 2 Nusselt number as a function of wavenumber for various degree of internal heating. Nusselt number is evaluated by calculating the ratio of the convective to conductive gradients at the surface.

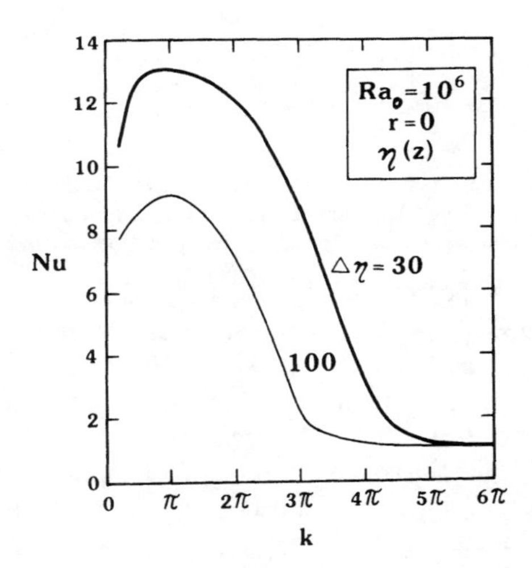

Figure 3 Nusselt number versus k for a depth-dependent viscosity. Rayleigh number is based on the surface viscosity. See eqn (15) for the viscosity law.

is supported by geoid anomaly data (Richards and Hager, 1984; Hager,1984). From considerations of the neutral curves in the case of the depth-dependent viscosities, we can explain the gradual decrease of Nu to 1.0, the conductive value, at higher wavenumbers. An increase of the viscosity would suppress the higher modes according to linear theory. We find it interesting, that the characteristics of the $Nu(k)$ curve, derived from the mean-field method reflect in fact the shape of the neutral curve (Vincent, 1985), based on linear stability analysis.

2.4 Boussinesq results for variable viscosity

Over the past few years, there have been numerous investigations of the role played by variable viscosity in mantle convection. But the differences in the dynamics between temperature ($\eta=\eta(T)$) and temperature and pressure dependent ($\eta=\eta(T,p)$) viscosities seem to have gone unappreciated for the most part,except in the work of Fleitout and Yuen (1984a) and Yuen and Fleitout (1985). This may have to do with the popular use of temperature-dependent, viscous fluids in laboratory experiments (e.g. Richter et al., 1983). The subtle differences between $\eta(T)$ and $\eta(T,p)$ in geodynamics have recently been discussed by Yuen (1986).

We illustrate these points in the next two figures.

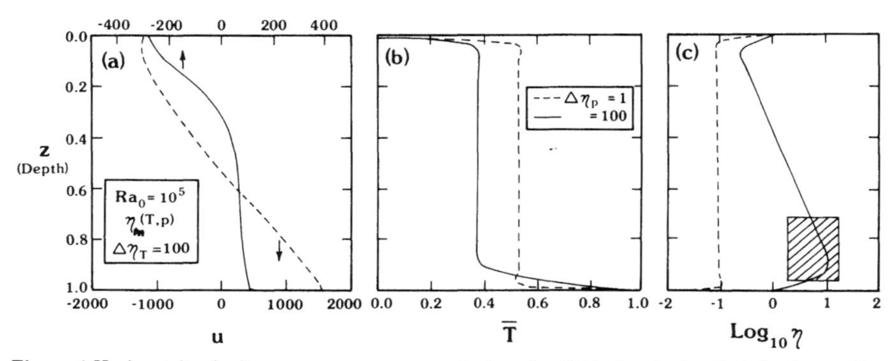

Figure 4 Horizontal velocity, mean-temperature and viscosity fields for the "molle" rheology, given by eqn (12). Solid curve denotes presence of both temperature and pressure dependences in the rheology, while dashed curve represents just a temperature-dependent viscosity. Hatched area indicates the viscosity lid.

In figs. 4 and 5 respectively the "molle", η_M, and the "duro", η_D rheologies are shown, where a temperature-induced $\Delta\eta_T$ is applied. Solid curves represent the presence of a pressure-dependent component with $\Delta\eta_p$=100, while dashed curves denote purely temperature dependent viscosities. We observe that the velocities and internal temperature fields are lower in the case of $\eta=\eta(T,p)$. The viscosity structure of $\eta(T,p)$ has two distinct features, each of which has important implications for mantle dynamics. The first is the presence of a low viscosity channel beneath the surface, which is commonly called the asthenosphere. For $\eta(T)$ there does not exist an asthenosphere. The second feature, indicated by the hatched area is the high viscosity lid right above a low viscosity zone at the base of the convecting layer. In the mantle this region can become even more viscous by an accumulation of subducted slab material. Thus it is important to assess the dynamical effects of this high viscosity lid on plumes emerging from the D'' - layer (Yuen and Peltier, 1980; Loper and Stacey, 1983).

Another interesting property, which is caused by variable viscosity convection, deals with the strong focusing of stress at the top thermal boundary layer (Fowler, 1985; Quareni et al., 1986). These stress boundary layers are embedded within this thermal boundary layer. In the case of $\eta(T,p)$ a second stress-boundary layer is found to exist at the bottom. There may be some important consequences on earthquake-faulting from this type of

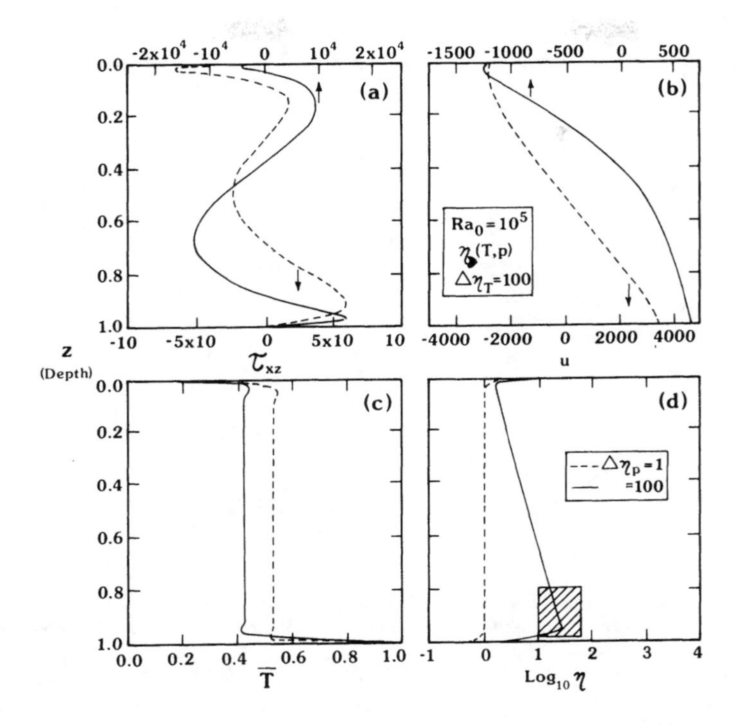

Figure 5 Shear-stress, horizontal velocity, temperature and viscosity fields for a "duro" rheology, given by eqn (13). Dashed and solid curves denote respectively $\eta_D(T)$ and $\eta_D(T,p)$ viscosities. Hatched area is the viscosity lid above the core-mantle boundary for a temperature- and pressure dependent viscosity.

concentrated basal shear stress acting on the elastic lithosphere (Melosh, 1977). In fig. 6 we show the effects of increasing depth dependence $\Delta\eta_p$ on the Nu vs. k relationship for a temperature and pressure dependent rheology, $\eta_D(T,p)$. Heat transfer is decreased by the increase in $\Delta\eta_p$ but not to the same extent as observed for a purely depth dependent viscosity (cf. fig. 3). There is a small shift of k_{max} towards higher values with larger $\Delta\eta_p$. The behavior of $Nu(k)$ for large k is different from a depth dependent viscosity.

Studies of thermal history have relied principally on heat transfer relationships drawn for constant viscosity, steady state convection for aspect- ratio one cells (Turcotte and Oxburgh, 1967; Olson and Corcos, 1980; Jarvis and Peltier, 1982). The relationship $Nu \propto Ra^{0.33}$ has taken on great popularity among geophysicists engaged in parameterized convection (e.g. Sharpe and Peltier, 1979; Schubert et al., 1979). However, some recent studies (Christensen, 1984, 1985; Fleitout and Yuen, 1984a) have shown that the power law exponent β relating Nu to Ra may, in fact, become quite small, close to 0.1, as the interior Rayleigh number Ra_T, based on the viscosity of the averaged temperature $\overline{T}$ and, for a pressure-dependent viscosity, the mid-depth (z=0.5), is raised to a sufficiently large value greater than, say, 10^5 (Christensen, 1984,1985; Quareni et al., 1985). These results were based primarily on the "molle" rheology. In fig. 7 we investigate the effects of the "duro"

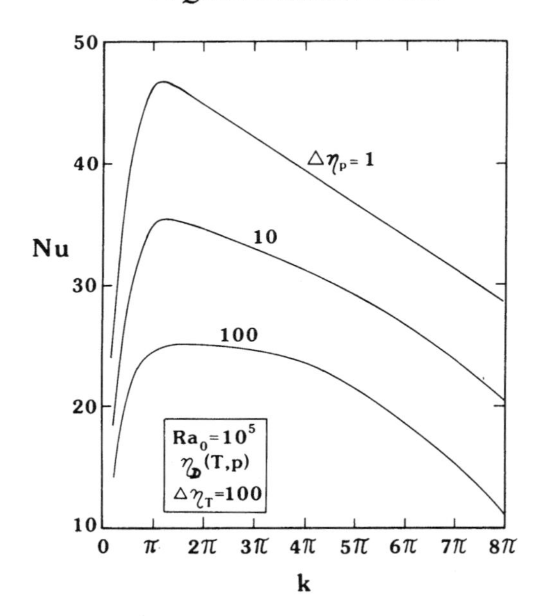

Figure 6 *Nu* versus *k* for "duro" rheology. Various viscosity contrasts due to pressure dependence have been considered. The surface viscosity is used in calculating the Rayleigh number. A viscosity contrast of 100 due

rheology on the flattening of the Nu vs. Ra_T curves for various values of Ra_0, the Ra based on the surface viscosity. The amount of pressure- induced viscosity contrast is kept the same at $\Delta\eta_p=100$. For three values of Ra_0, 10^4, 10^5 and 10^6 we then vary the amount of thermally-induced viscosity contrast $\Delta\eta_T$, which is indicated in powers of 4 by the dots on the curves.

It is found that flattening occurs most easily at low values of Ra_0. The curves are found to flatten for both types of rheology, η_D and η_M. There are little differences in the power law exponent between the two types of variable viscosity. This story, however, is not the same when effects from equation of state are brought into play (Quareni and Yuen, 1986).

3. Mean-field techniques: Anelastic, compressible equations

The influences of compressibility on mantle convection have, until now, not been studied extensively, save for the work of Jarvis and McKenzie (1980). Yuen et al. (1987) have recently demonstrated the importance of incorporating variable viscosity and equation of state in mantle convection and, in particular, the sensitivity of the equation of state parameters in controlling mantle geotherms for temperature and pressure dependent rheology. In this section we will set down the steady-state, mean field equations for variable viscosity, compressible convection, as applied to the mantle. We will also discuss the effects from truncating certain terms in the mean-field equations for the purpose of assessing their relative importances.

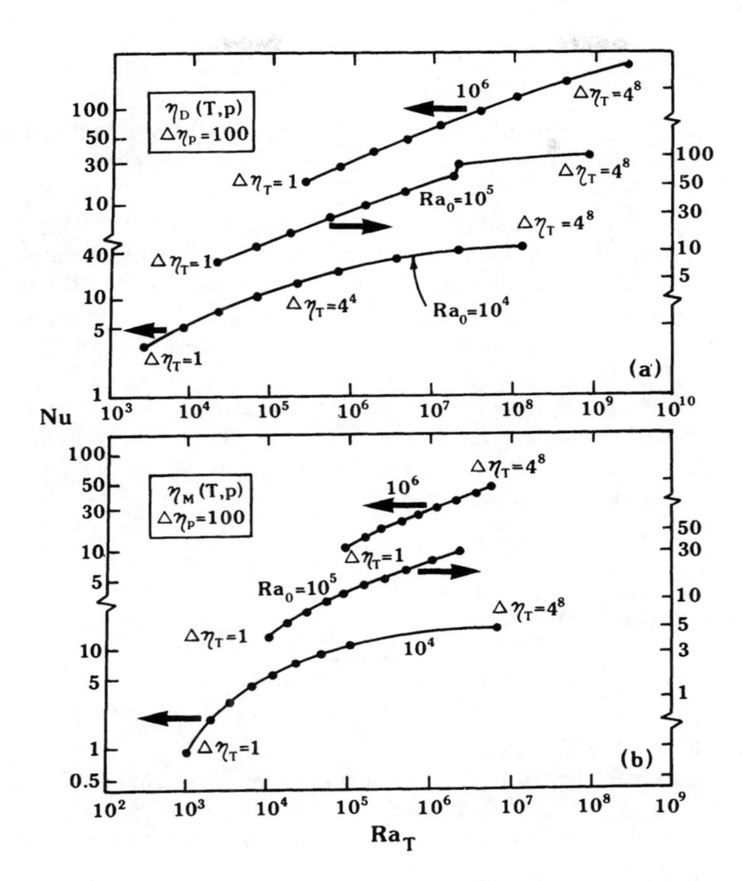

Figure 7 *Nu* versus the effective Rayleigh number Ra_T for both "duro" (panel a) and "molle" (panel b) rheologies. Ra_T is based on using a viscosity based on the averaged temperature and the depth in the middle, z=0.5. Viscosity contrasts due to thermal variations in powers of 4 are used and are given next to the curves.

3.1 Mathematical formulation: Full mean-field equations

We shall begin by setting down the appropriate steady-state conservation equations in the full form. They may be written as

$$\nabla.(\rho \underline{u}) = 0 \tag{18}$$

$$-\nabla P + \rho \underline{g} + \nabla.\tau_{ij} = 0 \tag{19}$$

$$\rho c_p \underline{u} . \left[\nabla T - \frac{\alpha T}{\rho c_p} \nabla P \right] = \nabla .(k_0 \nabla T) + \rho H + \tau_{ij} \frac{\partial u_i}{\partial x_j} \tag{20}$$

where α is the the thermal expansion coefficient and c_p is the specific heat at constant pressure. The total pressure field P consists of both the hydrostatic and the dynamical component π.

This set of equations was first given by Jarvis and McKenzie (1980) for constant viscosity fluids. It is based on the anelastic approximation for compressible fluids $d\rho/dt = 0$ (Ogura and Phillips, 1962; Gough, 1969). Hence, the name "anelastic convection" is sometimes given to mantle convection when compressibility is taken into account. We now need an equation of state $\rho(T,P)$ to complete the set of equations. The problem of equation of state in the mantle is fraught with great uncertainties, especially in regard to the amount of anharmonic contributions under high temperature conditions in the deep mantle (Mulargià and Boschi, 1980). We will follow the track of Jarvis and McKenzie where an expansion about an adiabatic state is used to get the Adams-Williamson relation (Birch,1952) given by the non-dimensional form

$$\frac{1}{\rho} \frac{d\rho}{dz} = \frac{\alpha g d}{\gamma C_p} \tag{21}$$

with the Grueneisen parameter γ meaning

$$\gamma = \frac{\alpha K_s}{\rho C_p} \tag{22}$$

where K_s is the adiabatic bulk modulus.

This depth dependent equation of state, akin to the Lane-Emden equation used in astrophysics (Chandrasekhar, 1957), can further be generalized by including the functional dependences of γ and α on ρ from the seismic equation of state (Anderson, 1987). In this case, it would take the canonical form

$$\frac{1}{\rho} \frac{d\rho}{dz} = f(\rho) \tag{23}$$

where $f(\rho)$ is an analytical function of ρ (Anderson, 1987). For the simple power law dependences of α, C_p and γ on ρ, simple analytical expressions of $\rho(z)$ can easily be derived. For constant properties eqn. (21) may be integrated over z to give

$$\rho(z) = \rho_0 \exp\left[\frac{D}{\gamma} z \right] \tag{24}$$

where we have designated $D = \alpha g d / c_p$ as the dissipation number and ρ_0 is the value of the reference density at the surface. In this work we will treat all of the equation of state parameters as constants and, hence, use eqn. (24) exclusively. Eqns. (18) to (20) can be non-dimensionalised as in the Boussinesq equations. The density is scaled by ρ_0 and the temperature by

$$\bar{T} = \frac{T - T_0}{T_D} \tag{25}$$

where T_0 is the absolute temperature at the upper boundary and T_D is the temperature drop across the layer for fixed bottom temperature and is equal to $\rho_0 H d^2/k_0$ in the case of a fixed bottom heat flux. We note that in convection with adiabatic heating T_0 is an intrinsic parameter to the hydrodynamical problem (Jarvis and McKenzie, 1980).

A mass flux stream function is used to satisfy the solenoidal vector $\rho(z)\mathbf{u}$ in eqn. (18). The vertical and horizontal velocities w and u are related to W in the Boussinesq formulation by

$$w = \frac{1}{\rho(z)} W(z) f(x) \tag{26a}$$

$$u = \frac{1}{k^2 \rho(z)} W'(z) \frac{df(x)}{dx} \tag{26b}$$

where the prime denotes derivative with respect to z and $f(x)$ is the planform function.

For deriving the mean-field equations for variable viscosity, compressible convection, we make use of an identity originally used in deriving an expression for the vorticity ω in variable viscosity, Boussinesq convection (Turcotte et al., 1973). For viscosity varying only with depth, the Poisson equation for $\eta\omega$ reads

$$\nabla^2(\eta\omega) = \frac{\eta'}{\eta}\frac{\partial \pi}{\partial x} - 2\eta''\frac{\partial w}{\partial x} - \rho Ra_\infty \frac{\partial \theta}{\partial x} \tag{27}$$

with

$$\omega = \left[\frac{\rho'}{\rho^2} W' - \left(\frac{1}{\rho}\right)(W'')\right]\frac{1}{k^2}\frac{df}{dx} - \frac{W}{\rho}\frac{df}{dx} \tag{28}$$

Substituting eqn. (28) into (27) and making use of the Laplacian operator for the particular Fourier mode with wavenumber k, we may derive a fourth order linear differential equation for W when $\rho(z)$ takes the form given by eqn. (24). The compressible momentum equation now reads

$$\begin{aligned} &\frac{d^4W}{dz^4} + 2\frac{d^3W}{dz^3}\left[\frac{1}{\eta}\frac{d\eta}{dz} - 2\frac{D}{\gamma}\right] + \frac{d^2W}{dz^2}\left[\frac{1}{\eta}\frac{d^2\eta}{dz^2} - 5\frac{D}{\gamma}\frac{1}{\eta}\frac{d\eta}{dz} + 5\frac{D^2}{\gamma^2} - 2k^2\right] \\ &+ \frac{dW}{dz}\left[3\frac{D^2}{\gamma^2}\frac{1}{\eta}\frac{d\eta}{dz} - \frac{D}{\gamma}\frac{1}{\eta}\frac{d^2\eta}{dz^2} - 2k^2\frac{1}{\eta}\frac{d\eta}{dz} - 2\frac{D^3}{\gamma^3} + 4k^2\frac{D}{\gamma}\right] \\ &+ k^2W\left[\frac{1}{\eta}\frac{d^2\eta}{dz^2} + k^2 + \frac{D}{\gamma}\frac{1}{\eta}\frac{d\eta}{dz} - \frac{2}{3}\frac{d^2}{\gamma^2}\right] + k^2\exp\left(2\frac{D}{\gamma}z\right)Ra_\infty\frac{\theta}{\eta} = 0 \end{aligned} \tag{29}$$

As in the Boussinesq case,we can similarly eliminate terms involving η''' by introducing a new component Y_1, given by

$$Y_1 = W''' + \left[\frac{\eta'}{\eta} - \frac{4D}{\gamma} \right] W'' - \frac{\eta'}{\eta}\frac{D}{\gamma} W' + \frac{\eta'}{\eta} k^2 W \tag{30}$$

An equation, similar to eqn. (3), can be written in terms of γ, W and its derivatives. This equation has been used in our calculations because of its numerical ease over eqn (29).

For compressible convection the mean-field energy equations now contain both adiabatic heating, which is scaled by D, and viscous dissipation Φ, which is scaled by two parameters D/Ra_∞ and D/γ. They are given by

$$\frac{d^2T}{dz^2} - \frac{d}{dz}(W\theta) + DW\theta + \Phi \tag{31}$$

$$\frac{d^2\theta}{dz^2} - k^2\theta + D(T+T_0)W - W\frac{dT}{dz} \tag{32}$$

where the mean-field expression for viscous dissipation is:

$$\Phi = \frac{D}{Ra_\infty}\frac{\eta}{\rho^2}\left[\left[2W' + \frac{D}{\gamma}W \right] + k^2 \left[W + \frac{D}{\gamma}\frac{W'}{k^2} - \frac{W''}{k^2} \right]^2 + \left[\frac{1}{3}\frac{D}{\gamma}W \right]^2 \right] \tag{33}$$

Inspection of eqn (33) readily reveals a potential maximum of the dissipative heating function for γ sufficiently small. We shall address in the next section this question concerning the threshold value of small γ for which viscous dissipation becomes significant.

The compressible mean-field equations with variable viscosity now consist of eqns (29) through (33),which constitute an eighth order, non-linear system. The mechanical and thermal boundary conditions are not altered by the presence of compressibility in the deep interior. The compressible equations are more complicated than the Boussinesq set and require about three times longer computational time, using the same numerical methods outlined in Quareni et al. (1985). They may take as long as 20 seconds on the CRAY-2 per steady-state run.

3.2 Mean-field equations: truncated sets.

The purpose of this subsection is to assess the influences of dropping certain terms in the momentum equation (27), in order to provide some guidelines as to similar truncated versions for two-dimensional solutions. Eqn (27) shows that the product $\eta\omega$ is generated by horizontal gradients of temperature and the dynamical pressure. Jarvis and McKenzie (1980) have dropped the pressure gradient term in the momentum equations because of computational considerations. For the parameter range examined, they found that the solutions of the truncated version

$$\nabla^2\omega = Ra_\infty \frac{\partial T}{\partial x} \tag{34}$$

provides a reasonable approximation to the vorticity.

We will also neglect the pressure gradient and study the effects of this truncation upon variable viscosity as well. Eqn (27) now becomes

$$\nabla^2(\eta\omega) = -\rho Ra_\infty \frac{\partial \theta}{\partial x} - 2\eta'' \frac{\partial w}{\partial x} \tag{35}$$

and we can simplify eqn (29) to

$$\frac{d^4W}{dz^4} + \frac{d^3W}{dz^3}\left[\frac{2\eta'}{\eta} + \frac{3D}{\gamma}\right] + \frac{d^2W}{dz^2}\left[\frac{\eta''}{\eta} + 4\frac{D}{\gamma}\frac{\eta'}{\eta} + \frac{3D^2}{\gamma^2} - 2k^2\right]$$

$$+\frac{dW}{dz}\left[\frac{2D^2}{\gamma^2} - 2k^2 - \frac{D}{\gamma}\frac{\eta''}{\eta} + \frac{D}{\gamma}(3k^2 - \frac{D^2}{\gamma^2})\right]$$

$$+k^2W\left[\frac{\eta''}{\eta} - \frac{2D^2}{\gamma^2}\frac{\eta'}{\eta} + k^2 + \frac{D^2}{\gamma^2}\right] + \frac{k^2}{\eta}\exp(2\frac{D}{\gamma}z)Ra_\infty\theta = 0 \tag{36}$$

The energy equations, (31) and (32) are unchanged. We shall designate this approximation the "first level of truncation" in what follows.

Next we neglect terms in the momentum equation in order to render symmetric the operator of the vorticity-like equation, eqn (35). Having a symmetric operator at hand would lead to a greater ease in computation for the finite- element formulation. For this reason we would like to the errors inferred by an approximation within the mean-field framework. From eqn (28) we see that neglecting the term with the first derivative in W would allow one to write a Boussinesq-like vorticity ω_1, in terms of W. This relationship is given by:

$$\left[\frac{d^2}{dz^2} - k^2\right] W = -\rho k^2 \omega_1 \tag{37}$$

with the second level of truncation we find the momentum equation becoming:

$$\nabla^2(\eta\omega_1) + 2\eta''\frac{\partial w}{\partial x} + \rho Ra_\infty \frac{\partial \theta}{\partial x} = 0 \tag{38}$$

For constant viscosity eqns (37) and (38) constitute a system of Poisson equations in terms of ω_1 and W. The mean-field momentum equation with the second-level truncation is given by

$$\frac{d^4W}{dz^4}+\frac{d^3W}{dz^3}\left[\frac{2\eta'}{\eta}+\frac{2D}{\gamma}\right]+\frac{d^2W}{dz^2}\left[\frac{\eta''}{\eta}-2\frac{D}{\gamma}\frac{\eta'}{\eta}+\frac{D^2}{\gamma^2}-2k^2\right]$$

$$+2k^2\frac{dW}{dz}\left[\frac{D}{\gamma}-\frac{\eta'}{\eta}\right]+k^2\left[\frac{\eta''}{\eta}+2\frac{D}{\gamma}\frac{\eta'}{\eta}+k^2-\frac{D^2}{\gamma^2}\right]W+k^2\frac{\rho^2}{\eta}Ra_\infty\theta=0 \quad (39)$$

We can define another variable Y_2 which will help in getting rid of the second derivatives of the viscosity. It is given by:

$$Y_2=\frac{d^3W}{dz^3}+\frac{\eta'}{\eta}\left[\frac{d^2W}{dz^2}+k^2W\right] \quad (40)$$

The second-level mean-field compressible equations consist then of eqns (32), (39) and (40). They are much easier to solve than than the full set.

For constant and variable viscosities we have conducted calculations using the complete and the two truncated sets of mean-field equations. The purpose of this undertaking is to determine the accuracies of the truncated equations relative to the complete set.

Tables 1 and 2 summarize the results in terms of the deviations of the Nusselt number and averaged temperature. In the case of constant viscosity the effects

Table 1: first-order truncation

Rheology	$-\Delta Nu$ (%)	$-\delta\langle T\rangle$ (%)
$Ra_\infty = 10^7$		
constant	0.052	0.13
$\eta(T)$, $Q^* = 50$	0.39	0.037
$\eta(T,p)$, $Q^* = 30, V^* = 3$	5.2	0.49
$Ra_\infty = 10^8$		
constant	0.12	0.073
$\eta(T)$, $Q^* = 50$	0.34	0.031
$\eta(T,p)$, $Q^* = 50$, $V^* = 4$	10.5	1.13

$\gamma = 2$, $D = 0.5$, r=0, aspect-ratio = 1
$\Delta T = 3000$ K, $T_0 = 1100$ K
Q^* has unit of kcal/mole
V^* has unit of cm^3/mole

of the two levels of truncation are observed to be minimal at most 0.5 % at $Ra_\infty=10^9$. The results are worst for temperature and pressure dependent viscosity. Errors of around 10 % are committed by the first level truncation, while up to a 40 % deviation in the Nusselt number can be caused as a result of using the second level truncated equations. Greater inaccuracies are found for higher Rayleigh numbers.

Table 2: second-order truncation		
Rheology	$-\Delta Nu$ (%)	$-\delta\langle T\rangle$ (%)
$Ra_\infty = 10^7$		
constant	0.060	0.27
$\eta(T)$, $Q^* = 50$	2.2	0.075
$\eta(T,p)$, $Q^* = 30$, $V^* = 4$	15	0.51
$Ra_\infty = 10^8$		
constant	0.044	0.27
$\eta(T)$, $Q^* = 50$	3.4	0.049
$\eta(T,p)$, $Q^* = 30$, $V^* = 4$	19	1.0
$Ra_\infty = 10^9$		
constant	0.054	0.22
$\eta(T)$, $Q^* = 50$	7.0	0.037
$\eta(T,p)$, $Q^* = 30$, $V^* = 4$	37	1.4
Same physical parameters and units as in table 1		

In fig. 8 we show for one of the worst cases the differences in the temperature and velocity fields associated with temperature and pressure dependent viscosity convection in the presence of both base-heating (r=0) and bi-modal (r=5) heating configurations. The temperature curves are very similar, while greater differences are found for the velocity structures at the top of the convecting region. The curves for constant viscosity at this Rayleigh number (Ra_∞=10^8) are indistinguishable.

We show the effects of truncation on the behavior of the $Nu(k)$ curves in fig. 9 for a difficult case involving temperature and pressure dependent viscosity. It is seen that the effects of truncation are to decrease the Nusselt numbers. The shape of the curves are preserved. What is important, is that the wavenumbers associated with the maximal heat transfer are not changed as a result of these approximations. For this Rayleigh number we found that the three curves for constant viscosity lie practically on top of one another. Another aspect of this comparison concerns the ability of the truncated equations in picking up the physics captured by the complete mean-field equations. A good example of this is depicted in fig. 10 where we show that the three approaches can essentially portray the phenomenon of a rise in the interior temperature (Yuen et al., 1987), when the compressibility of the medium is increased sufficiently (conversely decreasing the Grueneisen parameter).

4. Geophysical considerations

As mentioned in the introduction to this paper, one of the advantages of the mean-field method is the ability to sweep rapidly over parameter space and thus learn about the

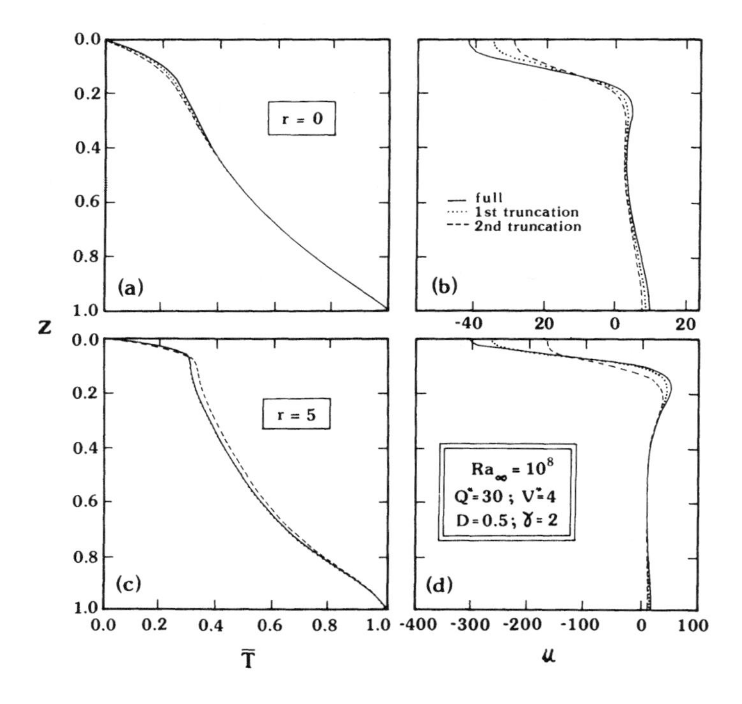

Figure 8 Mean temperature and horizontal velocity for full, first and second order truncated mean-field equations. The physical parameters are provided in the figure. Rayleigh number is based on the asymptotic viscosity given in eqn (13).

interesting consequences arising from new types of new types of non-linear interactions, such as adiabatic heating and viscous dissipation, compressibility and viscous heating, etc. We will discuss in this section the interestingly novel results found with the mean-field method for the incompressible limit ($\gamma \rightarrow \infty$, D finite) and the compressible regime (both D and γ finite).

4.1 Adiabatic and viscous heating in the Boussinesq limit

There are intrinsically two new dimensionless parameters in the compressible equations; D and D/γ. Adiabatic and viscous heating can still be included within the Boussinesq framework by taking the limit $D/\gamma \rightarrow 0$, while keeping D to a finite value. This type of extended Boussinesq expansion was employed by Christensen and Yuen (1985) in studying the effects of endothermic phase transitions on inducing layered convection. Studying the problem for the limit of zero compressibility ($\gamma = \infty$) would also shed light on the relative contributions made by the compressible physics.

It was pointed out by Stacey (1977) using Lindemann scaling arguments of melting, that the lower mantle temperature profile might be superadiabatic because of the

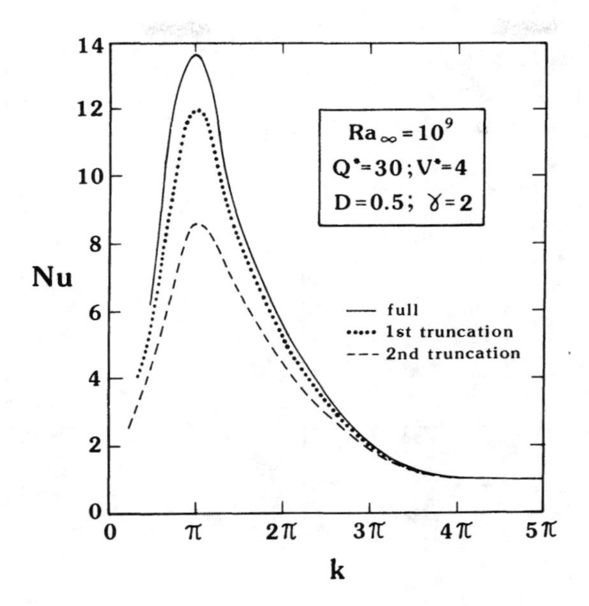

Figure 9 Nusselt number as a function of k for the full and the truncated equations. Parameters can be found in the figure.

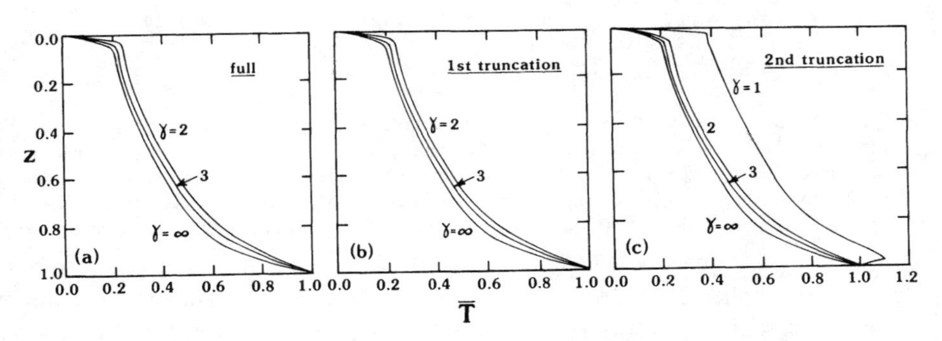

Figure 10 Temperature field for the full and truncate equations. A suite of Grueneisen parameters has been considered. A "duro" rheology has been employed with Q^*=30 kcal/mole, V^*=4 cm^3/mole, Ra_∞=10^9, r=0 and D=0.5. An aspect-ratio one cell is assumed.

temperature and pressure dependences of the viscosity. Quareni (1986a) have quantified this hypothesis in a precise way by solving the mean-field equations for variable viscosity in the extended Boussinesq limit ($D \neq 0$). The results for the temperature and viscosity fields associated with $\eta_D(T,p)$ are shown in fig. 11, where the key controlling variable is the activation volume V^*. We consider two types of bottom thermal conditions: (1.) a specified heat flux, $\mu = 0.5$, for panels (a) and (b); (2.) a constant bottom temperature without any internal heating for panels (c) and (d). It is observed that too large an activation

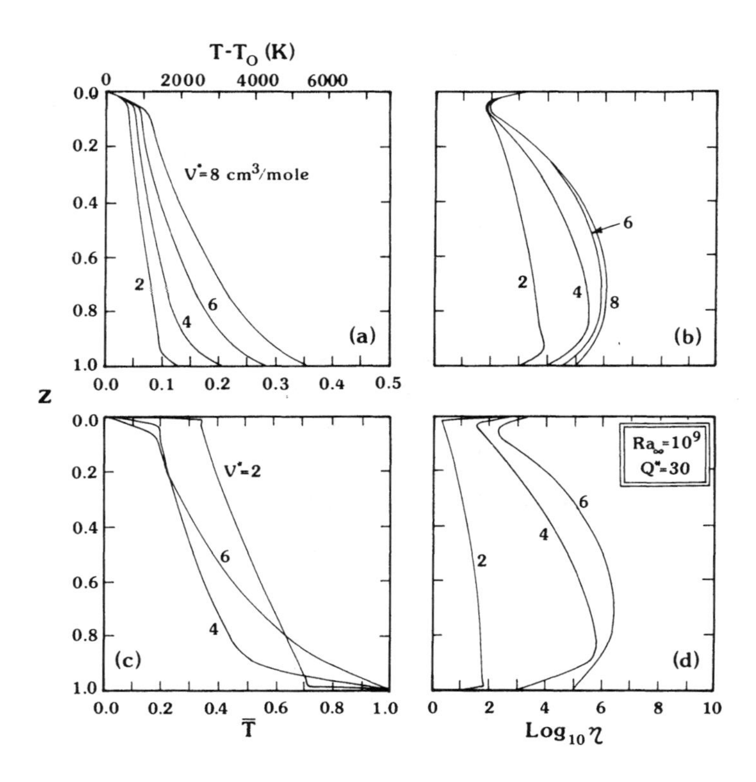

Figure 11 Effects of varying the activation volume on the temperature and viscosity structures. An activation energy of 30 kcal/mole has been used. The surface temperature T_0 is set to 1100 K within the framework of the extended Boussinesq approximation (e.g. Christensen and Yuen, 1985). In panels (a) and (b) the bottom heat flux μ=0.5 (see eqn (9)). A constant bottom temperature boundary condition is assumed for panels (c) and (d). The dissipation number D is 0.5 in all calculations and D/γ is zero. An aspect-ratio one cell is taken.

volume , greater than 5 cm^3/mole, would create a superadiabatic interior. For the purely base-heated case, even an activation volume as low as 4 cm^3/mole would produce a fair viscosity profile with a marked degree of stratification, which is inconsistent with postglacial rebound data for a Maxwell rheology (Yuen et al., 1982; Peltier, 1983).

In this connection it is of paramount interest to point out that recent experimental evidences for perovskite seem to indicate extremely low dependences of the viscosity on both temperature and pressure. (Knittle and Jeanloz, 1987; Heinz and Jeanloz, 1987). Extremely low activation volumes, less than 2 cm^3/mole, are suggested by these preliminary experimental results.

The role played by the magnitude of internal heating in elevating the interior temperature by virtue of non-linear coupling between radioactive heat sources and variable viscosity was first quantitatively established by Quareni et al. (1986). The interior temperature and the core-mantle temperature are in fact shown to be very sensitive to increased amounts of internal heating. This tendency is shown in fig. 12 for $\eta_D(T,p)$ and

$Ra_\infty = 10^9$. We have varied H from 10^{-12} W/kg to 6.10^{-12} W/kg, a value characteristic of the mean heat generation in the mantle (Schubert et al., 1980). It is clear that for H greater than 10^{-12} W/kg the interior temperatures (greater than 3500 K) are above the solidus of perovskite (Heinz and Jeanloz, 1987).

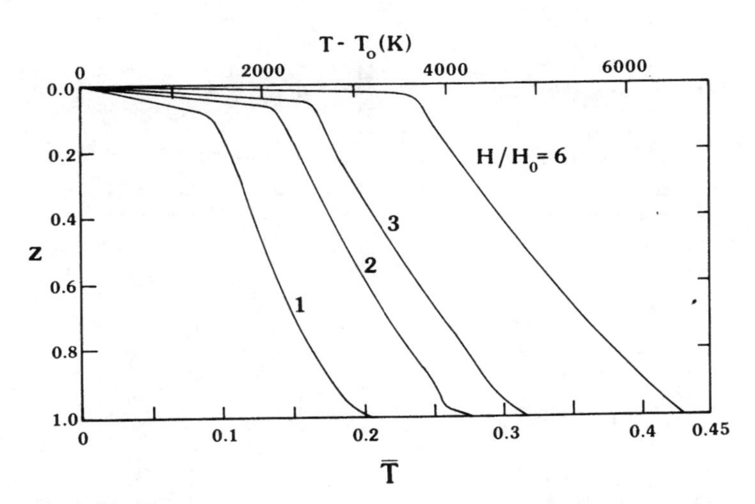

Figure 12 Temperature as a function of internal heating for a fixed bottom heat flux (μ=0.5) boundary condition. Other parameters are Ra_∞=10^9 , Q^*=50 kcal/mole, V^* = 4 cm ^{3}/mole, D =0.5, D/γ=0 and H_0=10^{-12} W/kg. T is rescaled with respect to the basal conductive temperature with internal heating. See eqn (25).

The temperature of the core-mantle boundary would exceed existing estimates, between 4000 and 5000 K, based on melting temperatures of potential core materials (Brown et al., 1984; Mulargià and Boschi, 1980). These results illuminate the potential importance of using rheology and melting temperature constraints to place bounds on the amount of radiogenic heating in the mantle. Effects of time dependence may change these amounts at most 40 to 50 %.(Quareni et al,. 1985). Current models of heat source distribution may need seriously to be reevaluated, since H is found to be a factor of at least five smaller than the averaged estimates based on the surface best-flow.

Next we study the influences of varying the activation energy Q^* for the fixed, heat flux boundary condition (μ=0 and μ=1). A low value of activation volume (V^*=2 cm^3/mole) is used in view of the recent inferences drawn from melting temperatures. We found, as in Fleitout and Yuen (1984a), that the interior temperature increases with Q^*. Viscosity decreases with larger activation energies. The viscosity is nearly constant for Q^* greater than 40 kcal/mole, while greater amounts of interior viscosity stratification are produced by low activation energies. In models, which are heated entirely from within, the temperatures at the CMB are too high for Q^* greater than 40 kcal/mole. This figure makes clear the fact that there are two regimes in so far as the variation in Q^* is concerned. For this low value of V^*, solutions greater than about 40 kcal/mole exhibit similar temperature and viscosity fields. The transitional value of Q^* is about 35 kcal/mole.

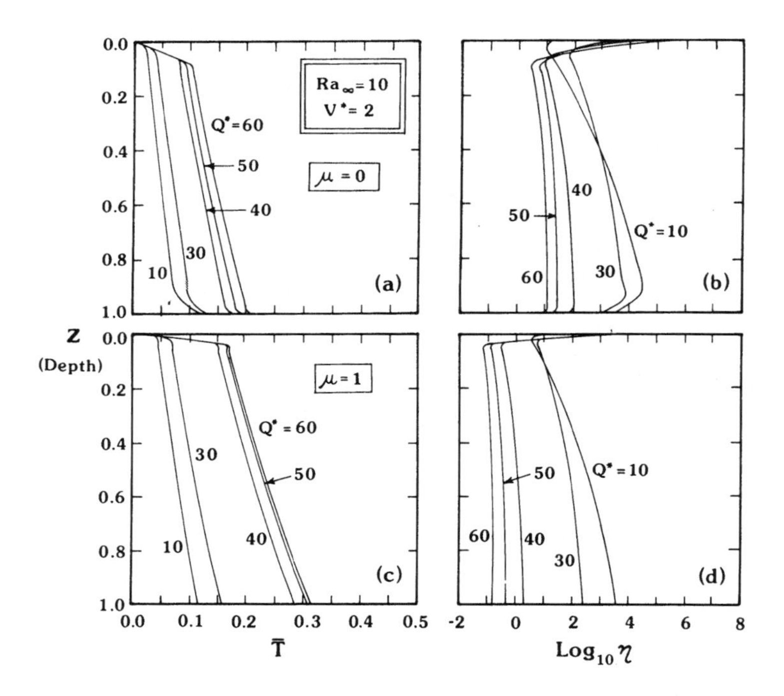

Figure 13 Influences of activation energy on the temperature and viscosity profiles. The bottom thermal condition is based on a given heat flux, denoted by μ. An activation volume of 2 cm^3/mole is used throughout; D/γ=0 and D =0.5. Other parameters can be found in the figure.

Variation of the dissipation number D is the theme of fig. 14, where a constant heat flux condition ($\mu = 0$) is applied throughout. Considerations of melting temperature would clearly reject models with too large a value in D, greater than D=0.8. Recent predictions of α based on seismic equation of state (Anderson, 1987) would reduce D considerably in the lower mantle. On this basis values of D greater than 0.6 seem high for the deep mantle. These variable viscosity results contrast those from constant viscosity in that the vigor of convection increases with D giving rise to higher interior temperatures, lower viscosities and higher velocities. Too small a value of D would result in a significant viscosity stratification across the mantle.

Examination of the viscosity profiles in both figs. 13 and 14 would suggest that it is not difficult to attain a nearly isoviscous mantle by fine-tuning the relevant parameters, i.e. increasing D and Q^* and decreasing V^*. However, one must take heed of the melting curve constraint. Thus this situation resembles problems in non-linear programming, where the solution space must satisfy many constraints simultaneously. These constraints exist by virtue of the increasingly complicated but realistic physics being built into the model.

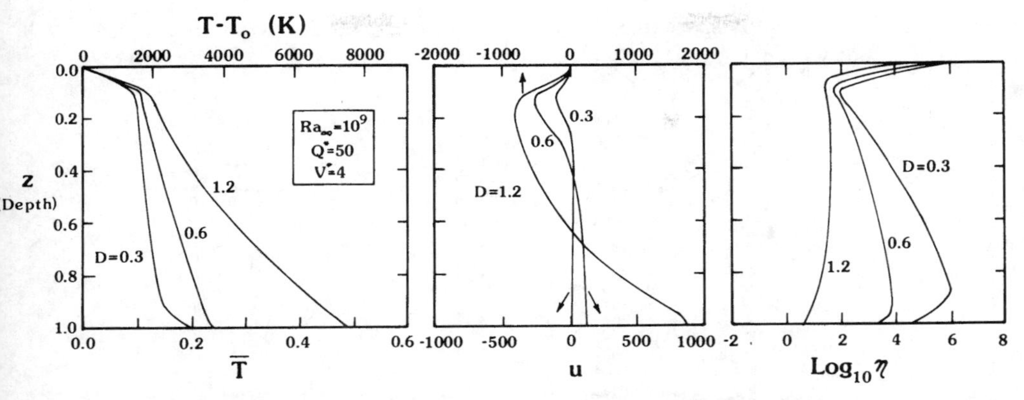

Figure 14 Effects of varying the dissipation parameter in the adiabat on the thermo-mechanical structure. A fixed heat flux at the bottom is provided (μ=0). The creep parameters are Q^*=50 kcal/mole and V^* = 4 cm^3/mole. T_0 is the same as in the other figures, i.e., T_0=1100 K. D/γ=0 is assumed.

4.2 Effects of compressibility in variable viscosity convection

In this section we will devote our attention to the importance of compressibility, when coupled with variable viscosity, in modifying the geotherms. The dissipation term of the energy equation, when viewed by the rate of change of internal energy (Batchelor, 1967), is given by:

$$\Phi = 2\frac{\eta}{\rho}\left[\dot{e}_{ij}\dot{e}_{ij} - \frac{1}{3}\left(\nabla.\underline{u}\right)^2\right] \tag{41}$$

Viscous dissipation reduces to the first term involving the strain-rate product in the extended Boussinesq regime (*D/γ=0 but $D\neq$0*). The second term Φ_B is due to dissipative heating caused by volumetric changes, when compressibility is accounted for. It may be scaled as

$$\Phi_B \approx \eta u_c{}^2\frac{\rho''}{\rho^3} \tag{42}$$

where u_c is a characteristic velocity. For the Adams-Williamson equation of state with constant properties (eqn (24)), this dissipation term goes like

$$\Phi_B \approx \frac{\eta u_c{}^2 D^2}{\gamma^2 \exp(\frac{Dz}{\gamma})} \tag{43}$$

In the limit of infinite compressibility (γ= 0), Φ_B goes to zero from inspection of eqn (43), but for some intermediate values of γ it takes on a maximum value. One would then expect some modification of the thermal profiles from intense dissipative heating, as this stationary point of $\Phi_B(\gamma)$ is approached. Yuen et al. (1987) found that significant contributions to the thermal profile from enhanced Φ_B take place at small values of γ. For constant and temperature dependent viscosities this critical value of γ_c is about ten times smaller than that for common materials, whereas for temperature and pressure dependent viscosities the value of γ_c lies within geophysically relevant values.

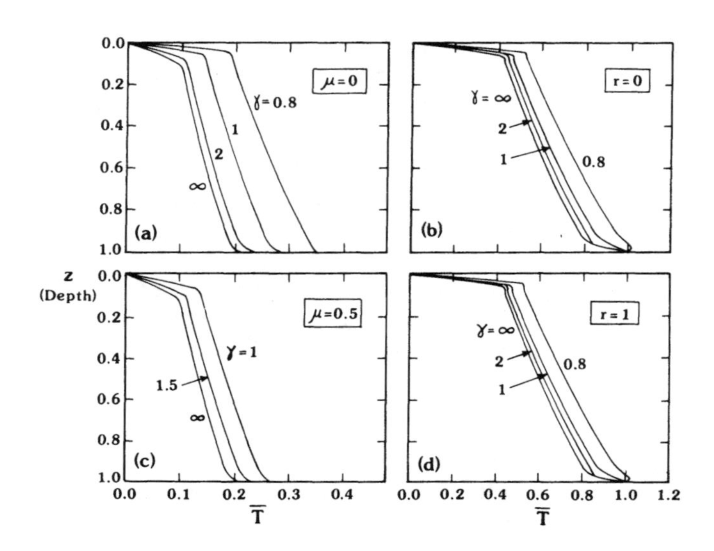

Figure 15 Effects from increasing the compressibility (decreasing the Grueneisen parameter) on the thermal profiles for a temperature and pressure dependent viscosity. Both types of heating are assumed. (1.) fixed temperature in panels (b) and (d). (2.) fixed heat flux in panels (a) and (c). Other parameters are Q^*=50 kcal/mole, V^*=4 cm^3/mole and D=0.5. A unity aspect-ratio cell is assumed.

Fig. 15 shows this type of non-linear phenomenon for two different heating configurations in which a temperature and pressure dependent viscosity is used. A dissipation number D=0.5 is used in all cases. As γ is reduced to values around one, the temperature field becomes heated up globally.

For a temperature and pressure dependent viscosity, the viscosity has a maximum near the bottom (z=0), in this case one sees that a stationary point of Φ_B exists for γ_c close to D. We have, in fact, found an empirical relationship that

$$\frac{D}{\gamma_c} \approx C_1 \tag{44}$$

where C_1 lies within 0.1 and 1.0, depending on the heating configuration and the amount of internal heating. Results from fig. 15 suggest that the relationship expressed by eqn (44) holds.

Next we focus our attention on effects due to the dissipation number D, which may range between 0.6 and 1.2, for whole mantle convection. These variations arise from the large uncertainties in the values of the physical parameters of the deep mantle. The results are shown in fig. 16 for temperature and pressure dependent viscosity with Q^*=50 kcal/mole and V^*=4 cm^3/mole. For low values of D (D=0.3), appropriate for upper mantle

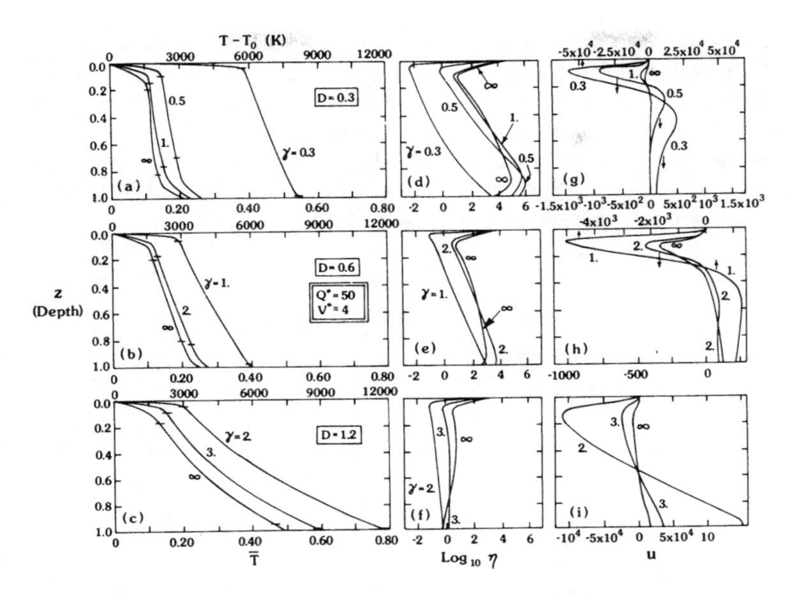

Figure 16 Temperature, horizontal velocity and viscosity profiles for temperature and pressure dependent compressible convection. Same creep parameters are used as in the previous figure. A unity aspect-ratio roll is used. Velocity is non-dimensionalized with respect to the thermal diffusion speed. T_0 is the same as in fig. 14.

convection, γ_c is smaller than 1. This would imply that compressibility effects are not important for mantle materials if the dissipation number D is sufficiently small. On the other hand, raising D to an extreme value of 1.2 causes γ_c to increase to around three, the upper limit of the Grueneisen parameter for the mantle, corresponding to a 50 % density variation. In these cases the interior temperatures for realistic values of γ exceed 5000 K, which could suggest that D cannot be so high in order to avoid large-scale melting. Reducing D to 0.6, keeps the interior below 4000 K. We emphasize here that the temperature estimates are based in cartesian formulation and that effects of sphericity (Olson, 1981; Zebib et al., 1980) would reduce these interior temperatures close to a factor of two in the case of constant viscosity. As the effects of variable viscosity on the interior temperature of Boussinesq , spherical-shell convection have not been studied at all, it is difficult to estimate its impact for our situation. Nonetheless, the physics of varying D should be clear. That is, if we insist that γ lies within bounds imposed by mineral physics, then the dissipation number D cannot be allowed to float freely, but is rather tightly constrained by the non-linear coupling between $\eta(T,p)$ and equation of state in the dissipative heating term. For preferable values of D and γ, i.e. $D=0.6$ and γ around 2, the total density changes by only 35 %, and the viscosity profile varies by about an order of magnitude between the upper and the lower mantle, in accordance with estimated geoid anomalies (Hager et al., 1985). By contrast, a nearly isoviscous mantle is predicted only for models in which D is regarded to be anomalously high. For small values of D and γ around 1, the viscosity contrast between the upper and lower mantle is greater than two orders of magnitude.

In the models with modest viscosity variation (panel h) it is seen that there is greater vigor of motion in the upper mantle, while in the hot viscous cases, (panels f and i) convective stirring is distributed uniformly across the mantle. For a given D, we find that shear heating becomes prominent when D/γ exceeds a threshold value of 0(.1), see eqn (44). The density variations for temperature and pressure dependent viscosity models are all reasonable, between 30 to 50 %. This threshold quantity does not change much if we impose a constant temperature boundary condition at the bottom.

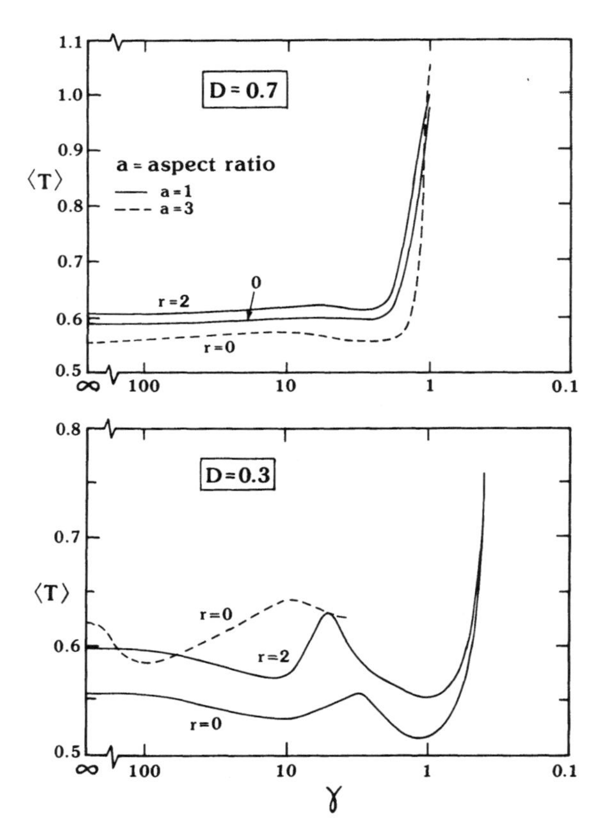

Figure 17 Averaged temperature as a function of increasing compressibility (decreasing Grueneisen parameter) for temperature and pressure dependent viscosity. Same creep parameters are used as in fig. 15. A temperature contrast of 4000 K is imposed for a purely base-heated configuration. Two aspect-ratio cells have been used: one (solid curve) and three (dashed curve). The Rayleigh number is 10^9 and T_0 is 1100 K.

Fig. 17 summarizes the effects of increasing the compressibility (or decreasing γ) on the averaged temperature of T(z) for $\eta(T,p)$. Both the bimodal heating configuration and purely basal heating (r=0) situation are shown for two different values of D. We observe that $<T>$ is higher for a wider aspect-ratio when D is decreased to 0.3. The interior naturally is hotter with internal heating (r=2). What is striking are the predictions for intense shear heating, when eqn (44) is satisfied. This is clear by the shift of steep temperature climb to lower values of γ, as D diminishes. It is important to pursue this question of melting within the context of a time-dependent model.

There is now a growing consensus that large aspect-ratio cells with intermittent boundary layer instabilities may be the most likely form in mantle convection.(Hansen ,1986; Yuen et al,. 1986; Christensen, 1987; Machetel and Yuen, 1987). It is thus important to consider larger aspect-ratios than one.

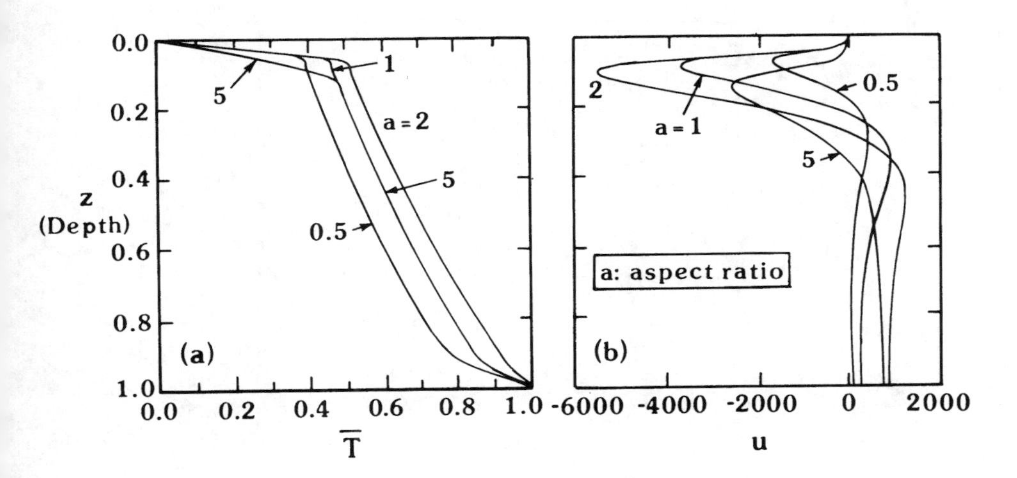

Figure 18 Influences of varying the aspect-ratio on temperature and pressure dependent compressible convection. The same heating configuration as in fig. 17 is used. Other parameters are D=0.5, γ=1, Q^*=50 kcal/mole, V^*=4 cm^3/mole, Ra_∞=10^9 and T_0=1100 K.

In fig. 18 we study the effects of varying the aspect-ratio a on the temperature and velocity profiles for $\eta(T,p)$ convection with a fixed temperature of 5100 K at the bottom. The temperature field reaches a maximum for aspect-ratio two, then decreases again with larger aspect-ratios. The top thermal boundary layer of large aspect-ratios (a=5) is thicker. Velocity fields exhibit a maximum for a=2, owing to its hotter interior, which serves to decrease the viscosity. It is of paramount importance to carry out calculations for large aspect-ratio boxes using with the full 2-D compressible equations, in light of these recent findings on the interesting dynamics of large aspect-ratio flows in the Boussinesq framework.

It is of physical interest to inquire as to the differences in the dissipation function for the various rheologies, since it has been found that compressible shear heating plays a much stronger role for $\eta(T,p)$ than for either $\eta(T)$ or constant viscosity. In fig. 19, we have plotted the quantity Φ'/Ra versus depth for constant viscosity (panel a), temperature dependent viscosity (panel b), solely pressure dependent viscosity (panel c), temperature and pressure dependent viscosity with Frank-Kamenetzky approximation (panel e) and

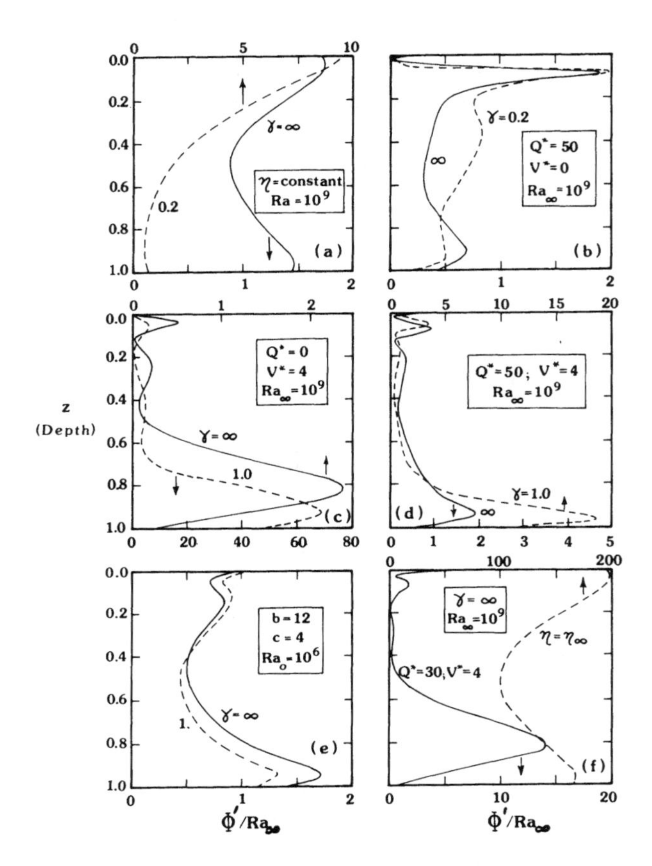

Figure 19 Dissipation heating function for various types of rheologies. An aspect-ratio one cell is used. A constant bottom temperature condition is imposed with ΔT=4000 K.

temperature and pressure dependent viscosity (panels d and f). The dimensionless quantity $\Phi'(z)$, which is a measure of the viscous heating function within the mean-field theory, is defined by

$$\Phi'(z) = 2\rho(z)\eta(z)\left[\dot{e}_{ij}\dot{e}_{ij} - \frac{1}{3}\left(\nabla.\underline{u}\right)^2\right] \tag{45}$$

Some interesting physics are unveiled by the display of this major source of non-linearity in the compressible problem. Firstly, dissipation is not focused for constant viscosity convection, except for extremely small values of the Grueneisen parameter (see panels a and f). Secondly, dissipation is extremely strong only in the upper mantle for temperature dependent viscosity (see panel b). Most important are the effects introduced by having pressure dependence in the creep law. There one notices immediately that the peak of the dissipation function is shifted to the lower mantle where the viscosity tends to be maximal.

It is this particular property that enables us to derive the analytical prediction for the onset of enhanced shear-heating, given by eqn (43), where $\exp(Dz/\gamma)$ can be approximated with $\exp(D/\gamma)$. This enhancement of shear-heating in the lower mantle is still present in the Boussinesq approximation (see panel f). Since this source of non-linearity may be important for the dynamics of "plumes" arising out of the D″-layer and the potential of melting pockets in plumes, every effort should be made to measure the pressure dependence of lower mantle constituents which will help to address this problem. Already we have seen some encouraging experimental efforts towards this end (Heinz and Jeanloz, 1987; Knittle and Jeanloz, 1987).

In the mean-field equations the viscous dissipation term is not balanced entirely by adiabatic heating in the equation for the mean temperature (eqn 31), since part of the adiabatic term also appears in the temperature equation for the fluctuating component , eqn. (32). An important feature of the full compressible equation is that the global rate of dissipative heating is equal to the global rate at which work is done by the fluid in adiabatic volume changes (Jarvis and McKenzie, 1980). We find, in fact, that in some cases the integral of the dissipative heating exceeds the integral of the adiabatic heating term $DW\theta$ in the mean-field energy equation. This imbalance would lead to an enhanced surface heat flux, since the excess amount represents a heat source. One can observe this behavior in fig. 15, where the surface heat fluxes are generally larger than the specified ones at the bottom of the layer, especially for values of γ approaching the critical. In panels b and d of fig. 15 viscous heating is so intensive that the heat actually flows into the core, clearly an indication of a potentially nonequilibrium situation. We display this inherent imbalance between viscous dissipation and adiabatic heating in fig. 20, by showing the relative distributions of both heating functions for various rheologies and bottom thermal conditions. The global comparison would entail the differences in the areas under each of the curves.

We find, as in Jarvis and McKenzie (1980) that viscous dissipation has a tendency to be strongly localized in space, while adiabatic heating is a relatively smooth function. In some cases the global rate of shear can be twice the amount produced by adiabatic heating. Whether or not this imbalance, owing to the very nature of the mean-field approximation, would alter the findings reported above on the effects of large compressibility on the thermal structure must await future efforts with the full set of compressible equations.

In their work on compressible convection Jarvis and McKenzie have focused only on the fixed heat flux boundary condition at the bottom. In the Nu versus Ra relationship they found a power law index β of 0.25 for Ra about 10^5. Fig. 21 shows Nu as a function of Ra for constant viscosity convection, where both a fixed bottom temperature (solid line) and a fixed flux (dashed line) condition have been applied. A big difference in β is found for the two boundary conditions. We have employed eqns (16) and (17) respectively for the fixed temperature and constant flux boundary conditions in calculating Nu.

For fixed bottom temperature β is found to be 0.33, close to the Boussinesq mean-field result of β=0.35 (Quareni et al., 1985). However , for the fixed flux, β is observed to to decrease with increasing vigor of convection. For T_0=500 K, β is 0.23 for $Ra=10^5$, and decrease to 0.087 at $Ra=10^9$. At a higher surface temperature T_0=1100 K, β is 0.21 for $Ra=10^5$ and is reduced to 0.048 at $Ra=10^{10}$. The mean-field results are in agreement with the 2-D results of Jarvis and McKenzie for the fixed flux condition and for modest Ra.

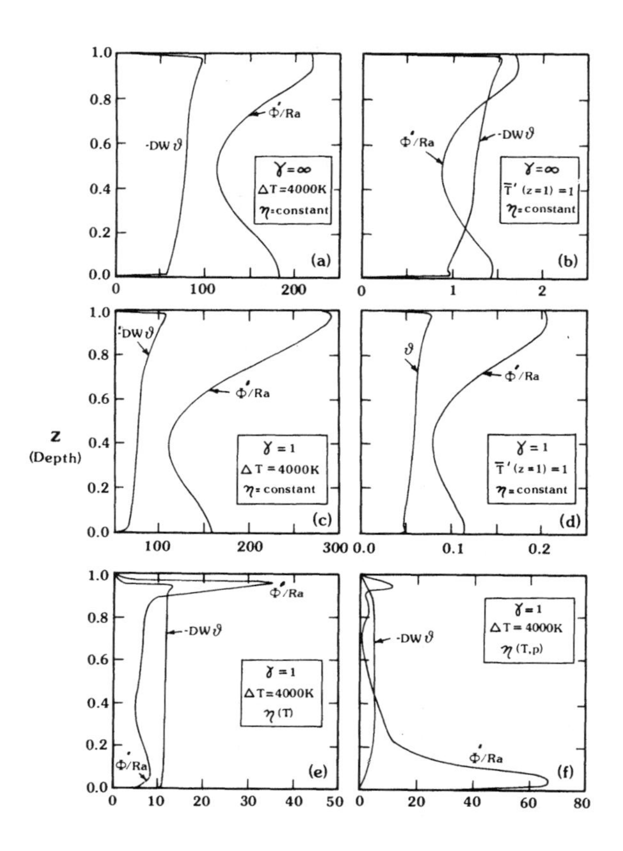

Figure 20 Comparison between viscous dissipation and adiabatic heating for various rheologies and different bottom thermal conditions. A Rayleigh number of 10^9 is used throughout and D=0.5 has been employed. An aspect-ratio one cell is taken. Creep parameters are Q^*=50 kcal/mole and V^*=4 cm^3/mole for all the variable viscosity cases.

Two-dimensional compressible convections are warranted in order to verify this potential change in β.

The effects of variable viscosity are shown in fig. 22, where a "duro" temperature and pressure dependent viscosity is used. Both aspect-ratio one (solid and dashed curves) and aspect-ratio three (solid and dotted and dotted lines) cells have been considered. These results further reinforce the importance of bottom thermal conditions on heat transfer properties associated with compressible convection (Quareni and Yuen, 1986). From the fixed temperature condition β is found to be 0.25 for aspect-ratio one and decreases to 0.22 for aspect-ratio three cell. The power law indexes for the constant flux boundary condition are extremely low: β=0.038 for aspect-ratio one and β=0.021 for aspect-ratio three . As the constant temperature boundary condition may seem the more appropriate one at the CMB, from considerations of the geodynamo (Mollett, 1984; Stevenson et al., 1983) and the

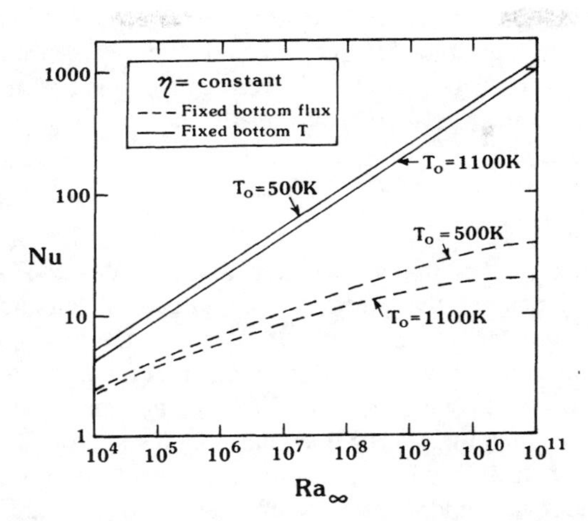

Figure 21 Nusselt number versus Rayleigh number for constant viscosity compressible convection. Two different surface temperatures have been considered as well as two different basal thermal boundary conditions. One notes the flattening of the Nusselt number curve for the fixed heat flux condition. The aspect-ratio is one.

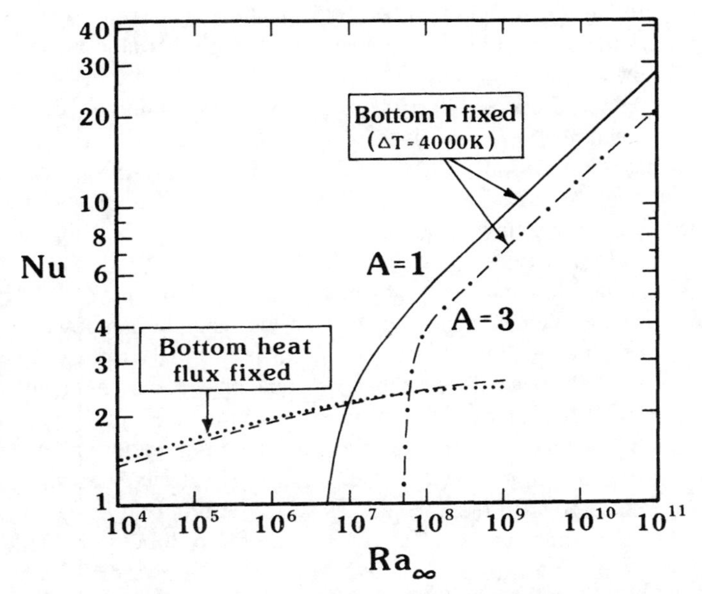

Figure 22 *Nu* versus Ra_∞ for temperature and pressure dependent compressible convection. Creep parameters are Q^* =40 kcal/mole and V^* =4 cm^3/mole. Other parameters include D=0.5, γ=1 and T_0 =1100 K. Both aspect-ratio one (solid and dashed) and three (dotted and solid-dotted) cells are employed.

melting temperature of candidate core material (Williams et al., 1987), heat transfer coefficients useful for parameterized convection may, in fact, not need to be revised substantially downwards. On the other hand, dynamical effects of sphericity may be of importance in governing heat transfer properties for compressible, variable viscosity convection, as the interior temperatures for spherical convection are much lower than those produced in a cartesian geometry, at least in the Boussinesq approximation (Zebib et al., 1980).

5. Summary and perspectives

The use of the mean field method in mantle convection provides a valuable tool to study in a rapid and economical manner the averaged thermal and mechanical properties of the mantle. Their proper understanding and limitations are hence paramount for a variety of problems in geodynamics. Here we summarize the main points of this work, suggest directions for further work and try to put the results into some perspective.

The studies of the mean-field method have shown the feasibility of deriving numerically estimates of scaling coefficients for the dynamical properties, such as heat transfer, in which analytical boundary layer solutions are not available. This approach has made scaling predictions, in variable viscosity (Quareni et al., 1985), double diffusive convection (Spera et al., 1986; Clark et al., 1987) and, in this work, compressible convection. Mean-field modelling has, in fact, called attention to the different dynamics caused by the temperature and pressure dependent viscosities in convection processes (Fleitout and Yuen, 1984a; Yuen and Fleitout, 1985), to the existence of a stress-boundary layer embedded within the thermal boundary layer for strongly variable viscosity (Quareni et al., 1986b), and to the existence of thin chemical strands lodged within thermal boundary layers (Spera et al., 1986; Clark et al., 1987) for high Lewis number, double-diffusive convection in magma chambers. Comparison of mean-field results with two-dimensional calculations for thermal convection (Quareni et al., 1985; Olson, 1987) has shown that, although the Nusselt numbers can behave a discrepancy of up to 50 %, the averaged interior temperatures, the power law exponent β and the wave number dependence are in good agreement, within 5 %, between the two methods.

Mean-field method also has helped to advance our understanding of convection with many interacting non-linear elements such as variable viscosity, adiabatic heating, internal heating, viscous heating and equation of state. From our preliminary results (Quareni et al., 1986a and Yuen et al., 1987) we have seen how each of these components within the context of a dynamical model can exert such a strong influence upon the other so that definite constraints may be placed on the range of plausible parameter values such as (1.) activation volume in the lower mantle (2.) the amount of internal heating and (3.) the dissipation number associated with the adiabat. These zeroth order understandings, derived from mean-field, pose challenges for the next generation of two-dimensional calculations in either the steady-state or the time averaged context. There are, of course, many shortcomings, inherent with the single-mode mean-field method, such as predicting transitions in time dependent convection or describing phenomenon requiring strong lateral heterogeneities as the operating mechanism, like subduction.

A natural extension of the mean-field approach to the spherical shell problem should be the next step and work has already been initiated by us in this direction (Zhang and Yuen, 1987). Further understanding is needed to assess the impact of sphericity on these non-linear couplings, among the physical parameters, which are already found to be in a balanced state. There is the possibility that even tighter constraints can be placed by introducing this element of geometrical realism.

With the development of spectral transform codes in mantle convection (Vincent ,1985, 1987) we have at our disposal the ability to determine in a quantitative fashion the relative distribution of the dominant horizontal modes participating in strongly time dependent convection (Vincent and Yuen, 1987). A knowledge of the modal partitioning function will be an important first step on the road towards building a multimode algorithm for the purpose of understanding time dependent convection by the non-linear dynamics approach (Bergé et al., 1984)

Acknowledgments

This research has been supported by N.A.S.A. grant NAG 5-770 and NSF grant EAR-8511200. Computing has been done on the CRAY-XMP12 at Bologna. We thank discussions with Alain Vincent and F. Mulargià. We are grateful for preprints sent to us by Don Anderson and Ray Jeanloz.

References

Anderson, D.L., 1987. A seismic equation of state, II. Shear properties and thermodynamics of the lower mantle, *Phys. Earth Planet. Int.*, **45**, 307-323.

Batchelor, G.K., 1967. *An Introduction to Fluid Dynamics*, Chap. 3, Cambridge Univ. Press.

Birch, F., 1952. Elasticity and constitution of the Earth's interior, *J. Geophys. Res.*, **57**, 227-286.

Bergé, P. Pomeau, Y. and Vidal, Ch., 1984. *L'ordre dans le chaos*, Herman, Paris, 353 pp.

Brown, J.M., Ahrens, T.J. and Shampine, D.L., 1984. Hugoniot data for pyrrhotite and Earth's core, *J. Geophys. Res.*, **89**, 6041-6048.

Buckmaster, J.D. and Ludford, G.S.S., 1984. *Theory of laminar flames*, Cambridge University Press, Cambridge, 266 pp.

Chandrasekhar, S., 1957. *An Introduction to the Study of Stellar Structure*, Dover Press, New York.

Chandrasekhar, S., 1961. *Hydrodynamic and Hydromagnetic Stability*, Oxford Clarendon Press.

Chan, S.K., 1971. Infinite Prandtl Number Turbulent Convection, *Studies in Appl. Math.*, **50**, 13-49.

Christensen, U.R., 1984. Heat transfer by variable viscosity convection and implications for the Earth's thermal evolution, *Phys. Earth Plan. Inter.* , **35**, 264-282.

Christensen, U.R., 1985. Thermal evolution models for the Earth, *J. Geophys. Res.*, **90**, 2995-3007.

Christensen, U.R. and Yuen, D.A., 1985. Layered convection induced by phase transitions, *J. Geophys. Res.*, **90**, 10291-10300.

Christensen, U.R., 1987. Time dependent convection in elongated Rayleigh-Bénard cells, *Geophys. Res. Lett.*, **14**, 220-223.

Clark, S., Spera, F.J. and Yuen, D.A., 1987. Steady-state double-diffusive convection in magma chambers heated from below, in *Magmatic Processes: Physiochemical Priciples*, ed. by B.O. Mysen, the Geochemical Society, Special Publ. No. 1, 289-305.

Fleitout, L. and Yuen, D.A., 1984a. Steady-state secondary convection beneath lithospheric plates with temperature and pressure dependent viscosity, *J. Geophys. Res.*, **89**, 7653-7670.

Fleitout, L. and Yuen, D.A., 1984b. Secondary convection and the growth of the oceanic lithosphere, *Phys. Earth Planet. Int.*, **36**,391-412.

Fowler, A.C., 1985. Fast thermoviscous convection, *Stud. Appl. Math.*, **72**, 189-219.

Golubitsky, M. and Schaeffer, D., 1985. *Singularities and groups in bifurcation theory*, Vol. 1, Springer Verlag, New York.

Gough, D.O., 1969. The anelastic approximation for thermal convection, *J. Atm. Sci.*, **26**, 448-456.

Gough, D.O., Spiegel, E.A. and Toomre, J., 1975. Modal equations for cellular convection, *J. Fluid Mech.*, **68**, 695-719.

Hager, B.H., 1984. Subducted slabs and the geoid: constraints on mantle rheology and flow, *J. Geophys. Res.*, **89**, 6003-6017.

Hager, B.H., Clayton, R.W., Richards, M.A., Comer, R.P. and Dziewonski, A.M., 1985. Lower mantle heterogeneity, dynamic topography and the geoid, *Nature*, **313**, 514-545.

Hansen, U., 1986. On the influence of the aspect-ratio on time dependent thermal convection, *E.O.S.*, **67**, no. 44, 1195.

Hansen, U., 1987. Zeitabhaengige Phaenomenon in Thermische und Doppelt diffusiver Konvektion, Ph. D. thesis, Geophysics, Univ. Cologne, F.R. Germany.

Heinz, D.L. and Jeanloz, R., 1987. Measurement of the melting curve of $Mg_{.9}Fe_{.1}SiO_3$ at lower mantle conditions and its geophysical implications, *J. Geophys. Res.*, in press.

Herring, J.R., 1963. Investigation of problems in thermal convection, *J. Atm. Sci.*, **20**, 325-338.

Jacobsen, S.B. and Wasserburg, G.J., 1979. The mean age of mantle and crustal reservoirs, *J. Geophys. Res.*, **84**, 7411-7427.

Jarvis, G.T. and McKenzie, D.P., 1980. Convection in a compressible fluid with infinite Prandtl number, *J. Fluid Mech.*, **96**, 515-583.

Jarvis, G.T. and Peltier, W.R., 1982. Mantle convection as a boundary-layer phenomenon, *Geophys. J. R. Astr. Soc.*, **68**, 385-424.

Kimura, S., Schubert, G. and Straus, J.M., 1986. Route to chaos in porous-medium thermal convection, *J. Fluid Mech.*, **166**, 305-324.

Koch, M. and Yuen, D.A., 1985. Surface deformations and geoid anomalies over single and double layered convection systems, *Geophys. Res. Lett.*, **12**, 701-704.

Knittle, E. and Jeanloz, R., 1987. The activation energy of silicate perovskite, in press, ed. by M. Manghani and Y. Sonyo, U.S. - Japan Conference on High Pressure Research, American Geophysical Union, Washington, D.C..

Krause, F. and Raedler, K.-H., 1980. *Mean-field magneto-hydrodynamics and dynamo theory*, Pergamom Press, Oxford.

Loper, D.E. and Stacey, F.D., 1983. The dynamical and thermal structure of deep mantle plumes, *Phys. Earth Plan. Int.*, **33**, 304-317.

Machetel, P. and Yuen, D.A., Infinite Prandtl number spherical shell convection, in this volume.

McLaughlin, J.B. and Orszag, S.A., 1982. Transition from periodic to chaotic thermal convection, *J. Fluid Mech.*, **122**, 123-142.

Melosh, H.J., 1977. Shear stress at the base of a lithospheric plate, *Pure and Appl. Geophys.*, **115**, 429-439.

Mihahjan, J.M., 1962. A rigorous exposition of the Boussinesq approximations applicable to a thin layer of fluid, *Astroph. J.*, **3**, 1126-1133.

Mollett, S., 1984. Thermal and magnetic constraints on the cooling of the Earth, *Geophys. J. R. Astr. Soc.*, **76**, 653-666.

Mulargià, F. and Boschi, E., 1980. The problem of the equation of state in the Earth's interior, in *Physics of the Earth's Interior*, 337-361, Proceedings of the Enrico Fermi International School of Physics, Course 78, ed. by A.M. Dziewonski and E. Boschi, North Holland Co..

Ogura, Y. and Phillips, N.A., 1962. Scale analysis of deep and shallow convection in the atmosphere, *J. Atm. Sci.*, **19**, 173-179.

Olson, P.L. and Corcos, G.M., 1980. A boundary layer model for mantle convection with surface plates, *Geophys. J. R. Astr. Soc.*, **62**, 195-219.

Olson, P.L., 1981. Mantle convection with spherical effects, *J. Geophys. Res.* , **86**, 4881-4890.

Olson, P.L., 1987. A comparison of heat transfer laws for mantle convection at very high Rayleigh numbers, *Phys. Earth Plan. Int.*, in press.

Peltier, W.R., 1983. Constraints on deep mantle viscosity from LAGEOS acceleration data, *Nature*, **304**, 434-436.

Quareni, F. and Yuen, D.A., 1984. Time dependent solutions of mean-field equations with applications for mantle convection, *Phys. Earth Plan. Int.*, **36**, 337-353.

Quareni, F., Yuen, D.A., Sewell, G. and Christensen, U.R., 1985. High Rayleigh number convection with strongly variable viscosity: A comparison between mean-field and two-dimensional solutions, *J. Geophys. Res.*, **90**, 12, 633-12, 644.

Quareni, F., Yuen, D.A. and Saari, M.R., 1986a. Adiabaticity and viscosity in deep mantle convection, *Geophys.*

Res. Lett., **13**, 38-41.

Quareni, F., Koch, M. and Yuen, D.A., 1986b. Stress fields produced by mantle convection: Effects of variable viscosity, *Terra Cognita Vol. 6*, no. 2, 321.

Quareni, F. and Yuen, D.A., 1986. Heat transfer characteristics in compressible convection,*E.O.S.*, **67**, no. 44, 1194.

Richards, M.A. and Hager, B.H., 1984. Geoid anomalies in a dynamic Earth, *J. Geophys. Res.*, **89**, 5987-6002.

Richter, F.M., Nataf, H.C. and Daly, S.F., 1983. Heat transfer and horizontally averaged temperature of convection with large viscosity variations, *J. Fluid Mech.*, **129**, 173-192.

Roberts, P.H., 1967. Convection in horizontal layers with internal heat generation. Theory, *J. Fluid Mech.*, **30**, 33-49.

Sammis, C.G., Smith, J.C., Schubert, G. and Yuen, D.A., 1979. Viscosity-depth profile of the Earth's mantle: effects of polymorphic phase transitions, *J. Geophys. Res.*, **82**, 3747-3760.

Schubert, G., Cassen, P. and Young, R.E., 1979. Subsolidus convective cooling histories of terrestrial planets, *Icarus*, **38**, 192-211.

Schubert, G., Stevenson, D.J. and Cassen, P., 1980. Whole planet cooling and the radiogenic heat sources contents of the Earth and the moon, *J. Geophys. Res.*, **85**, 2531-2543.

Sharpe, H.N. and Peltier, W.R., 1978. Parametrized mantle convection and the Earth's thermal history, *Geophys. Res. Lett.*, **5**, 737-740.

Spera, F.J., Yuen, D.A., Clark, S. and Hong, H.-J., 1986. Double-diffusive convection in magma chambers: single or multiple layers?, *Geophys. Res. Lett.*, **13**, 153-156.

Stacey, F.D., 1977. A thermal model of the Earth, *Phys. Earth Plan. Int.*,**8**, 282-286.

Stevenson, D.J., Spohn, T. and Schubert, G., 1983. Magnetism and thermal evolution of the terrestrial planets, *Icarus*, **54**, 466-489.

Toomre, J., Gough, D.O. and Spiegel, E.A., 1982. Time dependent solutions of multimode convection equations, *J. Fluid Mech.*, **125**, 99-122.

Turcotte, D.L. and Oxburgh, E.R., 1967. Finite amplitude convective cells and continental drift, *J. Fluid Mech.*, **28**, 29-42.

Turcotte, D.L., Torrance, K.E. and Hsui, A.T., 1973. Convection in the Earth's mantle, in *Meth. Comp. Phys.*, **13**, 431-451, Academic Press, New York.

Vincent, A.P., 1985. Apport des methodes spectrales aux problemes de convection en geophysique interne, 3rd cycle Ph. D. thesis, Astronomy and Space Physics, Univ. Paul Sabatier at Toulouse, France.

Vincent, A.P., 1987. Spectral methods in thermal convection, Minnesota Supercomputer Institute Report, Minneapolis, MN, 70 pp.

Vincent, A.P. and Yuen, D.A., 1987. Spectral methods in time-dependent chaotic mantle convection, *Minnesota Supercomputer Institute Report*, Minneapolis, MN, 120 pp.

Williams, Q., Jeanloz, R., Bass, J., Svendsen, B. and Ahrens, T.J., 1987. The melting curve of iron to 2.5 Mbar: First experimental constraint on the temperature at the Earth's center, *Science*, **236**, 181-182.

Yamaguchi, Y., Chang, C.J. and Brown, R.A., 1984. Multiple buoyancy-driven flows in a vertical cylinder heated from below, *Phil. Trans. R. Astr. Soc. Lond.*, A **312**, 519-552.

Yuen, D.A. and Peltier, W.R., 1980. Mantle plumes and the thermal stability of the D''-layer, *Geophys. Res. Lett.*, **9**, 625-628.

Yuen, D.A., Sabadini, R. and Boschi, E.V., 1982. Viscosity of the lower mantle as inferred from rotational data, *J. Geophys. Res.*, **87**, 10, 745-10, 762.

Yuen, D.A. and Fleitout, L., 1985. Thinning of the lithosphere by small-scale convective destabilization, *Nature*, **313**, 125-128.

Yuen, D.A., Weinstein, S.A. and Olson, P.L., 1986. High-resolution calculations of large aspect-ratio convection, *E.O.S.*, **67**, no. 44, 1194.

Yuen, D.A., 1986. Variable viscosity makes waves, *Nature*, **323**, 669.

Yuen, D.A., Quareni, F. and Hong, H.-J., 1987. Dissipative heating in compressible mantle convection: effects from equation of state and rheology, *Nature*, **326**, 67-69.

Zebib, A., Schubert, G. and Straus, J.M., 1980. Infinite Prandtl number thermal convection in a spherical shell, *J. Fluid Mech.*, **97**, 257-277.

Zhang, S. and Yuen, D.A., 1987. Deformation of the core-mantle boundary induced by spherical-shell compressible convection, *Geophys. Res. Let.*, in press.

FRANCESCA QUARENI, Instituto di Geofisica, Facoltà di Scienze, Unversità degli Studi, Viale Berti Pichat 8, 40127 Bologna, Italy.
DAVID A. YUEN, Minnesota Supercomputer Institute and Department of Geology and Geophysics, University of Minnesota, Minneapolis MN 55455, USA.

Chapter 12

Infinite Prandtl number spherical-shell convection

Philippe Machetel and David A. Yuen

This work presents an overview of numerical simulations of thermal convection for constant viscosity, infinite Prandtl number fluids in a spherical shell, with mantle convection being the main application. Using high-resolution grids on a supercomputer Cray-2, we have monitored the transitions from steady state to the onset of oscillatory time-dependent convection. This occurs at a Rayleigh number which is around 30 times the critical for an inner to outer radii of .62. Additional bifurcations are found with increasing strength of convection. This process culminates in chaotic convection. Analysis of the spatial correlation function of the time-dependent signals shows that the dimensionality of this chaotic attractor, at about 60 times the critical, is around 2.8 and resembles a low dimensional fractal. A large scale circulation, dominated by the degree n=2 component, is found to coexist with aperiodic boundary layer instabilities, mainly starting from the bottom. Spectral analysis of the power associated with the thermal anomalies reveals an upward cascade of energy from n=2 to n=4 to 6 at the bottom boundary. This last signature agrees well with recent seismic findings at the core-mantle boundary.

1. Introduction.

Thermal convection has become truly a working hypothesis for earth scientists. Over the past few years, convincing evidences from seismic tomography (Dziewonski et al., 1977, Woodhouse and Dziewonski, 1984; Dziewonski, 1984) have demonstrated definitively the dynamic nature of the Earth's interior. There is indeed correlation between the long wavelength component of the geoid anomalies (Crough and Jurdy, 1980) and the seismic velocity anomalies (Masters et al., 1982). These phenomena can be attributed to lateral

N.J. Vlaar, G. Nolet, M.J.R. Wortel & S.A.P.L. Cloetingh (eds), Mathematical Geophysics, 265-290.

thermal differences due to thermal convection (Busse, 1983). Moreover, the patterns observed cannot be explained by simple steady-state cellular convection (Turcotte and Oxburgh, 1967). In order to understand better the relationship between these inferred images and the dynamics of mantle convection, it is very important now for geophysicists to delve into the subject matter of thermal convection more intensively than before, especially regarding the real nature of time-dependent convection.

The basic physics of thermal convection has been studied extensively in the laboratory (Krishnamurti, 1970 a and b; Busse and Whitehead, 1971; Krishnamurti, 1973; Whitehead and Parsons, 1978; Dubois et al., 1981; Nataf et al., 1981; Koster and Müller, 1984, Carrigan, 1982; Olson, 1984; and Hart et al., 1986) and theoretically by linear or weakly non-linear theory (Pekeris, 1935; Chandrasekhar, 1961; Busse, 1967; Busse, 1975; Busse and Riahi, 1982; Riahi et al., 1982). With the introduction of more powerful computers we have witnessed the gradual use of numerical methods in mantle convection (Torrance and Turcotte, 1971; McKenzie et al, 1974, Houston and Debremaecker, 1974, Schubert and Young, 1976; Jarvis and McKenzie, 1980; Jarvis and Peltier, 1982; Christensen, 1984; Hansen and Ebel, 1984; Jarvis, 1984 and 1985; Fleitout and Yuen, 1984; Christensen and Yuen, 1984). These studies have brought new insights into various aspects of thermal convection as applied in the Earth's mantle, which can be considered as an extremely viscous fluid. These include the onset of time-dependent convection, effects of rheology and aspect-ratio of the flow. One aspect which has not been emphasized as much is the influence from sphericity on the convective circulations. Part of this neglect may have stemmed from the long-lasting idea of upper-mantle convection (Richter, 1973). However, there is now growing acceptance of a dynamical lower-mantle, especially in view of the seismic tomographic images. Thus it is of extreme importance to get away from the cartesian constraint and to investigate the mantle convection in a more realistic geometry, i.e. in a spherical shell. There have been few studies of spherical-shell, finite-amplitude convection in an infinite-Prandtl number fluid as the Earth's mantle (Schubert and Young, 1976; Schubert, 1979; Schubert and Zebib, 1980; Zebib et al., 1980; Zebib et al., 1983; Machetel and Rabinowicz, 1985; Zebib et al., 1985; Machetel et al., 1986; Machetel and Yuen, 1986 and 1987). In this paper we will endeavour to summarize the principal results obtained thus far in this still growing field of research. The equations and numerical methods will be discussed in section 2. Next we will study the influences of internal-heating and of the Rayleigh number in the weakly non-linear regime. Section 4 will see the development of time dependence in the steady-state solutions, the subsequent transition to weak turbulence and the interpretation of these results in light of modern concept of non-linear dynamics (Bergé et al., 1984, Schuster, 1984). We will then examine in the spectral domain the time-dependent thermal anomalies in section 5. The results are summarized in section 6. We conclude in section 7 with a discussion of the limitations of these rudimentary models.

2. Theoretical and numerical approach.

The dimensionless equations governing the convective motion of an infinite Prandtl number isoviscous fluid in the Boussinesq approximation, where the bottom temperature is kept constant, are the mass conservation equation (1), the momentum equation (2) and the energy equation (3).

$$\nabla\cdot(\mathbf{U}) = 0 \tag{1}$$

$$Ra_b(T-T_0)\mathbf{e}_r + \nabla\cdot(2\sigma) = \nabla P \tag{2}$$

$$\frac{\partial T}{\partial t} + \mathbf{U}\cdot\nabla T = \nabla^2 T + \frac{Ra_i}{Ra_b} \tag{3}$$

where **U** is the velocity field, T the temperature, $\mathbf{e}_r$ the unit radial vector outward oriented, σ the viscous stress tensor, P the pressure, t the time. Ra_b is the Rayleigh number based on the difference ΔT between the temperature T_o on the outer boundary and the temperature T_i on the inner boundary. Ra_i is the Rayleigh number based on the rate of internal heating H.

$$Ra_b = \frac{g\,\alpha\Delta T d^3}{K\nu} \tag{4}$$

$$Ra_i = \frac{g\,\alpha H d^5}{kK\nu} \tag{5}$$

As the constant temperature boundary condition is more appropriate for the core mantle boundary, we have used this non-dimensionalization scheme with two Rayleigh numbers. The gravity acceleration g will be considered as radially oriented and constant in the mantle, $d = R_o - R_i$ is the thickness of the convective layer (R_o and R_i are respectively the radius of the outer surface and the radius of the internal surface), ν is the kinematic viscosity, K is the thermal diffusivity, and k the thermal conductivity.

The scale factors used for the distance, the time, the velocity, and the temperature are respectively d, d^2/K, K/d, ΔT.

In axisymmetrical geometry, equation (1) allows to express the velocity field in term of a stream function ψ, with :

$$U_r = \frac{1}{r^2\sin\theta}\frac{\partial\psi}{\partial\theta} \tag{6}$$

$$U_\theta = \frac{-1}{r\sin\theta}\frac{\partial\psi}{\partial r} \tag{7}$$

The vorticity $\boldsymbol{\omega}$ is given by the one component vector:

$$\boldsymbol{\omega} = r\sin\theta\,\nabla\times(\mathbf{U}) \tag{8}$$

The stream function equation (9) is obtained from the equations (6), (7) and (8).

$$D^2(\psi) = -\omega \tag{9}$$

$$D^2(\,) = \frac{\partial^2(\,)}{\partial r^2} + \frac{1}{r^2}\frac{\partial^2(\,)}{\partial\theta^2} - \frac{\cot\theta}{r^2}\frac{\partial(\,)}{\partial\theta} \tag{10}$$

Where ω is the only non-zero component of the vorticity vector for the two-dimensional flow. The equation of vorticity (11) is obtained by taking the curl of equation (2), by expressing the viscous stress in term of velocity (Batchelor, 1967), and by replacing the velocity by the vorticity.

$$D^2(\omega) = Ra_b \sin\theta \frac{\partial T}{\partial \theta} \tag{11}$$

With $\xi = Ra_i / Ra_b$ measuring the strength of internal heating, the energy equation becomes:

$$\frac{\partial T}{\partial t} = \nabla^2 T - \mathbf{U}.\nabla T + \xi \tag{12}$$

The equations (9), and (11) have been solved using alternate direction implicit methods (A.D.I.) (Roache, 1972). The temperature equation has been solved using a semi-implicit scheme. The time step δt is based on the Courant criterion :

$$\delta t = \frac{\delta r}{V_{rm}} \tag{13}$$

where δr is the distance between two equidistant radial levels and V_{rm} the maximal radial velocity.

In this paper, the brackets will denote averaging over a spherical surface, double brackets averaging over the spherical surface and the radial direction, and the overbars will denote time averaging. In order to monitor the temporal development of the flow, we have computed the following quantities :

- The Nusselt number averaged on the spherical surface $<Nu(r,t)>$ which reflects the global heat transport :

$$<Nu(r,t)> = \frac{(1-\eta)^2}{2\eta} \int_0^{\pi} \left[U_r T - \frac{\partial T}{\partial r} \right] r^2 \sin\theta \, d\theta \tag{14}$$

- The shell volume averaged Nusselt number $<<Nu(t)>>$:

$$<<Nu(r,t)>> = \frac{1}{d} \int_{R_i}^{R_o} <Nu(r,t)> dr \tag{15}$$

- The instantaneous horizontally averaged temperature profile $<T(r,t)>$:

$$<T(r,t)> = \frac{1}{2} \int_0^{\pi} T(r,\theta,t) \sin\theta \, d\theta \tag{16}$$

- The time averaged mean temperature profile $\overline{<T(r)>}$:

$$\overline{<T(r)>} = \int_{t_1}^{t_2} <T(r,t)> dt \Big/ \int_{t_1}^{t_2} dt \tag{17}$$

This has been done only for the higher Rayleigh number computations. In these cases t_1 is situated well beyond the initial transient, spin-up period, t_2 is the time at the end of computation.

- The thermal perturbation $T_p(r,\theta,t)$ from the instantaneous mean temperature profile:

$$T_p(r,\theta,t) = T(r,\theta,t) - <T(r,t)> \tag{18}$$

- The Legendre spectral decomposition $\tau_n(r,t)$ of the thermal perturbation :

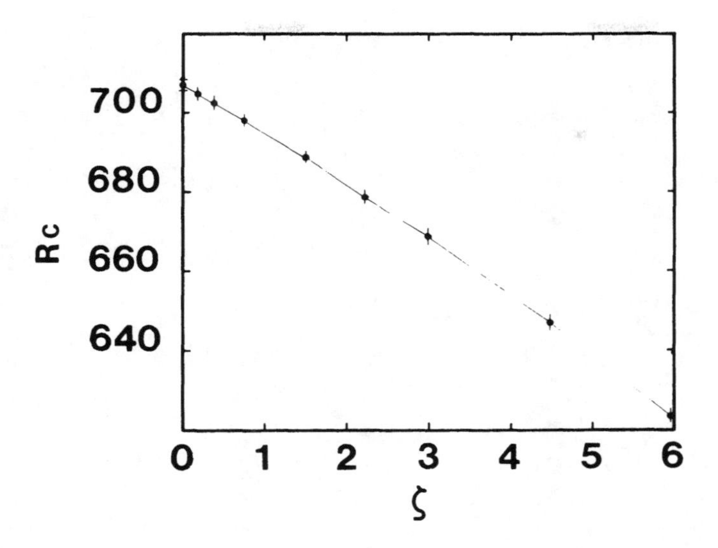

Figure 1.
Critical Rayleigh number R_c, for the P40 solution, as a function of the internal heating rate ξ.

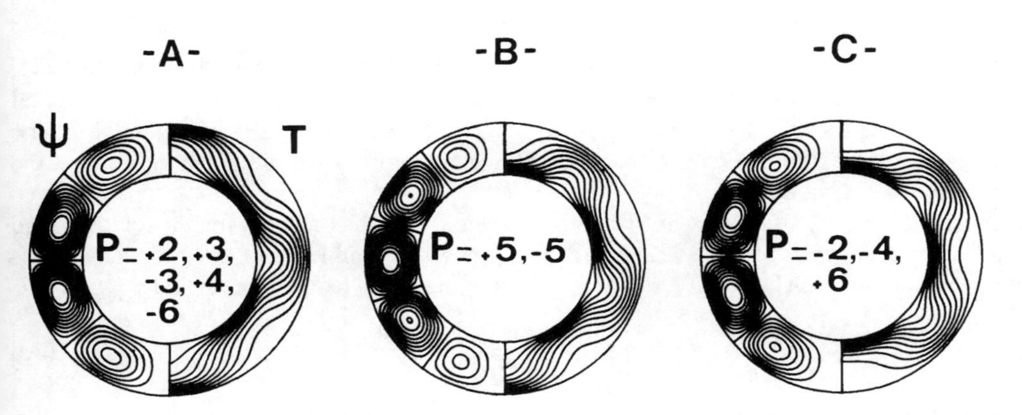

Figure 2. Different possible steady state solutions for $Ra_b = 1.28R_c$ and $\xi = 0$. As initial condition, the conductive temperature profile has been perturbed by P (Eqn. 20) with $n = 2$ to 6 and with both positive and negative amplitude.

$$\tau_n(r,t) = \frac{2n+1}{2}\int_0^{\pi} T_p(r,\theta,t)P_n(\cos\theta)\sin\theta \, d\theta \tag{19}$$

The number of grid points used for the computations ranges from 10×49 for the lower Rayleigh number cases to 241×737 for the higher Rayleigh number, time-dependent calculations, which were done on the Cray-2 computer.

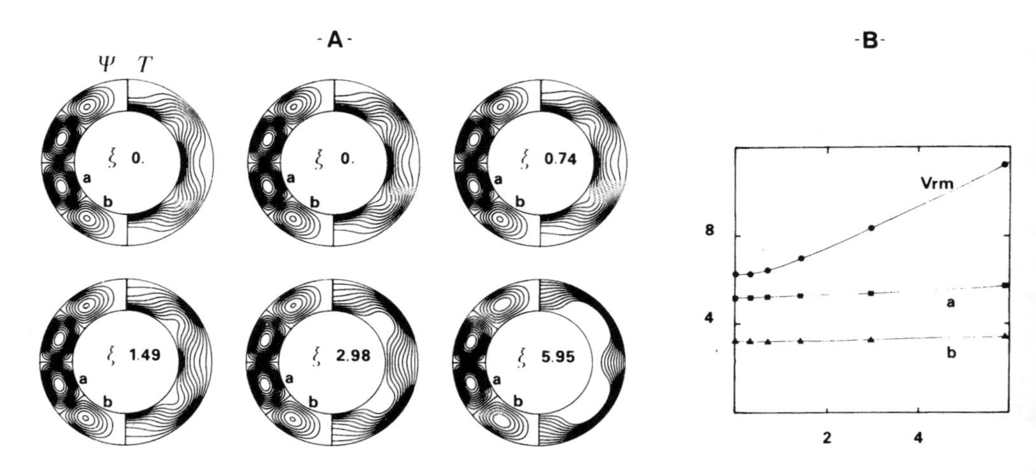

Figure 3. Influence of the internal heating rate ξ on the structure of the steady M40 solution computed for Ra_b = 1.28 R_c. Inset A displays the stream functions and the temperature fields, inset B shows the variation of the maximum radial velocity V_{rm} in the solution and the maximum stream function inside the cells with internal heating (letters refer to inset A).

3. Weakly non-linear regime.

The onset of convection is characterized by a critical value R_c of the Rayleigh number Ra_b (Chandrasekhar, 1961). For a fluid partly heated from within, R_c depends on the internal heating rate. As it is shown in Fig. 1, R_c can be considered as a linear function of the internal heating rate ξ. Henceforth, all the Rayleigh numbers will be given in terms of the ratio to the critical one R_c.

Equation (12) is non-linear. Therefore, several solutions can be found, for the same Rayleigh number and the same internal heating rate. The influence of the initial conditions on the final solution has been tested by perturbing the initial temperature field by P:

$$P(n,r,\theta) = \pm\, 0.1 \sin\left\{ \frac{-\pi}{\ln(\eta)} \ln\left[\frac{(1-\eta)}{\eta} r \right] \right\} \cos(n\,\theta) \tag{20}$$

Thus, we imposed the modes $n=1$ to 6 with both positive and negative amplitudes at the beginning of computation. The different steady solutions found, for a geometrical aspect ratio η= .62, ξ= 0, and Ra_b= 1.28 R_c, are shown in Fig.2 . In these conditions, three solutions called P40, P50, and M40 are stable. P40 solution (M40 solution) is displaying four cells along a meridian great circle with rising (sinking) currents at the poles and at the equator (Figure 2-A) (see figure 2-C for M40 solution). The third solution P50 (Figure 2-B) displays five cells along a meridian. These results are in good agreement with previous studies which have already shown that the most unstable mode at the onset of convection is strongly dependent on the geometrical aspect ratio, i.e. inner to outer radii, and that several solutions are possible for the same physical conditions (Chandrasekhar, 1961; Young, 1974; Busse, 1975; Zebib et al., 1980).

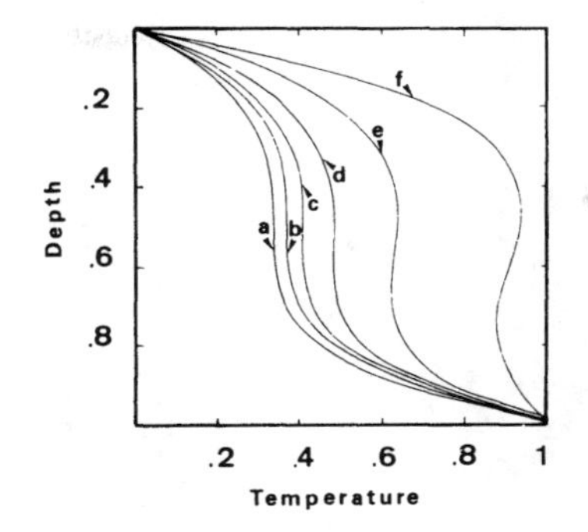

Figure 4.
Influence of the internal heating rate ξ on the spherically averaged temperature <T(r)> for $Ra_b = 1.28\ R_c$. a : $\xi = 0$, b : $\xi = 0.37$, c : $\xi = 0.74$, d : $\xi = 1.49$, e : $\xi = 2.98$ and f : $\xi = 5.95$.

Figure 3-A shows that the internal heating rate ξ does not have, for constant Ra_b, a strong effect on the lateral extent of the cells. The maximum radial velocity in the cells (Fig. 3-B) increases by about 80 % while the mass flux of the cells, given by the maximum stream function, remains approximately constant. On the other hand, the temperature field is strongly affected by the internal heating rate. The higher internal heating rate cases are characterized by large zones of stably stratified interiors (Fig. 3-A). This effect influences the spherically averaged temperature $<T(r)>$ inside of the convective layer. The increase of $<T(r)>$ with ξ can be seen on Fig. 4. A value of $\xi = 5.95$ is able to generate an inversion in the mean temperature profile : the mean temperature is higher in a range of depth from $z = .4$ to $z = .6$ than deeper down. The increase of ξ induces a thinning of the upper boundary layer and a thickening of the lower one. The presence of internal heating reinforces the discrepancy between the lower and the upper boundary layers and will likely play an important role in the generation process of boundary layer instabilities at higher Rayleigh numbers.

Since the increase of the internal heating rate ξ seems to have only a weak influence on the lateral extent of the cells, we now consider cases with $\xi = 0$. Fig. 5-A (top) displays the steady M40 solutions found for Ra_b ranging from R_c to approximately 8 R_c. The lateral extent of the equatorial cells is increasing at the expense of the polar ones. This evolution which favors the equatorial cells (analogous to the rolls in plane geometry) at the expense of the polar cells (analogous of the hexagonal cells) can be explained by looking at the results in cartesian geometry. For Rayleigh numbers near the critical, rolls and hexagonal circulations are equally likely to exist but, when the Rayleigh number is increased rolls become then the preferential structure (Busse, 1967; Busse and Whitehead, 1971). If Ra_b is increased more than approximately 7 to 8 R_c then a transition from a four cells steady state to a two cells steady state P20 occurs. Fig. 5-A displays the aspect of the streamlines before and after this transition. This kind of transition occurs for different geometrical aspect ratii $\eta = .62$ (Fig. 5-A (top)) and for $\eta = .56$ (Fig. 5-A (bottom)). Fig. 5-B shows that the sum of the maximum stream function inside the cells and the Nusselt number $<<Nu>>$ are less after the transitions than before (Machetel and Rabinowicz, 1985). The level of Rayleigh

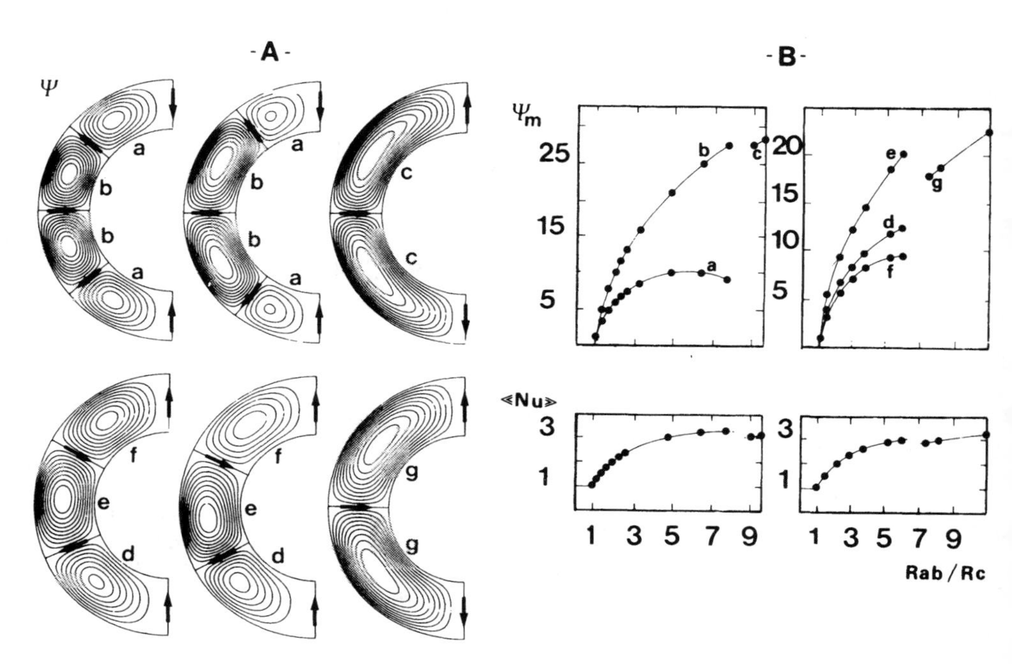

Figure 5.
Influence of the increase of the Rayleigh number Ra_b on the structure of the steady M40 solution. Part A displays the stream function computed, for $\xi = 0$, for $\eta = .62$ (top) and $\eta = .56$ (bottom). The Rayleigh numbers are, for $\eta = .62$, $Ra_b = 1.00\ R_c$ (left), $Ra_b = 6.42\ R_c$ (middle) and $Ra_b = 10.59\ R_c$ (right). For $\eta = .56$, they are : $Ra_b = 1.01\ R_c$ (left), $Ra_b = 3.66\ R_c$ (middle), and $Ra_b = 8.06\ R_c$ (right). Part B gives the evolution of the maximum stream function in the cells and of the Nusselt number $<<Nu>>$ for $\eta = .62$ (left) and $\eta = .56$ (right) (letters refer to part A).

number for which the transition to P20 solution occurs is dependent on the value of η. This transition has also been observed by Young (1974) for a Prandtl number equal to 5 and $\eta = .6$ and by Zebib et al. (1985) for $\eta = .5$. These last authors found that the P20 is the only axisymmetrical stable solution for Rayleigh numbers greater than approximately 40 R_c. The existence of the P20 solution has been tested in three-dimensional geometry by Machetel et al. (1986). For low Rayleigh numbers the same axisymmetrical solutions P40, P50 and M40 are found among several other co-existing three-dimensional solutions. When the Rayleigh number is increased a general trend to become 'polygonal' solutions is observed. In this context, the same transition from a M40 to a P20 solution is observed for the same Rayleigh number range.

These results indicate that the P20 solution is able to play a particularly special role among all the other solutions of spherical convection. From a geophysical point of view, this solution displays strong resemblances to the structure of the geoid anomalies (Crough and Jurdy, 1980) and to the structure of seismic velocity anomalies (Master et al., 1982). To check whether this planform can exist in the Earth's mantle conditions, it is of interest to study its behavior for higher Rayleigh numbers.

4. From steady-state to chaotic convection.

In the previous section results for low Rayleigh numbers were presented. However, in order to understand better the dynamical features in the lower mantle, as revealed by seismic tomography, it is more useful to explore higher Rayleigh number convection. The onset of time dependence is an issue which has not been explored too much in the context of mantle convection. Numerical simulations provide the only available tools in understanding the detailed dynamical interaction of flows in a spherical shell. It came therefore, as a surprise to us, that, for spherical shell convection with $\eta = .62$, the transition to oscillatory time-dependent convection occurs at Rayleigh numbers only around 30 times supercritical, with $l = 2$ cells being the dominant mode (Machetel and Yuen, 1986). This whole question of time-dependent convection certainly merits greater attention in the future.

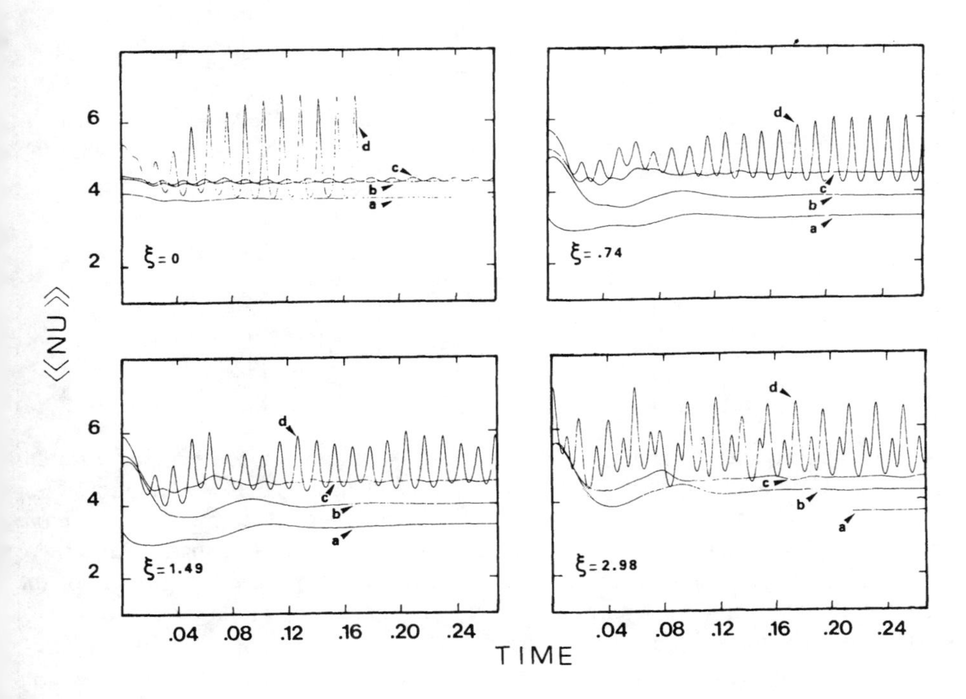

Figure 6.
Temporal evolution of the Nusselt number $\langle\langle Nu \rangle\rangle$ for different internal heating rate ξ and different values of the Rayleigh number Ra_b (curves a to d in each panel). Table 1 the values of supercritical Rayleigh number used.

In Fig. 6, we display the evolution of the Nusselt number $\langle\langle Nu(t) \rangle\rangle$ for various value of the internal heating rate ξ. Table 1 gives the supercritical Rayleigh numbers Ra_b used in these calculations.

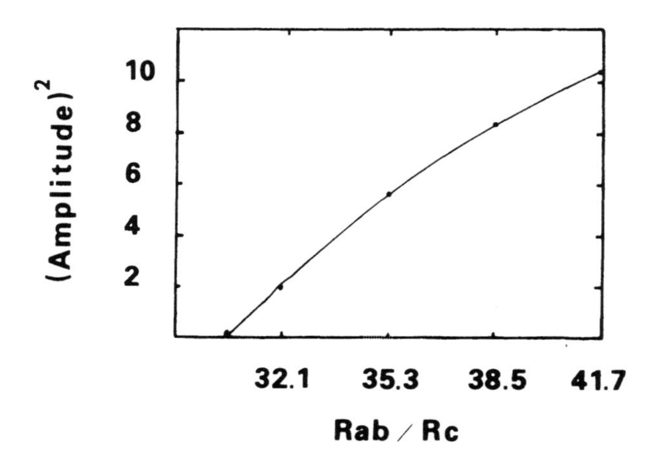

Figure 7.
The evolution of the square of the amplitude of the temporal evolution of <<*Nu* (*t*)>> is shown to be proportional to the excitation strength, defined by Ra_b/R_c, near the onset of time dependent convection. This property is characteristic of a Hopf bifurcation (Bergé et al., 1984).

Table 1

	$\xi = 0.$	$\xi = .74$	$\xi = 1.49$	$\xi = 2.98$
a	19.3	9.8	9.9	10.2
b	28.9	16.3	16.5	17.0
c	30.5	29.3	29.6	23.8
d	38.5	39.2	39.5	40.7

For low Rayleigh numbers (curves a, Fig. 6), steady solutions are reached for different values of ξ. <<*Nu*>> reached equilibrium very quickly during these calculations. For higher Ra_b, the influence of the overshoot at the beginning of computation is more important and <<*Nu*>> require more time to reach the equilibrium. The shape of the curve A-b in Fig. 6 is characterized by decaying oscillations which mark the beginning of the evolution and disappear upon reaching the steady state. For slightly higher Ra_b, (curves c) steady states are no longer attained and periodic behavior seems to prevail for all cases. Finally it can be observed that, at Ra_b greater than approximately 38 times R_c, the solutions display strong and oscillatory type of evolutions. For $\xi = 0$, and $\xi = .74$ the solutions are periodic whereas they are quasi-periodic for $\xi = 1.49$ and $\xi = 2.98$ (Machetel and Yuen, 1986).

Two characteristics have let us to conclude that the onset of time dependence take place, via a Hopf bifurcation mechanism (Bergé et al., 1984), at $Ra_b = R_{td}$. First, in the neighbourhood of the transition, the amplitude of the time dependent oscillations of <<*Nu*>> is proportional to $(Ra_b - R_{td})^{1/2}$ (Fig. 7). Additionally, the period of these oscillations, to first order, does not depend on the distance $Ra_b - R_{td}$.

Since the internal heating rate does not seem to have a strong influence on the onset of time dependent convection, we focus our attention on the case $\xi = 0$. Figure 8 shows the

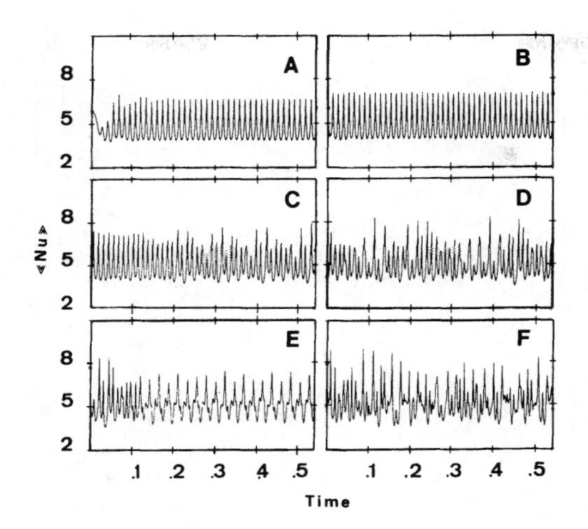

Figure 8.
Temporal evolution of the Nusselt number $<<Nu(t)>>$ for f = 0. A : $Ra_b = 38.5\ R_c$, B : $Ra_b = 41.3\ R_c$, C : $Ra_b = 44.9\ R_c$, D : $Ra_b = 48.2\ R_c$, E : $Ra_b = 51.4\ R_c$, F : $Ra_b = 57.8\ R_c$ (we now employ the same convention in lettering for Figures 9 to 14).

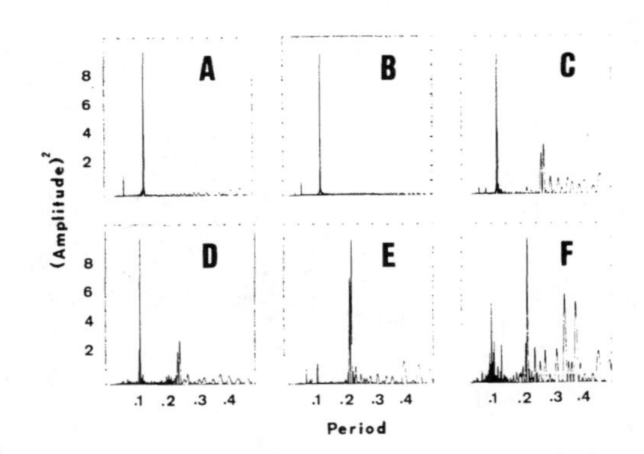

Figure 9.
Spectral analysis of the temporal evolution for Fig. 8. The transient part of the signal has been removed (see text). A, B, C, D, E, F follow the same convention as in fig. 8.

temporal evolution of $<<Nu>>$ for Rayleigh numbers ranging from 38.5 R_c to 57.8 R_c. For Ra_b = 38.5 R_c, and Ra_b = 41.3 R_c (Figure 8 A and B) the evolution of $<<Nu>>$ is clearly periodic. For Ra_b = 44.9 and Ra_b = 48.2 a modulation, which is characteristic of a quasi-periodic signal, appears in the temporal signatures of $<<Nu>>$. In the case Ra_b = 51.4 R_c, after a spin-up time of approximately .15 the evolution seems more regular while, for Ra_b = 57.8 the signal becomes more complicated again. Fig. 9 displays the spectra of these evolutions. These spectra have been calculated by removing the transient part of the signal at the beginning of computation. Over thirty overturns have been employed in the time series. Table 2 gives the transient time t_c, and the associated periods.

Table 2

Ra_b	t_c	T_1	T_2	T_3	T_2/T_1	T_3/T_1
38.5	.10	.013118	---	---	---	---
41.3	.10	.012489	---	---	---	---
44.9	.20	.011843	.027404	---	2.31	---
48.2	.07	.011544	.024928	---	2.16	---
51.4	.15	.011340	.022836	.044444	2.01	---
54.6	.40	.010 ??	.022 ??	.041 ??	2.2	4.1
57.8	.15	.010031	.021983	??	2.19	---

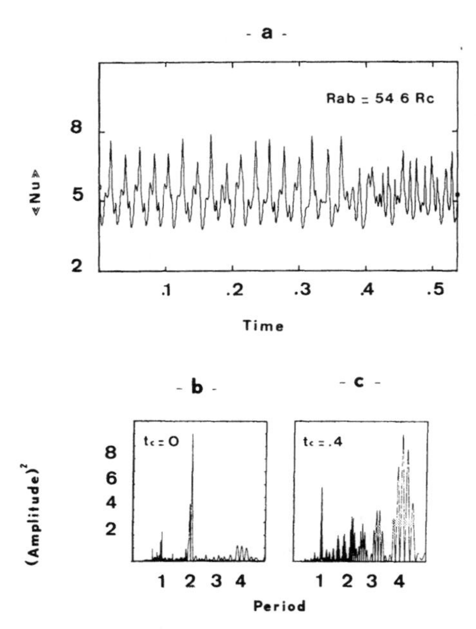

Figure 10.
The top panel displays the temporal evolution of the Nusselt number $<<Nu(t)>>$ for $Ra_b = 54.6\ R_c$. The curve shows two different behaviors : the first (t less than 0.4) is similar to those of curve E in fig. 8, the second (t greater than 0.4) is similar to those of curve F in fig. 8. The bottom panels display the spectral analysis corresponding to this time series. The analysis has been conducted without cutting the transient part of the signal (left panel) and by cutting this transient part (right panel).

For $Ra_b = 38.5\ R_c$ and $Ra_b = 41.3\ R_c$, the time histories are characterized by a single frequency. The presence of the harmonics of the fundamental period is due to the difference in the shape between the top and the bottom of the curves of the figure 8 A and B. For $Ra_b = 44.9\ R_c$, a second incommensurate period appears in the spectrum (Figure 9 C). The increase of the Rayleigh number decreases the periods. As they are not always commensurate, the ratio T_2/T_1 falls from 2.31 to 2.16. For $Ra_b = 51.4$ this ratio is almost equal to 2. This phenomenon of phase-locking can explain this particular feature of Figure 8 E (Guckenheimer, 1986).

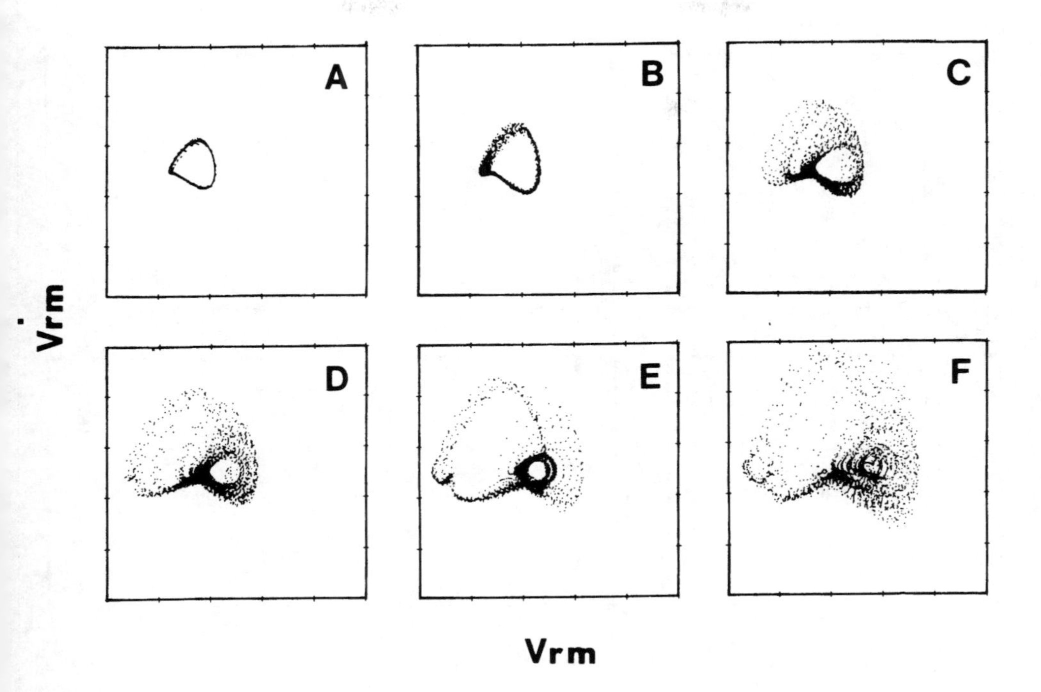

Figure 11.
Projection in the plane ($\dot{V}_{rm}$, time derivative of V_{rm}) of the phase space of the time dependent solutions. A, B, C, D, E, F follow the same convention as in fig. 8.

Fig. 10-A displays the temporal evolution of <<*Nu*>> for Ra_b = 54.6. It is difficult to ascertain whether the first portion of the evolution, similar to those of the case Ra_b = 48.2 R_c, can be considered to be a transient phenomenon. A sudden change then appears at a time approximately 0.4 the conductive time across the shell. The second part of this evolution has a very different shape. Fig. 10-B shows the spectrum of this solution computed without cutting the 'transient' part of the evolution, Fig. 10-C displays the spectrum computed on the second portion of the time history. The spectrum 10-B has the same aspect than those of the case Ra_b = 51.4 R_c. On the other hand, the spectrum computed on the second part of the signal displays a chaotic-like behavior. That is to say, numerous periods have non-negligible contributions. But, even if the time series used for this last spectrum is too short for determining accurate periods, the qualitative properties of the spectrum are likely to be preserved. In the case Ra_b = 57.8 R_c, the transient evolution is short again and the chaotic- like behavior appears both on the temporal evolution of *Nu* (Figure 8-F) and in the spectrum (Figure 9-F).

In the last decade there have been important advances made in understanding order in chaos from concepts of non-linear dynamics (Ruelle and Takens, 1971; Newhouse et al., 1978; Bergé et al., 1984, Guckenheimer, 1986). Phase space diagrams provides a powerful

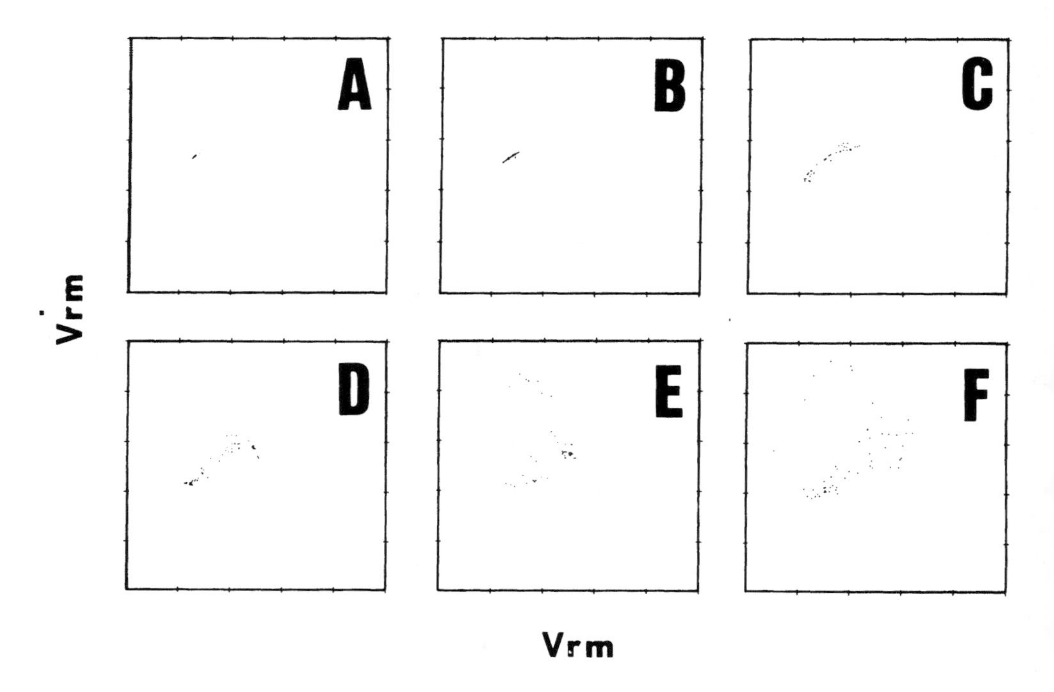

Figure 12.
Poincaré section of the time dependent solutions obtained from phase space projection of figure 11 by selecting situations where the Nusselt number <<$Nu(t)$>> is increasing and equal to the time averaged <<Nu>> (similar to eq. 17). A, B, C, D, E, F follow the same convention as in fig. 8.

tool for describing the behavior chaotic systems, since one can visualize the complicated behavior in geometric form. We present, in Fig. 11 a projection of the phase space on the plane (V_{rm} , $\dot{V}_{rm}$), where V_{rm} is the maximum radial velocity and $\dot{V}_{rm}$ its time derivative, for the same series presented in Fig. 8. One observes the increased complexity in the orbits, as the Rayleigh number is increased. It is well known that qualitatively different trajectories can be distinguished by their Poincaré sections (Bergé et al., 1984, Guckenheimer, 1986). By slicing the attractor with a plane and plotting successive points of intersection, one obtain a trajectory for the Poincaré cross-section. This plane is defined by setting one of the variables, here <<$Nu(t)$>>, to be equal to the time averaged Nusselt number $\overline{<<Nu>>}$. We choose the Poincaré sections to be the points where <<$Nu(t)$>> increases with time. The corresponding Poincaré sections for the phase trajectories in Fig. 11 are shown in Fig. 12. In the case of monochromatic evolutions, for Ra_b = 38.5 R_c and Ra_b = 41.7 R_c, the trajectories in the phase space reach limit cycles (Figure 11 A and B). As the solutions are periodic, the trajectories cross the phase space plane defined by the temporal evolution of <<Nu>> and $\overline{<<Nu>>}$ in the same conditions of velocity. Therefore the Poincaré section is restricted, in these cases, to a single 'dot'. When an evolution is characterized by two periods, the trajectories in the phase space are confined to a torus (Schuster, 1984). Each period being related to the motion along one of the axis of the torus. Poincaré sections for $Ra_b = 44.9\ R_c$ and $Ra_b = 48.2\ R_c$ are characteristic of non-commensurate periods. The points are approximately disposed along a curve to draw a continuous segment (Figure 12 C and D). In the case of commensurate periods, a continuous curve can no longer be distinguished. The points are rearranged in widely separated areas discontinuously (Figure 12-E). The Poincaré section of Figure 12 F ($Ra_b = 57.8\ R_c$) does not have a simple shape, the points are scattered over the surface. In this case, the dimension of the attractor in the phase space is greater than 2. This property can reveal a three-periodic system, for which the attractor will be a torus of dimension 3, as well as a chaotic system for which the attractor will have a fractal dimension between 2 and 3 (Schuster, 1984; Bergé et al., 1984, Guckenheimer, 1986). Indeed, because of the contraction of surface, due to the viscous dissipation, the phase space of a dissipative flow must have a volume equal to 0 (Bergé et al., 1984). Therefore, the dimension of the attractor in the phase space must be less than 3. However, if the dimension of the attractor is equal to 2, this property of surface contraction will force the trajectories to reach a torus where the determinism of the equations does not allow the trajectories to cross themselves. If they did, this would then mean that the solutions could display several evolutions from the same initial conditions. In this case the divergence of all trajectories could not exist. Therefore, the dimension of the attractor has to be greater than 2. Display of the trajectories in 3-D should reveal these strange attractors much more lucidly.

The task of the scientific investigator is to reconstruct the dimension of the attractor from limited data. An efficient technique, introduced by Grassberger and Procaccia (1983), made it possible to construct a phase space and look for chaotic attractors. Simply put, one looks at a time series in the phase space and treats the measured values at fixed time delays. This is done by considering a spatial correlation $C(r)$, r measuring the distance on the attractor in the phase space, between points of a time series $X(t)$ of a dynamical system in its phase space. The expression of the correlation function is :

$$C(r) = \lim_{m \to \infty} \frac{1}{m^2} \sum_{i,j} H(r - |\mathbf{x}_i - \mathbf{x}_j|) \tag{21}$$

Where the summation has to be done for $i = 1$ to m and $j = 1$ to m, m being the number of points of the time serie $X(t)$. In eqn. (21), H is the Heaviside function, $\mathbf{x}_i$ and $\mathbf{x}_j$ are the p-component vectors derived (see eqn.23) by sampling the trajectories of $X(t)$. In our case, m is equal to several hundred and $X(T)$ is equal to $\langle\langle Nu(t)\rangle\rangle$. For small distances r, $C(r)$ behaves as a power of r :

$$C(r) \approx r^s \tag{22}$$

Where s is closely related to the fractal dimension D of the attractor (Grassberger and Proccacia, 1983). Basically, this method entails the computation of the correlation function $C(r)$ by increasing the dimension of the test space. This is accomplished by building up the p-component vectors $\mathbf{x}$:

$$\mathbf{x}_i = (X(t_i), X(t_i + \tau), \ldots, X(t_i + (p-1)\tau)) \tag{23}$$

Where τ is the arbitrary but fixed time lag. When the dimension p of the test space is increased, the asymptotic slope s of the representation $\ln(C(r))$ versus $\ln(r)$ gives the

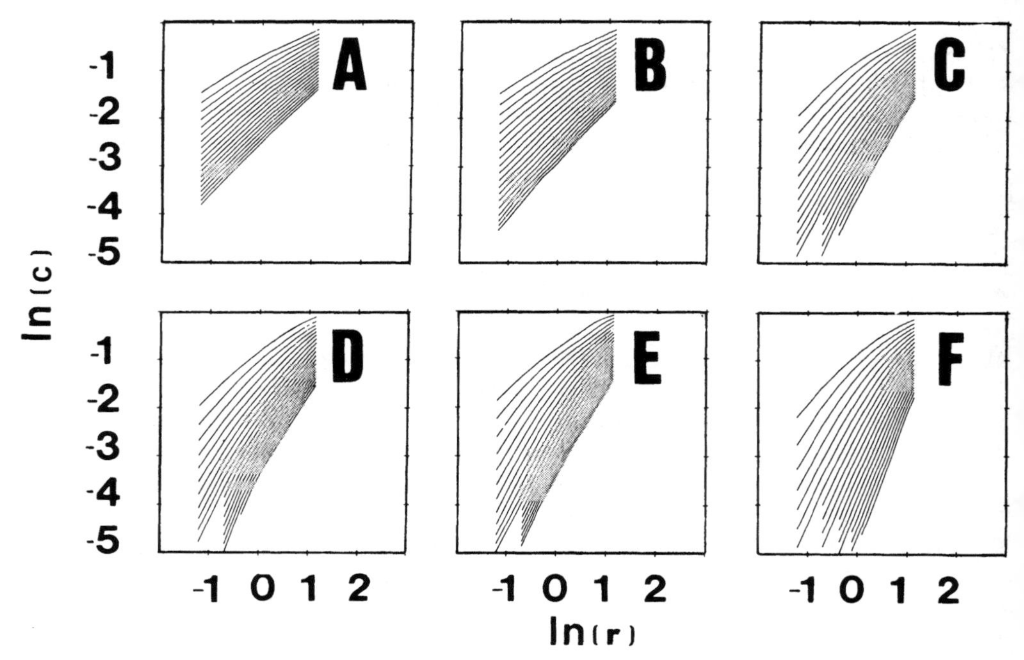

Figure 13.
Determination of the size of the attractor of the solution in the phase space. The characteristic relation of the fractal dimension leads to $\ln(C(r)) = s \ln(r)$ (Grassberger and Proccacia, 1983). The asymptotic value of the slope yields the dimension D of the attractor (see text). A time series consisting of four hundred points, sampled well beyond the transient time, and a time lag s = .0001 have been used to compute the correlation function. A, B, C, D, E, F follow the same convention as in fig. 8.

dimension D of the attractor. Fig. 13 displays the curves $\ln(C(r))$ versus $\ln(r)$ and Fig. 14 gives the asymptotic evolution of s with the dimension of the test space p. In the case of mono-periodic evolution, the size of the attractor is.1 (Fig. 13 and 14, A and B). This result is not surprising because the trajectories in the phase space are confined to a limit cycle. In the case of quasi periodic evolution with commensurate or non-commensurate frequencies the dimension of the attractor is 2 (Fig. 13 and 14, C,D and E). The trajectories describe a torus in the phase space. But, in the case of chaotic circulation, the asymptotic dimension reached by the curves of Fig. 13 F is approximately 2.7 (Fig. 14 F). This result is in good agreement with an experimental determination of the fractal dimension 2.8 for the strange attractor in Rayleigh-Bénard convection by Malraison et al. (1983).

The presence of this strange attractor for Ra_b = 57.8 R_c confirms that convection is chaotic. This new idea of using fractal analysis will be important for furthering the understanding of time-dependent convection, as there may exist a relationship between the dimension of the underlying attractor and the controlling variables of convection, such as the physical and rheological parameters. Fractals will also help geophysicists in appreciating the variablility of mantle flow processes.

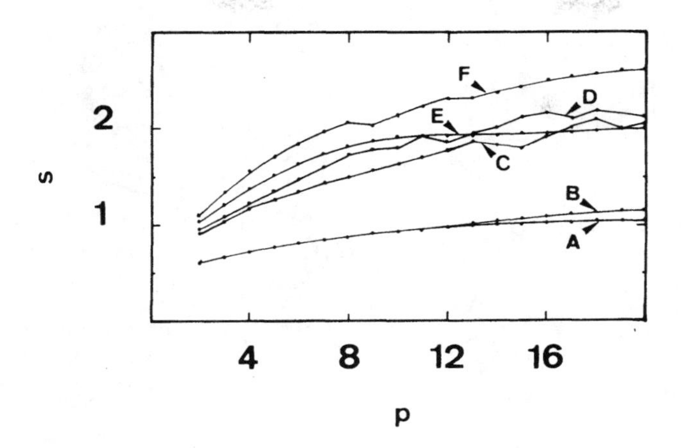

Figure 14.
Evolution of the slope s of $\ln(C(r))$ versus the size p of the test space. Curves A, B, C, D and E reach integer dimension for the attractor while curve F reaches a fractal dimension which is characteristic of strange attractor (Bergé et al., 1984; Schuster, 1984). A, B, C, D, E, F follow the same convention as in fig. 8.

5. Spherical harmonic analysis of the time dependent solutions.

In this section, we will focus on the geophysical interpretation of the spectral contents of the non-linear regime, more characteristic of the Earth's mantle, since there is a recent rapid growth of global geophysical data bases which provide descriptions of actual mantle heterogeneities, dominated by low order spherical harmonics. At the same time, there is also mounting experimental and theoretical interests in understanding the transition to large scale flow in thermal convection at high Rayleigh number (Krishnamurti and Howard, 1981; Carrigan, 1982; Howard and Krishnamurti, 1986; Hansen, 1986; Hansen, 1987; Vincent, 1986; Yuen et al, 1986a).

It is important to note that, even for chaotic or weakly turbulent convection at a couple hundred times supercritical, the boundary layer instabilities are not able to break the global circulation with $n = 2$ dominating and to generate smaller cells with the more expected aspect ratio one cells. The presence of thin hot and cold tongues in these simulations may explain certain features in tectonics, such as lithospheric doubling (Vlaar, 1983). This phenomenon is displayed in the series of isotherm histories in Fig. 15. These experiments were performed at Rayleigh numbers smaller than what is commonly expected in mantle convection, between 10^6 and 10^7. However, if the pressure-dependences of thermal expansivity and thermal conductivity (Birch, 1952; Anderson, 1987) are accounted for, then it is conceivable that the effective Rayleigh number of the mantle can be $O(10^2)$ to $O(10^3)$ supercritical. In this light we can regard these solutions as possible reasonable mantle convection models.

In order to insure the time-dependent solutions be properly resolved (Marcus, 1981), we have employed a sufficiently fine grid for this purpose. The computations reported below have been performed with two successive fine meshes : one with 121 points in the radial direction and 369 points along the horizontal for a Rayleigh number 100 times the

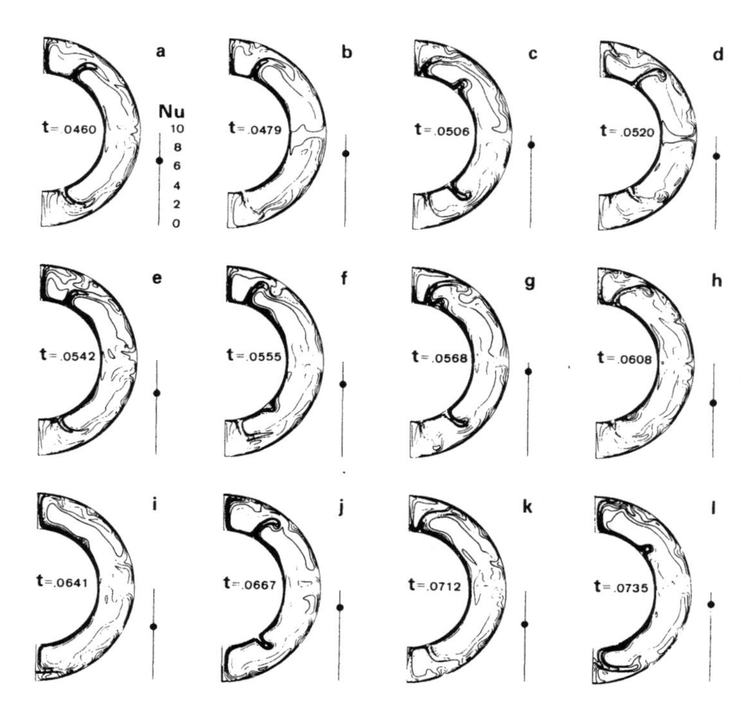

Figure 15.
Temporal evolution, for $Ra_b = 200\ R_c$, of the temperature field and of the Nusselt number <<$Nu(t)$>>. The chaotic-like pattern is confirmed. No regular trends can be observed on the Nusselt number, but, in spite of boundary layer instabilities, the global circulation remains with a P20 shape.

critical, the second one with 241 in the radial and 737 along the horizontal for a Rayleigh number 200 times surpercritical (Machetel and Yuen, 1987).

In both cases, we found that the circulation is time-dependent and is characterized by a large scale $n = 2$ component (P20 solution), which has persisted for timescales greater than the age of the Earth. Aperiodic boundary layer instabilities have a greater tendency to erupt at the bottom surface than at the top. This imbalance is due to the higher temperature gradient of the bottom boundary layer. In any cases, these unstable events can not destroy the large-scale circulation and produce equi-dimensional cells. The evolution of the surface Nusselt number <$Nu(r,t)$> is plotted given adjacent to each panel in Fig. 15. The numerical noise induced by truncation (computations were done on a Cray-2 computer with a word length of 64 bits) combined with the intrinsic time-dependence of the flow helps to maintain the basic asymmetry between the upper and lower portions of the shell in Fig. 15.

Fig. 16 displays the characteristic fluctuations of the mean temperature <$T(r,t)$> and the overall time-averaged mean temperature $\overline{<T(r)>}$ for Rayleigh numbers of $O(10^2)$. Although the mean temperature profiles are seen to fluctuate randomly, it is important to emphasize that the time-averaged fields reach quickly an equilibrium level (Fig. 16 bottom panels). The temperature profiles are shown for both $Ra = 100\ R_c$ (Fig. 16-A) and $Ra =$

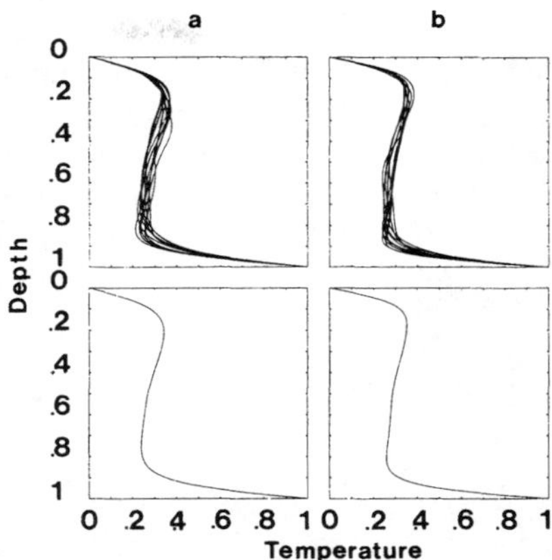

Figure 16.
Instantaneous spherically averaged temperature profile $<T(r,t)>$ (top panels) and time averaged mean temperature profile (bottom panels) for $Ra_b = 100\, R_c$ (left panels) and $Ra_b = 200\, R_c$ (right panels).

200 R_c (Fig.16-B). We observe that in these time-dependent flows, the thermal boundary layers display the same asymmetry as in the steady state results of Figure 4 and also by Zebib et. al. (1980 and 1985) and by Olson (1981). What is particularly interesting is the existence of a stably stratified interior which would be absent for the cartesian geometry at this level of supercritical Rayleigh number (Jarvis and Peltier, 1982). In fact, asymptotic steady state theory also predicts a stably-stratified interior because of the spherical geometry (Olson, 1981). This aspect may contribute to certain regions in the lower mantle being subadiabatic.

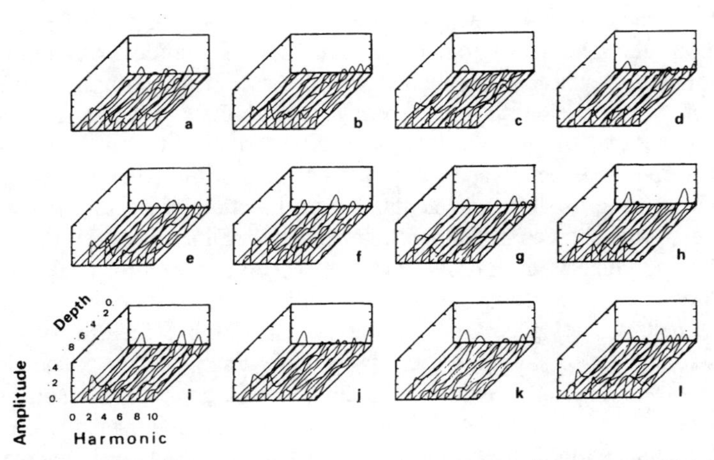

Figure 17. Temporal evolution of the depth-dependent spherical harmonic spectrum of the thermal anomalies corresponding to isotherms from fig. 15.

The effects of internal heating may contribute additional tendencies to the profile being subadiabatic at the base of the mantle (see results for low Rayleigh number in Fig. 4). It is important therefore, to separate out the relative contributions to subadiabaticity from sphericity and radioactivity under high Rayleigh number situations. It may be then feasible to place constraints on the amount of radioactivity in the lower mantle from the degree of subadiabaticity allowed from seismology.

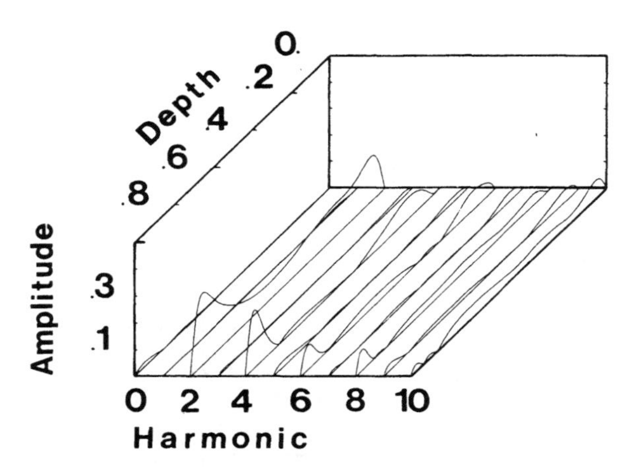

Figure 18.
Spectrum of the depth dependent time averaged thermal anomalies.

Signatures derived from convective calculations when cast in the spectral domain have been used, within the context of steady-state cartesian model (Jarvis, 1985), to make strong inferences upon the lateral heterogeneous structure of mantle convection. In a time dependent situation, spectral components are likely to vary strongly with depth as a consequence of short-wavelength boundary layer instabilities (see Fig. 15). In Fig. 17, we construct the power spectrum of the temperature perturbation $\tau_n(r,t)$ in order to illustrate explicitly, as a consequence of aperiodic thermal instabilities, the cascading of spectral components, from long to short wavelengths, as a function of depth and time. It is quite clear that there are differences in the dominant spectral content between the various levels. Especially of importance is the observation that the higher spectral peaks, just above the bottom, are not always concentrated in the fundamental mode $n = 2$, as would be predicted by a steady-state spectral analysis (Jarvis, 1985), in which the dominant spectral component is found not to shift in wavenumber with depth. In this connection, it is very interesting to note that recent seismic evidences (Morelli and Dziewonski, 1987; Creager and Jordan, 1987; Giardini et al, 1987) suggest that the dominant seismic power is shifted from $n = 2$ to higher harmonics, $n = 4$ and 6, near the core-mantle boundary. Thus these observations accord well with the spectral predictions of our globally time-dependent convection models. These results based on constant viscosity should be even more valid for temperature-dependent viscosity which has been proved able to generate violent boundary layer instabilities (Yuen and Peltier, 1980).

In Fig. 18 we show the spectral decomposition of the overall time averaged thermal anomalies associated with $Ra = 200\ R_c$. We observe that in this case the cascading of thermal energy upward into the lower boundary layer is quite evident in that the spectrum of the interior of the convective layer is strongly dominated by the $n = 2$ component while near the bottom the $n = 4$ and the $n = 2$ component have almost the same strength. Clearly more work on the spectral contents of time-dependent convection and the relative contributions of the dominant modes is needed, since non-linear transfer of modal energies is facilitated by the interaction between large-scale shear flow and thin transient plumes (Howard and Krishnamurti, 1986).

6. Summary of results.

Table 3 summarizes the principal scenarii for base heated spherical convection and the Rayleigh number ranges where transitions occur. If the presence of internal heating induce a lowering of the critical Rayleigh number (Figure 1), it seems to have a moderate influence on the onset of time dependence at least for $\xi \leq 3$ (Fig. 6 and Table 1). On the other hand, internal heating rate has a strong influence on the mean temperature profile and on the thickness of the boundary layers (Fig. 4). This phenomenon is certainly able to bring significant changes at high Rayleigh numbers and the influence of ξ needs to be studied carefully, also with an eye for subadiabacity.

Table 3

Conduction	Convection				
	Steady state convection		Time dependent convection		
	Polygonal Axisym. Bimodal	Polygonal	Periodic	Quasi-Periodic	Chaotic
Ra / R_c : 1	7-15	30	41-45	52-55	$\rightarrow 200$

In the range R_c to approximately 15 R_c, constant viscosity base heated convection can display polygonal solutions, axisymmetrical solution with toroidal cells, and bimodal solution (Machetel et al., 1986). For higher Rayleigh numbers transitions from axisymmetrical solution with toroidal cells and from bimodal solutions lead to polygonal convective circulations. The axisymmetrical solution P20, displaying only two polar cells is, in fact, the simplest polygonal circulation. This solution which possesses strong resemblances with the geoid anomalies (Crough and Jurdy, 1980) and with the seismic velocity anomalies (Masters et al., 1982; Dziewonski et al., 1977; Dziewonski, 1984) becomes time-dependent at a Rayleigh number about 30 times supercritical. Afterwards an oscillatory flow with one frequency prevails. At about 41 R_c, a second period, not commensurate with the first, appears. With further increase of Ra_b, the two periods T_1 and T_2 decrease with T_2 decreasing faster than T_1. Then at $Ra_b = 51.4\ R_c$ the two frequencies are now commensurate. A cursory examination of the temporal evolution of the Nusselt number (Fig. 8) or of the spectrum (Fig. 9) could possibly mislead one to the wrong conclusion that a reverse transition from a quasi-period to a mono-periodic flow takes place. But the attractor of the solution in the phase space still remains a torus whose dimension is equal to 2. When the Rayleigh number is increased the bi-periodic regime is unstable and the solution becomes now chaotic. From these results, it is, in fact, not possible to ascertain whether the chaotic regime appears as a consequence of a bi-periodic flow, as in the Curry and Yorke scenario (Curry and Yorke, 1977) or due to the appearance of a third non-commensurate period (Ruelle and Takens, 1971). The fractal dimension of this strange attractor at the onset of chaos has been estimated to be around 2.8.

Finally, it is of extreme importance to emphasize that persistence of large-scale global circulation of P20 nature even at high Rayleigh numbers. Further work is needed to determine the evolution at higher Rayleigh numbers about $O(10^3)$ the critical value. Within this global flow there appear aperiodic boundary layer instabilities which are not able to tear up the large scale flow. Of potential interest to lithospheric dynamicists are the possibilities of thin hot tongues running against cold boundary layers, reminiscent of lithospheric doubling (Vlaar, 1983). These thermal instabilities induce a shift from degree 2 to higher degrees in the spherical harmonic decomposition of the thermal anomalies near the bottom. This last finding is in a good qualitative agreement with recent seismological studies near the core-mantle boundary.

7. Discussion.

In this work we have given an overview of the development of spherical-shell convection, which is still at a stage of infancy. Nevertheless, our results have demonstrated quite clearly the importance of time dependent convection and the chaotic nature inherent in mantle flow processes. The onset of the transition to large scale flow, with $n = 2$ mode dominating, represents a relatively new manifestation of secondary transitions in mantle convection. This phenomenon definitely has direct geophysical relevance, especially in view of the recently acquired seismic and geoid anomaly data bases. These findings, however, are derived from a certain class of simple models. As such, they have limitations which we should bear in mind.

Firstly, the physical conditions associated with mantle convection are not well known. We are still not sure about the mineralogical composition, the amount of internal heating, and the style of convection (layered versus whole mantle convection). Partial melting and phase transitions induce mineralogical stratification of the mantle able to modify the structure of convection (Christensen and Yuen, 1985). A thermal coupling may occur trough this stratification to modify deeper convection (Anderson and Bass, 1986). The question of whole-mantle or layered mantle convection will likely find an answer by studying the thermodynamical and topographical properties of the major mantle discontinuities. The internal heating rate influences strongly the temperature inside of the mantle and the thickness of the thermal boundary layers. This effect, combined with the stress and temperature dependence of the viscosity has been shown able to generate violent rising mantle plumes with short horizontal wavelengths (Yuen and Peltier, 1980). Thus, all of these questions remain open ones and would still require a great deal of research for even a partial solution.

Secondly, the creep laws of lower mantle constituents are not well known at all. Inferences of lower mantle viscosity from glacial isostasy may be contaminated with the effects of transient lower mantle rheology (Yuen et al., 1986-b). There may also be additional contaminating effects from compressibility for inferring deep mantle viscosity with Boussinesq models.

Third, there may be problems related to numerical computations. In the case of time-dependent convection. It is important to check the accuracy of the solution with finer resolution, since the resolution required for highly time-dependent flows is much finer than for steady state solutions. For Rayleigh number about 200 times the critical, Yuen et al. (1986-a) have found that one should use at least 100×300 grid points for describing the time dependent behavior associated with aspect-ratio three convection. Marcus (1981) has shown that too coarse a grid could lead to spurious time dependent behavior. This same phenomenon has recently been confirmed by Kimura et al. (1986) for porous medium, time- dependent convection. In the course of solving the governing equations, different levels of approximations are typically invoked, e.g. Boussinesq approximation, which may obscure the appearance of interesting time-dependent phenomena, geometrical restrictions, or simplified boundary conditions. In this connection, the mean field methods (Quareni and Yuen, this issue) may provide a rapid and economical access to the averaged thermal properties of the Earth's interior with more physics being built into the model.

Finally, measurement of convective structure in the mantle relies on indirect evidences, such as geoid anomalies, topography or seismic velocity anomalies. All of these are based, in turn, on certain assumptions, which are related directly to our still meagre knowledge of the physical properties of the Earth's interior. In order to promote progress in realistic modelling of mantle convection in a spherical shell, it is important for geodynamicists to interact more frequently with mineral physicists. The recent work by Anderson (1987) on the seismic equation of state in the lower mantle is certainly a right step toward this direction.

Acknowledgments

This research has been supported by C.N.R.S. grant "A.T.P. dynamique des fluides géophysiques et astrophysiques", N.A.S.A grant NAG 5-770 and N.S.F. grant EAR-8511200. Supercomputing has been done on the Cray-2 of the Minnesota Supercomputer Institute. Data reduction of the results has been done on the C.D.C. computer of the "Centre National d'Etudes Spatiales" in Toulouse. We thank Richard Walsh, Minnesota Supercomputing Center, Inc., for technical assistance.

References

Anderson, D.L. and Bass, J.D., 1986. Transition region of the Earth's upper mantle, *Nature*, **320**, 321-328.

Anderson, D.L., 1987. A seismic equation of state II. Shear properties and thermodynamics of the lower mantle, *Phys. Earth Plan. Int.*, **45**,307-323.

Batchelor, G.K., 1967. *An introduction to fluid dynamics*, Cambridge University Press, 615 pp.

Bergé, P., Pomeau, Y. and Vidal, Ch., 1984. *L'ordre dans le chaos*, Herman, Paris, 353pp, (in French).

Birch, F., 1952. Elasticity and constitution of the Earth's interior, *J. Geophys. Res.*, **57**, 227-286.

Busse, F.H., 1967. The stability of finite amplitude cellular convection and its relation to an extremum principle, *J. Fluid Mech.*, **30**, 625-649.

Busse, F.H. and Whitehead, J.A., 1971. Instabilities of convection rolls in a high Prandtl number fluid, *J. Fluid Mech.*, **47**, 305-320.

Busse, F.H. and Whitehead, J.A., 1974. Oscillatory and collective instabilities in large Prandtl number convection, *J. Fluid Mech.*, **66**, 67-69.

Busse, F.H., 1975. Patterns of convection in spherical shells, *J. Fluid Mech.*, **72**, 67-85.
Busse, F.H. and Riahi, N., 1982. Patterns of convection in spherical shells. Part 2, *J. Fluid Mech.*, **123**, 283-301.
Busse, F.H., 1983. Quadrupole convection in the lower mantle, *Geophys. Res. Lett.*, **10**, 285-288.
Carrigan, C.R., 1982. Multiple-scale convection in the Earth's mantle : a three dimensional study, *Science*, **215**, 965-967.
Carrigan, C.R., 1985. Convection in an internally heated, high Prandtl number fluid : a laboratory study, *Geophys. Astr. Fluid Dyn.*, **32**, 1-21.
Chandrasekhar, S., 1961. *Hydrodynamic and hydromagnetic stability*, Oxford, Clarendon press.
Creager, K.C. and Jordan, T.H., 1986. Aspherical structure of the core mantle boundary from PKP travel times, *Geophys. Res. Lett.*, **13**, 1497-1500.
Christensen, U.R., 1984. Convection with pressure and temperature dependent non-Newtonian rheology, *Geophys. J. R. Astr. Soc.*, **77**, 343-384.
Christensen, U.R. and Yuen, D.A., 1984. The interaction of a subducting lithospheric slab with a chemical or phase boundary, *J. Geophys. Res.*, **89**, 4389-4402.
Christensen, U.R. and Yuen, D.A., 1985. Layered convection induced by phase transitions, *J. Geophys. Res.*, **90**, 10291-10300.
Crough, S.T. and Jurdy, D.M., 1980. Subducted lithosphere, hotspots, and the geoid, *Earth Plan. Sci. Lett.*, *48*, 15-22.
Curry, J.H. and Yorke, J.A., 1977. A transition from Hopf bifurcation to chaos : computer experiments with map on R2, *Lecture notes in Mathematics*, **668**, 48-60, Springer-verlag.
Dubois, M., Bergé, P. and Croquette, V., 1981. Etude de regimes convectifs instationnaires à l'aide des diagrammes de Poincaré, *C. R. Acad. Sc. Paris*, **293**, 409-411.
Dziewonski, A.M., Hager, B.H. and O'Connell, R.J., 1977. Large scale heterogeneities in the lower mantle, *J. Geophys. Res.*, **82**, 239-255.
Dziewonski, A.M., 1984. Mapping the lower mantle: determination of lateral heterogeneity in P velocity up to degree and order 6, *J. Geophys. Res.*, **89**, 5929-5952.
Ellsworth, K., Schubert, G. and Sammis, C.G., 1985. Viscosity profile of the lower mantle, *Geophys. J. R. Astr. Soc.*, **83**, 199-214.
Fleitout, L.M. and Yuen, D.A., 1984. Steady-state, secondary convection beneath lithospheric plates with temperature and pressure dependent viscosity, *J. Geophys. Res.*, **84**, 9227-9244.
Giardini, D., Li, X.H. and Woodhouse, J.H., 1987. Three dimensional structure of the Earth core from splitting in free oscillation spectra, *Nature*, **325**, 405-411.
Grassberger, P. and Procaccia, I., 1983. Characterization of strange attractors, *Phys. Rev. Lett.*, **50**, 346-349.
Guckenheimer, J., 1986. Strange attractors in fluid : another view, *Ann. Rev. Fluid Mech.*, **18**, 15-31.
Hansen, U. and Ebel, A., 1984. Numerical and dynamical stability of convection cells in the Rayleigh number range 10 3 - 8x10 5, *Ann. Geophys.*, **2**, 291-302.
Hansen, W., 1987. *Zur Zeitabhängigkeit der thermischen Konvektion im Erdmantel und der doppelt-diffusiven Konvektion in Magmakammern*, PhD thesis, Geophysics, Univ. Cologne, F.R.Germany.
Hansen, U., 1986. On the influence of the aspect ratio on time dependent thermal convection, *E.O.S.*, **67**, 1195.
Hart, J.E., Toomre, J., Deane, A.E., Hurlburt, N.E., Glatzmaier, G.A., Fichtl, G.H., Leslie, F., Fowlis, W.W. and Gilman, P.A., 1986. Laboratory experiments on planetary and stellar convection performed on spacelab 3, *Science*, **234**, 61-64.
Houston, M.H. and Debremaecker, J.C., 1974. ADI solution of free convection in a variable viscosity fluid, *J. Comput. Phys.*, **16**, 221-239.
Howard, L.N. and Krishnamurti, R., 1986. Large scale flow in turbulent convection. A mathematical model, *J. Fluid Mech.*, **170**, 385-410.
Jarvis, G.T. and McKenzie, D.P., 1980. Convection in a compressible fluid with infinite Prandtl number, *J. Fluid Mech.*, **96**, 515-583.
Jarvis, G.T. and Peltier, W.R., 1982. Mantle convection as a boundary layer phenomenon, *Geophys. J. R. Astr. Soc.*, **68**, 385-424.
Jarvis, G.T., 1984. Time-dependent convection in the Earth's mantle, *Phys. Earth Plan. Int.*, **36**, 305-327.
Jarvis, G.T., 1985. The long-wavelength component of mantle convection, *Phys. Earth Plan. Int.*, **40**, 24-42.
Jarvis, G.T. and Peltier, W.R., 1986. Lateral heterogeneity in the convecting mantle, *J. Geophys. Res.*, **91**, 435-451.
Kimura, S., Schubert, G. and Straus, J.M., 1986. Route to chaos in porous medium thermal convection, *J. Fluid*

Mech., **166**, 305-324.

Koster, J.N. and Müller, U., 1984. Oscillatory convection in vertical slots, *J. Fluid Mech.*, **139**, 363-390.

Krishnamurti, R., 1970-a. On the transition to turbulent convection. Part1. The transition from the two to three-dimensional flow., *J. Fluid Mech.*, **42**, 295-307.

Krishnamurti, R., 1970-b. On the transition to turbulent convection. Part 2. The transition to time-dependent flow., *J. Fluid Mech.*, **42**, 309-320.

Krishnamurti, R., 1973. Some further studies on the transition to turbulent convection, *J. Fluid Mech.*, **60**, 285-303.

Krishnamurti, R. and Howard, L.N., 1981. Large scale flow generation in turbulent convection, *Proc. Nat. Acad. Sci.*, **78**, 1981-1985.

Machetel, P. and Rabinowicz, M., 1985. Transitions to a two mode axisymmetrical spherical convection: application to the Earth's mantle, *Geophys. Res. Lett.*, **12**, 227-230.

Machetel, P., Rabinowicz, M. and Bernardet, P., 1986. Three-dimensional convection in spherical shells, *Geophys. Astr. Fluid Dyn.*, **37**, 57-84.

Machetel, P. and Yuen, D.A., 1986. The onset of time-dependent convection in spherical shells as a clue to chaotic convection in the Earth's mantle, *Geophys. Res. Lett.*, **13**, 1470-1473.

Machetel, P. and Yuen, D.A., 1987. Chaotic axisymmetrical spherical convection and large scale mantle circulation, *Earth Plan. Sci. Lett.*, in press.

Malraison, B., Atten, P., Bergé, P. and Dubois, M., 1983. Dimension d'attracteurs étranges : une détermination expérimentale en régime chaotique de deux systèmes chaotiques, *C. R. Acad. Sc. Paris*, **297**, 209-214 (also in *Journal de Physiques lettres*).

Marcus, P.S., 1981. Effects of truncation in modal representations of thermal convection, *J. Fluid Mech.*, **103**, 241-255.

Master, G., Jordan, T.H., Silver, P.G. and Gilbert, F., 1982. Aspherical Earth structure from fundamental spheroidal mode data, *Nature*, **298**, 609-613.

McKenzie, D.P., Roberts, J.M. and Weiss, N.O., 1974. Convection in the Earth's mantle : toward a numerical simulation, *J. Fluid Mech.*, **62**, 465-538.

Morelli, A. and Dziewonski, A.M., 1987. Topography of the core-mantle boundary and lateral homogeneity of the liquid core, *Nature*, **325**, 678-682.

Nataf, H.C., Froidevaux, C., Levrat, J.L. and Rabinowicz, M., 1981. Laboratory convection experiments : effects of lateral cooling and generation of instabilities in the horizontal boundary layers, *J. Geophys. Res.*, **85**, 6143-6154.

Newhouse, S., Ruelle, D. and Takens, F., 1978. Occurrence of strange axiom A attractors near quasi periodic flows on Tm, m ≥ 3, *Commun. Math. Phys.*, **64**, 35-40.

Olson, P.L., 1981. Mantle convection with spherical effects, *J. Geophys. Res.*, **86**, 4881-4890.

Olson, P.L., 1984. An experimental approach to thermal convection in a two layered mantle, *J. Geophys. Res.*, **89**, 11293-11302.

Pekeris, C.L., 1935. Thermal convection in the interior of the Earth, *Mon. Not. Roy. Astr. Soc., Geophys. Suppl.*, **3**, 343-367.

Quareni, F. and Yuen, D.A., 1987. Mean field methods in mantle convection, this issue.

Riahi, N., Geiger, G. and Busse, F.H., 1982. Finite Prandtl number convection in spherical shells, *Geophys. Astr. Fluid Dyn.*, **20**, 307-318.

Richter, F.M., 1973. Convection and the large scale circulation of the mantle, *J. Geophys. Res.*, **78**, 8735-8745.

Roache, P.J., 1972. *Computational fluid dynamics*, Hermosa Publishers, 434 pp.

Ruelle, D. and Takens, F., 1971. On the nature of turbulences, *Commun. Math. Phys.*, **20**, 167-192.

Schubert, G. and Young, R.E., 1976. Cooling the Earth by whole mantle subsolidus convection : A constraint on the viscosity of the lower mantle, *Tectonophysics*, **35**, 201-214.

Schubert, G., 1979. Subsolidus convection in the mantles of terrestrial planets, *Ann. Rev. Earth Plan. Sci.*, **7**, 289-342.

Schubert, G. and Zebib, A., 1980. Thermal convection of an internally heated infinite Prandtl number fluid in a spherical shell, *Geophys. Astr. Fluid Dyn.*, **15**, 65-90.

Schuster, H.G., 1984. *Deterministic chaos : an introduction*, Physik Verlag, Weinheim, F.R. Germany.

Torrance, K.E. and Turcotte, D.L., 1971. Structure of convection cells in the mantle, *J. Geophys. Res.*, **76**, 1154-1161.

Turcotte, D.L. and Oxburgh, E.R., 1967. Finite-amplitude convective cells and continental drift, *J. Fluid Mech.*,

28, 29-42.
Vincent, A.P., 1986. Spectral modelling of plumes in multiprocessing supercomputers, *E.O.S.*, **67**, 1195.
Vlaar, N.J., 1983. Thermal anomalies and magmatism due to lithospheric doubling and shifting, *Earth and Planet. Sci. Lett.*, **65**, 322-330.
Whitehead, J.A. and Parson, B., 1978. Observations of convection at Rayleigh numbers up to 760,000 in a fluid with large Prandtl number, *Geophys. Astr. Fluid Dyn.*, **9**,201-217.
Woodhouse, J.H. and Dziewonski, A.M., 1984. Mapping the upper mantle : Three dimensional modeling of Earth structure by inversion of seismic waveforms, *J. Geophys. Res.*, 89,5953-5986.
Young, R.E., 1974. Finite amplitude convection in a spherical shell, *J. Fluid Mech.*, **63**, 695-721.
Yuen, D.A. and Peltier, W.R., 1980. Mantle plumes and the thermal stability of the D" layer, *Geophys. Res. Lett.*, **9**, 625-628.
Yuen, D.A., Weinstein, S.A. and Olson, P.L., 1986-a. High resolution calculations of large aspect-ratio convection, *E.O.S.*, **67**, 1194.
Yuen, D.A., Sabadini, R., Gaspareni, P. and Boschi, E., 1986-b. On transient rheology and glacial isostasy, *J. Geophys. Res.*, **91**, 11420-11438.
Zebib, A., Schubert, G. and Straus, J.M., 1980. Infinite Prandtl number convection in a spherical shell, *J. Fluid Mech.*, **97**, 257-277.
Zebib, A., Schubert, G., Dein, J.L. and Paliwal, C., 1983. Character and stability of axisymmetric thermal convection in spheres and spherical shells, *Geophys. Astr. Fluid Dyn.*, **23**, 1-42.
Zebib, A., Goyal, A.K. and Schubert, G., 1985. Convective motions in a spherical shell, *J. Fluid Mech.*, **152**, 39-48.

PHILIPPE MACHETEL, GRGS/CNES, 18 Avenue E.Belin, 31055 Toulouse, France.
DAVID A. YUEN, Minnesota Supercomputer Institute and Department of Geology and Geophysics, University of Minnesota, Minneapolis MN 55455, USA.

Chapter 13

Lateral heterogeneity and the geoid: the importance of the surface kinematic constraints

A.M. Forte and W.R. Peltier

We apply the Green function technique to describe the internal load induced deformation of 3-D, self-gravitating, spherical shells of incompressible Newtonian fluid consisting either of a single constant-viscosity shell or of two adjacent shells of different viscosity. Using this method we derive closed form, analytic expressions for the kernel functions connecting the internal lateral heterogeneity of density to the horizontal divergence of the surface flow and to the non-hydrostatic geoid. Although the geoid constrains only the ratio of the upper and lower mantle viscosities, the horizontal divergence field constrains their absolute values. We find that both surface divergence and geoid fields are best fit with only a factor of eight viscosity increase at 1200 km depth. We point out, however, that the coupling of poloidal and toroidal flow in the Earth's mantle, which is required to understand surface velocity spectra, will most probably allow us to reduce the viscosity increase required by the geoid data.

1. Introduction

Current inferences of the spherically symmetric viscosity distribution in the Earth's mantle rely upon internal loading schemes which relate the seismically inferred lateral density variations in the Earth to the large-scale undulations of the non-hydrostatic geoid (e.g., Hager, 1984). These internal loading calculations all seem to require an increase in the viscosity at the 670 km seismic discontinuity which is considerably greater than the viscosity increase required by glacial isostatic adjustment analyses (e.g., Peltier, 1982). In

N.J. Vlaar, G. Nolet, M.J.R. Wortel & S.A.P.L. Cloetingh (eds), Mathematical Geophysics, 291-323.

this report we will present an alternative formulation of the internal loading problem and we will later argue that a likely reason for the discrepancy between the viscosity inferences obtained from post-glacial rebound data and isostatic geoid anomalies is that the Newtonian viscous flow models used in current internal loading analyses are unable to satisfy the surface kinematic constraints provided by the observed tectonic plate velocities.

The mathematical method most commonly used to relate geophysical surface observables, such as the geoid, to lateral density variations in the mantle is the propagator matrix technique (e.g., Richards and Hager, 1984; Ricard et al., 1984). In this report we will show that the Green function technique employed by Parsons and Daly (1983) is a mathematically straightforward alternative to the propagator matrix schemes allowing one to readily derive analytic, closed-form, expressions for the kernels which connect lateral density heterogeneity at depth to the surface observables. The Green functions considered by Parsons and Daly (1983) were derived for a 2-D cartesian geometry where the effects of self-gravitation were ignored; here we will show how the Green function method may be extended to 3-D, spherical, self-gravitating shells of constant viscosity and also to the case of shells consisting of two layers having different viscosities. The Green function method will be employed to derive expressions for the kernel functions describing the non-hydrostatic geoid and the horizontal divergence of the convective flow field produced at the Earth's surface by lateral density variations in the mantle.

Previous attempts to model the non-hydrostatic geoid (e.g., Hager, 1984) have ignored the importance of the observed surface plate kinematics and the constraints they provide on the viscosity structure of the mantle. The equipartition of kinetic energy between the poloidal and toroidal components of the observed surface-plate velocity field is likely to have a direct bearing on our ability to properly model the geoid. Despite the complications produced by the poloidal-toroidal coupling of the convective flow in the mantle we will demonstrate, using the simple physical models that will be presented, that on the whole the observed large-scale plate motions are those expected to exist on the basis of the seismic tomographic inferences of the internal density heterogeneity. This last result provides strong and convincing support for the hypothesis that the large-scale tectonic plate motions are the surface expression of deep-seated convection in the Earth's mantle.

In section 2 of the report we will briefly summarize the main features of the observed tectonic plate velocities. In section 3 we present our extension of the Green function method which is then employed to infer the contrast between the viscosities of the upper and lower mantles. In section 4, we discuss the trade-off we believe to exist between the conversion of poloidal to toroidal flow and the strength of the viscosity contrast required by the observed non-hydrostatic geoid. Finally, in section 5 of this report, we provide our main conclusions.

2. Present-day tectonic plate velocities

Employing the Wilson-Morgan hypothesis of fixed hot spots, Minster and Jordan (1978) have obtained a model for the absolute, present-day, tectonic plate velocities which are depicted in Fig. 1. We shall refer to this velocity field as the "observed surface-plate velocities". From the observed surface-plate velocities we have calculated the spherical harmonic coefficients of two scalars, the horizontal divergence and the radial vorticity, which completely characterize the tectonic plate motions:

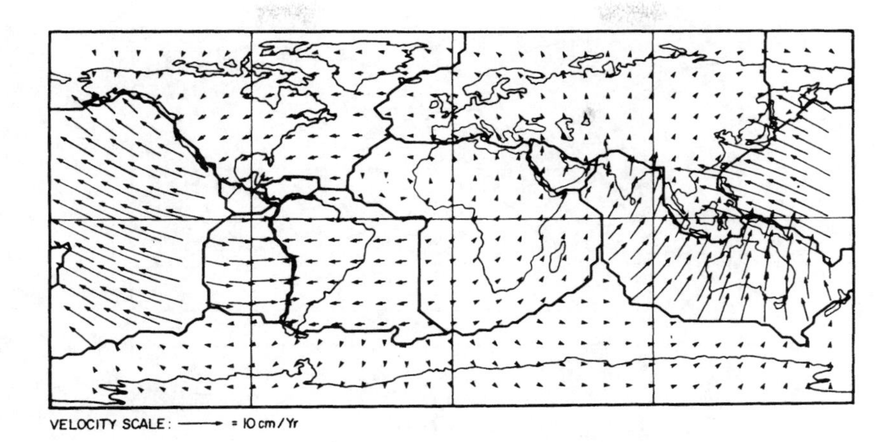

Figure 1. Surface plate velocity field constructed from the absolute angular velocity vectors of Minster and Jordan (1978) in the "hot spot" frame.

$$\nabla_H \cdot \mathbf{v} = \sum_{l,m} D_l^m \, Y_l^m(\theta,\phi) \tag{1a}$$

$$\hat{\mathbf{r}} \cdot (\nabla \times \mathbf{v}) = \sum_{l,m} V_l^m \, Y_l^m(\theta,\phi) \, . \tag{1b}$$

In equation (1) $\mathbf{v}$ is the observed surface-plate velocity field and D_l^m and V_l^m are, respectively, the harmonic coefficients of the horizontal divergence and radial vorticity.

An alternative scalar representation of the surface kinematics, previously employed by Hager and O'Connell (1978), is obtained by describing the observed surface-plate velocities in terms of poloidal and toroidal components:

$$\mathbf{v} = \sum_{l,m} S_l^m \, \hat{\mathbf{r}} \times \mathbf{\Lambda} \, Y_l^m(\theta,\phi) + \sum_{l,m} T_l^m \, \mathbf{\Lambda} \, Y_l^m(\theta,\phi) \, . \tag{2}$$

In equation (2) we have employed the angular momentum operator $\mathbf{\Lambda} = \mathbf{r} \times \nabla$ (Backus, 1958) and S_l^m and T_l^m are, respectively, the poloidal and toroidal scalars. A description of the surface kinematics in terms of the two scalars appearing in (1) is completely equivalent to a description in terms of the scalars appearing in (2) since one may readily show that

$$S_l^m = \frac{aD_l^m}{l\,(l+1)} \tag{3a}$$

$$T_l^m = \frac{-aV_l^m}{l\,(l+1)} \, , \tag{3b}$$

where a is the radius of the Earth.

The harmonic coefficients D_l^m and V_l^m (tabulated in Forte and Peltier, 1986) have been used to synthesize the maps of horizontal divergence and radial vorticity presented in Fig. 2. From this figure it is evident that the horizontal divergence scalar is necessary to represent two of the types of plate boundary which are observed in nature: the ridges and

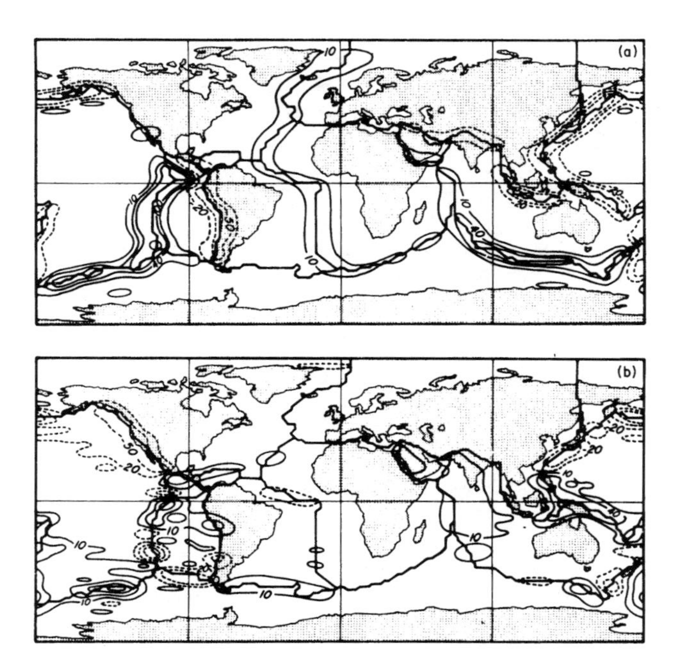

Figure 2. (a) Surface divergence up to degree and order 32. The contour interval is 30 x 10^{-9} rad/yr. and the individual contour levels are in units of 10^{-9} rad/yr. The dashed contour lines indicate negative divergence. (b) Radial vorticity to degree and order 32. The contour interval and units are as in (a). The dashed contour lines indicate negative vorticity (i.e., clockwise circulation).

trenches. The radial vorticity scalar is necessary to represent the third type of plate boundary: the transform fault.

The partitioning of kinetic energy between the poloidal and toroidal components of the observed surface-plate velocities may be determined by calculating their degree variances, σ_l:

$$\sigma_l \text{ (poloidal)} = \left[\sum_{m=-l}^{l} S_l^m S_l^{m*} \right]^{1/2} \tag{4a}$$

$$\sigma_l \text{ (toroidal)} = \left[\sum_{m=-l}^{l} T_l^m T_l^{m*} \right]^{1/2} , \tag{4b}$$

where the asterisk denotes complex conjugation. These degree variances are shown in Fig. 3(a) where one observes that there is an almost exact equipartition of kinetic energy between the poloidal and toroidal components; this result was first obtained by Hager and O'Connell (1978) who did not further consider its importance except in the context of their Stokes flow extrapolation to depth of the observed surface-plate velocities. In a spherical shell of chemically uniform fluid with physical properties that vary with radius only one

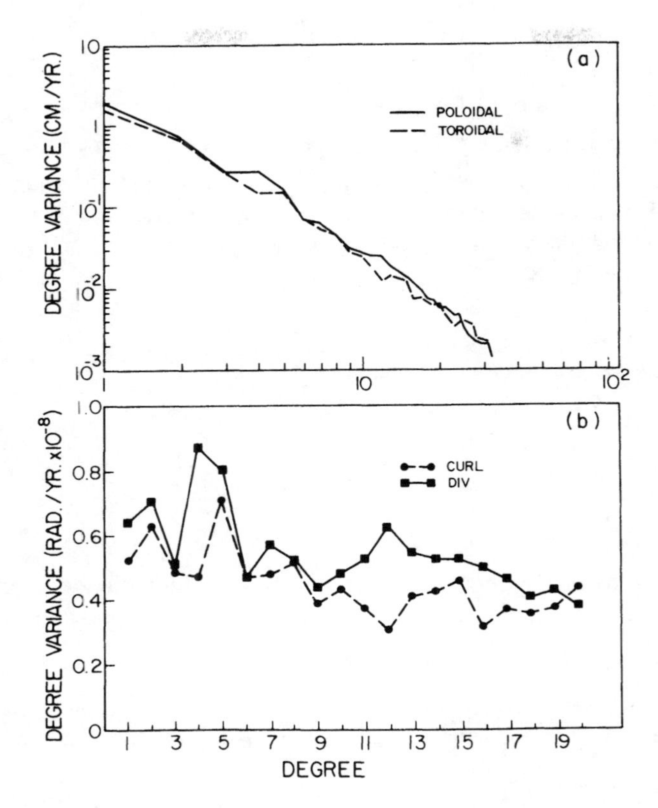

Figure 3. (a) Degree variance of the poloidal and toroidal components of the surface velocity field. (b) Degree variance of the horizontal divergence and radial vorticity fields derived from the surface plate velocities.

expects thermally induced buoyancy forces to produce poloidal flow only. The factors that are most probably responsible for generating the required large flux of energy from poloidal to toroidal flow are the extreme lateral variations of viscosity in the mantle (e.g. Hager and O'Connell, 1978) as well as the presence of chemically differentiated, continental crust which is buoyant and therefore cannot be subducted (Peltier, 1985). In Fig. 3(b) we show the degree variances of the horizontal divergence and radial vorticity scalars and it will be noted in particular that the divergence spectrum is characterized by a well-defined peak at $l = 4, 5$.

3. Predicting surface motions and the geoid from seismically inferred mantle heterogeneity

3.1 Green function for a constant-viscosity mantle

The Boussinesq hydrodynamic equations which we consider describe the conservation of mass and momentum in a viscous fluid:

$$\nabla \cdot \mathbf{u} = 0 \,, \tag{5}$$

$$\rho_o \mathbf{g}_1 + \rho_1 \mathbf{g}_o - \nabla P_1 + \eta \, \Delta \mathbf{u} = 0 \,. \tag{6}$$

The quantities with subscript 1 are perturbations to the hydrostatic reference state which is denoted by the subscript o. The inertial force terms have been ignored in equation (6) since the Earth's mantle is characterized by an essentially infinite Prandtl number. We will adopt the sign convection that the perturbed body force $\mathbf{g}_1$ appearing in the self-gravitation term $\rho_o \mathbf{g}_1$ in (6) is given by

$$\mathbf{g}_1 = \nabla \, \phi_1 \,, \tag{7}$$

where ϕ_1, the perturbed gravitational potential, then satisfies

$$\Delta \phi_1 = - 4\pi G \, \rho_1 \,. \tag{8}$$

The body force $\mathbf{g}_o$ is simply given by

$$\mathbf{g}_o = - \frac{g_o}{r} \, \mathbf{r} \,, \tag{9}$$

where g_o is the (approximately) constant gravitational acceleration in the Earth's mantle.

Equation (5) shows that $\mathbf{u}$ is solenoidal and may therefore be represented as

$$\mathbf{u} = \nabla \times \Lambda p + \Lambda q \tag{10}$$

where p and q are, respectively, the poloidal and toroidal scalars (Backus, 1958). Substitution of results (7), (9), and (10) into equation (6) and application of $\nabla\times$ to the resulting equation yields

$$\Lambda \, \frac{\rho_1 g_o}{r} + \eta \, \nabla \times \Lambda \, \Delta q - \eta \, \Lambda \, \Delta^2 p = 0 \,. \tag{11}$$

If one now applies the operator $\Lambda\cdot$ to equation (11) and uses the result $\Lambda \cdot (\nabla \times \Lambda) = 0$ (Backus, 1958) then the following is obtained:

$$\eta \, \Lambda^2 \Delta^2 p = \Lambda^2 \, \frac{\rho_1 g_o}{r} \,; \tag{12}$$

where the operator $\Lambda^2 = \Lambda \cdot \Lambda$ is characterized by the property

$$\Lambda^2 \, Y_l^m(\theta,\phi) = - \, l\,(l+1) \, Y_l^m(\theta,\phi) \tag{13}$$

and $Y_l^m(\theta,\phi)$ is the complex spherical harmonic function which is normalized such that

$$\frac{1}{4\pi} \int_S Y_l^m(\theta,\phi) \, Y_s^{t*}(\theta,\phi) \, dS = \delta_{ls} \, \delta_{mt} \,, \tag{14}$$

where S denotes integration over the surface area of the unit sphere. If one expands the quantities p and ρ_1, in (12), in terms of spherical harmonics and uses results (13) and (14) then the following important equation is obtained:

$$D_l^2\, p_l^m(r) = \frac{g_o}{\eta}\, \frac{(\rho_1)_l^m(r)}{r}\ , \tag{15}$$

where $p_l^m(r)$ and $(\rho_1)_l^m(r)$ are the radially-varying spherical harmonic coefficients of the scalars p and ρ_1 respectively. D_l^2 is the transformed biharmonic operator Δ^2 and D_l is defined as

$$D_l = \frac{d^2}{dr^2} + \frac{2}{r}\frac{d}{dr} - \frac{l(l+1)}{r^2}\ .$$

Now since $\mathbf{r}\cdot\mathbf{\Lambda} = 0$, it is evident from equation (11) that the toroidal scalar is governed by the following equation:

$$\eta\, \mathbf{\Lambda}^2\, \Delta q = 0\ ,$$

and expanding q in terms of spherical harmonics this last result becomes:

$$D_l\, q_l^m(r) = 0\ . \tag{16}$$

Equations (15) and (16) show that when the dynamic viscosity, η, is assumed constant then lateral density variations will only excite a poloidal flow field; this is also true for the more general case in which the viscosity is an arbitrary function of radius as shown, for example, by Arkani-Hamed and Toksöz (1984).

The poloidal flow Green function, $p_l^m(r,r')$, is found by solving

$$D_l^2\, p_l^m(r,r') = \delta(r - r')\ , \tag{17}$$

where $\delta(r)$ is the Dirac delta function and r' is the radius at which the δ-function load is placed. The Green function will depend on the boundary conditions assumed at the Earth's surface ($r{=}a$) and at the core-mantle boundary (CMB; $r{=}b$). If one assumes that, on the whole, the tectonic plates are participating in the large scale flow in the mantle then a free-slip boundary condition at $r{=}a$ is suggested. We will also assume a free-slip condition at the CMB. The derivation of the poloidal flow Green function which satisfies these boundary conditions is presented in Appendix A. Once the Green function has been obtained it is clear from equations (15) and (17) that the poloidal scalar will be given by

$$p_l^m(r) = \frac{g_o}{\eta} \int_b^a \frac{(\rho_1)_l^m(r')}{r'}\, p_l^m(r,r')\, dr'\ . \tag{18}$$

The toroidal flow present in the observed surface plate velocities obviously poses a problem when attempting to predict the plate motions using the poloidal scalar of equation (18). We believe however that the conversion of poloidal to toroidal flow occurs largely near the Earth's surface where the lateral variations in rheology and chemistry are the most extreme. This suggests then that density heterogeneity below the lithosphere will mostly excite poloidal flow thus maintaining the validity of equation (15). We found moreover (Forte and Peltier, 1986) that a strong correlation exists between the horizontal divergence and the seismically inferred lateral heterogeneity in the upper mantle at degrees 2 and 4; this suggests that there is a linear relationship between these two fields and that it should therefore be possible to model the horizontal divergence at these two degrees with a spherically symmetric viscosity model.

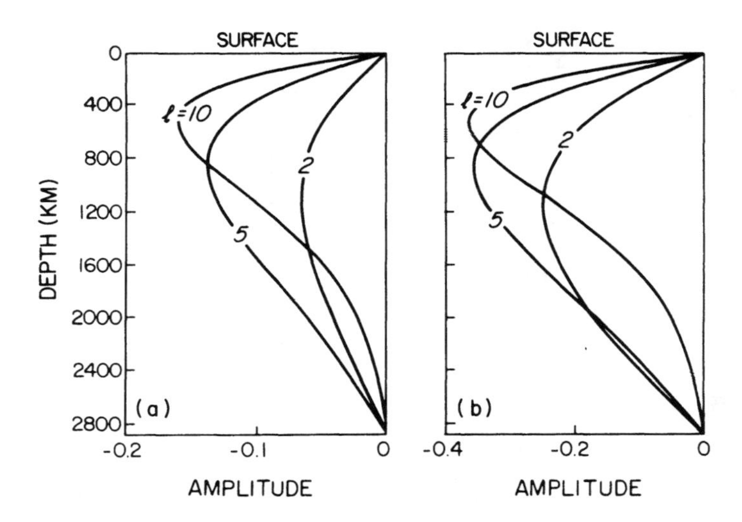

Figure 4. Kernels for an incompressible, homogeneous mantle with free-slip conditions at the surface and CMB.: (a) Horizontal divergence kernels (b) Geoid kernels for a self-gravitating mantle.

The kernel function, $S_l(r')$, which allows us to predict the spherical harmonic coefficients of the horizontal divergence at $r=a$ from the lateral density variations in a constant-viscosity mantle is derived in Appendix A where we show

$$(\nabla_H \cdot \mathbf{u})_l^m(r=a) = \frac{g_o}{\eta} \int_b^a S_l(r')\,(\rho_1)_l^m(r')\,dr' \ . \tag{19}$$

In Fig. 4(a) we show the function $S_l(r')$ for several degrees. The density perturbation, $(\rho_1)_l^m(r)$, is obtained from models M84C of Woodhouse and Dziewonski (1984) and L02.45 of Dziewonski (1984) where we are assuming that the lateral variations in seismic wave speeds are caused solely by lateral temperature variations:

$$(\rho_1)_l^m = \begin{cases} \dfrac{(\partial\rho/\partial T)}{2v_s(\partial v_s/\partial T)}\,[\delta(v_s^2)]_l^m\,, & 25\text{ km} \le z \le 670\text{ km} \\[2ex] \dfrac{(\partial\rho/\partial T)}{(\partial v_p/\partial T)}\,[\delta(v_p)]_l^m\,, & 670\text{ km} \le z \le 2895\text{ km} \end{cases} \tag{20}$$

The temperature derivatives we use in (20) were obtained by Anderson et al. (1968) from laboratory analyses of spinel. Forte and Peltier (1986) show that the temperature derivatives of spinel are probably reasonable estimates of the actual values. The degree correlations (e.g. O'Connell, 1971) between the observed horizontal divergence and the divergence predicted using (19) are shown in Table 1 where one observes that excellent correlations are obtained at $l=2$ and 4 as expected.

Table 1 Degree correlations between predicted surface divergence for a homogeneous mantle and observed surface divergence

Degree, l	Correlation Coefficient	Significance Level, α
l=2	.80	$90\% < \alpha < 95\%$
l=3	.30	$\alpha < 80\%$
l=4	.68	$95\% < \alpha < 98\%$
l=5	.28	$\alpha < 80\%$

The straightforward manner with which the kernel functions for the non-hydrostatic geoid are derived is a good illustration of the flexibility offered by the Green function method. The harmonic coefficients of the interior gravitational potential produced only by the internal density heterogeneity in the mantle are:

$$(U_{int})_l^m(r) = \frac{4\pi G}{2l+1} \int_b^a r'^2 \frac{r_<^l}{r_>^{l+1}} (\rho_1)_l^m(r')\, dr' , \tag{21}$$

where $b \le r \le a$, $r_< = \min(r, r')$, and $r_> = \max(r, r')$. The harmonic coefficients of the internal potential produced by the deflected surface boundary at $r=a$ are:

$$(U_a)_l^m(r) = \frac{4\pi Ga}{2l+1} (\rho_o - \rho_w) \left[\frac{r}{a} \right]^l \delta a_l^m , \tag{22}$$

where δa_l^m are the harmonic coefficients of the deflection of the Earth's surface, $\delta a(\theta,\phi)$, from its reference position and we have assumed that the mantle of density ρ_o is overlain by ocean with density ρ_w. The expression for the internal potential produced by the deflected CMB is:

$$(U_b)_l^m(r) = \frac{4\pi Gb}{2l+1} (\rho_c - \rho_o) \left[\frac{b}{r} \right]^{l+1} \delta b_l^m , \tag{23}$$

where ρ_c is the density of the outer core and δb_l^m are the harmonic coefficients describing the deflection of the CMB. We expect that the deflections of the phase-change boundaries in the mantle are likely to contribute significantly to the Earth's gravitational potential but in the absence of information regarding these deflections we will state that the total perturbed potential, ϕ_1, is simply given by:

$$(\phi_1)_l^m(r) = (U_{int})_l^m(r) + (U_a)_l^m(r) + (U_b)_l^m(r) . \tag{24}$$

The surface deflection $\delta a(\theta,\phi)$ may be determined from the condition that the σ_{rr} component of the stress tensor will be continuous at the bounding surface:

$$\delta a(\theta,\phi) = \frac{P_1(r{=}a)}{g_o(\rho_o - \rho_w)} - \frac{2\eta}{g_o(\rho_o - \rho_w)} \left[\frac{\partial u_r}{\partial r} \right]_{r=a} , \tag{25}$$

where P_1 is the non-hydrostatic pressure and $u_r = \hat{\mathbf{r}}\,\mathbf{u}$. The deformed CMB will also be described by (25) except that ρ_w is replaced by ρ_c and all quantities are evaluated at r=b. In Appendix A we provide complete details of the derivation of the self-gravitating geoid kernels, $G_l(r')$, which allow us to predict the geoid:

$$(Ge)_l^m = \frac{3}{(2l+1)\bar{\rho}} \int_b^a G_l(r')\,(\rho_1)_l^m(r')\,dr' , \tag{26}$$

where $(Ge)_l^m$ are the harmonic coefficients of the predicted geoid and $\bar{\rho}$ is the average density of the Earth. The behaviour of the function $G_l(r')$ is shown in Fig. 4(b) where it is observed that the kernels all have a negative sign. The negative sign of these kernels is due to the dominant negative contribution of the flow-induced deflection of the surface boundary compared to the weaker positive contribution of the internal lateral density variations (e.g. Hager, 1984). A comparison of equations (19) and (26) shows that, in contrast to the predicted horizontal divergence field, the predicted non-hydrostatic geoid produced by lateral density variations in a constant-viscosity mantle will not depend on the viscosity.

In Table 2 we present the degree correlations between the observed, GEM10B, non-hydrostatic geoid of Lerch et al. (1979) (filtered by removal of the hydrostatic flattening determined by Nakiboglu (1982)) and the geoid predicted using (26). From an analysis of the degree correlations between the observed, non-hydrostatic geoid and the seismically inferred mantle heterogeneities we found (Forte and Peltier, 1986) that the kernels describing the degree-2 and 3 geoid should have maximum (negative) amplitudes in the lower mantle and negligible amplitude in the upper mantle while for $l \geq 4$ the kernels should generally have maximum (positive) amplitudes in the mid-mantle and negligible amplitudes elsewhere. An examination of the shapes of the geoid kernels shown in Fig. 4(b) will then reveal why the correlations in Table 2 are generally poor.

Table 2 Degree correlations between predicted geoid for a homogeneous mantle and observed non-hydrostatic geoid

Degree, l	Correlation Coefficient	Significance Level, α
l=2	-.01	$\alpha < 80\%$
l=3	.53	$\alpha < 80\%$
l=4	-.51	$80\% < \alpha < 90\%$
l=5	.60	$\alpha \approx 95\%$

3.2 Green function for a two-layer mantle

A simple Earth model that permits one to determine the effects of viscosity stratification is an incompressible mantle divided into two layers having different viscosities. We continue to assume that a whole-mantle convective flow exists and we do not therefore consider the case in which the depth of the viscosity jump also coincides with a chemical discontinuity which occurs if there is a layered convective circulation.

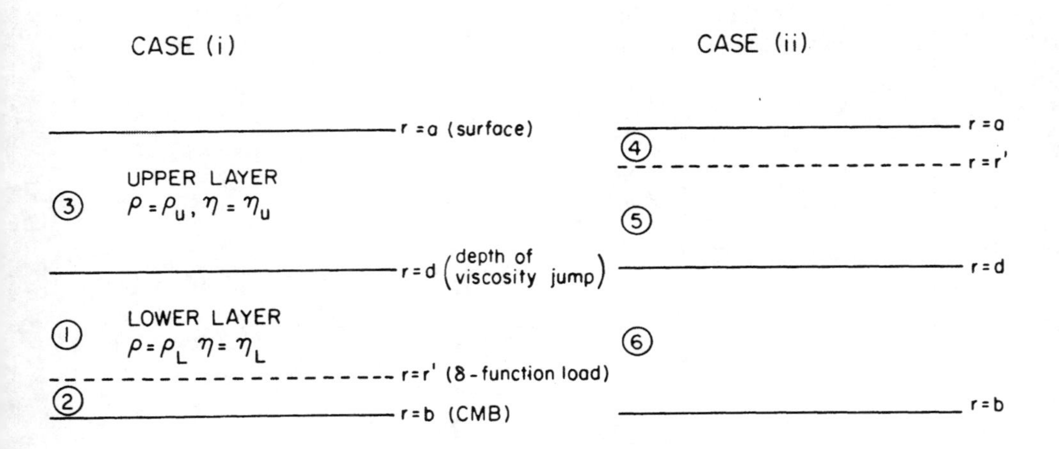

Figure 5. Schematic diagram of the geometry for the two-layer Green function.

The dynamic viscosity in each layer is assumed constant and thus equation (15) is valid in each layer. Since the delta-function loads may be placed in either layer the problem consists of treating two cases: case (i) - find the poloidal flow Green function for a δ-function load in the lower layer, and case (ii) - find the poloidal flow Green function for a δ-function load in the upper layer. The Green function for case (i) will thus satisfy the following equation:

$$D_l^2(p_L)_l^m(r,r') = \begin{cases} 0 & d < r < a \\ \delta(r-r') & b < r < d \end{cases}, \tag{27}$$

while the Green function for case (ii) satisfies:

$$D_l^2(p_u)_l^m(r,r') = \begin{cases} \delta(r-r') & d < r < a \\ 0 & b < r < d \end{cases}. \tag{28}$$

where $r=d$ is the radius at which the viscosity jump occurs. It is evident from equations (27) and (28) that the mantle will in each case consist of three regions which are shown and numbered for reference in Fig. 5. The poloidal flow Green function will again satisfy free-slip boundary conditions at the Earth's surface and at the CMB. From equations (15), (27), and (28) one may verify that the poloidal scalar is given by:

$$p_l^m(r) = \frac{g_o}{\eta_u} \int_d^a \frac{(\rho_1)_l^m(r')}{r'} (p_u)_l^m(r,r')\, dr'$$

$$+ \frac{g_o}{\eta_L} \int_b^d \frac{(\rho_1)_l^m(r')}{r'} (p_L)_l^m(r,r')\, dr' \,, \tag{29}$$

where η_u and η_L are, respectively, the viscosities of the upper and lower layers. If we let, for example, $(p_5)_l^m(r,r')$ denote the Green function $(p_u)_l^m(r,r')$ in the region $d < r < r'$ (i.e. layer 5 in Fig. 5) then the numbering scheme in Fig. 5 allows us to rewrite (29) as:

$$p_l^m(r) = \frac{g_o}{\eta_u} \int_r^a \frac{(\rho_1)_l^m(r')}{r'} (p_5)_l^m(r,r')\, dr'$$

$$+ \frac{g_o}{\eta_u} \int_d^r \frac{(\rho_1)_l^m(r')}{r'} (p_4)_l^m(r,r')\, dr' \tag{30a}$$

$$+ \frac{g_o}{\eta_L} \int_b^d \frac{(\rho_1)_l^m(r')}{r'} (p_3)_l^m(r,r')\, dr', \text{ for } d < r < a$$

and

$$p_l^m(r) = \frac{g_o}{\eta_u} \int_d^a \frac{(\rho_1)_l^m(r')}{r'} (p_6)_l^m(r,r')dr'$$

$$+ \frac{g_o}{\eta_L} \int_r^d \frac{(\rho_1)_l^m(r')}{r'} (p_2)_l^m(r,r')\, dr' \tag{30b}$$

$$+ \frac{g_o}{\eta_L} \int_b^r \frac{(\rho_1)_l^m(r')}{r'} (p_1)_l^m(r,r')dr', \text{ for } b < r < d$$

The complete details of the Green function derivation for the two-layer mantle will be found in Appendix B. In this Appendix we also derive analytic expressions for the kernel function, S_l, describing the horizontal divergence:

$$(\nabla_H \cdot \mathbf{u})_l^m(r=a) = \frac{g_o}{\eta_u} \int_b^a S_l(r';\eta_L/\eta_u,d)\, (\rho_1)_l^m(r')dr' \,. \tag{31}$$

From (31) it is evident that although S_l depends on the ratio of the viscosities the multiplicative factor, η_u, implies that the amplitude of the predicted horizontal divergence depends on the absolute value of the viscosity in each layer. In contrast, the predicted non-hydrostatic geoid will depend only on the ratio of the viscosities of the two layers:

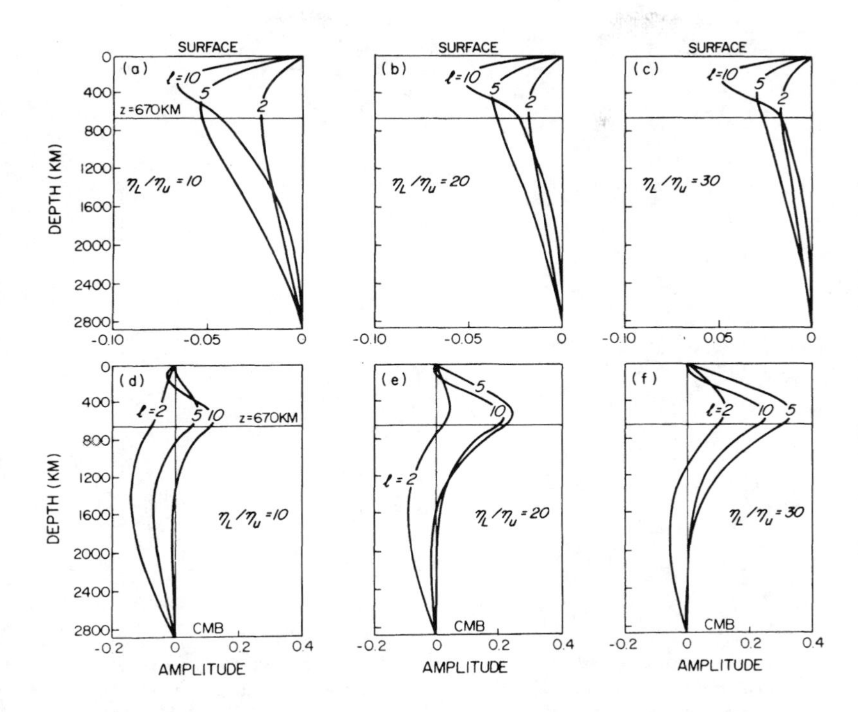

Figure 6. Kernels for an incompressible, two-layer mantle with free-slip conditions at the surface and CMB and viscosity jump at z=670 km. (a), (b), and (c) Horizontal divergence kernels; (d), (e) and (f) Geoid kernels for a self-gravitating mantle.

$$(Ge)_l^m = \frac{3}{(2l+1)\bar{\rho}} \int_b^a G_l(r';\eta_L/\eta_u,d)\,(\rho_1)_l^m(r')dr' \ , \tag{32}$$

where G_l is the geoid kernel, derived in Appendix B, for the self-gravitating, two-layer mantle.

The behaviour of the kernel functions describing the horizontal divergence and the geoid for the case d = 5701 km (i.e. depth z = 670 km) are shown in Fig. 6 for several viscosity ratios. A comparison of Figs. 4 and 6 shows that the effect of increasing the viscosity at depth has "split" the divergence and geoid kernels: the longest wavelength geoid (l = 2,3) is now more sensitive to lateral density variations in the lower mantle while the horizontal divergence is most sensitive to lateral density variations in the upper mantle. One also observes that the geoid kernels in Fig. 6 become increasingly positive as the viscosity of the lower layer increases since the mantle becomes "stiffer" thus diminishing the negative contribution to the geoid provided by the surface boundary deflection; this allows the positive contribution of the internal density anomalies to the geoid to become dominant (Hager, 1984). In Fig. 7 we show profiles of the divergence and geoid kernels for the case d = 5171 km (z = 1200 km). The kernel functions shown in Figs. 6 and 7 show that the predicted geoid will be very sensitive to changes in both the viscosity ratio and the

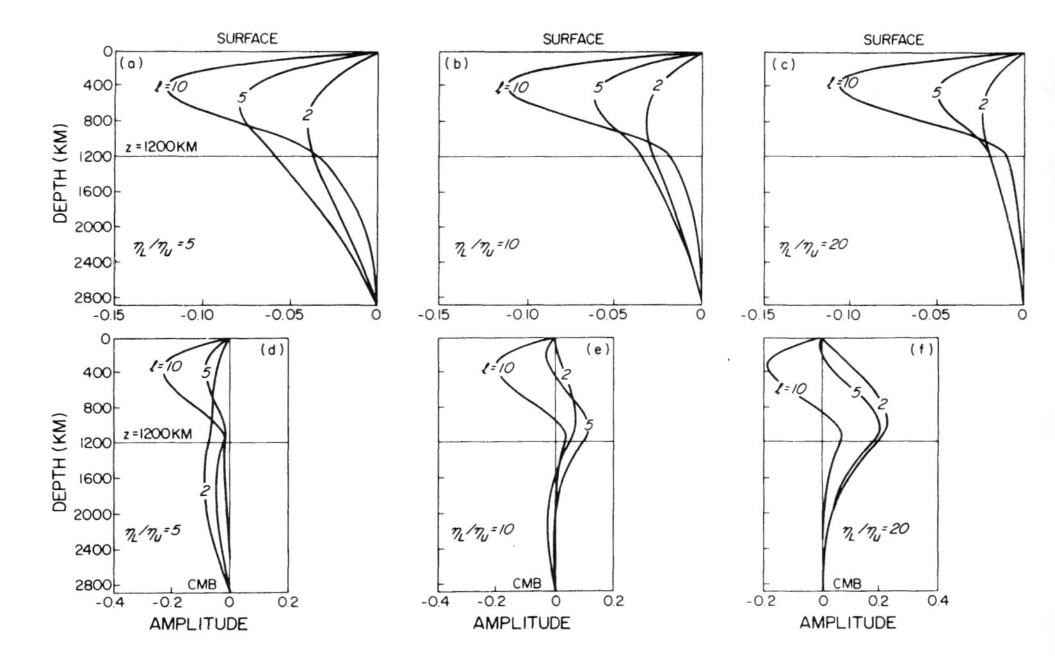

Figure 7. Kernels for an incompressible, two-layer mantle with free-slip conditions at the surface and CMB and viscosity jump at z=1200 km.: (a), (b) and (c) Horizontal divergence kernels; (d), (e) and (f) Geoid kernels for a self-gravitating mantle.

depth of the viscosity jump while the predicted horizontal divergence will be relatively insensitive to these changes.

The degree correlations of the geoid predicted using (32) and the observed non-hydrostatic geoid are presented in Fig. 8. When the boundary depth is at z = 670 km the best correlations for degrees 2 and 3 occur when $16 < \eta_L/\eta_u < 18$ but the correlation at l = 4 is very poor in this range while the correlation for l = 5 has the "wrong" sign. The best fit, at all degrees, is obtained when z = 1200 km since for $\eta_L/\eta_u = 8$ the correlations for l = 2-5 are all positive and those at l = 2,3 and 5 are more than 90% significant. In Fig. 9 we show the degree variances of the predicted geoid.

We noted previously that the profiles of the horizontal divergence kernels are relatively insensitive to changes in the viscosity ratio and the depth of the viscosity jump. The degree correlations of the horizontal divergence predicted using (31) and the observed divergence are therefore expected to be fairly constant and this is confirmed in Fig. 10. The most appropriate viscosity ratio, η_L/η_u, and radius of the boundary, d, must then be determined by fitting the predicted geoid to the observed non-hydrostatic geoid. An objective measure of fit which includes the information provided by both the degree correlations and the degree variances is the root mean square (rms) error, δ_{rms}, defined by:

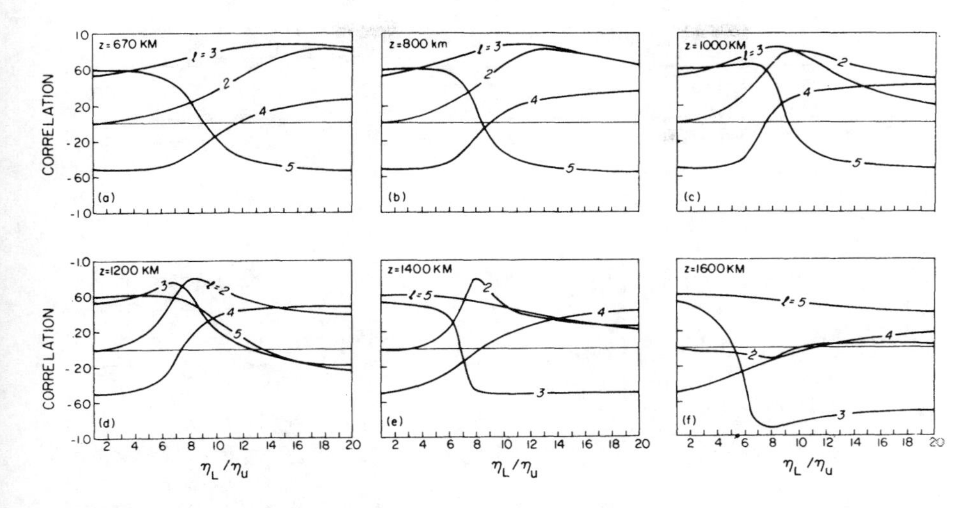

Figure 8. Degree correlations of corrected, non-hydrostatic, geoid (Nakiboglu, 1982) obtained from the GEM10B geopotential model (Lerch et al., 1979) with the predicted geoid, as a function of viscosity contrast and depth of viscosity jump.

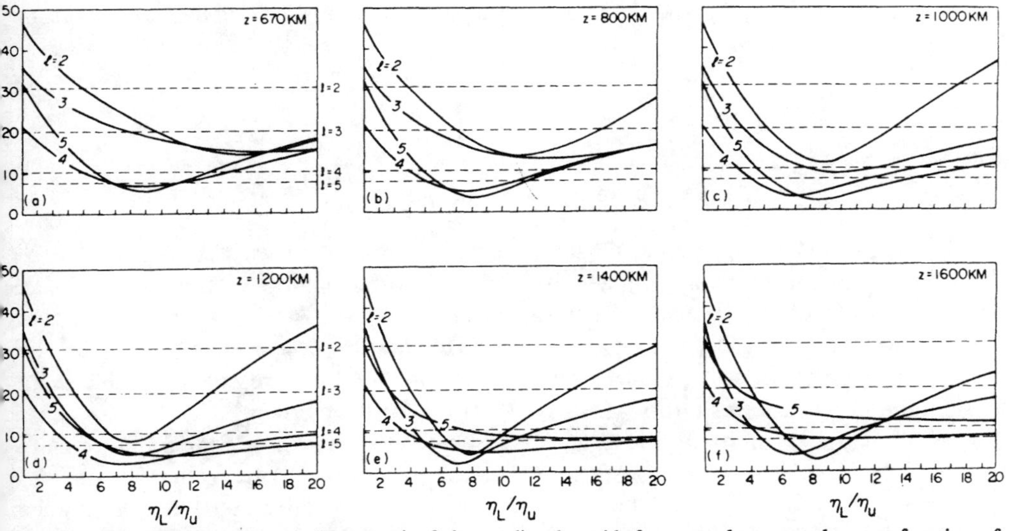

Figure 9. Degree variances (in metres) of the predicted geoid, for a two-layer mantle, as a function of viscosity contrast and depth of viscosity jump. The dashed horizontal lines are the degree variances of the corrected, non-hydrostatic, GEM10B geoid.

$$\delta_{rms}^2 = \frac{1}{4\pi} \int_o^{2\pi} \int_{-1}^{1} \left[G_p(\theta,\phi) - G_o(\theta,\phi) \right]^2 d\cos\theta \, d\phi \, , \tag{33}$$

where $G_p(\theta,\phi)$ and $G_o(\theta,\phi)$ are, respectively, the predicted and observed non-hydrostatic

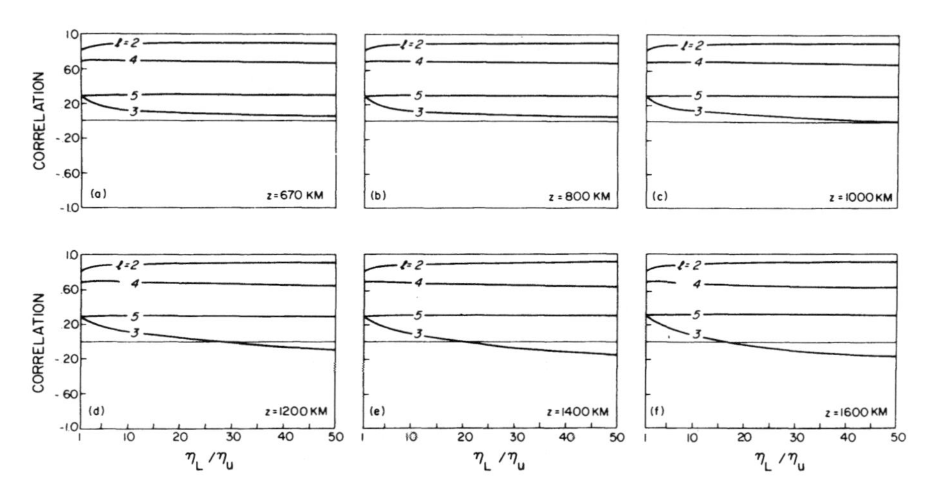

Figure 10. Degree correlations of the observed and predicted surface divergence fields for a two-layer mantle, as a function of viscosity contrast and depth of viscosity jump.

geoids. Expanding both G_p and G_o in terms of spherical harmonics one then obtains from (33),

$$\delta_{rms}^2 = \sum_l \delta_l^2 , \tag{34}$$

where,

$$\delta_l^2 = (\sigma_p)_l^2 + (\sigma_o)_l^2 - 2\rho_l \, (\sigma_p)_l \, (\sigma_o)_l$$

In (34), $(\sigma_p)_l$ and $(\sigma_o)_l$ are the degree variances of the predicted and observed non-hydrostatic geoid, respectively, while ρ_l is the degree correlation between these two fields. When the depth of the boundary between the layers is at z = 670 km, the rms errors obtained for the case η_L/η_u = 18 (which maximizes the degree correlations) and for the case η_L/η_u = 6 (which best matches all the degree variances) are presented in Table 3. In Table 4 are shown the rms errors when z = 1200 km for the cases η_L/η_u = 8 (maximizes degree correlations) and η_L/η_u = 4 (matches the degree variances). A comparison of these two tables show that a sort of "trade-off" exists in that essentially equivalent fits (as measured by δ_{rms}) to the observed non-hydrostatic geoid are obtained by either placing the boundary at z = 670 km with η_L/η_u = 18 or by placing it deeper, at z = 1200 km, which then requires a lower viscosity ratio of η_L/η_u = 8. The slightly better fit to the observed non-hydrostatic geoid in the range l = 2-5 when z = 1200 km and η_L/η_u = 8 is entirely due to the much improved fits at l = 4 and 5 compared to when z = 670 km. A more graphic illustration of the extent to which the predicted geoid matches the observed non-hydrostatic geoid for the case z = 1200 km and η_L/η_u = 8 is provided in Fig. 11.

Now that the optimum viscosity ratio and boundary depth have been determined, the absolute value of the mantle viscosity may be found by matching the kinetic energy of the

Table 3 Rms errors between predicted and observed non-hydrostatic geoid for the case z = 670 km.

$\eta_L/\eta_u = 18$

Degree, l	$(\sigma_p)_l$ (metres)	ρ_l	δ_l (metres)
l=2	15.8	.82	19.8
l=3	15.2	.87	10.0
l=4	13.6	.25	14.8
l=5	16.3	-.52	21.2

$\delta_{rms} = 34.1$ m

$\eta_L/\eta_u = 6$

Degree, l	$(\sigma_p)_l$ (metres)	ρ_l	δ_l (metres)
l=2	28.0	.13	38.7
l=3	22.8	.67	17.5
l=4	8.8	-.48	16.4
l=5	9.7	.53	8.6

$\delta_{rms} = 46.3$ m

predicted, purely poloidal, surface flow field to the kinetic energy of the observed surface plate velocities which consist of both poloidal and toroidal components. We include the kinetic energy of the toroidal field since we assume that it is produced by a flux of energy out of the poloidal field. The amplitude of the predicted horizontal divergence (or, equivalently, of the predicted poloidal scalar; cf. (3a)) depends on both η_L/η_u and η_u and to illustrate this dependence in a single diagram we have plotted in Fig. 12 a quantity called $\log_{10}(\sigma_R)$ which is defined by:

$$\log_{10}(\sigma_R) = \log_{10}\left[\frac{\sigma_l(\eta_L/\eta_u)}{\sigma_l(\eta_L/\eta_u{=}1)}\right], \tag{35}$$

where $\sigma_l(\eta_L/\eta_u)$ is the degree of the predicted horizontal divergence for a given value of η_L/η_u. The quantity σ_R defined in (35) will be independent of η_u since this viscosity appears merely as a multiplicative factor in (31) and thus $\log_{10}(\sigma_R)$ is only a function of η_L/η_u. The observational constraints on η_u (and hence η_L) are introduced by considering

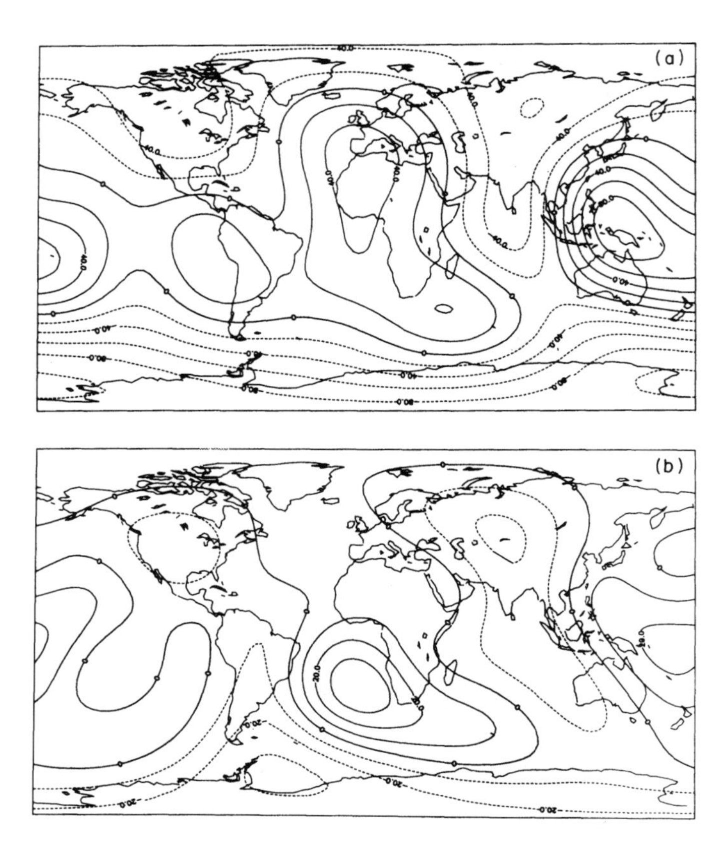

Figure 11. (a) The corrected, non-hydrostatic, GEM10B geoid synthesized from harmonics in the range $2 \leq l \leq 5$. The contour interval is 20 metres. (b) The non-hydrostatic geoid predicted using the geoid kernel for a two-layer, self-gravitating mantle with a factor of eight viscosity increase at a depth of 1200 km. The contour interval is 10 metres.

$$\log_{10}(\sigma_o) = \log_{10}\left[\frac{\sigma_l(\text{observed})}{\sigma_l(\eta_l/\eta_u\!=\!1)}\right], \tag{36}$$

where σ_l(observed) is the total kinetic energy (i.e. poloidal + toroidal), at each degree, of the observed surface plate velocities. The quantity σ_o, defined in (36), is directly proportional to η_u and is independent of η_L/η_u. The horizontal dashed lines in Fig. 12 are plots of $\log_{10}(\sigma_o^*)$ for $\eta_u^* = 1 \times 10^{21}$ Pa s. When an appropriate value for η_L/η_u has been found then the value of η_u is determined by equating $\log_{10}(\sigma_R)$ with $\log_{10}(\sigma_o)$ and in Fig. 12 this is equivalent to measuring, for a particular degree, the distance as measured along the ordinate axis of the point (determined by η_L/η_u) on a $\log_{10}(\sigma_R)$ curve from the $\log_{10}(\sigma_o^*)$ line ; if this distance is called d then the value of η_u is $\eta_u = 10^d \times \eta_u^* = 10^d \times 10^{21}$ Pa s. Now, since the best degree correlations between the predicted and observed

Table 4 Rms errors between predicted and observed non-hydrostatic geoid for the case z = 1200 km.

$\eta_L/\eta_u = 8$

Degree, l	$(\sigma_p)_l$ (metres)	ρ_l	δ_l (metres)
l=2	8.4	.78	24.6
l=3	5.0	.65	17.1
l=4	3.0	.13	10.3
l=5	5.8	.53	6.6

δ_{rms} = 32.3 m

$\eta_L/\eta_u = 4$

Degree, l	$(\sigma_p)_l$ (metres)	ρ_l	δ_l (metres)
l=2	22.1	.11	35.7
l=3	14.6	.59	16.3
l=4	7.8	-.47	15.5
l=5	13.3	.62	10.5

δ_{rms} = 43.5 m

horizontal divergence fields are at l = 2 and 4, we will match the predicted and observed kinetic energies at these same two degrees in the manner just described. We find then that for the case z = 670 km and η_L/η_u = 18 the value of η_u determined from the degree-2 matching is 1.47 x 10^{21} Pa s which agrees well with η_u = 1.55 x 10^{21} Pa s determined from the degree-4 matching. When z = 1200 km and η_L/η_u = 8, the value of η_u determined from l = 2 is 2.31 x 10^{21} Pa s which again agrees well with η_u = 2.26 x 10^{21} Pa s obtained from l = 4.

The predicted and observed horizontal divergence fields are well-correlated at l = 2 and 4 as noted previously. A more pictoral illustration of the agreement between these two fields is given in Fig. 13 where we show maps of the predicted and observed horizontal divergence synthesized from harmonics in the range l = 2-5. The agreement between these maps is quite good and the places where the biggest mismatch occur are North America, Asia and Australia; these regions are incorrectly shown to be subducting in the predicted divergence map since our simple model does not incorporate the effects of lateral variations in chemistry which produce buoyant continental masses.

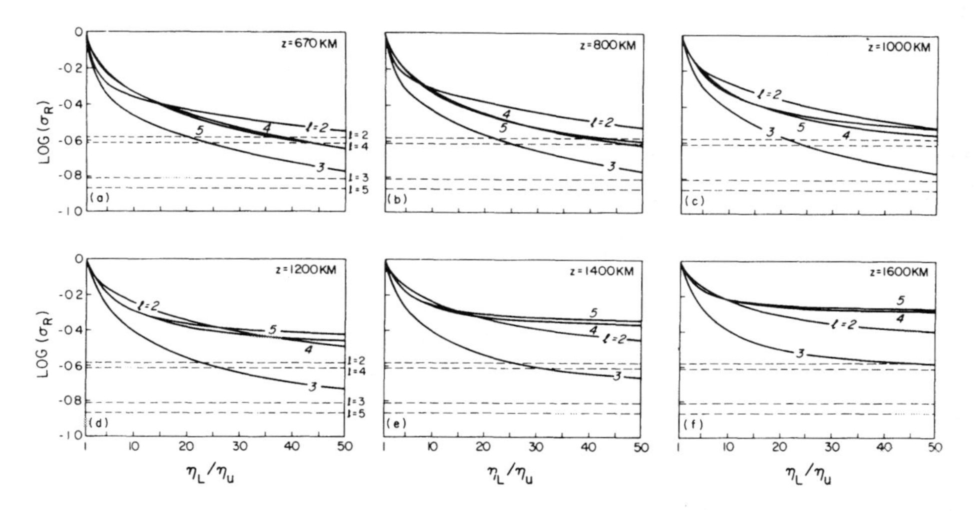

Figure 12. Degree variances of the predicted surface divergence field, for a two-layer mantle, as measured by log (σ_R) (cf. equation (35)) as a function of viscosity contrast and depth of viscosity jump. The dashed horizontal lines are the degree variances of the observed surface divergence field, as measured by log (σ_o) (cf. equation (36)).

4. The importance of poloidal-toroidal coupling

The agreement between the predicted and observed fields in the maps shown in Figs. 11 and 13 is encouraging and it illustrates the approximate validity of the very simple viscous flow models that we have used. It is certain that an improved fit to the observed surface fields may be made by increasing the accuracy and resolution of the seismically inferred lateral heterogeneity models that we have used. However, the most serious problem with our spherically symmetric viscosity models is that lateral density variations will only force a purely poloidal flow field. We are therefore unable to reproduce the observed equipartition of kinetic energy between the poloidal and toroidal flows at the Earth's surface. Toroidal flows may be generated by allowing the viscosity to vary laterally thus producing a coupling between poloidal and toroidal scalars (Forte and Peltier, in preparation). We expect that this coupling will be strongest in the lithosphere-asthenosphere where the lateral variations of rheology and chemistry are the most extreme.

An Earth model with poloidal-toroidal coupling is expected to produce geoids that are quite different from those presented in this report. The predicted geoid depends almost entirely on the fine balance achieved between the opposing contributions from the surface topography and the internal mass anomalies. The good geoid correlations obtained with the two-layer model are a direct result of the reduced surface topography produced by increasing the viscosity of the lower layer. If one introduces lateral viscosity/chemistry variations then we expect that a similar reduction in topography might be achieved since we assume that the toroidal flow that is generated in the near-surface region at the expense of the interior poloidal flow will itself produce little or no surface topography while the topography supported by the now weaker, near-surface, poloidal flow will be diminished.

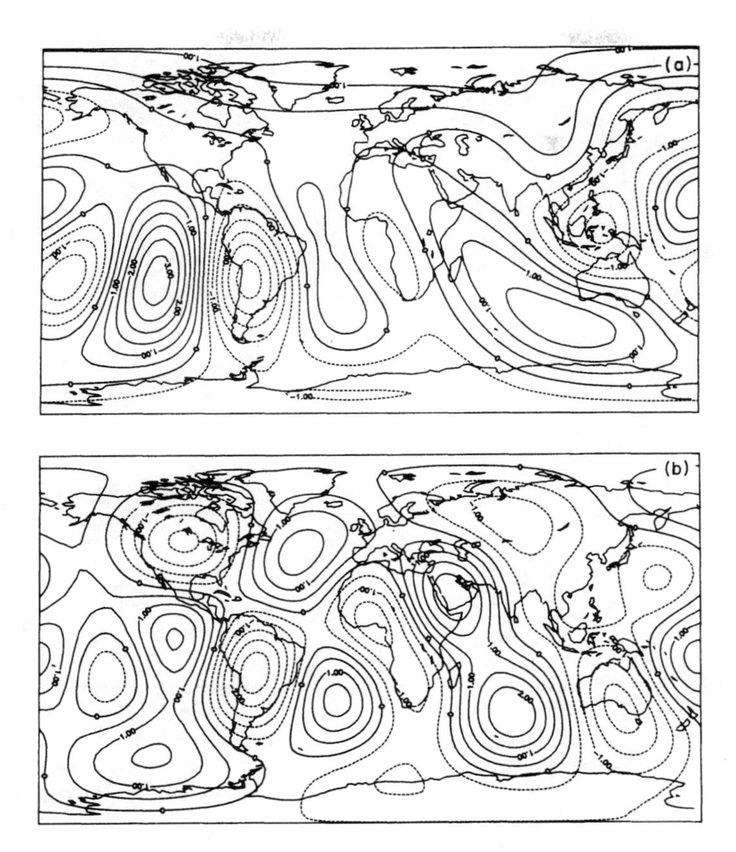

Figure 13. (a) The normalized, observed horizontal divergence of the surface plate motions synthesized from harmonics in the range $2 \leq l \leq 5$. This field has been normalized by dividing by the total degree variance of the actual field for the range $2 \leq l \leq 5$ such that the total degree variance of the field shown in the map is exactly unity. The contour interval is 0.5 (dimensionless units). (b) The normalized horizontal divergence predicted using the divergence kernel for a two-layer mantle with a factor of eight viscosity increase at a depth of 1200 km. This field has also been normalized as in (a). The normalized, predicted divergence field depends only on η_L/η_u and is independent of η_u (cf. equation (31)) thus facilitating a direct comparison with the field shown in (a). The contour interval is 0.5 (dimensionless units).

This suggests then that one may model the geoid with smaller viscosity increases at depth by introducing lateral viscosity/chemistry variations in the near surface region. The trade-off between poloidal-toroidal coupling and the strength of the viscosity contrast in the two-layer model may be illustrated in a somewhat ad hoc manner by defining a "poloidal conversion factor", α. We may then try to reproduce the effect of poloidal-toroidal conversion near the Earth's surface by multiplying the poloidal scalar, $p_l^m(r=a)$, by α thus simulating a $(1-\alpha^2) \times 100\%$ flux of energy to a toroidal flow which we assume contributes little or nothing to the geoid. In Fig. 14(a) and (d) we show, for example, that a great improvement in the l = 2 geoid correlation is obtained when $\alpha = 0.875$. In Table 5 we

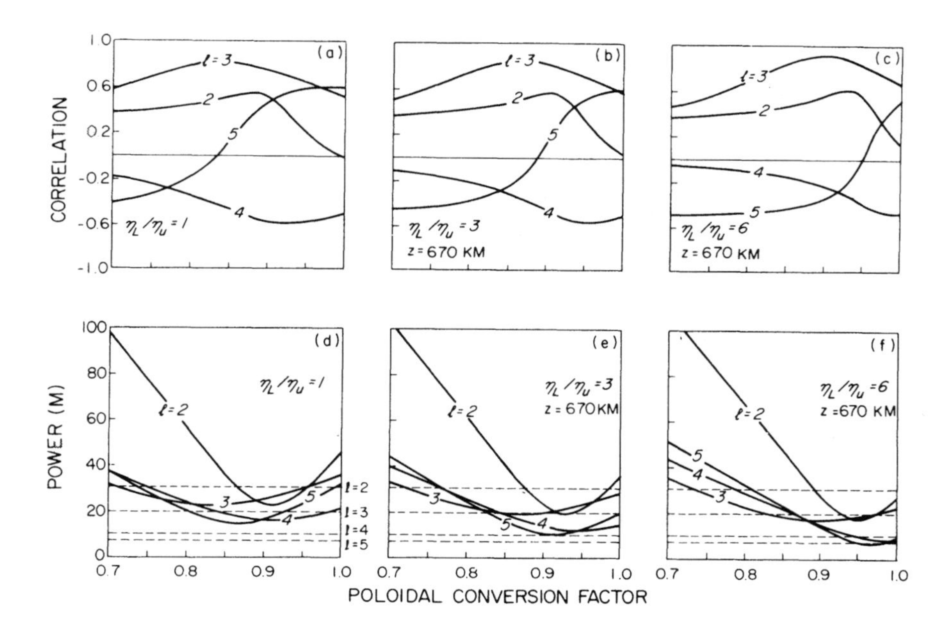

Figure 14. Behaviour of the predicted geoid when the effect of poloidal-toroidal coupling at the surface is simulated via a poloidal conversion factor. A conversion factor, α, mimics a $(1 - \alpha^2) \times 100\%$ flux of energy to the toroidal field: (a), (b) and (c) Degree correlations of corrected, non-hydrostatic, GEM10B geoid with predicted geoid as a function of α; (d), (e) and (f) degree variances (in metres) of predicted geoid as a function of α. Dashed horizontal lines are degree variances of corrected, non-hydrostatic, GEM10B geoid.

show that the good match between the predicted and observed geoid at $l = 2$ for $\eta_L/\eta_u = 20$ is due to the reduction of the surface topography at this degree from 695 m to 538 m and that a similar reduction is achieved in a constant viscosity mantle with a 23% flux of energy to the toroidal field.

5. Concluding remarks

The Green function formalism provides a mathematically straightforward method for deriving analytic, closed-form, expressions for the kernel functions relating the horizontal divergence of the surface flow and the non-hydrostatic geoid to the seismic tomographic inferences of lateral density variations in the mantle. The kernel functions for a two-layer mantle were used to predict horizontal divergence and non-hydrostatic geoid fields which are in fairly good agreement with the corresponding observed fields. In particular, we have shown that the agreement between the predicted and observed horizontal divergence fields (cf. Fig. 13) is reasonably good thus providing strong support to the notion that the large-scale tectonic plate motions are the direct expression of the convective flow in the mantle driven by internal thermally induced buoyancy forces.

A central aim of this report has been to present a unified description of both the non-hydrostatic geoid and the surface plate kinematics as represented by the horizontal

Table 5 Degree variances of the predicted surface topography

$\alpha = 1.0, \eta_L/\eta_u = 1$

Degree, l	Degree variance (metres)
l=2	695
l=3	449
l=4	582
l=5	1,012

$\alpha = 1.0, \eta_L/\eta_u = 20$, z = 670 km

Degree, l	Degree variance (metres)
l=2	538
l=3	342
l=4	452
l=5	753

$\alpha = 0.875, \eta_L/\eta_u = 1$

Degree, l	Degree variance (metres)
l=2	542
l=3	371
l=4	486
l=5	856

divergence field. The predicted non-hydrostatic geoid depends only on the viscosity ratio in a two-layer Earth model but the predicted horizontal divergence instead depends on the absolute values of the viscosity in each layer. We have thus been able to establish preliminary estimates of the absolute value of the viscosity by ensuring that the balance between buoyancy forces and viscous dissipation is such as to produce a surface flow whose kinetic energy matches that of the observed surface plate velocities. In attempting to match the predicted and observed non-hydrostatic geoids we found that a trade-off exists between the required viscosity contrast and the depth of the viscosity jump in the two-layer model. We found further that the two-layer model providing the most successful fit to the observed non-hydrostatic geoid has a factor of eight viscosity increase at a depth of 1200 km; in this model the viscosity of the upper layer is $\eta_u = 2.3 \times 10^{21}$ Pa s while that of the lower layer is $\eta_L = 1.8 \times 10^{22}$ Pa s. The discrepancy between these preliminary viscosity estimates and those obtained from post-glacial rebound (e.g. Wu and Peltier, 1983) may have implications for the issue of transient rheology in the mantle (Peltier et al., 1986; Forte and Peltier, 1986).

The existence of strong toroidal flow in the observed surface plate velocities points to an important defect in Earth models with a spherically-symmetric viscosity distribution since such models are unable to produce any toroidal flow. In section 4 we argued that the generation of toroidal flow by a flux of energy out of the poloidal flow field is likely to produce surface topography of smaller amplitude than that produced in the absence of this effect. Since the predicted geoid is very strongly dependent on the predicted surface topography we then believe that incorporation of the effects of poloidal-toroidal coupling may allow us to fit the predicted geoid to the observed geoid with a smaller viscosity increase at depth in the mantle thus providing a possible reconciliation with the small viscosity contrast at the 670 km seismic discontinuity required by glacial isostatic adjustment analyses. If such a reconciliation were not possible then it may be necessary to argue that the viscosity inferences obtained from glacial isostatic adjustment analyses represent transient values rather than steady-state values (Peltier et al., 1986).

Appendix A: Homogeneous, incompressible mantle

A.1 *Derivation of the Green function*

The required Green function must satisfy equation (17); that is,

$$D_l^2\, p_l^m(r,r') = \delta(r-r')\,. \tag{A1}$$

When $r \neq r'$, (A1) is the biharmonic equation whose solution is easily found; therefore, the solution of (A1) is split in two parts: a solution defined for the region $r' < r \leq a$ and a solution for the region $b \leq r < r'$. These two solutions must then be joined at $r=r'$ by applying the appropriate matching conditions. The biharmonic equation,

$$D_l^2\, p_l^m(r,r') = 0\,,$$

is a fourth order differential equation whose solution will consist of the sum of four linearly independent functions:

$$p_l^m(r,r') = A_1 r^l + \frac{B_1}{r^{l+1}} + C_1 r^{l+2} + \frac{D_1}{r^{l-1}}\,,\; r' < r \leq a$$

$$p_l^m(r,r') = A_2 r^l + \frac{B_2}{r^{l+1}} + C_2 r^{l+2} + \frac{D_2}{r^{l-1}}\,,\; b \leq r < r' \tag{A2}$$

In (A2) there are 8 coefficients which are found by applying the boundary conditions at $r=a$, $r=b$ and the matching conditions at $r=r'$. Two boundary conditions are obtained by requiring that the radial component of the fluid velocity, u_r, is zero at the deformed bounding surfaces of the fluid layer (which to first order accuracy may be taken as applying at the undeformed reference positions of the bounding surfaces):

$$u_r = \hat{\mathbf{r}} \cdot \mathbf{u} = \frac{1}{r} \Lambda^2 p \tag{A3}$$

Therefore, from (A3), one sees that if $u_r = 0$ for all θ and ϕ at both $r=a$ and $r=b$, then

$$p_l^m(r,r') = 0 \quad , \quad r=a,b \tag{A4}$$

Two more boundary conditions are obtained by requiring that the shear stresses, $\sigma_{r\theta}$ and $\sigma_{r\phi}$, are both zero at $r=a$ and $r=b$, and one may show (e.g., Chandrasekhar, 1961) this condition implies that

$$\frac{d^2}{dr^2} p_l^m(r,r') = 0 \quad , \quad r=a,b \tag{A5}$$

Three matching conditions are obtained by requiring that u_r, u_θ (and u_ϕ), and $\sigma_{r\theta}$ (and $\sigma_{r\phi}$) are continuous at $r=r'$; one can therefore show that:

$$p_l^m(r,r')\Big|_{r=r'^+} = p_l^m(r,r')\Big|_{r=r'^-} ,$$

$$\frac{dp_l^m(r,r')}{dr}\Big|_{r=r'^+} = \frac{dp_l^m(r,r')}{dr}\Big|_{r=r'^-} , \tag{A6}$$

$$\frac{d^2p_l^m(r,r')}{dr^2}\Big|_{r=r'^+} = \frac{d^2p_l^m(r,r')}{dr^2}\Big|_{r=r'^-} .$$

The final matching condition is obtained by integrating equation (A1) from $r=r'^-$ to $r=r'^+$ and using (A6) to obtain

$$\frac{d^3p_l^m(r,r')}{dr^3}\Big|_{r=r'^+} - \frac{d^3p_l^m(r,r')}{dr^3}\Big|_{r=r'^-} = 1 \tag{A7}$$

The eight conditions contained in equations (A4), (A5), (A6) and (A7) when applied to (A2) will yield expressions for the 8 coefficients:

$$A_1 = -\frac{D_1}{a^{2l-1}} \; , B_1 = -C_1\, a^{2l+3}$$

$$C_1 = \frac{1}{2(2l+3)(2l+1)} \; \frac{1}{r'^{\,l-1}} \; \frac{1-(r'/b)^{2l+3}}{1-(a/b)^{2l+3}} \tag{A8}$$

$$D_1 = \frac{1}{2(4l^2-1)} \; \frac{a^{2l-1}}{r'^{\,l-3}} \; \frac{1-(r'/b)^{2l-1}}{1-(a/b)^{2l-1}}$$

and

$$A_2 = -\frac{D_2}{b^{2l-1}},\ B_2 = -C_2\, b^{2l+3}$$

$$C_2 = -\frac{1}{2(2l+3)(2l+1)}\ \frac{1}{r'^{\,l-1}}\ \frac{1-(r'/a)^{2l+3}}{1-(b/a)^{2l+3}} \tag{A9}$$

$$D_2 = -\frac{1}{2(4l^2-1)}\ \frac{b^{2l-1}}{r'^{\,l-3}}\ \frac{1-(r'/a)^{2l-1}}{1-(b/a)^{2l-1}}$$

In order to obtain expressions for the kernel functions describing surface divergence and geoid height one requires expressions for the derivatives of the poloidal scalar which is given by equation (18), that is,

$$p_l^m(r) = \frac{g_o}{\eta}\left[\int_r^a \frac{(\rho_1)_l^m(r')}{r'}\,(p_2)_l^m(r,r')dr' + \int_b^r \frac{(\rho_1)_l^m(r')}{r'}\,(p_1)_l^m(r,r')dr'\right] \tag{A10}$$

where $(p_1)_l^m(r,r')$ is the Green function valid for $r' \leq r \leq a$ and $(p_2)_l^m(r,r')$ is the Green function valid for $b \leq r < r'$ (see equation (A2)). From expression (A10) one may readily show that the Green functions for the first, second, and third derivatives of $p_l^m(r)$ are the first, second, and third derivatives of $p_l^m(r,r')$ respectively.

A.2. *The surface divergence kernel*

Since $\nabla\cdot\mathbf{u} = 0$, then

$$\nabla_H\cdot\mathbf{u} = -\frac{1}{r^2}\frac{\partial}{\partial r}(r^2 u_r) \tag{A11}$$

Now, to first-order accuracy, $u_r = 0$ at $r=a$ and therefore from (A11) we get

$$\nabla_H\cdot\mathbf{u}(r=a) = -\frac{\partial u_r}{\partial r}\Big|_{r=a} \tag{A12}$$

If one substitutes (A3) into (A12), expands into spherical harmonics, and uses result (A4) then one can show that

$$(\nabla_H\cdot\mathbf{u})_l^m(r=a) = \frac{l(l+1)}{a}\ \frac{dp_l^m(r)}{dr}\Big|_{r=a}\ . \tag{A13}$$

Result (A13) proves equation (3a). From (A10) one gets

$$\frac{dp_l^m(r)}{dr}\Big|_{r=a} = \frac{g_o}{\eta}\int_b^a \frac{(\rho_1)_l^m(r')}{r'}\ \frac{d(p_1)_l^m(r,r')}{dr}\Big|_{r=a} dr' \tag{A14}$$

Upon substituting (A2) into (A14), and using (A8), one finds from (A13) that

$$(\nabla_H \cdot \mathbf{u})_l^m(r=a) = \frac{g_o}{\eta} \int_b^a S_l(r')\,(\rho_1)_l^m(r')\,dr' \ ,$$

where

$$S_l(r') = \frac{a^l\,l(l+1)}{2(2l+1)} \left[\frac{1}{r'^l}\,\frac{1-(r'/b)^{2l+3}}{1-(a/b)^{2l+3}} - \frac{1}{a^2 r'^{l-2}}\,\frac{1-(r'/b)^{2l-1}}{1-(a/b)^{2l-1}} \right] .$$

A.3. *The Geoid Kernel*

If one expands all quantities in equation (25) in terms of spherical harmonics then the following result is obtained

$$\delta a_l^m = \frac{1}{g_o\,(\rho_o - \rho_w)} \left[(P_1)_l^m(r=a) - 2\eta\,\frac{d\,(u_r)_l^m(r)}{dr}\Big|_{r=a} \right] . \tag{A15}$$

The non-hydrostatic pressure may be obtained from the $\hat{\theta}$-component of the momentum equation (6):

$$\frac{1}{r}\frac{\partial}{\partial\theta} P_1 = \eta \left[\Delta \mathbf{u}\right]_\theta + \rho_o\,\frac{1}{r}\frac{\partial \phi_1}{\partial \theta} \ ,$$

where one can show from (10), assuming a purely poloidal flow field, that

$$\left[\Delta \mathbf{u}\right]_\theta = -\frac{1}{r}\frac{\partial}{\partial\theta}\frac{\partial}{\partial r}\,r\,\Delta p \ ,$$

and thus if one expands all quantities in terms of spherical harmonics one can show that

$$(P_1)_l^m = -\eta\,\frac{d}{dr}\left[r D_l\, p_l^m(r)\right] + \rho_o\,(\phi_1)_l^m \ , \tag{A16}$$

where D_l is the transformed Laplacian operator in equation (15). Now, from (A3) it follows that

$$(u_r)_l^m(r) = -\frac{1}{r}\,l(l+1)\,p_l^m(r) \ . \tag{A17}$$

If (A16) and (A17) are substituted into (A15), and use is made of the boundary conditions (A4) and (A5), it follows that

$$\delta a_l^m = \frac{1}{(\rho_o - \rho_w)}\,X_l^m(r=a) + \frac{\rho_o}{g_o\,(\rho_o - \rho_w)}\,(\phi_1)_l^m(r=a) \ , \tag{A18}$$

where

$$X_l^m(r) = \frac{\eta}{g_o}\left[-r\,\frac{d^3}{dr^3} + \frac{3l(l+1)}{r}\,\frac{d}{dr} \right] p_l^m(r) \tag{A19}$$

Similarly, in the case of the core-mantle boundary deflection, δb_l^m, one can show that

$$\delta b_l^m = \frac{1}{(\rho_o - \rho_c)} X_l^m(r{=}b) + \frac{\rho_o}{g_o(\rho_o - \rho_c)} (\phi_1)_l^m(r{=}b) . \tag{A20}$$

If one now substitutes (A18) into (22), (A20) into (23), and combines the results in (24) then one may show that

$$(\phi_1)_l^m(r) = (U_{int})_l^m(r) + \frac{4\pi a G}{2l+1} \left[\frac{r}{a}\right]^l \left[X_l^m(r{=}a) + \frac{\rho_o}{g_o}(\phi_1)_l^m(r{=}a)\right]$$

$$- \frac{4\pi b G}{2l+1} \left[\frac{b}{r}\right]^{l+1} \left[X_l^m(r{=}b) + \frac{\rho_o}{g_o}(\phi_1)_l^m(r{=}b)\right] \tag{A21}$$

In equation (A21) it is clear that to find $(\phi_1)_l^m(r)$ one must know $(\phi_1)_l^m(r{=}a)$ and $(\phi_1)_l^m(r{=}b)$; this is easily accomplished by setting $r{=}a$ and $r{=}b$ in (A21) thus producing two simultaneous equations which may be solved to find $(\phi_1)_l^m(r{=}a)$ and $(\phi_1)_l^m(r{=}b)$ and having done this one may then show that

$$(\phi_1)_l^m(r) = (U_{int})_l^m(r) + \left[\frac{r}{a}\right]^l \left[(1-K_a)(1+K_b) + K_a K_b \left[\frac{b}{a}\right]^{2l+1}\right]^{-1}$$

$$\left\{\left[1 + K_b\left[1 - \left[\frac{b}{r}\right]^{2l+1}\right]\right] \left[K_a (U_{int})_l^m(r{=}a) + \frac{4\pi a G}{2l+1} X_l^m(r{=}a)\right] - \right.$$

$$\left. \left[\frac{b}{a}\right]^{l+1} \left[K_a\left[1 - \left[\frac{a}{r}\right]^{2l+1}\right] + \left[\frac{a}{r}\right]^{2l+1}\right] \left[K_b (U_{int})_l^m(r{=}b) + \frac{4\pi b G}{2l+1} X_l^m(r{=}b)\right]\right\} , \tag{A22}$$

where

$$K_a = \frac{4\pi a G}{2l+1} \frac{\rho_o}{g_o} \text{ and } K_b = \frac{4\pi b G}{2l+1} \frac{\rho_o}{g_o} . \tag{A23}$$

When expression (A19) is substituted into (A22) one may show that

$$(\phi_1)_l^m(a) = \frac{4\pi a G}{2l+1} \int_b^a G_l(r') (\rho_1)_l^m(r') \, dr' , \tag{A24}$$

where

$$G_l(r') = \left[1 - K_a + \frac{K_a K_b}{1+K_b} \left[\frac{b}{a}\right]^{2l+1}\right]^{-1} \left[\frac{r'}{a}\right]^{l+2} \left\{1 - \frac{K_a}{1+K_b} \left[\frac{b}{r'}\right]^{2l+1} \left[\frac{b}{a}\right] + \right.$$

$$\left. A_l(r') \left[1 - \frac{K_a}{1+K_b} \left[\frac{b}{a}\right]^{2l+2}\right] \left[\frac{a}{r'}\right]^{l+2} - B_l(r') \left[1 - \frac{K_a}{1+K_b} \left[\frac{b}{a}\right]\right] \left[\frac{b}{r'}\right]^{l+2}\right\} , \tag{A25}$$

and where

$$A_l(r') = -\frac{l(l+2)}{2l+1}\left[\frac{a}{r'}\right]^{l-2}\frac{1-(r'/b)^{2l-1}}{1-(a/b)^{2l-1}} +$$

$$\frac{(l+1)(l-1)}{2l+1}\left[\frac{a}{r'}\right]^{l}\frac{1-(r'/b)^{2l+3}}{1-(a/b)^{2l+3}} ,$$

$$B_l(r') = \frac{l(l+2)}{2l+1}\left[\frac{b}{r'}\right]^{l-2}\frac{1-(r'/a)^{2l-1}}{1-(b/a)^{2l-1}} -$$

$$\frac{(l+1)(l-1)}{2l+1}\left[\frac{b}{r'}\right]^{l}\frac{1-(r'/a)^{2l+3}}{1-(b/a)^{2l+3}} .$$

The non-hydrostatic geoid-height field is obtained from $\phi_1(r=a)$ by dividing by g_o and thus from (A24) one has

$$(Ge)_l^m = \frac{3}{2l+1}\,\frac{1}{\bar{\rho}}\int_b^a G_l(r')\,(\rho_1)_l^m(r')\,dr' ,$$

where $(Ge)_l^m$ is the harmonic coefficient of the geoid-height field, $\bar{\rho}$ is the average density of the Earth, and $G_l(r')$ is the geoid kernel given in equation (A25).

The effects of self-gravitation on the geoid are contained in the constants K_a and K_b which are defined in equation (A23). To obtain the geoid kernels, $G_l(r')$, for a non-self-gravitating mantle one simply sets both K_a and K_b to be zero in equation (A25); one may also note that this equivalent to letting $l \rightarrow \infty$.

Appendix B: Inhomogeneous (two-layer), incompressible mantle

B.1 *Derivation of the Green function*

As in equation (A2), the Green function $(p_i)_l^m(r,r')$, valid in region i (i = 1,2,3,4,5, and 6 - see Fig. 5 and equation (30)) is given by the expression

$$(p_i)_l^m(r,r') = A_i\, r^l + \frac{B_i}{r^{l+1}} + C_i\, r^{l+2} + \frac{D_i}{r^{l-1}} . \tag{B1}$$

For either case (i) or case (ii) (Fig. 5) equation (B1) shows there are 12 coefficients to be found; they are determined by applying the free-slip boundary conditions at $r=a$ and b, matching conditions at $r=r'$ (see Appendix A), and by applying appropriate matching conditions at $r=d$. Since we assume that a whole mantle flow exists the conditions we impose at $r=d$ are continuity of mass flux, normal stress, tangential stress, and tangential velocity:

$$\rho_u \; u_r \big|_{r=d^+} = \rho_L \; u_r \big|_{r=d^-} \,, \tag{B2a}$$

$$\sigma_{rr} \big|_{r=d^+} - \sigma_{rr} \big|_{r=d^-} = (\rho_L - \rho_u) \, g_o \, \delta d \,, \tag{B2b}$$

$$\sigma_{r\theta} \big|_{r=d^+} = \sigma_{r\theta} \big|_{r=d^-} \,, \tag{B2c}$$

$$u_\theta \big|_{r=d^+} = u_\theta \big|_{r=d^-} \,, \tag{B2d}$$

where ρ_u and ρ_L are, respectively, the densities of the upper and lower layers and δd is the deflection of the material interface $r=d$ from its reference level. A considerable simplification is made by assuming, as in Richards and Hager (1984), that $\rho_u = \rho_L$ (= 4.43 g/cm^3) since then we can ignore the deflection δd in equation (B2b). When the 12 boundary and matching conditions are applied to (B1) one may find all the coefficients; expressions for these coefficients may be found in Forte and Peltier (1986).

B.2 *The surface divergence kernel*

To determine the surface divergence field in a two-layered mantle we again use equation (A13) in Appendix A. The radial derivative of the poloidal scalar at $r=a$ is obtained from expression (30a):

$$\frac{dp_l^m}{dr}\bigg|_{r=a} = \frac{g_o}{\eta_u} \int_d^a \frac{d(p_4)_l^m(r,r')}{dr}\bigg|_{r=a} \frac{(\rho_1)_l^m(r')}{r'} \, dr' +$$

$$\frac{g_o}{\eta_L} \int_b^d \frac{d(p_3)_l^m(r,r')}{dr}\bigg|_{r=a} \frac{(\rho_1)_l^m(r')}{r'} \, dr' \tag{B3}$$

Using expression (B1) one can show, using (A13) with (B3), that

$$(\nabla_H \cdot \mathbf{u})_l^m(r=a) = \frac{g_o}{\eta_u} \int_b^a S_l(r';\gamma,d) \, (\rho_1)_l^m(r') \, dr' \,, \tag{B4}$$

where $\gamma = \eta_L / \eta_u$,

$$S_l(r';\gamma,d) = \begin{cases} p'(4) \,, & d \le r' \le a \\ \dfrac{1}{\gamma} \, p'(3) \,, & b \le r' \le d \end{cases}$$

where,

$$p'(i) = l(l+1)\left[-l\left(\frac{a}{r'}\right) A_i' + (l+1)\left(\frac{a}{r'}\right) B_i' \right.$$

$$\left. + (l+2)\left(\frac{a}{r'}\right)^l C_i' - (l-1)\left(\frac{r'}{a}\right)^{l+1} D_i' \right].$$

The quantities A_i', B_i', C_i', and D_i' are obtained by non-dimensionalizing the coefficients A_i, B_i, C_i, D_i according to:

$$A_i' = -a^{l-3} A_i ,\, B_i' = -\frac{1}{a^{l+4}} B_i ,\, C_i' = r'^{\,l-1} C_i ,\, D_i' = \frac{1}{r'^{\,l+2}} D_i .$$

B.3 *The geoid kernel*

In our two-layer model we have assumed, for simplicity, that the densities of the upper and lower layers are equal and therefore any deflection of the interface $r=d$ will not affect the perturbed potential ϕ_1; consequently, equation (A22) of Appendix A will also be valid for the two-layered Earth. The first task is thus to determine the kernels for the quantities $X_l^m(r=a)$ and $X_l^m(r=b)$. From equation (A19) we have

$$X_l^m(r=a) = \frac{\eta_u}{g_o}\left[-r\frac{d^3}{dr^3} + \frac{3l(l+1)}{r}\frac{d}{dr}\right] p_l^m(r)\Big|_{r=a} , \tag{B5a}$$

$$X_l^m(r=b) = \frac{\eta_L}{g_o}\left[-r\frac{d^3}{dr^3} + \frac{3l(l+1)}{r}\frac{d}{dr}\right] p_l^m(r)\Big|_{r=b} . \tag{B5b}$$

If one substitutes (30a) into (B5a) then the following expression is obtained:

$$X_l^m(r=a) = \int_b^a K_l^a(r')\,(\rho_1)_l^m(r')\,dr' , \tag{B6}$$

where,

$$K_l^a(r') = \begin{cases} p_1''(4) , & d \le r' \le a \\ \dfrac{1}{\gamma} p_1''(3) , & b \le r' \le d \end{cases}$$

where

$$p_1''(i) = 2\left[-l(l^2+3l-1)\left(\frac{a}{r'}\right) A_i' + (l+1)(l^2-l-3)\left(\frac{a}{r'}\right) B_i' + \right.$$

$$\left. l(l+1)(l+2)\left(\frac{a}{r'}\right)^l C_i' - l(l+1)(l-1)\left(\frac{r'}{a}\right)^{l+1} D_i' \right] .$$

The quantities, A_i', B_i', C_i', and D_i' appearing in the expression for p_1''(i) have the same definition as those in $p'(i)$. Similarly, if one substitutes (30b) into (B5b) it can be shown that

$$X_l^m(r=b) = \int_b^a K_l^b(r')\,(\rho_1)_l^m(r')\,dr' \,, \tag{B7}$$

where,

$$K_l^b(r') = \begin{cases} \gamma p_2''(6), & d \le r' \le a \\ p_2''(2), & b \le r' \le d \end{cases}$$

where

$$p_2''(i) = 2\left[-l(l^2+3l-1)\left[\frac{b}{r'}\right] A_i' + (l+1)(l^2-l-3)\left[\frac{b}{r'}\right] B_i' + l(l+1)(l+2)\left[\frac{b}{r'}\right]^l C_i' - l(l+1)(l-1)\left[\frac{r'}{b}\right]^{l+1} D_i' \right].$$

The quantities A_i', B_i', C_i', D_i' appearing in the expression for $p_2''(i)$ are now defined according to:

$$A_i' = -b^{l-3} A_i \,,\, B_i' = -\frac{1}{b^{l+4}} B_i \,,\, C_i' = r'^{\,l-1} C_i \,,\, D_i' = \frac{1}{r'^{\,l+2}} D_i \,.$$

If one sets $r=a$ in equation (A22), and substitutes (B6) and (B7) into the resulting equation, the following expression for the self-gravitating potential is obtained:

$$(\phi_1)_l^m(a) = \frac{4\pi a G}{2l+1} \int_b^a G_l(r';\gamma,d)\,(\rho_1)_l^m(r')\,dr' \,, \tag{B8}$$

where,

$$G_l(r';\gamma,d) = \left[\frac{r'}{a}\right]^{l+1} \left\{ 1 + \left[(1-K_a)(1+K_b) + K_a K_b \left[\frac{b}{a}\right]^{2l+1} \right]^{-1} \left[\left[1+K_b\left[1-\left[\frac{b}{a}\right]^{2l+1}\right]\right] \cdot \left[K_a + \left[\frac{a}{r'}\right]^{l+2} K_l^a(r') \right] - \left[\frac{b}{r'}\right]^{l+2} \left[K_b \left[\frac{b}{r'}\right]^{l-1} + K_l^b(r') \right] \right] \right\}, \tag{B9}$$

where the constants K_a and K_b are defined in equation (A23). The spherical harmonic coefficients of the non-hydrostatic geoid $(Ge)_l^m$ are directly obtained from (B8) by dividing by g_o:

$$(Ge)_l^m = \frac{3}{2l+1}\,\frac{1}{\bar\rho} \int_b^a G_l(r';\gamma,d)\,(\rho_1)_l^m(r')\,dr' \,,$$

where $\bar\rho$ is the average density of the Earth and $G_l(r';\gamma,d)$ is the dimensionless geoid kernel given in equation (B9).

References

Anderson, O.L., Schreiber, E., Lieberman, R.C. and Soga, N., 1968. Some elastic constant data on minerals relevant to geophysics, *Rev. Geophys. Space Phys.*, **6**, 491-524.

Arkani-Hamed, J., and Toksöz, M.N., 1984. Thermal evolution of venus, *Phys. Earth Planet. Int.*, **34**, 232-250.

Backus, G., 1958. A class of self-sustaining dissipative spherical dynamos, *Annals of Physics*, **4**, 372-447.

Chandrasekhar, S, 1961. *Hydrodynamic and Hydromagnetic Stability*, Oxford, University Press.

Dziewonski, A.M., 1984. Mapping the lower mantle: determination of lateral heterogeneity in P velocity up to degree and order 6, *J. Geophys. Res.*, **89**, 5929-5952.

Forte, A.M. and Peltier, W.R., 1987. Plate tectonics and aspherical earth structure: the importance of poloidal-toroidal coupling, *J. Geophys. Res.*, in press.

Hager, B.H., 1984. Subducted slabs and the geoid: constraints on mantle rheology and flow, *J. Geophys. Res.*, **89**, 6003-6015.

Hager, B.H. and O'Connell, R.J., 1978. Subduction zone dips and flow driven by the plates, *Tectonophysics*, **50**, 111-134.

Lerch, F.S., Klosko, S.M., Laubscher, C.E. and Wagner, C.A., 1979. Gravity model Improvement using GEOS 3 (GEM9 and GEM10), *J. Geophys. Res.*, **84**, 3897-3916.

Minster, J.B. and Jordan, T.H., 1978. Present-day plate motions, *J. Geophys. Res.*, **83**, 5331-5354.

Nakiboglu, S.M., 1982. Hydrostatic theory of the earth and its mechanical implications, *Phys. Earth Planet. Int.*,**28**, 302-311.

O'Connell, R.J., 1971. Pleistocene glaciation and the viscosity of the lower mantle, *Geophys. J. R. Astr. Soc.*, **23**, 299-327.

Parsons, B. and Daly, S., 1983. The relationship between surface topography, gravity anomalies, and the temperature structure of convection, *J. Geophys. Res.*,**88**, 1129-1144.

Peltier, W.R., 1982. Dyamics of the ice age earth, *Adv. Geophys.*,**24**, 1-146.

Peltier, W.R., 1985. Mantle convection and viscoelasticity, *Ann. Rev. Fluid Mech.*,**17**, 561-608.

Peltier, W.R., Drummond, R.A. and Tushingham, A.M., 1986. Post-glacial rebound and transient lower mantle rheology, *Geophys. J. R. Astr. Soc.*,**87**, 79-116.

Ricard, Y., Fleitout, L. and Froidevaux, C., 1984. Geoid heights and lithospheric stresses for a dynamic earth, *Ann. Geophys.*,**2**, 267-286.

Richards, M.A., and Hager, B.H., 1984. Geoid anomalies in a dynamic earth, *J. Geophys. Res.*,**89**, 5987-6002.

Woodhouse, J.H., and Dziewonski, A.M., 1984. Mapping the upper mantle: three-dimensional model of earth structure by inversion of seismic waveforms, *J. Geophys. Res.*,**89**, 5953-5986.

Wu, P., and Peltier, W.R., 1983. Glacial isostatic adjustment and the free air gravity anomaly as a constraint on deep mantle viscosity, *Geophys. J. R. Astr. Soc.*,**74**, 377-450.

A.M. FORTE and W.R. PELTIER, Department of Physics, University of Toronto, Toronto ON M5S 1A7, Canada.

Chapter 14

Lithospheric thickness, Antarctic deglaciation history, and ocean basin discretization effects in a global model of postglacial sea level change

W.R. Peltier

The global model of postglacial relative sea level variations which has been developed over the past decade is employed to investigate the constraints which it may be invoked to place on the timing of the deglaciation of West Antarctica. The analyses presented here confirm the suggestion of Wu and Peltier (1983) that the model of this event employed in Clark and Lingle (1979) may be simply modified to rectify the misfits between theory and observation that otherwise obtains at southern hemisphere sites. A large number of southern hemisphere relative sea level data are shown to require that the retreat of Antarctic ice was essentially contemporaneous with the retreat of Northern Hemisphere ice. Sensitivity tests are performed which demonstrate that this result is relatively insensitive to the discretization employed to represent the ocean basins; the only exception to this general rule corresponds to some coastal sites at which a trade-off is revealed between the delay of West Antarctic melting and the thickness of the lithosphere required to reconcile the observed local variations of relative sea level. At such sites, which are all in the far field of the ice sheets, some attention must be paid to the accuracy of the local finite element representation of the oceans. Otherwise ocean basin discretization effects in previously described integrations of this global model are shown to be small.

N.J. Vlaar, G. Nolet, M.J.R. Wortel & S.A.P.L. Cloetingh (eds), Mathematical Geophysics, 325-346.

1. Introduction

Over the past decade, beginning with the paper by Peltier (1974), a variety of different geophysical observations have been shown to be intimately related to the earth's dynamical response to the melting of the great continental ice sheets which last achieved their maximum volumes approximately 18,000 years ago. This last deglaciation event of the current ice age took approximately 10,000 years to complete and by approximately 7,000 years ago the coverage of the Earth's continental surface by ice had been reduced to that which characterizes the planet at present. Because of the extraordinary amount of water mass which had to be removed from the ocean basins to build these ice sheets, equivalent to a sea level fall of approximately 120 m, the dynamical effects induced by this surface loading event were such as to leave easily discernable evidence in the geological record in the form of relative sea level variations, the timing of which may be accurately constrained using ^{14}C dating methods. In addition to these data from Quaternary geomorphology, the memory of the deglaciation event is also recorded in free air gravity anomalies which are found today over the continental regions which were once ice covered, and in two anomalies of earth rotation. The latter consist firstly of the so-called nontidal acceleration of planetary rotation which was originally inferred on the basis of analyses of ancient eclipse data (e.g. Müller and Stephenson 1973) and which has recently been reconfirmed through analysis of laser ranging data for the LAGEOS satellite (Yoder et al. 1983, Rubincam 1984). The second rotational datum which has been shown to be unambiguously connected to the deglaciation event is the secular drift of the rotation pole with respect to the surface geography which was first revealed by the pzt data of the International Latitude Service (Vincente and Yumi 1969, 1970) and which has also recently been reconfirmed using Very Long Baseline Interferometry observations (Carter 1986). The geophysical theory which has been developed to relate all of these observations to the single cause of deglaciation, although based upon the viscoelastic field theory of Peltier (1974), has required the development of special additional ingredients for application to the understanding of each of these different physical signatures of the response.

In the case of the relative sea level data, methods had to be developed to enable an accurate prediction to be made of the time dependent separation of the surface of the solid earth from the surface of the ocean (the geoid). This required that careful account be taken of the perturbations of the gravitational potential induced by both the ice unloading and ocean loading components of the deglaciation process. Farrell and Clark (1976) showed how the methods of Peltier (1974) and Peltier and Andrews (1976) could be extended to produce a gravitationally self-consistent description of the isostatic adjustment process. This extension of the theory was first applied by Clark et al. (1978) and Peltier et al. (1978) who employed the deglaciation history tabulated in Peltier and Andrews (1976) called ICE-1 to implement the solution of the integral equation which describes relative sea level change in this extended theory (see equation 1 of the present paper). These initial calculations demonstrated that, although much of the relative sea level variability over the world's oceans could be explained with this theory, there were large systematic errors between theory and observation in several locations. In particular, predicted sea level variations in the southern oceans were shown to disagree sharply with the predictions of the ICE-1 melting chronology (which neglected any meltwater input from the deglaciation of Antarctica) by predicting the appearance of raised beaches 2 kyr prior to their actual

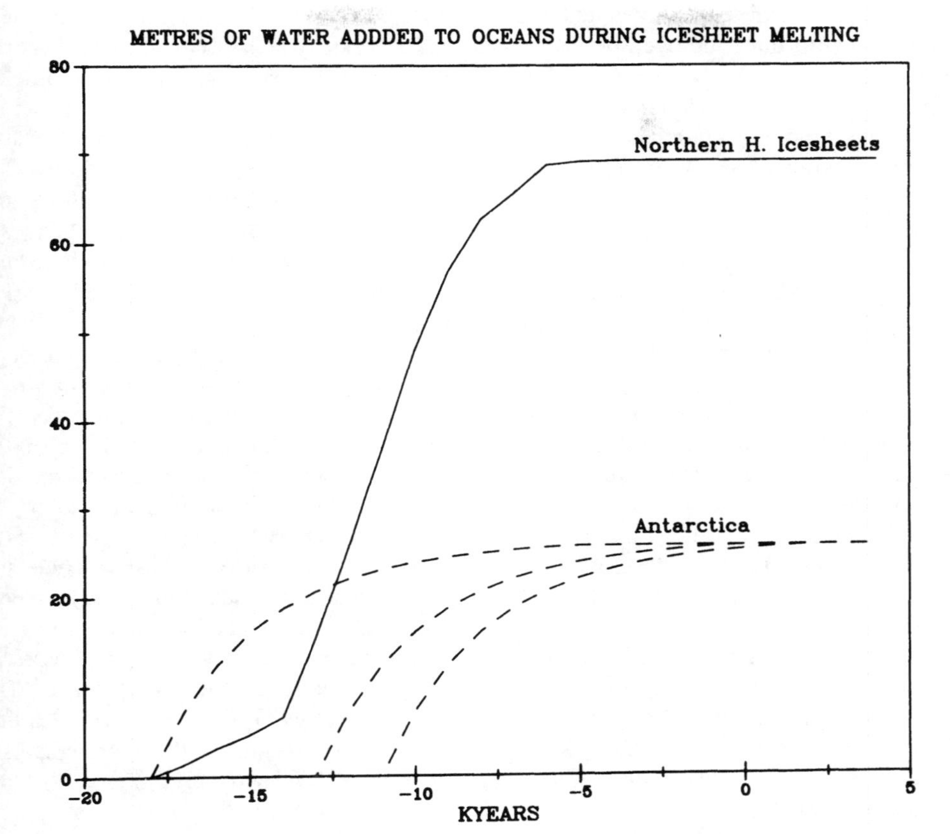

Figure 1. Contribution to the global eustatic rise of sea level due to the melting of the North American ice complexes of Laurentia and Fennoscandia (solid line) according to the ICE-2 melting chronology of Wu and Peltier (1983) and the contribution from Antarctica according to three different scenarios (the dashed lines). The earliest melting scenario shown is that from the ICE-2 chronology, the origins of whose Antartic component are in Hughes (1981). The two addition scenarios correspond to delays of the Antarctic component of 5 kyr and 7 kyr, the latter being the one preferred on the basis of the analyses presented in this paper.

appearance near 6 kbp (thousands of years before present). There were also large misfits between theory and observations in the regions immediately surrounding the main ice sheets. In Clark et al. the former of these misfits was corrected by delaying the melting of Northern Hemisphere ice by 2 kyr in marked violation of the ^{14}C controlled disintegration isochrones. Wu and Peltier (1983) showed that the latter misfits were a simple function of distance away from the ice margin and Peltier (1984, 1986) demonstrated that these misfits could be entirely removed by increasing the thickness of the continental lithosphere to a value somewhat in excess of 200 km. It was also suggested in Wu and Peltier (1983) that a more reasonable explanation of the misfits in the southern oceans (which Clark et al. had corrected by delaying Northern Hemisphere melting) might be simply to delay the melting of Antarctic ice which they tabulated in a model called ICE-2 whose Antarctic component was identical to that employed by Clark and Lingle (1979) which was itself based on the analysis reported in Hughes (1981). Although Wu and Peltier suggested that such a delay would provide a simple explanation of sea level misfits in the southern oceans, they did not experiment with such models. In this paper I will describe a new set of extensive numerical computations which have been performed to demonstrate the much improved fit to the global record of relative sea level change which this modified scenario allows and will comment on its implications with respect to current theories of paleoclimatic change.

Figure 1 shows the ice volume history for the separate Northern Hemisphere and Antarctic portions of the ICE-2 model with three separate versions of the deglaciation history of Antarctica shown as dashed lines. The Hughes (1981) model is the left-most of the dashed lines on the figure and corresponds to the assumption that Antarctic ice melted catastrophically beginning at 18 kbp. Also shown are versions of this history in which Antarctic melting is deplayed by 5 kyr and 7 kyr. As I will show in the present paper, the model with 7 kyr delayed Antarctic melting does in fact reconcile the misfits in the southern oceans very nicely when the northern hemisphere component of the melting chronology is fixed to that of the ICE-2 tabulation. This shift in the melting chronology of West Antarctic ice is also required to conform to the theory of the ice age cycle of glaciation and deglaciation developed in Peltier (1982), Peltier and Hyde (1984), Hyde and Peltier (1985, 1986) and Peltier (1986). In this version of the Milankovitch astronomical theory of paleoclimatic change, the basic 10^5 year cyclic oscillation of Northern Hemisphere ice volume is produced by the nonlinear interaction between the processes of ice sheet accumulation and flow and the process of glacial isostatic adjustment. The synchronous oscillation of Southern Hemisphere ice is supposed to be induced via coupling of the West Antarctic ice sheet to Northern Hemisphere events through the intermediary of the sea level variations induced by the Northern Hemisphere cryospheric volume fluctuations. Such strong coupling is expected because the West Antarctic ice sheet is a marine ice sheet whose base is well below sea level over much of its area. It is therefore highly susceptible to sea level variations.

Although the explicit illustration of the above described ideas will form the basis of the present paper it is important to keep simultaneously in view the other observational data which have also been successfully explained by the viscoelastic field theory of the isostatic adjustment process which is basic to the accurate prediction of relative sea level histories. For example, analyses of the free air gravity anomalies associated with the isostatic adjustment process has established (Peltier 1981, Peltier and Wu 1982, Wu and Peltier 1983, Peltier 1985, Peltier et al. 1986) that these data may be fit by the same almost uniform mantle viscosity model required by the sea level data only, apparently, if the mantle density stratification is assumed to be, at least in part, non-adiabatic. In fact it has been demonstrated in these articles that the almost isoviscous viscosity profile preferred by the RSL data is accepted as plausible by the free air gravity observations only if the density jumps associated with the seismic discontinuities at 420 and 650 km depth are assumed to behave in this fashion. As discussed in Peltier (1985) this is not necessarily incompatible with the interpretation of these discontinuities as equilibrium phase transformations since, to the extent that the phase boundaries are sharp (close to univariant), they will behave as chemical boundaries on the timescale of postglacial rebound. These analyses represent the first successful interpretation of the free air gravity data with a model that simultaneously reconciles RSL variability.

The explanations of the previously mentioned anomalies of earth rotation have also been successfully integrated into this same general picture. The interpretation of the LAGEOS observation of $\dot{J}_2$, which is directly related to the non-tidal acceleration of the rate of axial rotation, in terms of glacial isostasy, was in fact anticipated by the prediction of this effect in Peltier (1982) prior to the first published reports of the observation by Yoder et al. (1983). Detailed analysis of the LAGEOS data were described in Peltier (1983, 1985) and Peltier et al. (1986). As pointed out in Peltier (1985) the $\dot{J}_2$ observation is particularly useful because this physical datum is not particularly sensitive to either mantle non-adiabaticity or lithospheric thickness so that it serves as a particularly useful constraint upon the ratio of upper mantle to lower mantle viscosity. The above cited analyses also agreed with the RSL and free air gravity data in requiring a weak stratification of mantle viscosity with an asthenospheric value close to 10^{21} Pa s and a mesospheric value slightly higher but modestly so and near 2 x 10^{21} Pa s.

The second of the rotational data, the ongoing drift of the rotation pole towards eastern Canada at a rate near 0.95 (± 0.15) degrees per million years, is probably the most interesting phenomenon of all from a fundamental physical point of view. Since publication of the paper by Munk and Revelle (1952) it has generally been believed that no such wander of the rotation-pole with respect to the surface geography could occur so long as the planet was not actively subject to changes in its surface load. The Munk and Revelle analysis therefore led its authors to suggest that the observed secular drift of the pole was probably due to present day continued melting of Greenland and/or Antarctic ice. Their paper has continued to be invoked in favour of such an effect in very recent papers dealing with climatic change (e.g. Barnett 1981) even though the best present evidence suggests that both of these ice sheets are presently in equilibrium (e.g. Meier 1984). Reanalysis of this problem by Peltier (1982), Peltier and Wu (1983), Wu and Peltier (1984), and Peltier et al. (1986) has demonstrated the existence of a serious flaw in the theoretical analysis of Munk and Revelle, however, which is such as to invalidate their conclusion. As demonstrated in these papers, the homogeneous viscoelastic model of the earth employed by Munk and Revelle to compute the polar motion induced by surface loading has the extraordinary characteristic that the separate contributions to the forcing due to glacial isostasy, and to the centrifugal force associated with the changing rotation, exactly annihilate one another at any time when the surface load is steady. In terms of this model, therefore, no polar wander should be observed at a time like the present when this condition apparently holds. The above cited articles demonstrated, however, that when realistic radial stratification of the planet was taken into account then these two contributions to the forcing no longer cancelled one another and it became possible to explain the observed secular shift of the rotation pole with the same model of the radial viscoelastic structure of the planet as was shown to be required by the previously discussed isostatic adjustment data. The nature of the theoretical prediction, however, is such that the datum is strongly sensitive to all aspects of the radial viscoelastic structure, including lithospheric thickness and internal mantle buoyancy.

All of the analyses cited above have been based upon the assumption that the viscoelastic relaxation process could be described in terms of a Maxwell model of viscoelasticity in which the instantaneous response is Hookean elastic and the final response Newtonian viscous. Although some recent work on the inference of mantle viscosity on the basis of isostatic geoid anomalies (Hager 1984, Richards and Hager 1984, Forte and Peltier 1986) suggests that the convection process may be governed by a viscosity which is higher than that inferred through analyses of the glacial isostatic adjustment process, discussion in Peltier (1985, 1986) and Peltier et al. (1986) demonstrates that the implied importance of transient relaxation on the timescale of isostatic adjustment may be understood only if the elastic defect associated with the transient component is large. In this limit the general Burgers body description of the relaxation process degenerates to a Maxwell model governed by an effective viscosity $\nu_{eff} = \nu_1\nu_2/(\nu_1 + \nu_2)$ in which ν_1 is the steady state viscosity which governs the convection process and ν_2 is the viscosity which governs the timescale on which the steady state behaviour is approached. The work of Forte and Peltier (1986, 1987) rather strongly suggests that $\nu_2 \approx \nu_1$ and therefore that the viscosity which governs the isostatic adjustment process should be only a factor of about two lower than that which governs convection. In the present paper all of the analyses will therefore continue to employ the Maxwell representation of the viscoelastic relaxation process but we must keep in mind that the viscosity structure inferred from the sea level data is to be interpreted as an effective structure.

2. The global model of relative sea level change

As recently reviewed in detail in Wu and Peltier (1983), the variations of relative sea level $S(\theta,\phi, t)$ forced by a global deglaciation event may be computed as solutions to the integral equation:

$$S = \rho_I \frac{\phi}{g} * L_I + \rho_w \frac{\phi}{g} * S - \frac{1}{A_o} < \rho_I L_I * \frac{\phi}{g} + \rho_w S * \frac{\phi}{g} >_o - \frac{M_I(t)}{\rho_w A_o} \tag{1}$$

in which ρ_I and ρ_w are the densities of ice and water respectively, ϕ is the Green function for the perturbation of the gravitational field at the earth's surface induced by the addition onto the surface of a point load of 1 kg mass, g is the surface gravitational acceleration of the unperturbed spherical equilibrium configuration, A_o is the assumed constant area of the earth's oceans, the angle bracket $< >_o$ indicates integration over the surface of the oceans, and $M_I(t)$ is the assumed known mass loss history of the ice sheets, one example of the form of which (expressed in terms of equivalent ocean depth) was shown in Figure 1. In (1) the space-time convolution operation is represented by *. The sea level equation (1) is an integral equation because the unknown field $S(\theta,\phi,t)$ appears both on the left hand side and under the convolution integrals on the right hand side.

In order to solve (1) it must be discretized in both space and time. For all of the calculations which I have performed with (1) the temporal evolution of the field S has always been sampled at 10^3 year intervals as has the input field L_I which is the assumed known history of continental ice sheet disintegration. The latter is represented by

$$L_I(\mathbf{r},t) = \sum_{l=0}^{P} \delta L_l(\mathbf{r}) H(t - t_l) \tag{2}$$

in which H is the Heaviside step function and $\delta t = t_l - t_{l-1} = 10^3$ yrs. With this

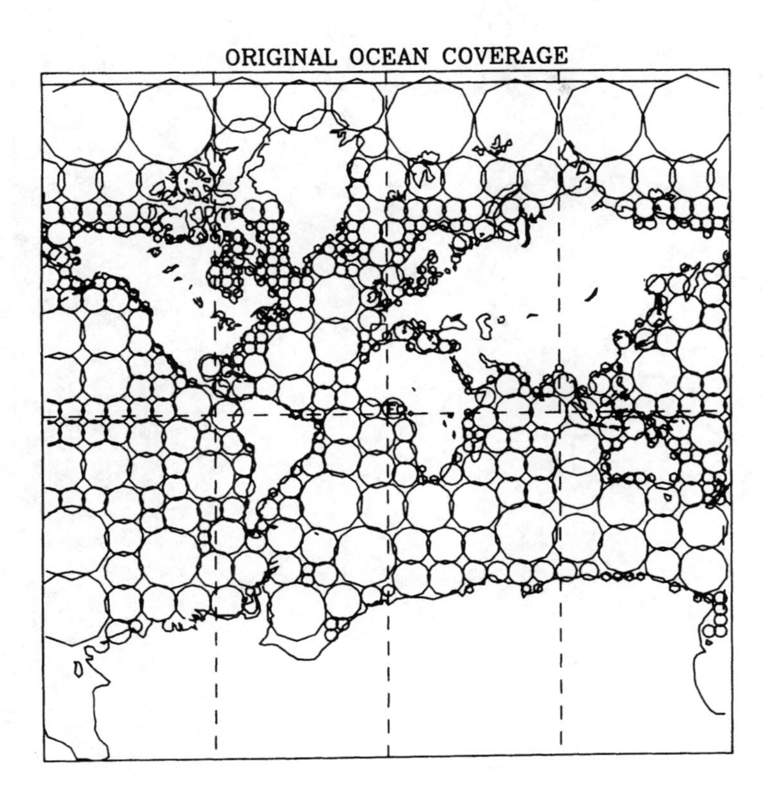

Figure 2. Finite element coverage of the oceans employed in the calculations of Peltier and Andrews (1976), Clark et al. (1978) and Wu and Peltier (1983).

discretization the sea level equation (1) becomes:

$$
\begin{aligned}
S(\mathbf{r},t) = & \iint_{ice} \rho_I L_I(\mathbf{r}',t)\, G^E\, (\mathbf{r}-\mathbf{r}')\, d^2r' \\
& + \iint_{ocean} \rho_w\, S\, (\mathbf{r}',t)\, G^E\, (\mathbf{r}-\mathbf{r}')\, d^2r' \\
& + \rho_I \sum_{l=0}^{P} \iint_{ice} \delta L_l\, (\mathbf{r}')\, G^{H,V}\, (\mathbf{r}-\mathbf{r}', t-t_l)\, d^2r' \\
& + \rho_w \sum_{l=0}^{P} \iint_{ocean} \delta S_l\, (\mathbf{r}')\, G^{H,V}\, (\mathbf{r}-\mathbf{r}', t-t_l)\, d^2r' \\
& - C(t)\,, \qquad (3)
\end{aligned}
$$

in which $\delta\, S_l$ is the increment in bathymetry at $t = t_l$ and the convolution of the Heaviside

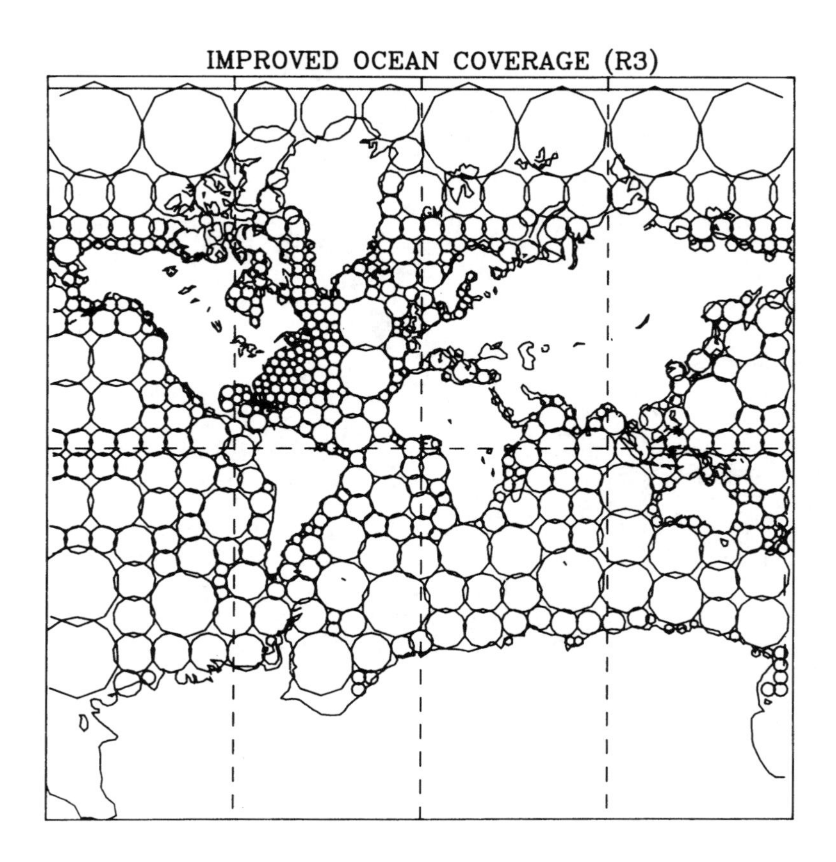

Figure 3. Refined finite element discretization of the ocean basins employed for the purposes of this paper. Comparison with the earlier model shown in Figure 2 will reveal that the refinement is confined to the western North Atlantic and, to a lesser degree, to the Western South Atlantic adjacent to the east coast of South America.

step functions in (1) with ϕ/gH produces $\frac{\phi}{g} * H = G^H = G^E + G^{H,V}$ for the kernels of the convolution integrals. The expansion of the Green function into the sum of elastic (G^E) and viscous $(G^{H,V})$ parts follows Peltier (1974). The function C(t) in (6) is given explicitly by

$$C(t) = \frac{1}{A_o}\Bigg[\iint_{ocean} d^2r \iint_{ice} \rho_I L_I (r',t)\, G^E (\mathbf{r}-\mathbf{r}')\, d^2r'$$

$$+ \iint_{ocean} d^2r \iint_{ocean} \rho_w\, S(\mathbf{r}',t)\, G^E (\mathbf{r}-\mathbf{r}')\, d^2r'$$

$$+ \iint_{ocean} d^2r \left\{ \sum_{l=0}^{P} \iint_{ice} \rho_I \delta L_l(\mathbf{r}')\, G^{H,V} (\mathbf{r}-\mathbf{r}', t-t_l)\, d^2r' \right\}$$

$$+ \iint_{ocean} d^2r \left\{ \sum_{l=0}^{P} \iint_{ocean} \rho_w \delta S_l(\mathbf{r}') G^{H,V} (\mathbf{r}-\mathbf{r}',-t-t_l)\, d^2r' \right\} \Bigg]$$

$$+ S_{EUS}(t)\,, \tag{4}$$

in which the "eustatic" water rise is defined as:

$$S_{EUS}(t) = \frac{\rho_I}{\rho_w A_o} \iint_{ice} L_I\, (\mathbf{r}',t)\, d^2r' \tag{5}$$

In order to complete the formulation of the numerical solution of (1) it remains to discretize the spatial dependence. In order to accomplish this what we do (Peltier and Andrews 1976, Peltier et al. 1978) is to divide the surface of the earth into a number of finite elements. The active elements cover all of the ocean surface and the glaciated regions of the continents. For the calculations which I shall discuss here, the ice covered portions of the surface are covered by 236 elements. Figure 1 of Wu and Peltier (1983) shows the discretization employed for North America and Europe and will not be reproduced here. The ocean basins are similarly covered with circular cap elements, two different versions of which are displayed in Figures 2 and 3. The former of these has the same density of coverage as was employed in the calculations of Wu and Peltier (1983), whereas the latter has a considerably more refined grid, particularly along the east coasts of North and South America.

If we denote the area of the jth active element by E_j and assume that the loads on each element are piecewise constant in time then equation (3) may be written in the following discrete form:

$$S_{ip} = \rho_w G^E_{ij} S_{jp} + \rho_I G^E_{ij} I_{jp} + \rho_w G^{H,V}_{ijpl}\, \delta S_{jl}$$

$$+ \rho_I\, G^{H,V}_{ijpl}\, \delta I_{jl} + \frac{\rho_I}{\rho_w A_o} E_j\, I_{jp}$$

$$- \frac{1}{A_o} \left\{ E_k \left[\rho_w G^E_{kj} S_{jp} + \rho_I G^E_{kj} I_{jp} + \rho_w G^{H,V}_{kplj}\, \delta S_{jl} + \rho_I G^{H,V}_{kplj}\, \delta I_{jl} \right] \right\} \tag{6}$$

In writing (6) the summation convention has been employed. For terms involving S_{jp} and δS_{jp}, j is summed over the finite elements in the ocean from 1 through N(= 470). For terms involving I_{jp} and δI_{jp}, j is summed over the elements making up the ice sheets from r through M(= 236). Terms involving δS_{jp} and δI_{jp} have p summed from 1 to P_o where P_o is the number of time intervals since deglaciation commenced. The matrices in (6) have the following definitions:

— $S_{ip} = S(\mathbf{r}_i, t_p)$, total relative sea level change at r_i by time t_p

— $I_{ip} = L_I(\mathbf{r}_i, t_p)$, total thickness of ice removed at r_i by time t_p

— $\delta S_{ip} = \delta S(\mathbf{r}_i, t_p)$, increment of seawater thickness at $\mathbf{r}_i$ for time $t_p = S_{ip} - S_{i(p-1)}$

— $\delta I_{ip} = \delta L_p(\mathbf{r}_i)$, increment of ice thickness at $\mathbf{r}_i$ for time t_p

— $G_{ij}^E = \int^{E_j}\int G^E(\mathbf{r}_i - \mathbf{r}')\, d^2r'$, the elastic response at $\mathbf{r}_i$ due to the jth element

— $G_{iplj}^{H,V} = \int^{E_j}\int G^{H,V}(\mathbf{r}_i - \mathbf{r}', t_p - t_l)\, d^2r'$, the viscous response at $\mathbf{r}_i$ for time t_p due to Heaviside loading of the jth element at time t_l.

It will be noted that the numbers G_{ij}^E and $G_{iplj}^{H,V}$ are computed on the basis of the assumption that each of the elements consists of a spherical cap as shown in Figures 2 and 3. This assumption greatly reduces the computational complexity since the matrix elements may be computed by interpolation in a sparse table constructed for a range of different cap radii which spans those employed in the model and for a range of distances inside and outside each of these caps. This is the main numerical trick employed in the model. The greatest advantage of this finite element technique over a spherical harmonics method is that one may refine the grid locally without paying a global price. This fact, coupled with the large errors associated with the application of a spherical harmonics technique due to Gibbs effect near the ice margins make this method optimal in the present application. In the new results to be described below, I will provide explicit examples of the magnitude of the discretization errors committed through the use of this method.

3. Illustrative results from the global model

In this section I will present results obtained from the global inversion of (6) which are intended to illustrate two different properties of the solutions, the first numerical and the second physical. First I shall discuss the impact on North American RSL predictions of the ocean grid refinement represented by the utilization of the finite element mesh shown in Figure 3 rather than that shown in Figure 2. Secondly, I shall describe a sequence of calculations using the refined ocean grid to investigate the impact of the different Antarctic melting scenarios described in Figure 1 on the relative sea level predictions for both the Northern and, more importantly, the Southern Hemispheres.

3.1 The influence of ocean grid discretization error on North American RSL predictions

Figures 4 and 5 together show a collection of 32 observed RSL curves represented by the solid crosses compared to the predictions of two models which differ from one another only in the finite element discretization employed for the oceans. The dashed lines are for the original ocean discretization whereas the solid lines are for the improved discretization. The main point which should be clear on the basis of these comparisons is that the influence of even rather substantial refinement of the ocean discretization upon the predicted relative sea level variations at sites near major ice masses is rather small and

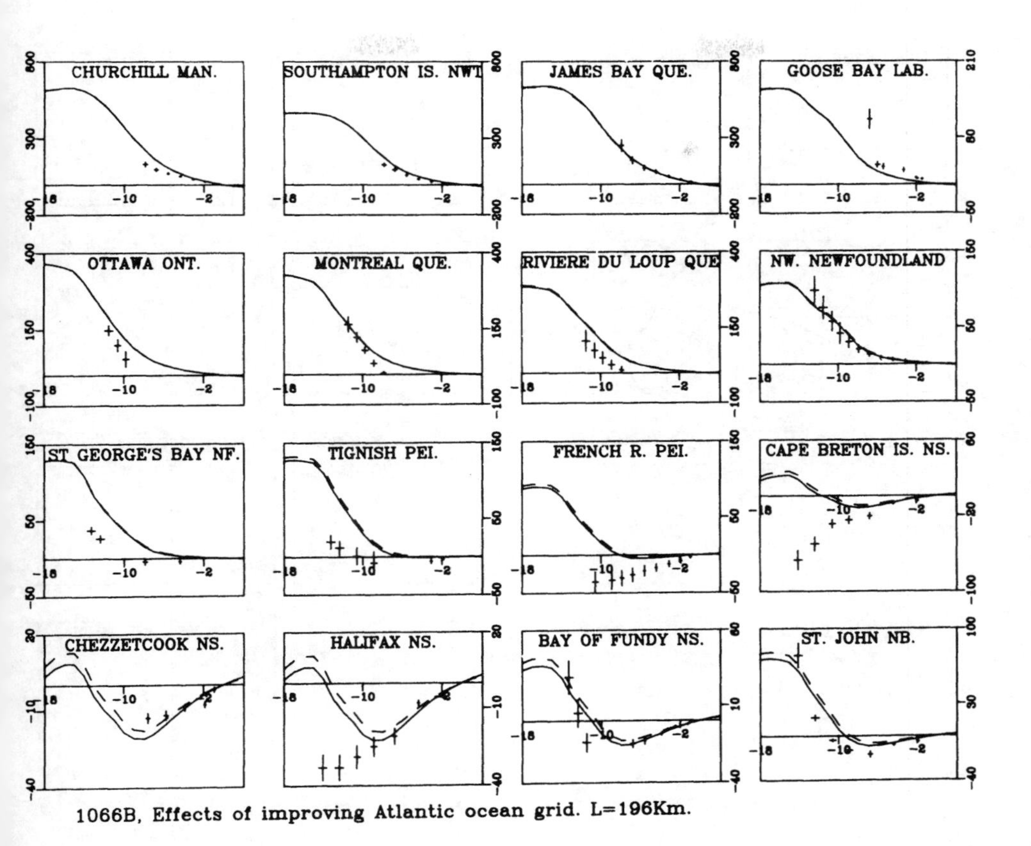

Figure 4. A comparison between predicted and observed (error bars) relative sea level data at sixteen North American sites illustrating the influence of the refinement of the finite element distribution in the Western North Atlantic represented by use of the mesh shown in Figure 3 (solid lines) rather than the mesh shown in Figure 2 (dashed lines). The elastic component of the earth model is 1066B, the lithospheric thickness is 196 km, the upper mantle viscosity is 10^{21} Pa s, and the lower mantle viscosity is 2×10^{21} Pa s.

restricted to positions near the ice margin such as Cape Breton Island, Chezzetcook, Halifax, and the Bay of Fundy in the Canadian province of Nova Scotia and to St. John, New Brunswick. Even at these locations, however, the differences are substantial only for the earliest times. Also modestly effected, and in the opposite sense, are sites in the Gulf of Mexico such as Matagorda Bay in Texas. The physical reason why the ocean discretization has so little influence at sites near the major ice sheets is that the response at such locations is strongly dominated by the deformation of the solid earth, and ocean loading effects play a relatively minor role.

All of the calculations displayed on these Figures, and all those to be discussed in the remainder of this article, have been based upon an earth model whose elastic structure is fixed to that of model 1066B of Gilbert and Dziewonski (1975), whose upper mantle has a

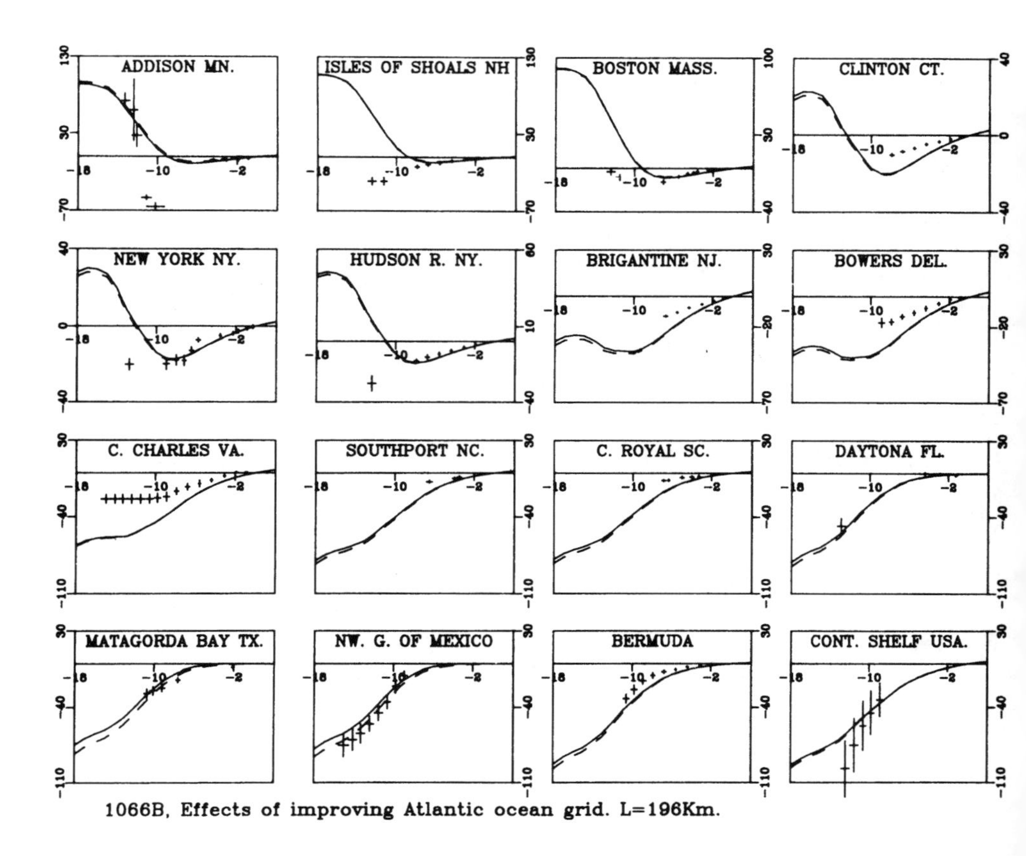

Figure 5 As in Figure 4 but for 16 additional North American relative sea level sites.

viscosity of 10^{21} Pa s and whose lower mantle has a viscosity of 2 x 10^{21} Pa s. This model has been shown to adequately reconcile all of the observation data discussed in the introduction to this paper. For the purpose of these initial calculations the thickness of the lithosphere has been fixed at 196 km. Effects due to the variation of this parameter on RSL data from the Southern Hemisphere will be discussed in a following sub-section.

3.2 The influence of the timing of Antarctic deglaciation

Figures 6, 7, and 8 show comparisons between observed and predicted relative sea level histories at 48 locations which are all well removed from the major ice sheets on Canada, Scandinavia, and Antarctica. At many of these sites, including Jonathan Point Belize, East Panama, Georgetown Guyana, Paramaribo Surinam, the Mekong Delta of Vietnam, the Huon Peninsula of Papua New Guinea, Townsville in Queensland Australia, Maackay Queensland, Moruya New South Wales, the Wariu Valley New Zealand, Christchurch, New Zealand, the East Caroline Islands, New Caledonia, and Oahu, Hawaii, there are substantial radio carbon dates from material of age up to 10 and in some cases more kyr.

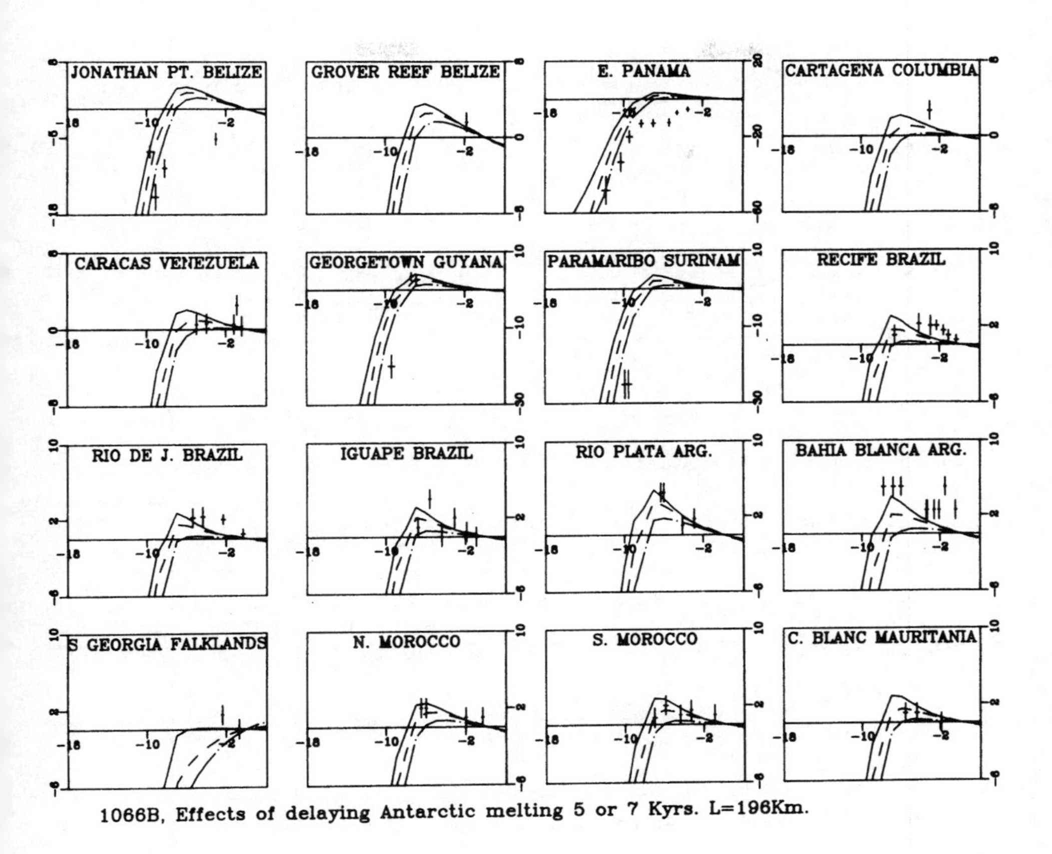

Figure 6. Comparison of predicted and observed (error bars) relative sea level histories at sixteen sites in the far field of the ice sheets illustrating the influence upon the predicted relative sea level histories of delaying the Antarctic component of the deglaciation chronology from that of the ICE-2 model listed in Wu and Peltier (1983). The solid curves are the predictions using the original melting chronology of ICE-2, the dashed curves are rsl predictions based upon a 5 kyr delay of the Antarctic component of melting, and the dash dotted curves are predictions employing a 7 ky delay of Antartic melting. The elastic structure of the earth model employed for the computations is 1066B, the lithospheric thickness is 196 km, the upper mantle viscosity is 10^{21} Pas , and the lower mantle viscosity is 2 x 10^{21} Pa s.

Such information provides extremely important constraints on the nature of the deglaciation history since the data substantially overlap the time in which ice sheet melting is still taking place. As will be clear by inspection of the fits to the observations, at every one of the above mentioned sites a marked improvement in the fit is achieved by delaying the melting of Antarctic ice by the seven thousand year maximum illustrated on Figure 1. In the absence of this alteration of the deglaciation history every predicted far field RSL curve is characterized by the appearance of raised beaches at 8 kbp rather than at 6 kyr bp as is observed, a misfit which was identified in Wu and Peltier (1983) and earlier in the articles of Clark et al. (1978) and Peltier et al. (1978). Many of the sites could tolerate even

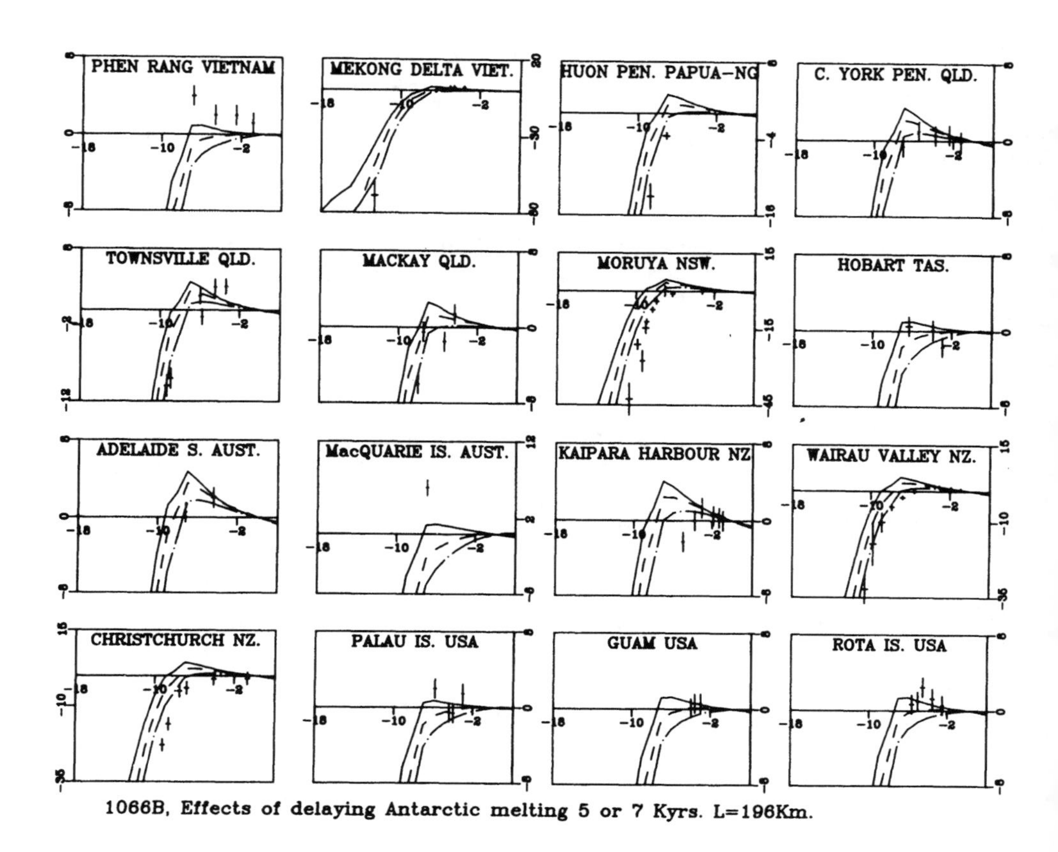

Figure 7. Same as Figure 6 but for 16 additional sites.

greater delay of the melting of Antarctic Ice than the 7 kyr maximum for which I have shown calculations on these Figures. It should also be kept in mind that it is only the Antarctic component of the deglaciation history which we are at liberty to adjust in this way since the Northern Hemisphere components of the melting chronology are very well constrained to the ICE-2 pattern by ^{14}C dates on material from terminal morraines.

Also evident by inspection of the comparisons shown on Figures 6-8, however, is the fact that this 7 kyr delay in the melting of Antarctic ice has disastrous effects on the fit to the RSL observations at a large number of other locations. For example at Recife, Rio De Janeiro, and Iguape Brazil, at Rio Plato, and Bahia Blanca, Argentina, in North and South Morroco, Cape Blane, Mauritania, and Rota Island, the South Gilbert Islands, New Caledonia, and Jarvis Island the delay of Antarctic melting completely eliminates the prediction of any raised beaches at all. Yet at every location raised beaches are actually observed. As we will demonstrate in the following subsection, much of the misfit induced at such sites when delayed Antarctic melting is assumed as simply rectifiable by thinning the lithosphere. This is one rather plausible explanation for the discrepancies at the above

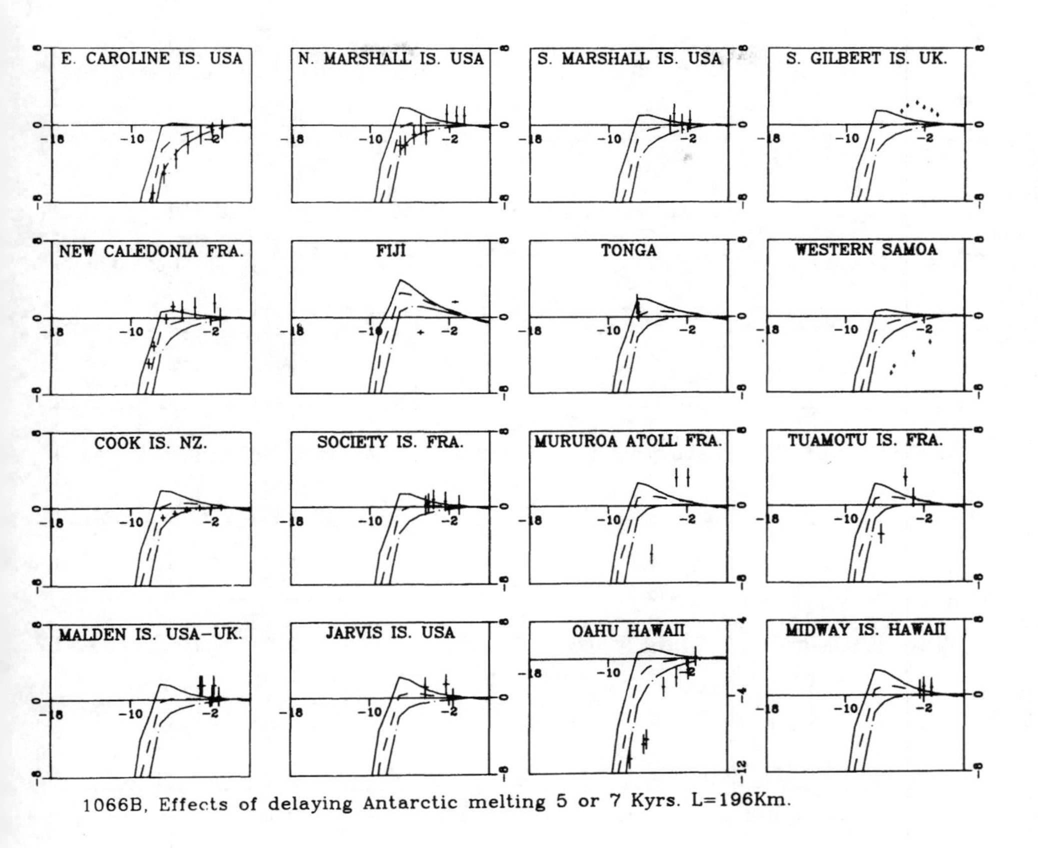

Figure 8. Same as Figure 6 but for 16 additional sites.

mentioned locations as the lithospheric thickness employed in the previously described calculations was near to but somewhat less than the value appropriate for the North American continental craton (e.g. Peltier 1984). It is certainly not appropriate, however, for the near surface structure of the earth in non-cratonic areas such as those whose RSL signatures we are concerned with there.

3.3 The influence of lithospheric thickness variations at far field RSL sites

Figures 9, 10, and 11 compare RSL predictions and observations at the same 48 sites discussed above in connection with the question of the preference of the data for a significant delay in the melting chronology of Antarctic ice. As will be clear by a careful inspection of all of these data, the effect of thinning the lithosphere to 71 km from 196 km is to increase the amount of emergence predicted for continental coastal sites in the far field of the ice sheets such as all of those along the east coast of South America including Recife, Rio De Janeiro and Iguape Brazil and Rio Plato, Argentina as well as at both North and South Morroco and Cape Blane, Mauritania and to thereby restore the good fit to the

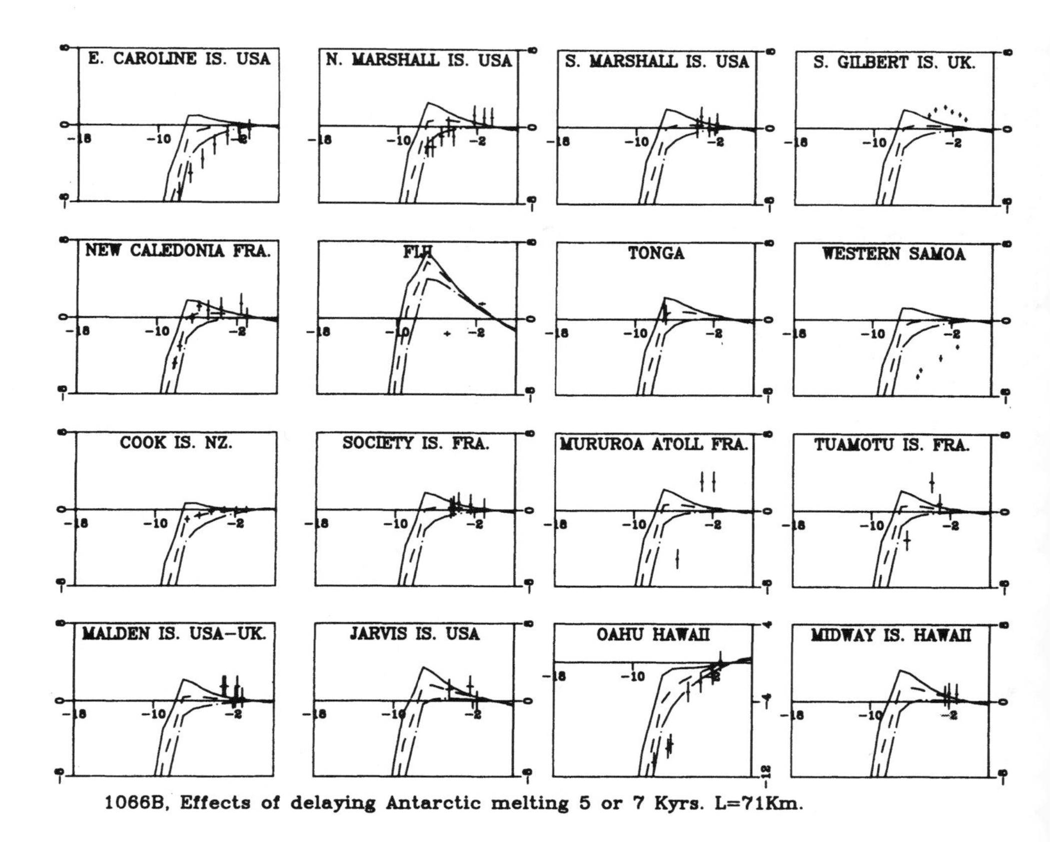

Figure 9. Comparison of predicted and observed (error bars) relative sea level curves at the same 16 sites shown on Figure 6 but with calculations employing an earth model whose lithospheric thickness is 71 km. As in Figure 6 the solid, dashed, and dash-dotted curves compare the predictions based upon the three scenarios of Antarctic melting illustrated in Figure 1.

data which was lost by delaying the melting of the Antartic ice sheet in order to remedy the misfits to the submergence curves extending back into the range of times in which deglaciation was still ongoing. It is therefore clear on the basis of the extensive new computations of relative sea level variability described here that this more reasonable scenario for southern hemisphere deglaciation is rather easily accommodated within the framework of the global model of relative sea level change if allowance is made for the trade-off between this effect and the effect of lithospheric thickness. The far field data do seem to require a much thinner and more conventional oceanic value for this parameter than that which has been found to be required to fit relative sea level data from the North American continent from sites beyond the margin of the Laurentian ice sheet. In order to close the cycle of new analyses described in this paper, we shall proceed to describe the influence of this new melting scenario for the Antarctic ice sheet upon RSL predictions for

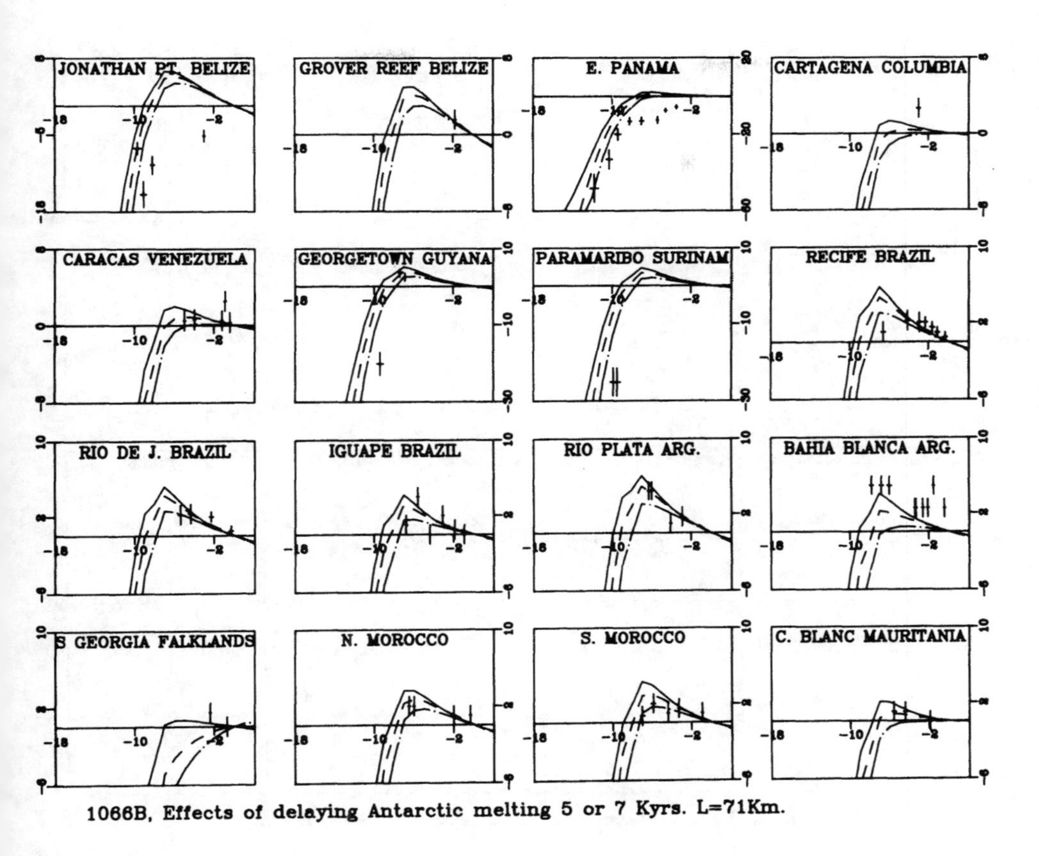

Figure 10. As in Figure 9 but for the same 16 sites as were employed in the thick lithosphere calculations illustrated on Figure 7.

sites in the near field of the Laurentian ice sheet.

3.4 Delayed Antarctic melting and North American relative sea level change

Figures 12 and 13 display predicted and observed relative sea level histories for 32 North American sites for the three model scenarios of Antarctic melting which were illustrated in Figure 1. The solid curve on each plate represents the prediction for the original scenario based upon the reconstruction of Hughes (1981) as employed by Clark and Lingle (1979) and tabulated in Wu and Peltier (1983). The dashed lines are predictions for a new model in which the Antarctic melting event is delayed by 5 kyr and the dash-dotted curve is for a 7 kyr delay.

Inspection of this suite of comparisons demonstrates a number of interesting aspects of the influence of the delayed southern hemisphere melting event on Northern Hemisphere sea level variations. The most apparent is that for sites within the margin of the ancient Laurentian ice mass like Churchill, Manitoba, James Bay, Montreal, and Riviere du Loup,

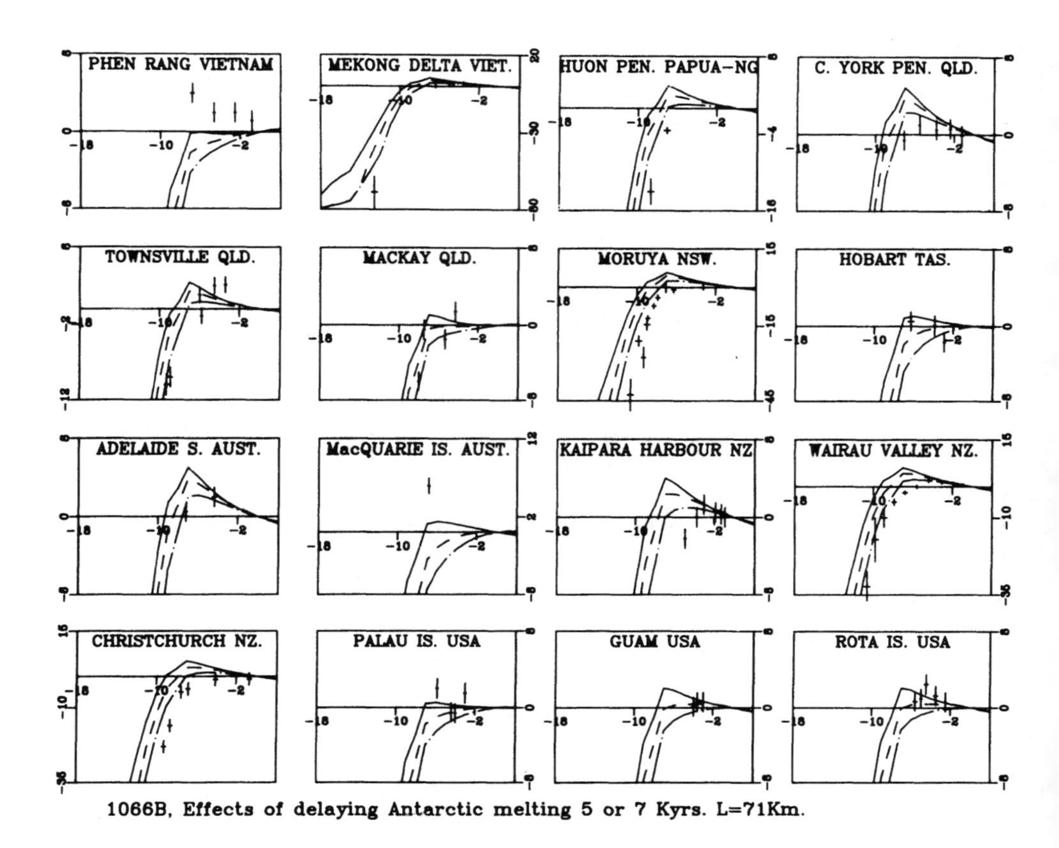

Figure 11. As in Figure 9 but for the same 16 sites as were employed in the thick lithosphere calculations illustrated on Figure 8.

Quebec, and Southampton Island in the North West Territories, the effect upon the RSL prediction is completely negligible. At such sites the response is so strongly dominated by the vertical motion of the solid earth that even for times within the period of active deglaciation the southern hemisphere variability has no noticeable influence.

This ceases to be the case as one moves progressively further away from the central Laurentide dome and across the location of the margin at ice age maximum. At these edge locations, the delayed melting of southern hemisphere ice diminishes the predicted uplift of the land at sites inside the margin and increases the predicted submergence at sites outside the ice sheet margin. However, at most locations the effect is significant only for ages in excess of about 7 kyr during the period of active deglaciation. For later times nearer the present, to which much of the ^{14}C data correspond, the effects are very small. Only at Halifax, Nova Scotia in Canada and at New York City and the Continental Shelf sites in the US does this influence appear to improve the comparison between the predicted and observed relative sea level data. At some other locations, such as Daytona, Florida, the

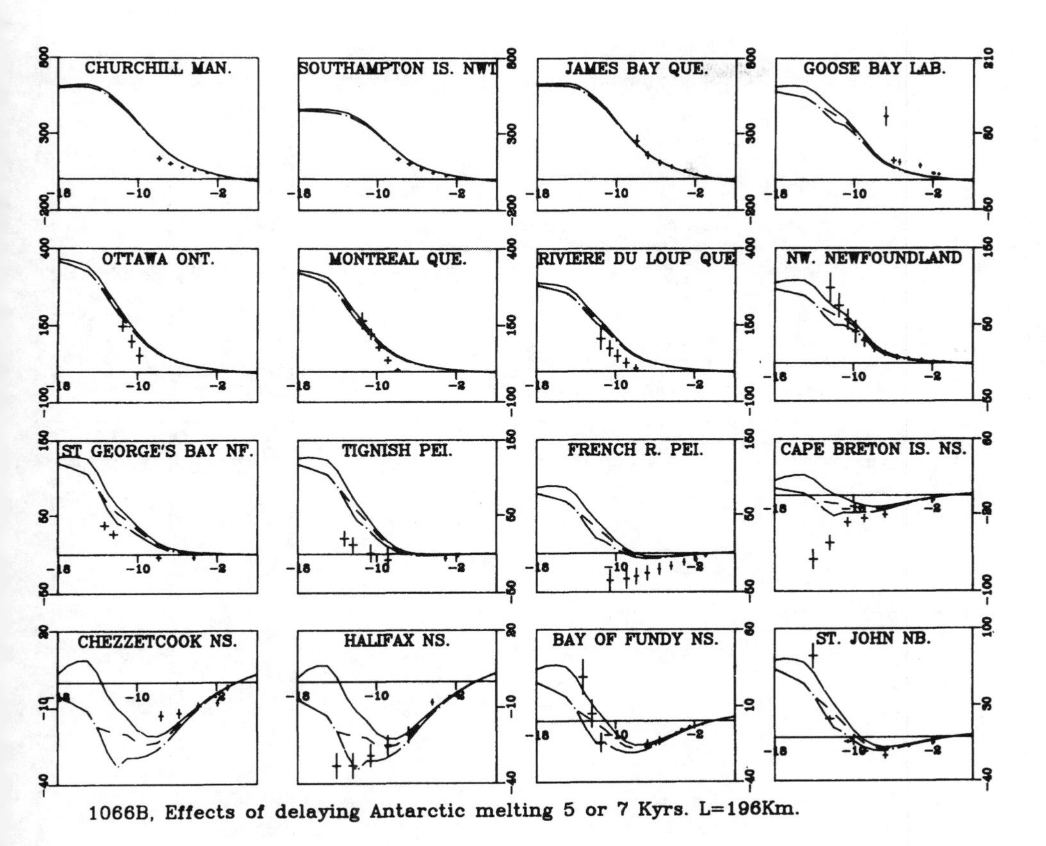

Figure 12. Comparisons of predicted and observed (error bars) relative sea level histories at 16 North American sites illustrating the influence of the delay of Antarctic melting upon the predictions at such near field locations. As in the previously described analyses, the computational results shown as solid, dashed, and dash dotted curves are predictions using the Antarctic melting chronology of ICE-2, and those with 5 ky and 7 kyr delays respectively. The point to note is that the influence of this variation is extremely small at sites with the margins of Laurentian ice and noticeable at most external sites only for times within the actual deglaciation phase which ended at about 7 kbp. Subsequent to this time, within which almost all of the ^{14}C data are found, the influence is small.

influence is somewhat adverse. Of greatest importance for our purposes here, however, is that the modification of the RSL prediction at sites near the crest of the glacial forebulge such as Clinton, Connecticut, Brigantine, New Jersey and Bowers, Delaware, is to further amplify the misfit whose sense would require a further increase of lithospheric thickness from the value of 196 km employed here. As first discussed in Peltier (1984), increasing the thickness of the lithosphere sharply diminishes the predicted submergence at sites near the forebulge crest. Since the model over predicts the submergence back to 10 kyr bp at such locations, it is clear that the 196 km thickness which has been employed for these calculations might have to be further increased unless errors in the deglaciation chronology

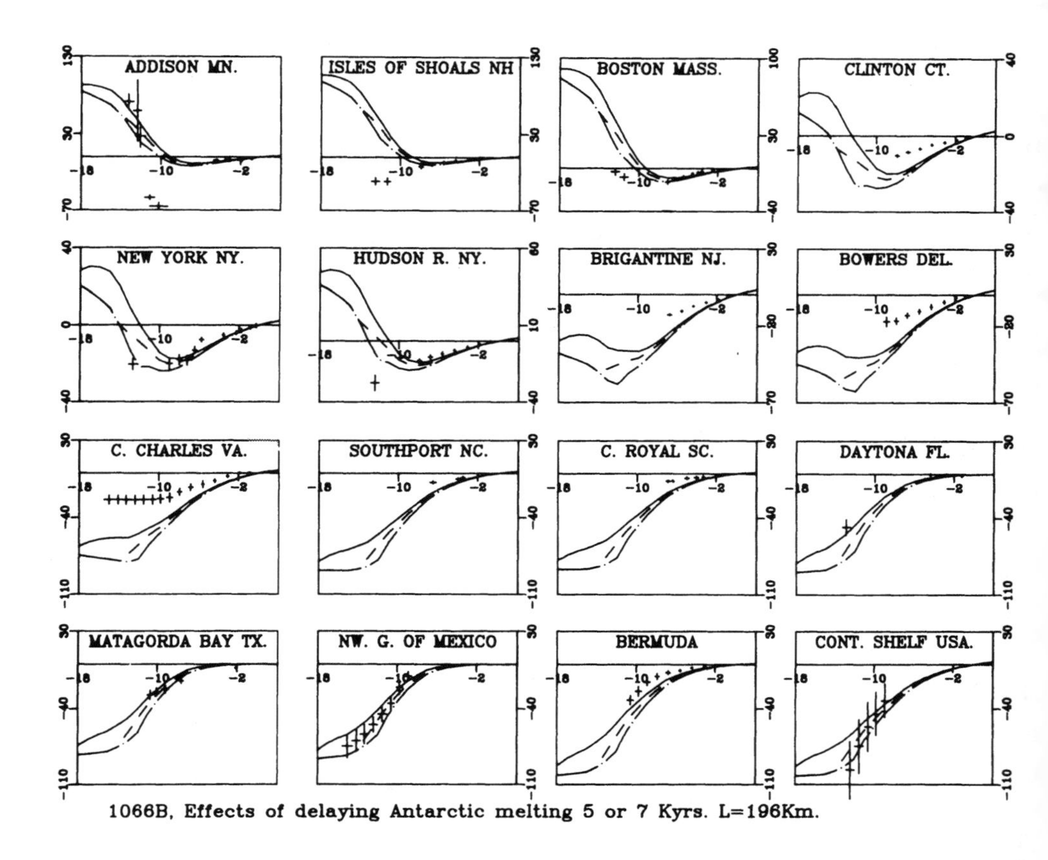

Figure 13. Same as in Figure 12 but for 16 additional sites.

were severely contaminating this inference.

4. Conclusions

The global model of postglacial relative sea level change which has been developed in the past decade of research, based upon the analysis of the response of a layered viscoelastic planet to surface load forcing in Peltier (1974), has proven to be an extremely useful vehicle in terms of which various elements of the radial viscoelastic structure may be quantitatively inferred. The global data base which we have assembled at the University of Toronto to use in this work now consists of Relative Sea Level data from almost 400 sites on the earth's surface. A large number of these data have been presented together here for the first time and employed to demonstrate the strong constraints which they may be employed to place, not only upon the radial viscoelastic structure of the planet, but also upon the deglaciation history itself. The analyses previously described have established that the melting history of Antarctica proposed by Hughes (1981) and employed by Clark and Lingle (1979) is not in accord with observations of relative sea level change. These data require that the West Antarctic ice mass must have disintegrated in phase with the disintegration of the Northern Hemisphere ice sheets as required in the theory of the ice age cycle developed in Peltier (1982, 1986), Peltier and Hyde (1984), and Hyde and Peltier (1985, 1987). Also demonstrated by the analyses reported here was the fact that the thickness of the lithosphere is a parameter to which most far field RSL data are quite sensitive. In regions distant from continental shield areas, whose lithospheres appear to have thicknesses greater than 200 km, the lithospheric thicknesses required by the RSL data are very substantially thinner, with values nearer 70 km being acceptable to most sea level sites.

This paper represents the first in a new series reporting analyses in which higher resolution finite element models than previously employed will be applied systematically to the optimum extraction of geophysical information from the very large RSL data base which we have now assembled. It will prove possible, I believe, to obtain a great deal of valuable new information on the lateral heterogeneity of the near surface viscoelastic structure of the planet from analyses of the sort which have been described here.

References

Barnett, J.P., 1983, Possible changes in global sea level and their causes, *Clim. Change*, **5**, (1), 15-38.

Carter, W.E., Robertson, D.S., Pyle, T.E. and Diamante, J., 1986, The application of geodetic radio interferometric surveying to the monitoring of sea level, *Geophys. J.R. Astr. Soc.*, **87**, 3-13.

Clark, J.A., Farrell, W.E. and Peltier, W.R., 1978, Global changes in postglacial sea level: A numerical calculation, *Quat. Res.*, **9**, 265-287.

Clark, J.A. and Lingle, C.S., 1979, Predicted relative sea-level changes (18,000 years BP to present) caused by late glacial retreat of the Antarctic ice sheet, *Quat. Res.*, **11**, 279-298.

Farrell, W.E. and Clark, J.A., 1976, On post-glacial sea level,*Geophys. J. R. Astr. Soc.*, **46**, 647-667.

Forte, A.M. and Peltier, W.R., 1986, Mantle convection and surface plate kinematics, in *The Lithosphere Asthenosphere System*, C. Froidevaux ed., AGU Press, Washington, D.C.

Forte, A.M. and Peltier, W.R., 1987, Mantle convection and a-spherical earth structure: the importance of poloidal-toroidal coupling, *J. Geophys. Res.*, **92**, 3645-3679.

Gilbert, F. and Dziewonski, A.M., 1975, An application of normal mode theory to the retrieval of structural parameters and source mechanisms from seismic spectra,*Philos. Trans. R. Soc.*, London, Ser. A **278**, 187-269.

Hughes, T.J., 1981, The last great ice sheets: a global view, in *The Last Great Ice Sheets*, eds. G.H. Denton and T.J. Hughes, Wiley, New York.

Hyde, W.T. and Peltier, W.R., 1985, Sensitivity experiments with a model of the ice age cycle: the response to harmonic forcing, *J. Atmos. Sci.*, **42**, 2170-2188.

Hyde, W.T. and Peltier, W.R., 1987, Sensitivity experiments with a model of the ice age cycle: the response to Milankovitch forcing, *J. Atmos. Sci.*, in press.

Meier, Mark F., 1984, Contribution of small glaciers to global sea level, *Science*, **226**, 1418-1421.

Müller, P.M. and Stephenson, F.R., 1975, The acceleration of the earth and moon from early observations, in *Growth Rhythms and History of the Earth's Rotation*, G.D. Rosenburg and S.K. Runcorn, eds., Wiley, New York, 459-534.

Munk, W.H. and Revelle, R., 1952, On the geophysical interpretation of irregularities in the rotation of the earth, Mon. Net. R. astr. Soc., *Geophys. Supp.*, **6**, 331-347.

Peltier, W.R., 1974, The impulse response of a Maxwell earth, *Rev. Geophys. Space Phys.*, **12**, 649-669.

Peltier, W.R., 1976, Glacial isostatic adjustment. II. The inverse problem, *Geophys. J.R. Astr. Soc.*, **46**, 669-706.

Peltier, W.R. and Andrews, J.T., 1976, Glacial isostatic adjustment. I. The forward problem, *Geophys. J.R. Astr. Soc.*, **46**, 605-646.

Peltier, W.R., Farrell, W.E. and Clark, J.A., 1978, Glacial isostasy and relative sea level: A global finite element model, *Tectonophysics*, **50**, 81-110.

Peltier, W.R., 1981, Ice Age geodynamics, *Ann. Rev. Earth Planet. Sci.*, **9**, 199-225.

Peltier, W.R., 1982, Dynamics of the ice age earth, *Advances in Geophysics*, **24**, 1-146.

Peltier, W.R. and Wu, Patrick, 1982, Mantle phase transitions and the free air gravity anomalies over Fennoscandia and Laurentia, *Geophys. Res. Lett.*, **9**, 731-734.

Peltier, W.R., 1983, Constraint on deep mantle viscosity from LAGEOS acceleration data, *Nature*, **304**, 434-436.

Peltier, W.R. and Wu, Patrick, 1983, Continental lithospheric thickness and deglaciation induced time polar wander, *Geophys. Res. Lett.*, **10**, 181-184.

Peltier, W.R., 1984, The thickness of the continental lithosphere, *J. Geophys. Res.*, **89**, 11303-11316.

Peltier, W.R., 1985a, The LAGEOS constraint on deep mantle viscosity: results from a new normal mode method for the inversion of viscoelastic relaxation spectra, *J. Geophys. Res.*, **90**, 9411-9421.

Peltier, W.R., 1985b, New constraints on transient lower mantle rheology and internal mantle buoyancy from glacial rebound data, *Nature*, **318**, 614-617.

Peltier, W.R., 1985c, Mantle convection and viscoelasticity, *Ann. Rev. Fluid Mech.*, **17**, 561-608.

Peltier, W.R., 1986a, Deglaciation induced vertical motion of the North American continent and transient lower mantle rheology, *J. Geophys. Res.*, **91**, 9099-9123.

Peltier, W.R., Drummond, R.A. and Tushingham, A.M., 1986, Post-glacial rebound and transient lower mantle rheology, *Geophys. J. R. Astr. Soc.*, **87**, 79-116.

Peltier, W.R. and Hyde, W.T., 1984, A model of the ice age cycle, in *Milankovitch and Climate*, A. Berger, J. Imbrie, J. Hays, G. Kukla and B. Saltzman eds., D. Reidel, Dordrecht.

Rubincam, D.P., 1984, Postglacial rebound observed by LAGEOS and the effective viscosity of the lower mantle, *J. Geophys. Res.*, **89**, 1077-1087.

Vincente, R.O. and Yumi, S., 1969, Co-ordinates at the pole (1899-1968), returned to the conventional international origin, *Publ. Int. Latit. Obs. Mizusawa*, **7**, 41-50.

Vincente, R.O. and Yumi, S., 1970, Revised values (1941-1961) of the co-ordinates of the pole referred to the CIO, Publ. Int. Latit. Obs. Mizusawa, **7**, 109-112.

Wu, Patrick and Peltier, W.R., 1983, Glacial isostatic adjustment and the free air gravity anomaly as a constraint on deep mantle viscosity, *Geophys. J. R. Astr. Soc.*, **74**, 377-450.

Wu, Patrick and Peltier, W.R., 1984, Pleistocene deglaciation and the earth's rotation: a new analysis, *Geophys. J. R. Astr. Soc.* **76**, 753-791.

Yoder, C.F., Williams, J.G., Dickey, J.O. Schultz, B.E., Eanes, R.J., Tapley, B.D., 1983, Secular variation of earth's gravitational harmonic J_2 coefficient from Lageos and nontidal acceleration of earth rotation, *Nature*, **303**, 757-762.

W.R. PELTIER, Department of Physics, University of Toronto, Toronto ON M5S 1A7, Canada.

Chapter 15

Non-uniqueness of lithospheric thickness estimates based on glacial rebound data along the east coast of North America

Masao Nakada and Kurt Lambeck

Variations in sealevels during the last 20000 years at sites along the eastern margin of North America have been widely used to estimate the viscosity of the Earth's mantle and the thickness of the elastic lithosphere. The sealevels at such sites are, however, quite sensitive to the details of the neighbouring ice load and models of these loads with a spatial resolution of 5° in latitude and 5° in longitude are inadequate. Instead, a spatial resolution of about 1° is required if lithospheric thickness is to be estimated from these data. Current ice models are not adequate and there appears to be a need to increase the ice load along the south eastern margin of the Laurentide ice sheet. The melting of the Antarctic ice sheet also introduces a non-negligible uncertainty into the models of sealevel along the eastern margin but differential changes in sealevel here are largely independent of this contribution. These uncertainties in the melting models make it impossible to reliably estimate the lithospheric thickness from relative sealevel variations observed at sites near the margins of the former ice sheets.

1. Introduction

That the lithosphere is an important concept in the study of the Earth's structure need hardly be emphasized. It is a layer that is capable of supporting and transmitting non-hydrostatic stresses on geological timescales thereby making plate tectonics possible. By acting as both a thermal and mechanical boundary layer it influences the nature and scale of convection in the mantle. But, by being a relatively rigid layer, it also masks much of the

N.J. Vlaar, G. Nolet, M.J.R. Wortel & S.A.P.L. Cloetingh (eds), Mathematical Geophysics, 347-361.

underlying mantle from closer examination and its presence adds considerable complexity to the study of the physical and chemical properties of the planet as a whole. Depending on the geophysical problem examined, or on the geophysical probing method used, the lithosphere has been defined in a variety of ways. The seismic definition is of a layer of high seismic velocities and low seismic attenuation overlying a layer of relatively low velocities and high attenuation. The thermal definition is a layer in which the transport of heat is predominantly by conduction, in contrast to the region below where this transport is achieved primarily by convection. A third definition is the mechanical lithosphere in which the layer is sufficiently strong to support and transmit substantial non-hydrostatic stress on geological timescales. This overlies a weak upper mantle that does not support stress-differences on these timescales and it can usually be modelled as a fluid. Apart from the magnitude difference, these seimic and mechanical definitions can be viewed as differing only in the frequency of the load cycle, this being of the order of 0.001 to 1 Hz at the seismic band and of the order of 10^{-14} Hz and lower at tectonic loading cycles. Because of increasing temperature with depth the stress relaxation time of the Earth decreases downwards and in the tectonic loading problem any non-hydrostatic stress in the lower lithosphere may relax faster than the time scales at which these stresses are generated by the loading cycle. The effective response of the layer is thereby controlled by that part of the lithosphere for which the stress relaxation time equals or exceeds the characteristic time constant of the load cycle. In contrast, the seismic loading cycle are of much shorter duration and the seismically defined lithosphere can be expected to be of greater thickness than the long-term mechanical lithosphere. Just what is this difference in thickness, is a function of the viscosity structure of the lithosphere and upper mantle. Furthermore, both thicknesses will be regionally variable, depending on the age of the layer and on its previous thermal and tectonic history.

Most studies of the mechanical lithosphere thickness have been carried out for load cycles of a duration of 10^6 years and longer but reliable estimates are difficult to obtain because of the need to know the details of the load cycles: often it is not possible to separate the effects of inadequacies in the information on the load from the estimated response (e.g. Lambeck, 1981). Load cycles of a duration of 10^4-10^5 years are provided by the glacial loading and unloading of the Earth and these may also be used to obtain information on lithospheric thickness (McConnell, 1968; Walcott, 1970; Nakiboglu and Lambeck, 1982; Peltier, 1984). Here, also, the problem of separating assumptions about the load cycle from estimates of the Earth's response presents a major problem because the evolution of the ice load is only partially known for the past 20000 years and only qualitatively known for earlier cycles of glaciation and deglaciation. Furthermore, the stress cycle is of a duration where the response of the mantle is more important, something that is not so for stress cycles of 10^6 years and longer duration. The question may well be asked whether it is possible to separate all of these factors from observations of the Earth's response to the late Pleistocene glacial unloading.

Peltier (1984) has examined the glacial loading problem and concluded from an examination of sealevels along the margin of North America that the lithospheric thickness is in excess of 200 km whereas seismic estimates of about 100 km have been suggested for the lithospheric thickness of the eastern United States (e.g. Cara, 1979). However, the response of the Earth at sites near the limits of the former ice sheet is not a good indicator of the Earth's lithospheric response for here the sealevels are very sensitive to the assumed ice models as well as to the viscosity of the mantle structure. Peltier's conclusion is sufficiently important, yet sufficiently uncertain, in view of the limited knowledge of the melting of the nearby Laurentide ice sheet, to warrant a re-examination of the problem of estimating lithospheric thickness from glacial rebound observations using regionally restricted observational data sets.

2. The sealevel equation

The equation describing the sealevel due to changes in normal surface loads associated with an exchange of mass between ice sheets and oceans has been given by Farrell and Clark (1976). Its solution for the sealevel change (at colatitude θ and longitude λ) can be written as (Nakiboglu et al, 1983, Nakada and Lambeck, 1987)

$$\zeta(\theta,\lambda:t) = \zeta_R(\theta,\lambda:t) + (Z_1 - \langle Z_1 \rangle) + (Z_2 - \langle Z_2 \rangle) \tag{1}$$

Here ζ_R is the response of sealevel on a rigid Earth and is given by

$$\zeta_R(\theta,\lambda:t) = \zeta_e(t)\left[1 + C_1(S_1 - \langle S_1 \rangle)\right] + C_2(S_2 - \langle S_2 \rangle) \tag{2}$$

where $\zeta_e(t)$ is the eustatic part of sealevel. Sealevel must be an equipotential surface at all times so that ζ_R is not constant over the oceans. These departures are given by S_1 and S_2 where the term S_1 describes the adjustment of the surface to the change in water volume and the term S_2 describes the sealevel response to the change in ice load. The brackets $\langle\ \rangle$ denote the mean value of the function enclosed and the two terms $\langle S_1 \rangle$ and $\langle S_2 \rangle$ ensure that mass is conserved. The constants C_1 and C_2 are defined as

$$C_1 = \frac{3\rho_w}{4\pi\bar{\rho}} \quad \text{and} \quad C_2 = \frac{3\rho_i}{4\pi\bar{\rho}}$$

and the S_1, S_2 are defined as

$$S_1(\theta,\lambda) = \iint_{\text{ocean}} \frac{d\omega}{2\sin\frac{\alpha}{2}}$$

$$S_2(\theta,\lambda,t) = \iint_{\text{ice}} \frac{\zeta_i(\theta',\lambda':t)}{2\sin\frac{\alpha}{2}} d\omega$$

Also

$$\zeta_e = \frac{-\rho_i}{A_0\rho_w} \iint_{\text{ice}} \zeta_i(\theta',\lambda':t)\, d\omega$$

In these definitions ρ_i, ρ_w and $\bar{\rho}$ are the densities of ice, water and the average Earth respectively. A_0 is the surface area of the ocean and $\zeta_i(\theta',\lambda':t)$ is the change in ice height at position (θ',λ') at time t. α is the angle between the point $P(\theta,\lambda)$ at which sealevel is evaluated and the position $P'(\theta', \lambda')$ of elements of the surface load. The terms Z_1 and Z_2 allow for the deformation of the Earth in response to the changes in ice and water loads with time. $Z_1 - \langle Z_1 \rangle$ represents the variation in sealevel associated with the loading and unloading of the crust by the ice sheets (without the meltwater being distributed in the oceans) and includes the effects of self-attraction of the ice and water. The term $Z_2 - \langle Z_2 \rangle$ represents the change in sealevel caused by the addition or substraction of the meltwater. The Z_1, Z_2 are defined by

$$Z_1 = C_2 \sum_{n=0}^{N_{max}} \iint_{ice} L^{-1}\left[(\tilde{k}_n - \tilde{h}_n)\zeta_i(s) \right] P_n(\cos\alpha) d\omega \tag{3}$$

$$Z_2 = C_1 a \sum_{n=0}^{N_{max}} \iint \left\{ \iint_{ice} L^{-1}\left[(\tilde{k}_n - \tilde{h}_n)\zeta \right] d\omega \right\} P_n(\cos\alpha) d\omega \tag{4}$$

The $P_n(\cos\alpha)$ are Legendre polynomials of degree n. L^{-1} denotes the inverse Laplace transform of quantities, denoted by $\tilde{}$, in the Laplace domain and s is the transform parameter of dimension $(\text{time})^{-1}$. The $\tilde{h}_n$ and $\tilde{k}_n$ are the n-th degree load Love numbers in the Laplace transform domain. Finally,

$$a = \frac{-\rho_i}{4\pi a_{00}\rho_w}$$

where a_{00} is the zero degree term in the spherical harmonic expansion of the ocean function. The observed quantities are the relative sealevels at time t, or

$$\Delta\zeta(\theta,\lambda:t) = \zeta(\theta,\lambda:t) - \zeta(\theta,\lambda:t_0) \tag{5}$$

where t_0 is the present time.

To obtain quantitative solutions of the sealevel change it is necessary to define in equation (1); (i) the geometry of the ice load through time, (ii) the shape of the oceans into which the meltwater flows during times of deglaciation or from which the meltwater is withdrawn during times of glaciation, and (iii) the Earth's mechanical response to the changing surface load. The principal numerical difficulties occur in the evaluation of the inverse Laplace transforms in equation (3) for realistic ice loads and in the integration of the integrals in (3), particularly of the integral of the function Z_2 over the ocean surface. See Nakada and Lambeck (1987) for a detailed discussion of these problems.

3. Model parameters

The Arctic ice loads adopted here are the models ICE 1 of Peltier and Andrews (1976) and the Arctic part of ICE 2 of Wu and Peltier (1983). We refer to these models as ARC 1 and ARC 2 respectively. These models describe the evolution of the ice sheets through time by a time series of heights of ice columns each of 5° in latitude and 5° in longitude areal coverage. An Antarctic 5° x 5° model has been derived from the difference of the Antarctic ice sheet 20000 years ago constructed by Denton and Hughes (1981) and the present ice sheet as given by Drewry (1983). Two melt histories have been assumed. In the first, the deglaciation of each column has, for want of further information, been assumed to be synchronous with the eustatic sealevel variation corresponding to the arctic model ARC 2. This is the Antarctic model ANT 1. Stuiver et al. (1981) and Broecker (1984) favour a model similar to this in which melting occurred primarily between 17000 and 6700 years ago with a significant fraction of the total melting having occurred after 12000 years ago. Nakiboglu et al. (1983) have also argued for such a melting history from an examination of sealevel variations in the southern hemisphere. Recent work (Nakada and Lambeck, in preparation) has confirmed the need for a substantial Antarctic meltwater component in the interval 12000 to 6000 years ago. Wu and Peltier (1983) introduced an Antarctic ice model in which much of the melting occurred before about 12000 years ago. This forms the basis for the second model ANT 2 in which the ice volume is the same as ANT 1 but for which the melting of each ice column is assumed to be synchronous with the eustatic sealevel variation of the Antarctic part of the original ICE 2 model of Wu and Peltier. These models are defined by ice columns of 5° x 5° areal extent and we introduce a further series of models in which these columns are linearly interpolated to produce a 1° x 1° description of the load.

The shape of the ocean is described by an ocean function O (which equals unity where there is ocean and which is zero elsewhere). The data base used to compute this function is a global 10' resolution of the coastlines. This 10' ocean function has been expanded to degree and order 180. We have previously noted that to obtain reliable coefficients out to this degree a 1° spatial resolution was inadequate (Nakada and Lambeck, 1987).

The Earth response model has the elastic moduli and density structure of the preliminary reference Earth model (PREM) of Dziewonski and Anderson (1981) and a uniform mantle viscosity η of 10^{21} Pa s. Other plausible viscosity models have been examined but the essential conclusions drawn in this paper remain independent of this choice. The mantle is overlain by a high (10^{25} Pa s) viscosity lithosphere of thickness H. Lateral variations in H and η are ignored although they may be important across continent-ocean boundaries.

4. Results

4.1 Solutions of the sealevel equation in the wave number domain

The characteristics of the solution of spatial and temporal sealevel change can be examined in the wavenumber domain by expanding the sealevels into spherical harmonic series and computing the power spectrum. To achieve these expansions, both the ocean function O and the ice load I are expanded as

$$O = \sum_n \sum_m \left[a_{nm} \cos m\lambda + b_{nm} \sin m\lambda \right] P_{nm}(\cos\theta)$$

$$I = \sum_n \sum_m \left[g_{nm} \cos m\lambda + h_{nm} \sin m\lambda \right] P_{nm}(\cos\theta) \qquad (6)$$

where the ice expansion coefficients g_{nm} and h_{nm} are functions of time. The equation (2) for sealevel on a rigid Earth is

$$\zeta(\theta,\lambda:t) = \zeta_e(t) + \sum_n \sum_m \left[A_{nm} \cos m\lambda + B_{nm} \sin m\lambda \right] P_{nm}(\cos\theta)$$

where

$$\left.\begin{matrix} A_{nm} \\ \\ B_{nm} \end{matrix}\right\} = \frac{4\pi}{2n+1} P_{nm}(\cos\theta) \left[\zeta_e C_1 \left\{ \begin{matrix} a_{nm} \\ \\ b_{nm} \end{matrix} \right\} + C_2 \left\{ \begin{matrix} g_{nm} \\ \\ h_{nm} \end{matrix} \right\} \right] \left\{ \begin{matrix} \cos m\lambda \\ \\ \sin m\lambda \end{matrix} \right.$$

The $P_{nm}(\cos\theta)$ are fully normalized associated Legendre polynomials. The power spectrum of this sealevel change is then

$$V_n^2(\zeta_R) = \left[\frac{4\pi}{2n+1} \right]^2 \left[(\zeta_e C_1)^2 V_n^2(O) + (C_2)^2 V_n^2(I) + 2\zeta_e C_1 C_2 V_n^2(O,I) \right] \tag{7}$$

where

$$V_n^2(O) = \sum_m (a_{nm}^2 + b_{nm}^2) \quad \text{and} \quad V_n^2(I) = \sum_m (g_{nm}^2 + h_{nm}^2)$$

are the power spectra of the ocean and ice functions and

$$V_n^2(O,I) = \sum_m (a_{nm} g_{nm} + b_{nm} h_{nm})$$

is the cross spectrum. Of the three terms in (7), the dominant one during the time of deglaciation is the ice spectrum and $V_n^2(\zeta_R)$ is approximately proportional to $V_n^2(I)$ at times of major growth or decay of the ice sheets. During Holocene time $V_n^2(I) = V_n^2(O,I) = 0$ and now $V_n^2(\zeta_R)$ is proportional to $V_n^2(O)$.

For an elastic Earth subjected to instantaneous unloading of the ice at time t^* the spectrum $V_n^2(\zeta_E)$ of sealevel change at $t > t^*$ is given by

$$V_n^2(\zeta_E) = (1 + k_n - h_n)^2 V_n^2(\zeta_R) \tag{8}$$

and for a viscoelastic Earth subjected to the same load

$$V_n^2(\zeta_{VE}) = (1 + k_n(t) - h_n(t))^2 V_n^2(\zeta_R) \tag{9a}$$

which can be approximated as

$$V_n^2(\zeta_{VE}) = \left[\frac{4\pi}{2n+1} C_2 \right]^2 (1 + k_n(t) - h_n(t))^2 V_n^2(I) \tag{9b}$$

where the viscoelastic Love numbers are now functions of time. The second approximate equality is for the time of deglaciation in the vicinity of the ice load. The function $f_n^2 = \left[1 + k_n(t) - h_n(t) \right]^2$ represents the power spectrum of the Earth's sealevel response to the step function loading. Figure 1 illustrates this spectrum for the above Earth model, but with different thicknesses of the lithosphere at 8000 years after the instantaneous

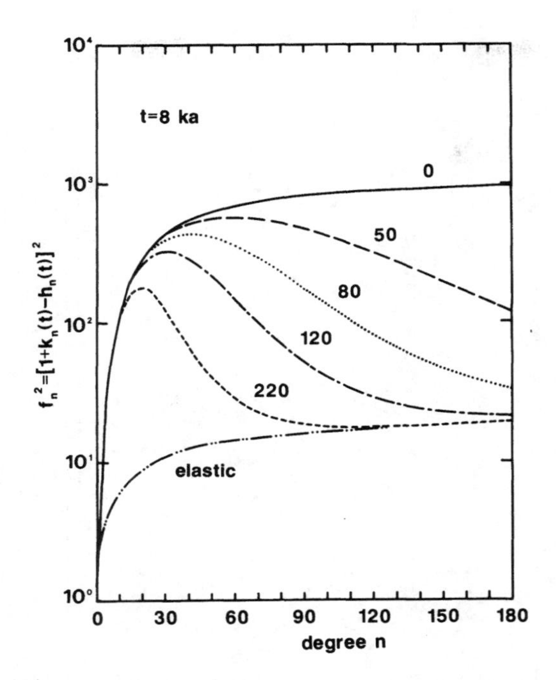

Figure 1. The Earth's response function $f_n{}^2$ for a Heavitree step function load applied at time t=0. The response is evaluated at time 8000 years later. The mantle has a uniform Newtonian viscosity of 10^{21} Pa s and is overlain with a very high viscosity lithosphere of thicknesses of 0, 50, 80, 120 and 220 km. The elastic solution is for an elastic mantle.

removal of the ice load. The elastic result is for the PREM elastic Earth model. For the Earth model without a lithosphere the function $f_n{}^2$ increases monotonically with increasing degree n but the introduction of the lithosphere reduces the Earth's response at high degree and $f_n{}^2$ converges on the elastic Earth value with the rate of convergence increasing with increasing lithospheric thickness.

Figure 2 illustrates the spectrum of the change in the ARC 1 ice sheet from the time of onset of melting to the end of melting. The spectrum marked 5° is based on the 5° area-mean ice heights. Very considerable power occurs at high degrees, a consequence of the inadequate stepwise spatial distribution of the ice load. This high wave number noise vanishes for the 1° interpolated ice load and the use of the 5° ice model may, through equation (9), induce considerable short wavelength variations in sealevel. If not observed, this could be suppressed by the introduction of a thick lithosphere in the Earth response model and, while it could lead to satisfactory solutions of the sealevel equation, the geophysical inference would be quite incorrect.

Figure 3 illustrates the sealevel spectrum defined by (9b) for the epoch 8000 years ago when the bulk of the arctic ice had disappeared. The results for both the 5° and the 1° ice models are illustrated for different lithospheric thicknesses. This figure illustrates clearly the consequences of using the 5° ice model: the 5° ice model with a 120 km thick

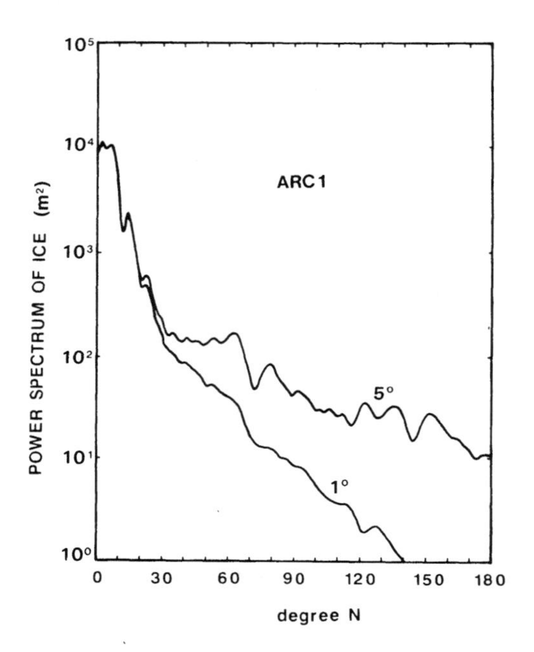

Figure 2 The power spectrum of the Arctic ice model ARC 1 based on the 5° and 1° spatial resolutions of the ice load. The spectrum corresponds to the total change in the load from 18000 years ago to the end of deglaciation.

lithosphere produces the same sealevel spectrum as the 1° ice model with a 50 km thick lithosphere. Clearly the 5° ice model is inadequate for infering geophysical parameters from observations of relative sealevel variations. The other conclusion that can be drawn from figure 3 is that the total power of the sealevel response at n ≥ 30 (i.e. $\sum_{n=30}^{\infty} V_n^2(\zeta)$), even with the 1° ice model, is significant and it appears that expansions of the ice functions to a high degree are required, particularly for sites near the edges of the ice sheets where the sealevel change is greater than these mean square values. Alternatively, convergence of the sealevel equation can be obtained by introducing Earth models with strong lithospheres but now the estimates of lithosphere thickness have no physical meaning.

4.2 Solutions for relative sealevel in the spatial domain

Sealevels are not well enough known to permit the spectra to be computed empirically through time and instead the computations need to be carried out in the spatial domain. In this section we consider sealevels at sites near the edge of the ice sheets and examine their sensitivity to the choice of lithospheric thickness and melting history models of the two polar regions. We adopt the uniform mantle viscosity model of 10^{21} Pa s favoured by Wu and Peltier (1983) but we emphasize that these sealevels are at least as sensitive to this choice as they are to the lithospheric thickness. The change in each ice column in the time

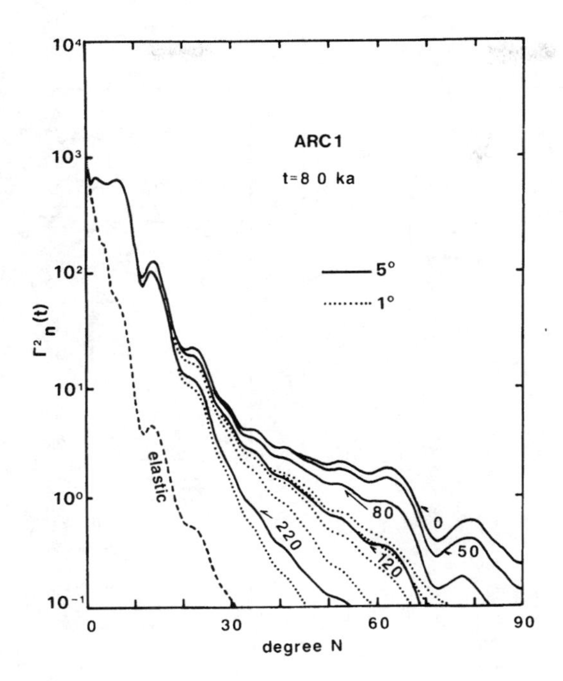

Figure 3 Power spectra of the sealevel response at 8000 years ago according to equation (9). The solid lines correspond to the 5° ice model whereas the dotted lines correspond to the 1° ice model.

interval t_j and t_{j+1} is assumed to be linear or

$$\zeta_i(\theta,\lambda:t) = a_j(\theta,\lambda)t + b_j(\theta,\lambda) \qquad t_j \leq t \leq t_{j+1}$$

for both the original 5° models and the interpolated 1° ice models. With this model, and with Peltier's (1974) analytical expressions for the Laplace transformed Love numbers, equations (1) to (4) give analytical solutions for the sealevel change whose only mathematical limitation is, that terms of the order $\zeta(C_1)^2$ have been ignored. As $C_1^2 \approx 2.10^{-3}$, this represents an error of 20 cm in sealevel change of 100 m and is negligible. The physical limitations of the model relate to the choice of ice models, ocean functions and the Maxwell viscoelastic assumption. Details of the solution are discussed by Nakada and Lambeck (1987).

The observed quantities are the sealevels for past epochs $\zeta(t)$ relative to present sealevel $\zeta(t_0)$, equation (50), or

$$\Delta\zeta = \zeta(t) - \zeta(t_0) = \sum_{n=0}^{N_{max}} \Delta\zeta_n \tag{10}$$

Figure 4 illustrates the $\Delta\zeta$ for an epoch 6000 years ago at two sites near the edge of the former Laurentide ice sheet. These model calculations are based on the ARC 1 5° (figure 4a) and the 1° (figure 4b) ice models with different degrees of truncation. The variable is

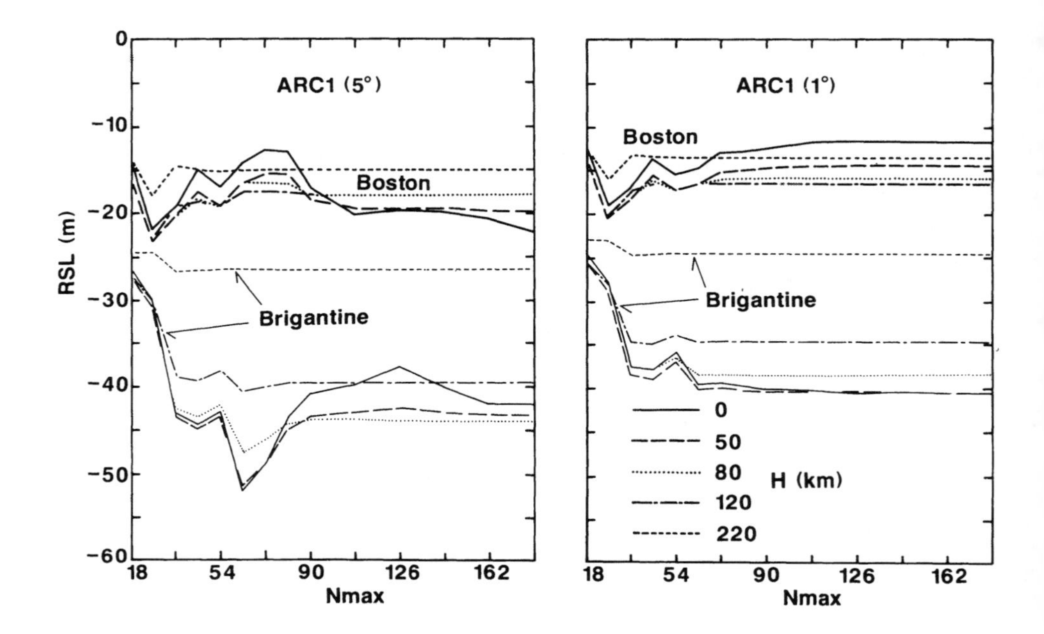

Figure 4 Relative sealevels, defined by equation (10), at 6000 years ago at Boston and Brigantine as a function of truncation degree N_{max}. The results on the left correspond to the 5° ice model, those on the right correspond to the 1° ice model.

the lithospheric thickness H. Only when a lithosphere greater than 220 km is introduced does the solution converge by N_{max} = 36 when the 5 ° ice model is used. The solutions based on the 1° ice load converge considerably more rapidly, even for models without a lithosphere, and at both sites convergence is achieved with $N_{max} \approx 90$. From figure 4 the error incurred at Boston by using the 5° ice model ranges from about 10 m for the zero thickness lithosphere to about 2 m for the 220 km thick lithosphere. This compares with the observed sealevel at 6000 years ago at 15 m below the present level (figure 5). Clearly the error is not insignificant at these sites. The need for the high resolution of the ice models was recognized by Quinlan and Beaumont (1981) in their glacial rebound calculations for eastern Canada.

Figure 5 illustrates the sealevel variations at Boston and Brigantine on the eastern Atlantic margin of the United States. These sites were used by Peltier (1984) to estimate the lithospheric thickness. Boston is located just inside the margin of the northern ice sheet at its maximum extent while the other site lies outside this limit. The theoretical sealevels are based on a uniform mantle viscosity of a nominal value of 10^{21} Pa s. The 1 deg. ice models are used throughout. The observations of relative sealevels are from Kaye and Barghoorn (1964) and Newman et al. (1980).

a

b

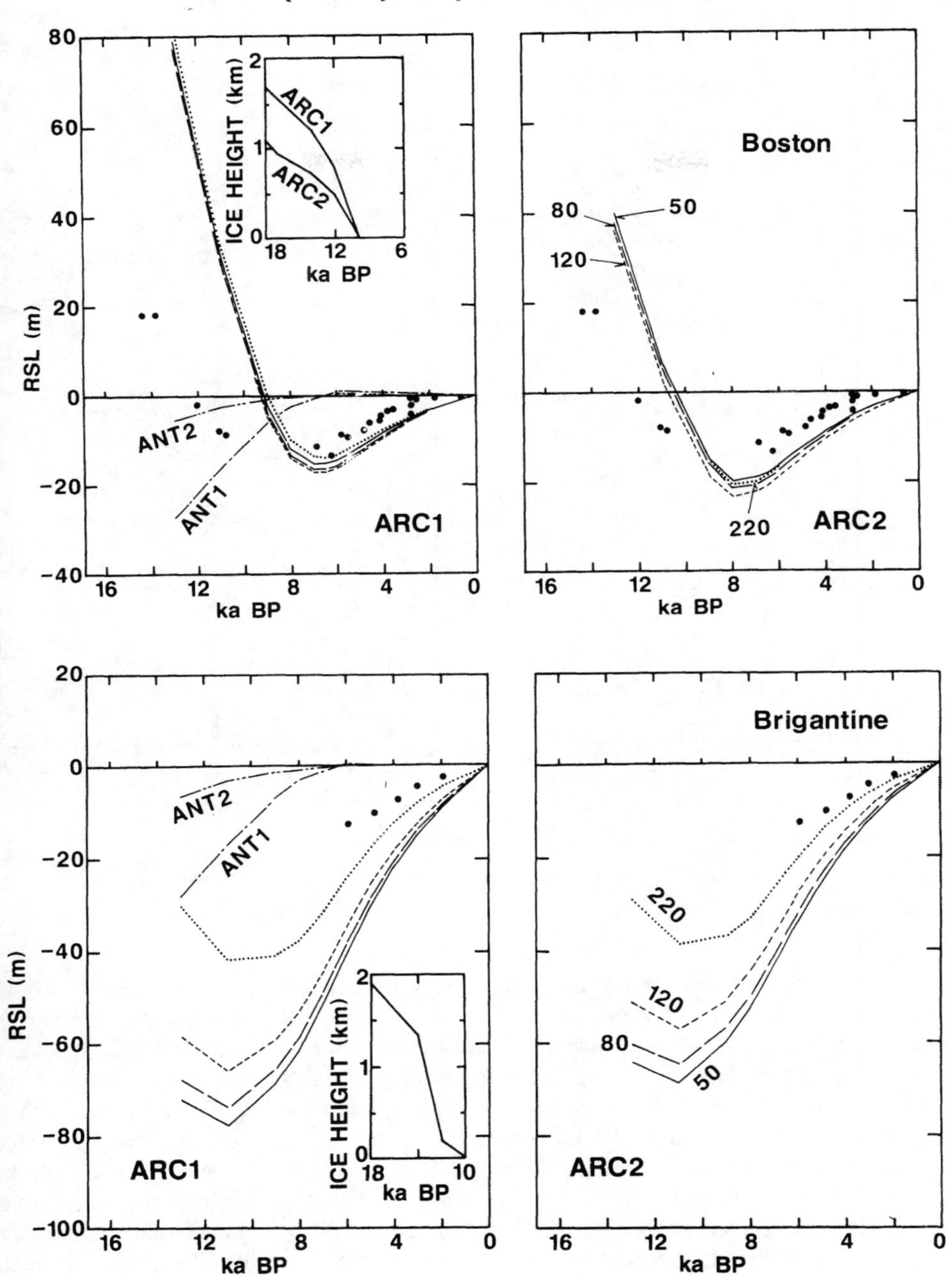

Figure 5 Comparison of observed relative sealevels with predicted values at (a) Boston, (b) Brigantine and (c) Chesapeake Bay. The mantle has a uniform viscosity of 10^{21} Pa s and the variable is the elastic thickness. The insets show the temporal variation in ice height near the site.

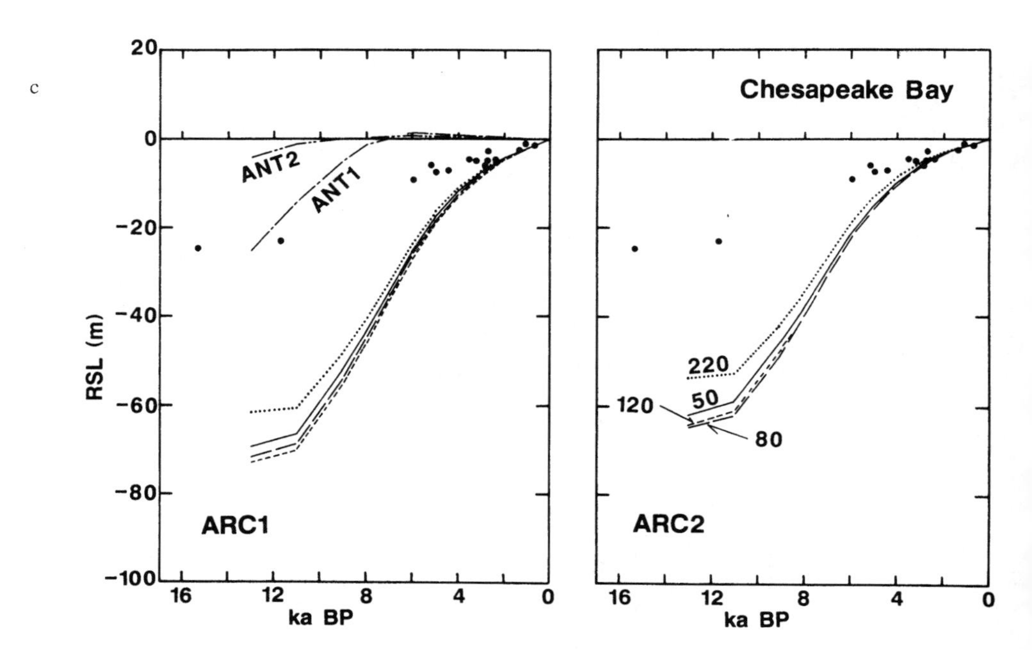

Figure 5 (Continued)

The contributions from both the Arctic and Antarctic ice sheets are considered. The latter contribution is quite insensitive to the choice of mantle viscosity and lithospheric thickness because the sites in question here are far from this ice load. A lithospheric thickness of 50 km is used to compute this contribution. The sealevels are, however, sensitive to the choice of melting history of the Antarctic ice sheet. The model with the early melting history (ANT 2), favoured by Wu and Peltier (1983), results in little meltwater being added after about 12000 years ago and the late glacial and Holocene sealevels are not significantly influenced. If, however, the melting model ANT 1, favoured by Stuiver et al. (1981) and Broecker (1984) is adopted then this meltwater does make a significant contribution, about 20 m at 12000 years ago and comparable to the sealevels at these sites at that time. In particular, the Antarctic contribution at this time is of opposite sign to that from the Arctic, and its neglect could have the consequence of concluding that Arctic melting occurred earlier than stated in the models.

As is evident from figure 4b, the Boston sealevels are relatively insensitive to the choice of lithospheric thickness. They are, however, sensitive to the choice of Arctic ice model, particularly during the early stages of deglaciation. This is a consequence of the changes in the two models in the immediate vicinity of Boston (see inset of figure 5a). Changes that may be minor from a glaciological viewpoint are of considerable consequence when it comes to the study of relative sealevel changes. Sealevels from sites such as Boston are unlikely to serve an important role in estimating the Earth's response.

The predicted differences in the Brigantine sealevels for the two ice models ARC 1 and ARC 2 are less than for Boston, simply because the two models are identical in the immediate vicinity of the site and the differences in sealevel reflect the variations in the ice models further away. The sealevels are strongly sensitive to the lithospheric thickness and the observations suggest a value of perhaps 250 km. But this estimate is only as good as the assumed mantle viscosity, and of course, the assumption that these ice models span the limits of plausible representations of the former ice loads. Sealevels at Chesapeake Bay are also quite similar for the two ice models but they do not agree well with the observations at about 12000 and 16000 years ago (see Newman et al., 1980). Here no amount of increasing the thickness of the lithosphere can explain the observations. The same is true at other sites in the neighbourhood of Brigantine and Chesapeake, for example, at Delaware and Virginia.

5. Concluding remarks

Sealevels at other sites along the eastern margin of the United States are discussed by Nakada and Lambeck (1987) but no firm conclusion was reached concerning the optimum combination of lithospheric thickness and mantle viscosity parameters. No one set of parameters adequately describes the sealevels from Maine to Miami, giving rise to the speculation that the ice models need further adjustment or that there is a component of tectonic motion at some of these sites. However, the required amount of tectonic uplift is such that it is more plausible to explain the discrepancy by the limitations of the ice models. What is required is a more substantial ice load near the south-eastern edge of the Laurentide ice sheet during the years from 18000 to 20000 years ago.

The main conclusion arising from this study is that sites near the margin of the former ice sheets are not reliable indicators of the Earth's response while uncertainties remain in the detailed melting histories of the ice sheets. Despite this, sites such as Boston have played an important role in estimating this response (Cathles, 1975; Peltier, 1984, 1985; Sabadini et al., 1985). Conclusions concerning the thickness of the lithosphere or the depth dependence on transient nature of the Earth's viscous response from such observations are therefore strongly dependent on the assumptions made in constructing the ice models.

The second principal conclusion drawn is that when the glacial rebound model is used to compute sealevel changes near the margins of the ice sheet there is a need to define the ice models with a high spatial resolution because the sealevel change is quite sensitive to the detailed loading of the crust in the neighbourhood of the site. The 5° representations are inadequate at such sites. Sites further away from the ice loads are less sensitive to the details of glaciation, as are sites near the center of the ice loads, but neither type of site is particularly sensitive to lithospheric thickness unless differential changes in sealevel at nearby sites are examined. Sites in the far field are sensitive to the gross melting history, for example, to the role played by the Antarctic ice sheets particularly when evidence on sealevels is available for the pre-Holocene period (Nakiboglu et al., 1983). By a careful selection of sites it does become possible in principal to separate some of these variables (e.g. Lambeck and Nakada, 1985) but generally the observational data remain inadequate to solve for both the ice model and Earth rheology parameters.

References

Broecker, W.S., 1984. Terminations, in *Milankovitch and Climate, part 2*, ed. by A.L.D. Berger, Reidel Publishing Company.

Cara, M., 1979. Lateral variations of S velocity in the upper mantle from higher Rayleigh modes, *Geophys. J. R. Astr. Soc.*, **57**, 649-670.

Cathles, L.M., 1975. *The Viscosity of the Earth's Mantle*, Princeton University press, New Jersey.

Denton, G.H. and Hughes, T.J. (eds.), 1981. *The Last Great Ice Sheets*, Wiley, New York.

Drewry, D.J. (ed.), 1982. *Antarctica: Glaciological and Geophysical Folio*, Scott Polar Res. Inst., Cambridge.

Dziewonski, A.M. and Anderson, D.L., 1981. Preliminary reference Earth model, *Phys. Earth Plan. Int.*, **25**, 297-356.

Farrell, W.E. and Clark, J.A., 1976. On postglacial sealevel, *Geophys. J. R. Astr. Soc.*, **46**, 647-667.

Kaye, C.A. and Barghoorn, E.S., 1964. Late Quaternary sea level change and crustal rise at Boston, Massachusetts, with notes on the Auto compaction peat, *Bull. Geol. Soc. Am.*, **75**, 63-80.

Lambeck, K., 1981. Flexure of the ocean lithosphere from island uplift, bathymetry and geoid heights: the Society Islands, *Geophys. J. R. Astr. Soc.*, **67**, 91-114

Lambeck, K. and Nakada, M., 1985. Holocene fluctuations in sea-level: Constraints on mantle viscosity and melt-water sources, *Proceedings of the Fifth International Coral Reef Congress, Tahiti*, vol. 3, 79-84.

Nakada, M. and Lambeck, K., 1987. Glacial rebound and relative sealevel variations: A new appraisal, *Geophys. J. R. Astr. Soc.*, **89**, in press.

Nakiboglu, S.M. and Lambeck, K., 1982. A Study of the Earth's response to surface loading with application to Lake Bonneville, *Geophys. J. R. Astr. Soc.*, **70**, 577-620.

Nakiboglu, S.M., Lambeck, K. and Aharon, P., 1983. Postglacial sealevels in the Pacific: Implications with respect to deglaciation regime and local tectonics, *Tectonophysics*, **91**, 335-358.

Newman, W.S., Cinguemani, L.J., Pardi, R.R. and Marcus, L.F., 1980. Holocene delevelling of the United States east coast, in *Earth Rheology, Isostacy and Eustacy*, ed. N. Morner, Wiley, New York.

Peltier, W.R., 1974. The impulse response of a Maxwell Earth, *Rev. Geophys. Space Phys.*, **12**, 649-669.

Peltier, W.R., 1984. The thickness of the continental lithosphere, *J. Geophys. Res.*, **89**, 11303-11316.

Peltier, W.L., 1985. New constraint on transient lower mantle rheology and internal mantle buoyancy from glacial rebound data, *Nature*, **318**, 614-617.

Quinlan, G. and Beaumont, C., 1981. A Comparison of observed and theoretical postglacial relative sealevel in Atlantic Canada, *Can. J. Earth Sci.*, **18**, 1146-1163.

Sabadini, R., Yuen, D.A. and Gasperini, P., 1985. The effects of transient rheology on the interpretation of lower mantle viscosity, *Geophys. Res. Lett.*, **12**, 361-365.

Stuiver, M., Denton, G.H., Hughes, T.J. and Fastoo, K.J.L., 1981. History of the marine ice sheet in west Antarctica during the last glaciation: a working hypothesis, in *The Last Great Ice Sheets*, ed. by Denton, G.H. and Hughes, T.J., Wiley, New York.

Walcott, R.I., 1970. Isostatic response to loading of the crust in Canada, *Can. J. Earth Sci.*, **7**, 716-727.

Wu, P. and Peltier, W.R., 1983. Glacial isostatic adjustment and the free air gravity anomaly as a constraint on deep mantle viscosity, *Geophys. J. R. Astr. Soc.*, **74**, 377-449.

MASAO NAKADA and KURT LAMBECK, Research School of Earth Sciences, Australian National University, P.O. Box 4, Canberra A.C.T. 2601, Australia.

Chapter 16

On the mechanics of plate boundary formation

S.A.P.L. Cloetingh and M.J.R. Wortel

We discuss the mechanics underlying the formation of new plate boundaries. Analysis of the relation between intraplate stress fields and lithospheric rheology leads to greater insight into the role that initiation of subduction and initiation of rifting play in the tectonic evolution of the lithosphere. Concentration of slab-pull forces is an effective mechanism for passive rifting of oceanic and continental lithosphere. Initiation of subduction at passive margins requires in general the action of external plate-tectonic forces, which will be most effective for young passive margins pre-stressed by thick sedimentary loads. Plate reorganizations occur predominantly by the formation of new spreading ridges, because stress relaxation in the lithosphere takes place much more efficiently through this process than through the formation of new subduction zones.

1. Introduction

Initiation of rifting and initiation of subduction are key elements in plate-tectonic schemes for evolution of the lithosphere. The importance of rifting processes is reflected in the considerable body of work devoted to this subject during the last decade (e.g. Palmason, 1982; Morgan and Baker, 1983), which has been motivated in part by the role of rifting in the formation of sedimentary basins. The mechanics of the initiation of new subduction zones has remained relatively unexplored (Cloetingh et al., 1982, 1983; Flinn, 1982) by comparison.

Advances in our understanding of lithospheric rheology (Goetze and Evans, 1979; Kirby, 1983) and in the modelling of intraplate stress fields (Wortel and Cloetingh, 1981, 1983; Cloetingh and Wortel, 1985, 1986) allow a unifying approach to the dynamics

N.J. Vlaar, G. Nolet, M.J.R. Wortel & S.A.P.L. Cloetingh (eds), Mathematical Geophysics, 363-387.

underlying the formation of new plate boundaries. In the present paper we concentrate on a comparison of the stresses and strengths in the lithosphere. In our studies we have followed a finite element approach, which allows us to take full account of depth-dependent rheological properties of the lithosphere and laterally varying forces. We investigate whether the stresses generated by the different geodynamic processes are capable of inducing lithospheric failure. Such an analysis will be shown to lead to a better understanding of the differing roles that initiation of subduction and rifting play in plate reorganization processes.

2. Intraplate stress fields and the rheology of the lithosphere

Lithospheric plates are in motion under the influence of plate-tectonic forces: the ridge push, which results from the elevation of the spreading ridge above the adjacent ocean floor and the thickening of the lithosphere with cooling (Hales, 1969); the drag exerted at the base of the lithosphere (Richter, 1973) due to viscous shear forces associated with the motion relative to the underlying asthenosphere; and the pull acting on the downgoing slab in a subduction zone (McKenzie, 1969). In addition to these plate-tectonic forces, there are also forces acting on the lithosphere which induce local stresses. A number of these mechanisms have been reviewed by Turcotte and Oxburgh (1976) and Cloetingh (1982). Among the locally induced stresses those associated with flexure and those induced by cooling of the lithosphere in near-ridge areas stand out in magnitude of stress level. In plates that are involved in collision or subduction processes, however, plate-tectonic forces dominate the stress field. This dominance is greater if there is a concentration of stresses due either to lithospheric age variations encountered along trench systems or to angular and curved geometry of plate boundaries (Wortel and Cloetingh, 1981; Cloetingh and Wortel, 1985).

In the first phase of modelling lithospheric stress fields resulting from plate-tectonic forces, models were tested against focal mechanism data of intraplate earthquakes to infer the relative and absolute importance of various possible driving and resistive forces (Solomon et al., 1975; Richardson et al., 1979). These and several other studies (Forsyth and Uyeda, 1975; Chapple and Tullis, 1977) resulted in the gross understanding that the ridge push and the slab pull were the main driving forces. Two developments, however, were necessary to set the stage for more advanced stress field modelling : 1. Recognition that the ridge push is not to be considered as a constant boundary force but as an integrated pressure gradient (Lister, 1975), which introduces the dependence of the force on the age of the lithosphere. 2. Deeper understanding of the variations in the subduction process of oceanic lithosphere and the associated variations in plate-tectonic forces (Vlaar and Wortel, 1976; Wortel and Vlaar, 1978; Molnar et al., 1979; Wortel, 1980; 1984). These studies demonstrated that age-dependent variations in length, depth and sinking rate affect the pull that a downgoing slab exerts on a converging plate.

By implementing these new features and insights in stress modelling Wortel and Cloetingh (1981, 1983, 1985) and Cloetingh and Wortel (1985, 1986) showed that the dynamical basis of their modelling procedure enabled the resulting stress field to be used to analyze, explain and even predict various deformational processes in lithospheric plates.

In the present study we elaborate on the tectonic effects of intraplate stress fields in the context of *rheological models* for the lithosphere. Stresses in the lithosphere are supported by its mechanically strong upper part, the rheological structure of which consists in its simplest form of two sections: a top section in the brittle regime, where strength increases rapidly with pressure according to Byerlee's law (Byerlee, 1968, 1978) , and a lower section where the effects of temperature dominate and stresses are limited by ductile flow (Goetze and Evans, 1979). Following Bodine et al. (1981) we take here the depth at which the ductile strength is 500 bar as the lower boundary of the mechanically strong part of the lithosphere (MSL). Studies of the flexural response of the *oceanic* lithosphere to seamount loading and to bending at trenches (Caldwell and Turcotte, 1979; Bodine et al., 1981), Seasat altimetry (McAdoo et al., 1985) and analysis of the depth distribution of oceanic intraplate seismicity (Wiens and Stein, 1983) all indicate an increase in thickness of the MSL from a few km near the spreading ridge to approximately 50 km at an age of 100 Ma, which is consistent with extrapolation of results of laboratory experiments on dry olivine by Goetze and Evans (1979). Similarly, according to predictions of rock-mechanics data the strength of the oceanic lithosphere increases according to a square root function of age.

The rheology of *continental* lithosphere involves many uncertainties, and is obviously more complex than the rheology of oceanic lithosphere. Independent evidence from widely different fields, however, suggests that the major rheological characteristics recognized for oceanic lithosphere apply also to continental lithosphere. The distribution of the depths and magnitudes of earthquakes in continental lithosphere is consistent with a depth-dependent rheology (Meissner and Strehlau, 1982; Sibson, 1983). Studies of the flexural rigidity of continental lithosphere (Kuznir and Karner, 1985) also agree with the existence of an MSL, the thickness of which is temperature-dependent and consistent with the extrapolation of laboratory data on crustal (Kirby, 1983; 1985) and mantle (Goetze and Evans, 1979) rocks. In the present paper we adopt an average model for continental rheology with a 15 km thick upper crust, 15 km lower crust and 120 km lithospheric mantle. Compared to its apparently minor influence on the rheology of oceanic lithosphere (McAdoo et al., 1985), the weakening effect of water on the rheology of continental upper crustal rocks is more substantial (Brace and Kohlstedt, 1980). Therefore, power-law creep parameters for quartzite (Shelton and Tullis, 1981), which mechanical behavior appears to be between that of wet and dry granite (Carter et al., 1981), are adopted to describe ductile flow in the upper crust. For the ductile rheology of the lower crust, we have adopted power-law creep parameters for diabase given by Shelton and Tullis (1981). Such a rheology with a relatively strong lower crust made up of mafic minerals, is supported by the occurrence of earthquakes in the lower crust in various continental rift zones (Shudofsky et al., 1987; Fuchs et al., 1987). Intraplate stresses must exceed the strength of the lithosphere in order to give rise to the formation of new plate boundaries. We shall elucidate this by comparing the outcome of the modelling of intraplate stresses with rheological models of the lithosphere. We shall then consider the implications for the initiation of rifting and subduction.

3. Initiation of rifting

In this section we discuss some mechanical aspects of the initiation of passive and active rifting (Turcotte and Emmerman, 1983, see Figure 1). The active rifting model has been strongly advocated in the early seventies (Sleep, 1971), especially in the context of studies on the subsidence of passive continental margins. Since then, the frequently observed absence of evidence for erosion and the widespread occurrence of listric normal faults on seismic profiles of passive margins gave rise in the late seventies to the increasing popularity of passive rifting models in the interpretation of subsidence of sedimentary basins (see e.g. Watkins and Drake, 1983). More recent studies (Hellinger and Sclater, 1983) have, however, shown that the occurrence or absence of erosion at the rift shoulders cannot be uniquely interpreted in favor of active or passive rifting mechanisms. In the arguments concerning passive and active rifting mechanisms, relatively little attention has been paid to the rheological properties of the lithosphere. Only recently has the role that thermo-mechanical properties play in the process of passive rifting been appreciated (e.g. Vierbuchen et al., 1983; Sawyer, 1985).

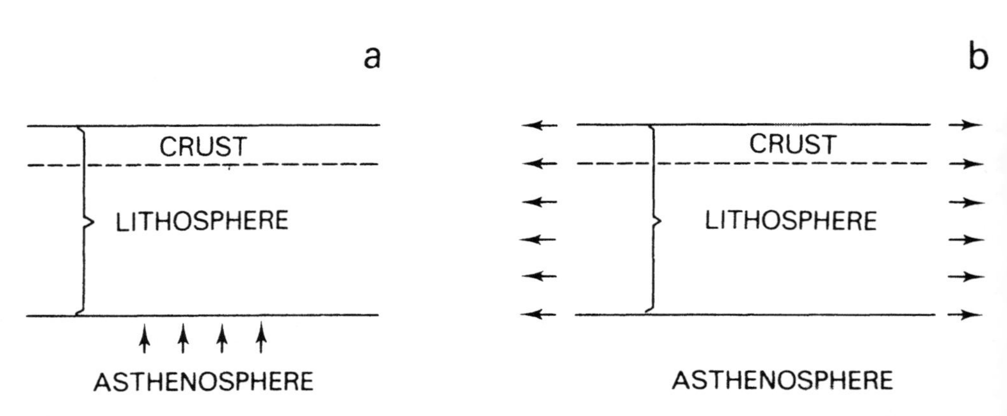

Figure 1. Schematic representation of mechanisms for active rifting and passive rifting. Forces are indicated by arrows. a. (left): active rifting. b. (right): passive rifting of lithosphere.

With a few exceptions (e.g. Cloetingh and Nieuwland, 1984; Houseman and England, 1986), studies of active rifting have been restricted to simplistic rheological models of the lithosphere, adopting either uniform viscoelastic, a purely viscous, or a uniform elastic rheology for the lithosphere. Here we concentrate on model calculations which we performed to investigate the interplay between the forces responsible for rifting of the lithosphere and its material properties. We investigate whether the stresses generated are capable of inducing failure and rifting in the lithosphere. We restrict our attention to the instantaneous response of the lithosphere to the applied forces. The subsequent evolution of the mechanically thinned lithosphere is a subject that falls outside the scope of the present paper.

3.1 Active rifting

Updoming of the lithosphere is caused by the effect of temperature perturbations on the density stratification of the upper mantle. Magnitudes of the temperature-induced buoyancy forces acting on the lower boundary of the lithosphere are based on thermal calculations such as reported by Detrick et al. (1981). We have incorporated in the models the influence of asthenospheric upwelling on the temperature distribution and thus on the rheology of the overlying MSL. In the Hawaiian case the thickness and strength of the MSL is reduced by the reheating process from values appropriate for 100 Ma old oceanic lithosphere (50 km, respectively 8 kbar) to values that vary from 20 to 30 km and 3.5 to 5.5 kbar, respectively. Figure 2 shows part of the finite element mesh used to calculate the response to uplift of rheological attenuated lithosphere. Details of the finite element calculations are given in Cloetingh and Nieuwland (1984). The results of the calculations are summarized in Figure 3. The Figure shows that even with the reduction in thickness and strength of the MSL in the region where uplift is taking place, the amounts of uplift required to induce lithospheric failure are generally in excess of geological estimates. For a comparison with oceanic swells, the selected case of Hawaii is well constrained by geological data. The width of the Hawaii swell is of the order of 400-600 km with a maximum uplift of 1.5 km (Detrick et al., 1981). According to our modelling, however, a 4 km uplift is necessary to induce lithospheric failure. Therefore, we suggest that active rifting alone is not sufficient to induce lithospheric failure in the Hawaiian case. In addition to uplift, tensional stresses associated with plate reorganization in the Pacific (Watts et al., 1980) seem to be required. This agrees with the mechanism for passive rifting worked out for the Farallon plate (Wortel and Cloetingh, 1981; see below).

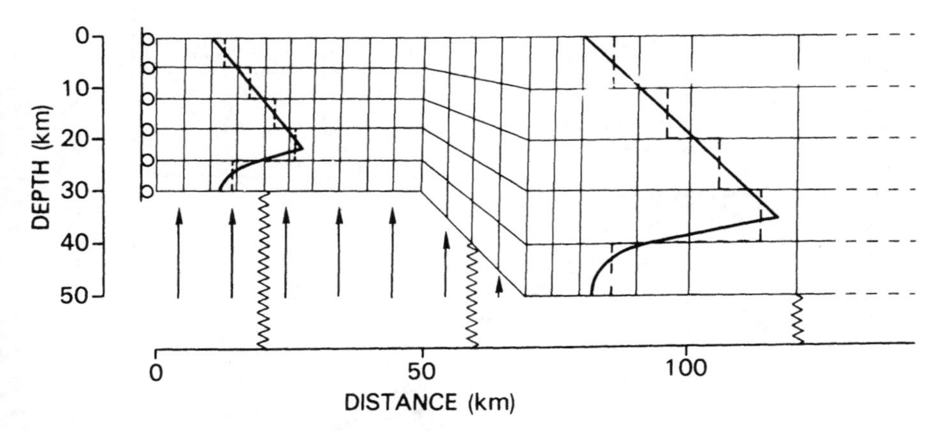

Figure 2. Central part of finite element mesh used for analysis of active rifting of a rheologically attenuated lithosphere. Note the differences in strength and thickness of the mechanically strong part of the lithosphere inside and outside the area of uplift.

Figure 3 also gives the results of the calculations carried out for the case of uplift of rheologically weakened continental lithosphere. As a result of its rheological layering, the effect of a temperature perturbation on continental lithosphere is more severe than the effect on oceanic lithosphere. Strong temperature perturbations will create minima in the lithospheric strength distribution near the boundaries between upper and lower crust and between lower crust and lithospheric mantle. This is due to differences in melt temperature of upper crustal, lower crustal, and mantle rocks. A strong temperature perturbation will shift the position of the brittle-ductile transition most effectively for rocks with a low melting point. This effect reduces the amount of uplift required to induce lithospheric failure in attenuated continental lithosphere considerably, relative to the uplift necessary to generate failure in oceanic lithosphere. Depending on the width of the dome, uplifts of 2-3 km are capable of inducing failure. These estimates are consistent (see Figure 3) with uplift data of major continental rift zones summarized by Withjack (1979) and Crough (1983).This applies in particular to data from the Baikal rift and the Rhinegraben systems. The calculated estimates are conservative in that we have assumed that the lithosphere is deformed at locations unaffected by earlier rifting events. Geological evidence for repeated rifting events, however, is abundant. Weakness zones created by a repetition of rifting events might provide favorable locations for the creation of new failure zones. On the other hand, evidence from rifted continental margins (Steckler and Watts, 1981) shows that faults associated with earlier break-up phases are locked and mechanically healed within a few Ma after the event if the source of heat is removed. Several other effects might further reduce the estimates of the amount of uplift and the forces required to induce lithospheric failure. These include erosion of continental lithosphere during elevation and the effect of gravitationally induced forces associated with uplift (Artyushkov, 1973). The stresses induced by the latter mechanism are a factor of five smaller than the flexural stresses associated with uplift and doming.

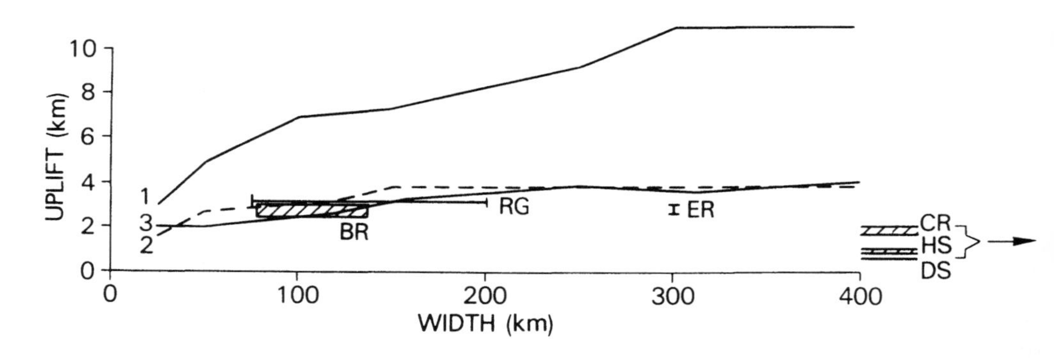

Figure 3. The uplift required to induce failure in the overlying lithosphere, when the effect of the temperature perturbation on the rheology is taken into account. Symbols (1) and (2) denote 100 Ma old oceanic lithosphere reheated to a solidus at depths of 50 km and 35 km, respectively. 3 denotes continental lithosphere thinned to 50 km (Lachenbruch and Sass, 1978). Included in the Figure are geological estimates of the geometry of the uplift for the Baikal rift (BR), the Rhine Graben (RG) (Withjack, 1979), the Ethiopian rift zone (ER), and the Hoggar Swell (HS), Darfur Swell (DS), Cape Verde Rise (CR) and Hawaii Swell (Crough, 1983).

In summary, modelling has shown that the lithosphere will not fail through doming unless this process is associated with weakening and thinning of its mechanically strong part. Such strong perturbations of the thermal structure of the lithosphere can be produced by lithospheric doubling and shifting (Vlaar, 1983). We have found that narrow domes of a few hundred km wide with an uplift of the order of 2 km are capable of inducing failure in rheologically weakened lithosphere. A comparison of the results of the model calculations with continental uplift data offers an explanation for the observation (Crough, 1983) that relatively few thermally induced domes generate new rifts. Particularly striking in this respect is that uplift of the order of 1 km at the Hoggar swell has not been followed by rifting, whereas uplift of the order of 2 km in the Ethiopian domes has (Crough, 1983).

3.2 Passive rifting

Wortel and Cloetingh (1981, 1983) proposed a new mechanism for passive rifting of oceanic plates and the subsequent inception of spreading centres under the influence of age-dependent plate-tectonic forces. This mechanism was based on a case study of the well-constrained plate-tectonic evolution of the eastern Pacific, in which the break-up of the Farallon plate into the Cocos plate and the Nazca plate at 25-30 Ma ago (Hey, 1977) played a central role. Using reconstructed geometries and lithospheric age distributions (Handschumacher, 1976) appropriate for the plate configuration just prior to break-up they were able to calculate the distribution of (age-dependent) ridge-push and slab-pull forces on the plate and, consequently, the induced stress field.

The calculated paleo-stress field (σ) is dominated by tensional stresses (see Figure 4). To induce failure in the vertical column of the MSL the integrated force $F = \int \sigma dz$ must exceed the integrated depth-dependent strength $A = \int \sigma_y \, dz$, where σ_y is the strength profile and z the depth. In Figure 5 we have plotted values for F corresponding with tensional stresses representative of large parts of the Farallon plate. A comparison of these data with (age-dependent) integrated strengths A (solid line) shows that the tensional stresses generated in the plate are capable of inducing its observed rupture and fragmentation. Since the dynamic situation of the Farallon plate and the resulting stress field at that time can, in itself, explain both location and timing of the inception of spreading , it is unnecessary to invoke hotspot activity as a mechanism for its fragmentation (Wortel and Cloetingh, 1981).

Also plotted in Figure 5 are the magnitudes of the integrated forces in the present-day Nazca plate (Wortel and Cloetingh, 1983). Figure 5 shows that the magnitude of the forces in the Nazca plate falls consistently below the integrated strength curve, for ages in excess of 30 Ma, which explains why the Nazca plate has not been subjected to the same fate as the Farallon plate. Note that only the youngest and therefore mechanically weakest part of the plate is near failure. In confirmation of this theory, work on the Nazca plate by Warsi et al. (1983) has shown that a fragmentation process is now active along the Mendana fracture zone, where Wortel and Cloetingh's (1983) model calculations predicted tensional stresses.

Cloetingh and Wortel (1985, 1986) modelled the state of stress in the Indo-Australian plate in order to investigate quantitatively observed variations in tectonic style. As noted by Bergman and Solomon (1985) intraplate earthquakes are a reliable indicator of the intraplate stress field because of the high level of intraplate seismicity in the Indo-Australian plate. Cloetingh and Wortel's (1985, 1986) modelled stress field gives a quantitative explanation for the high level of the stress field, the dominance of compression in the plate's interior (Bergman and Solomon, 1985), the selective occurrence of normal faulting seismicity parallel to its spreading centres (Stein et al., 1987), and the strong variations in observed stress directions. Furthermore, Cloetingh and Wortel (1985, 1986) showed that tensional stresses induced by slab-pull forces associated with subduction of old oceanic lithosphere at various segments of the convergent boundaries of the Indo-Australian plate are transmitted into the interior of the plate, where they affect lithosphere of both oceanic and continental character. This is particularly the case for eastern Australia

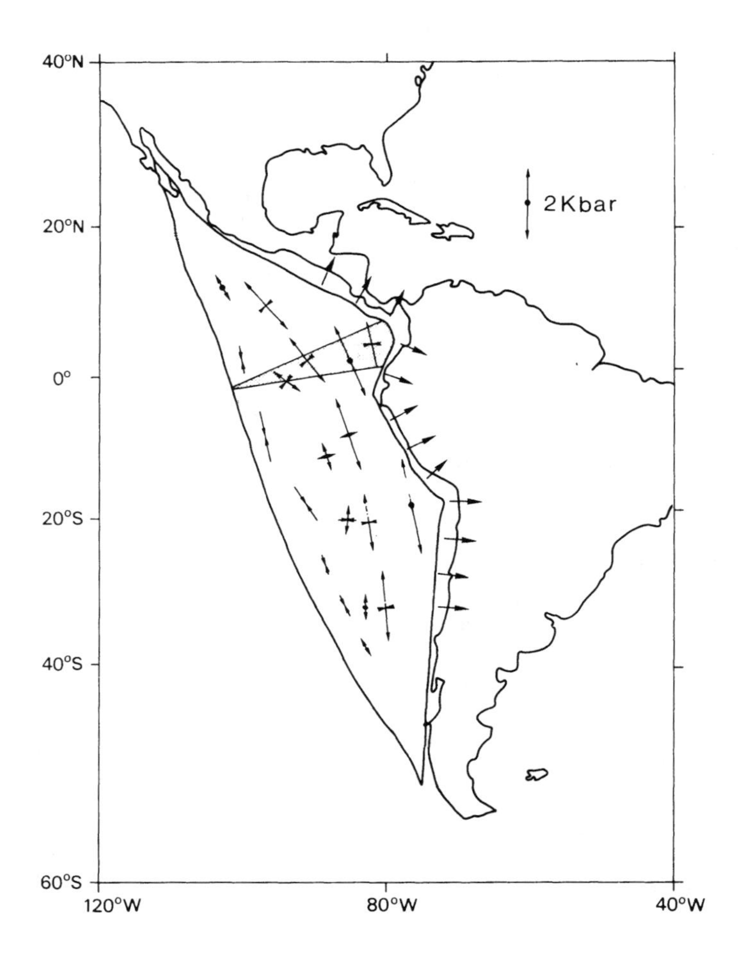

Figure 4. Calculated stress field in the Farallon plate under the reconstructed conditions of 30 Ma ago, prior to the break-up of the Farallon plate. The plotted principal horizontal non-lithostatic stresses are calculated for a 100 km thick elastic plate. The shaded area indicates the tensional zone which developed into the Cocos-Nazca spreading centre (Galapagos Rift). After Wortel and Cloetingh, 1981.

(see Figure 6) , an area characterized by recent volcanic activity, probably associated with a regional tensional stress regime (Duncan and McDougall, 1987). Furthermore, the measured continental flux of crustal Helium in the Great Artesian Basin of Queensland is consistent with fracturing of the crust under the influence of tensional stresses (Torgersen and Clarke, 1985).

The extent of continental thinning is a critical factor for its relative strength with respect to oceanic lithosphere (Steckler and Ten Brink, 1986, see Figure 7). In areas that have undergone earlier phases of lithospheric thinning such as rift zones and the hinge zones of passive continental margins, severely thinned continental lithosphere is probably substantially weaker than oceanic lithosphere and preferential rifting of continental

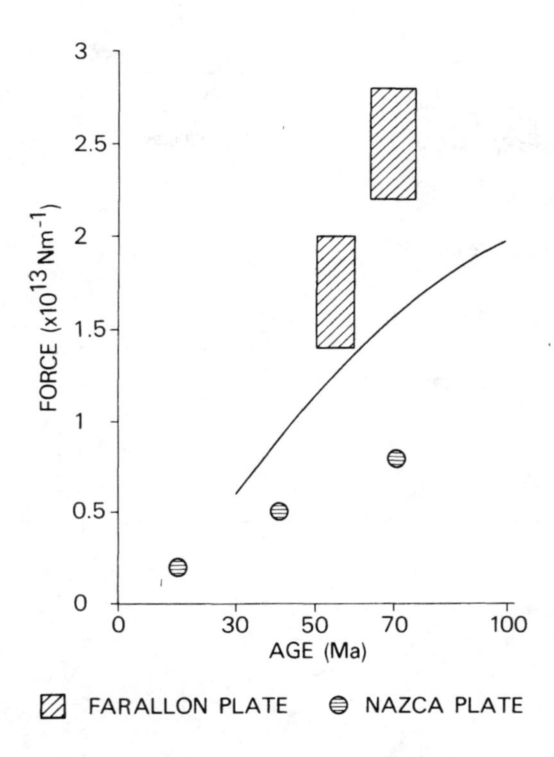

Figure 5. The force required to induce tensional failure in oceanic lithosphere with a depth-dependent rheology (Goetze and Evans, 1979) plotted as a function of lithospheric age (solid line). Boxes indicate stress levels prior to the break-up of the Farallon plate into the Nazca plates, calculated by Wortel and Cloetingh (1981). Circles indicate (intraplate) stress levels in the present-day Nazca plate (Wortel and Cloetingh, 1983). The values given are representative of large parts of the Farallon plate and Nazca plate, respectively.

lithosphere might occur (Vink et al., 1984; Steckler and Ten Brink, 1986). Thermo-mechanical models show that in this case stresses of the order of a few kilobars are required (Houseman and England, 1986; Cloetingh and Nieuwland, 1984). Fragmentation of plates under the influence of slab-pull forces is therefore seen to provide a mechanism for passive rifting of both oceanic and continental lithosphere. Due to its great lithospheric strength, rifting of old oceanic lithosphere is not expected to occur on a large scale. This is in agreement with the observation that, with two exceptions (Stein and Cochran, 1985; Mammericks and Sandwell, 1986), evidence for rifting of old oceanic lithosphere is absent. Whether continents in the interiors of plates break up under the influence of plate-tectonic forces is dependent on their specific thermo-mechanical structure, the position of the continental fragments in the plate relative to the surrounding trench systems, and the variation of the forces acting on each downgoing slab. While we do not wish to fully discuss here the stress regime in the overriding plate (see e.g. England and Wortel, 1980), we conjecture that variations in age-dependent slab-pull forces and associated stress fields of the type discussed here will prove to be of considerable importance in analysis of rifting

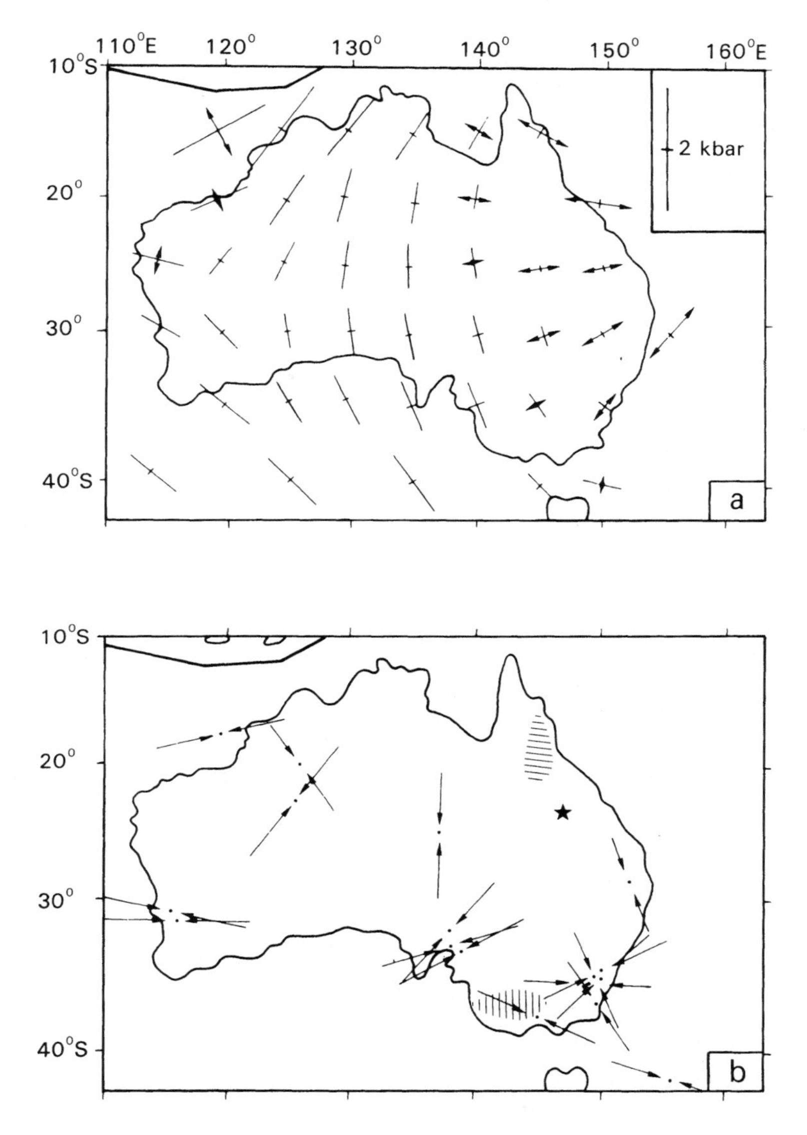

Figure 6. Regional stress field in the Australian continent and peripheral areas. a (top) : calculated stress field, after Cloetingh and Wortel (1986). b (bottom) : Stress orientation data from focal mechanism studies (Lambeck et al., 1984). Hatched areas mark the regions with Late Pliocene to present basaltic volcanism in northern Queensland and Victoria (Duncan and McDougall, 1987). The site of excessive He degassing (Torgersen and Clarke, 1985) is marked by a star.

processes in the overriding plate.

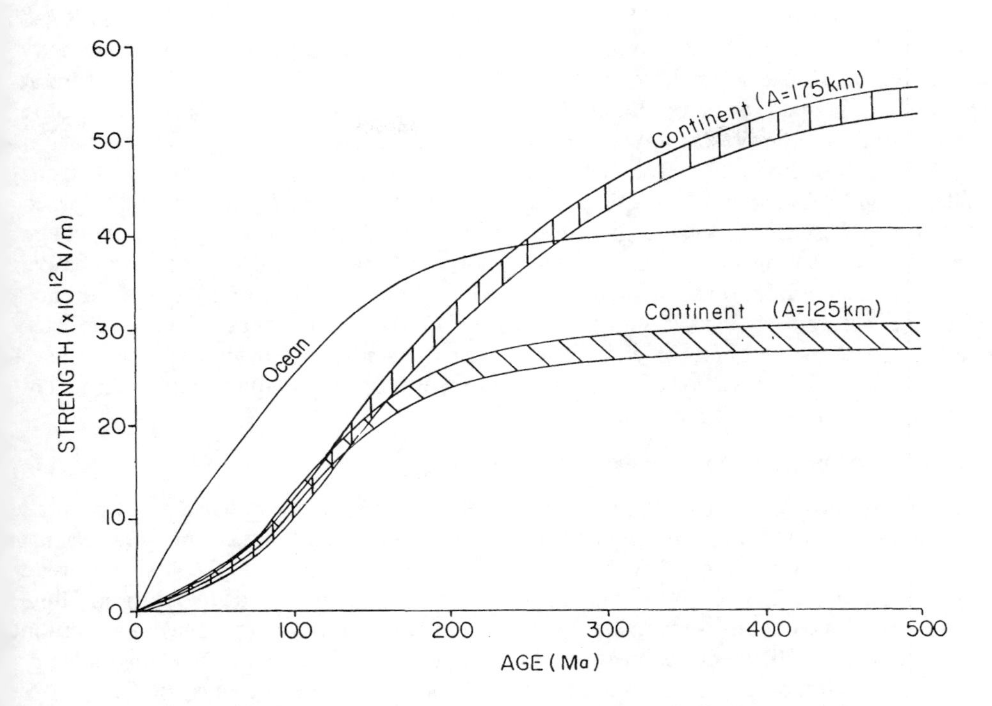

Figure 7. The force required to induce lithospheric failure in oceanic and continental lithosphere as a function of age. The curves are calculated adopting depth-dependent rheologies. Curves for continental lithosphere are calculated for two different thermal thicknesses A of continental lithosphere (after Steckler and Ten Brink, 1986).

4. Initiation of subduction zones

The mechanisms that initiate the formation of new subduction zones are not fully understood (Dickinson and Seely, 1979; Flinn, 1982; Turcotte and Schubert, 1982; Kobayashi and Sacks, 1985). This is in part because there are no obvious present-day examples of new trench formation. Dickinson and Seely (1979) summarize the possible mechanisms for initiation of oceanic lithosphere subduction and distinguish two classes: (1) plate rupture within an oceanic plate or at a passive margin and (2) reversal of the polarity of an existing subduction zone, possibly following a collision of an island arc with a passive margin (see also Mitchell, 1984). A third mechanism is initiation of subduction by inversion of transform faults into trenches (Uyeda and Ben-Avraham, 1972).

We deal here with mechanisms for the formation of new convergent plate boundaries rather than those for polarity reversal. Geological evidence for margins with widely different ages (e.g. Dewey, 1969; Cohen, 1982) support the thesis that passive continental margins in particular might be potential sites for the formation of convergent plate boundaries. Passive margins may, therefore, be expected to play a central role in the Wilson cycle of the opening and closing of oceans (Wilson, 1966).

Several authors (Turcotte and Schubert, 1982; Hynes, 1982) advocate a scenario based on the development of a major oceanic basin with old, and hence cold and gravitationally unstable (Oxburgh and Parmentier, 1977; Vlaar and Wortel, 1976), oceanic lithosphere at its continental boundaries before the basin can be closed. This seems to be more based on a comparison with the present-day size of the Atlantic Ocean rather than by an evaluation of geological observations. The same may be said of arguments for spontaneous foundering and subduction of oceanic lithosphere older than 200 Ma (Hynes, 1982), based on the current absence of oceanic lithosphere older than 200 Ma and the increase with age of the gravitational instability of oceanic lithosphere. The transformation of passive margins into active ones by spontaneous foundering of old unstable lithosphere is inhibited by the great strength of oceanic lithosphere (Kirby, 1980; McAdoo et al., 1985). We focus, therefore, on the state of stress at passive margins and investigate whether the stresses generated there are sufficiently high to induce lithospheric failure and subsequent transformation into an active margin.

4.1 Models for passive to active margin transition

Our model studies (Cloetingh, 1982; Cloetingh et al., 1983) have shown that, in general, flexure induced by sediment loading dominates the stress state at passive margins. In that work we have demonstrated that the continuing accumulation of sediments at passive margins leads to an increase in the induced stress level with the age of the margin. Finite element stress calculations were made for a passive margin in different stages of evolution (Figure 8) for two different sediment loading models using a depth-dependent rheology inferred from extrapolation of rock-mechanics data (after Goetze and Evans, 1979).

As a *reference* we adopted a sediment loading model in which the maximum thickness of the sedimentary wedge at passive margins followed a square-root-of-age relation (Turcotte and Ahern, 1977; Wortel, 1980). This model represents a fair average of the sediment loading histories and resulting thicknesses observed at passive margins (Southam and Hay, 1981). In the second model, the *full load* model, the entire loading capacity of oceanic lithosphere is taken up by the sediments. Sediment loading extends from the continental shelf to the continental rise, with maximum sediment thickness in the outer shelf/continental slope region. Lithospheric strength profiles combining the effect of pressure on brittle behaviour (Byerlee, 1968, 1978) and temperature- and stress dependence on ductile deformation from Goetze's flow laws for dry olivine (Goetze, 1978; Goetze and Evans, 1979) with an assumed strain-rate of $10^{-18}s^{-1}$ (characteristic for sedimentary basin development) were adopted (Cloetingh et al., 1982, 1984). Investigation of gravity anomalies at passive margins (Karner and Watts, 1982) has shown that the mechanical properties of oceanic lithosphere at passive margins are not essentially different from the rheological properties of "standard" oceanic lithosphere as inferred from studies of seamount loading. It should be noted that the models discussed here deal only with the gross features of passive margin evolution which are pertinent to the problem under consideration.

Figure 9 shows the stress field for the case of a reference load on a 100 Ma old passive margin. Differential stresses (σ_H - σ_V) are greatest at the transition from oceanic lithosphere to rift-stage lithosphere. In Figure 10 the stress maximum for a 100 Ma old passive margin is displayed as a function of depth, down to the base of the MSL. The

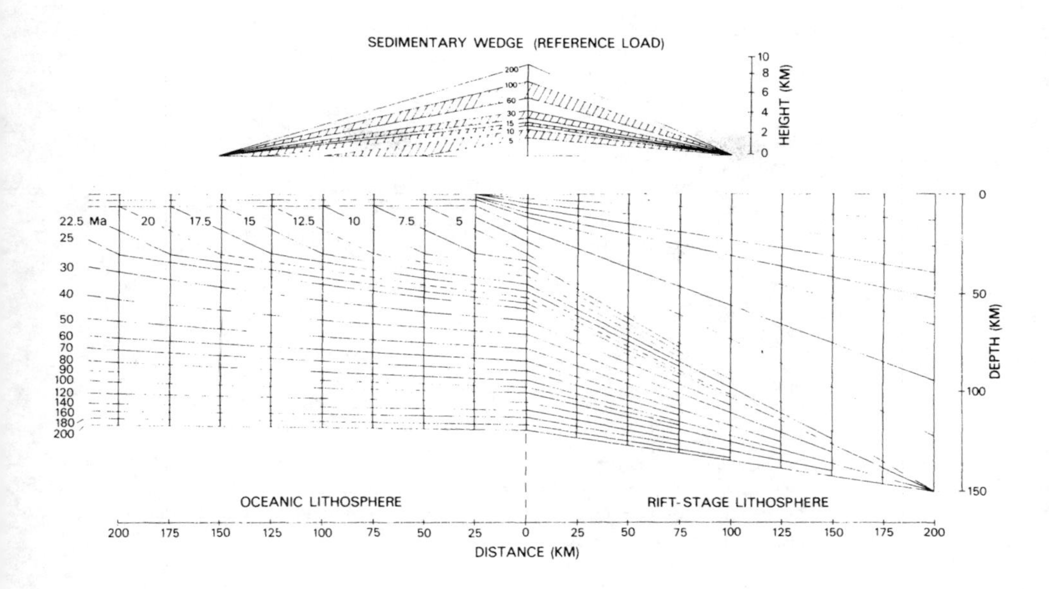

Figure 8. Central part of the finite element mesh employed for analysis of stress state on a passive margin in the interior of a plate not involved in collision or subduction processes. Layers of elements are incrementally added to the lithosphere, whose lower boundary is indicated at various time steps (specified in Ma). A comparison of the lower boundaries at 12.5 Ma and 20 Ma (given by dashed lines) illustrates the growth of the lithosphere in horizontal and vertical directions. The height and the width of the sedimentary wedge , corresponding to sedimentary loading according to the reference model on outer shelf, slope and rise, are given in kilometres at some selected time steps (specified in Ma).

lowermost and uppermost regions of the MSL fail due to the high stresses developed here. However, the main part of the MSL remains in the elastic state as stresses are too low to result in rupture of the lithosphere. The ratio of the maximum stress generated to the maximum strength is essentially independent of lithospheric age since load, strength and thickness of MSL all exhibit the same square-root-of-age behaviour due to the thermal evolution of the lithosphere.

The situation is very much different when the full loading capacity is taken up by the sediments. The surplus of sediment added to the reference load promote high stresses most effectively when deposited on a young (weak) passive margin. This is demonstrated in Figure 11 where we have plotted the ratio of A_y, corresponding to the area of the strength envelope in failure, and the total area A, corresponding to the integrated strength of the MSL. Figure 11 shows that full loading on passive margins with ages below 20 Ma leads to complete failure of the lithosphere. For ages in excess of 20 Ma the relative amount of failure decreases rapidly with age. Therefore, if subduction has not commenced at a passive margin during a youthful stage, continued aging alone will not result in conditions more favourable for the initiation of subduction. This explains the apparent paradox of why gravitationally unstable oceanic lithosphere at the margin of the Atlantic is not subject to subduction (see Kobayashi and Sacks, 1985).

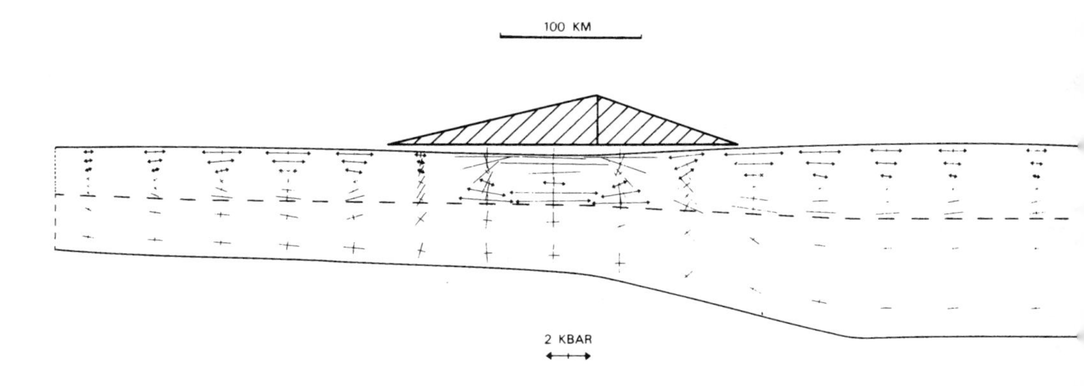

Figure 9. Stresses calculated for a passive margin at an age of 100 Ma based on the reference model of sediment loading. Flexure caused by sediment loading forms the dominant deformation mode at the margin. Principal stresses denoted by arrows are plotted in the undeformed configuration. Stresses are plotted only for the parts of the lithosphere where there is significant deformation.

These findings lead us to propose the modification of the classical sequence of the Wilson cycle concept shown in Figure 12, in which closure of a newly rifted basin occurs by initiation of subduction of young lithosphere, with implications for tectonics and volcanism which differ appreciably from those associated with deep subduction zones (Vlaar, 1983; Cloetingh et al., 1984; Vlaar and Cloetingh, 1984). In particular, it explains the absence of island arc volcanism in Wilsonian orogenies (e.g. Trumpy, 1982) and the emplacement of ophiolites, fragments of young gravitationally stable oceanic lithosphere, on the adjacent continent during the transformation of a passive margin, features that do not conform to the standard concepts of subduction of oceanic lithosphere and the long timespan of the classical form of the Wilson cycle. An interesting analogy for the destruction of passive margins of small oceanic basins exists in the form of marginal basins, whose evolution is characterized by a short timespan (less than 20 Ma, Taylor and Karner, 1983) between their creation and collapse. As pointed out by these authors, the tectonic setting of several back-arc basins suggests that they represent more than just a passive response to kinematic boundary conditions. Taylor and Karner (1983) therefore argue that neither the global nor local models adequately explain the conditions necessary for back-arc basin formation. They conclude that better understanding of the processes associated with the initiation of subduction and rifting is required.

As noted by Church and Stevens (1971), much of the geological evidence in collision orogens points to closing of smaller ocean basins rather than large oceans of the scale of the present- day Atlantic. Investigations of Alpine orogeny (Trumpy, 1982) have provided strong evidence for closure of small oceanic basins at an early stage after opening, as have studies of the evolution of the Appenines (Winterer and Bosellini, 1981). In fact, the Alpine basins, being characterized by young oceanic lithosphere, transcurrent faulting, extensive sediment loading, and a compressive tectonic regime, were ideally suited for the transformation of passive into active margins. Frequently, however, only indirect and

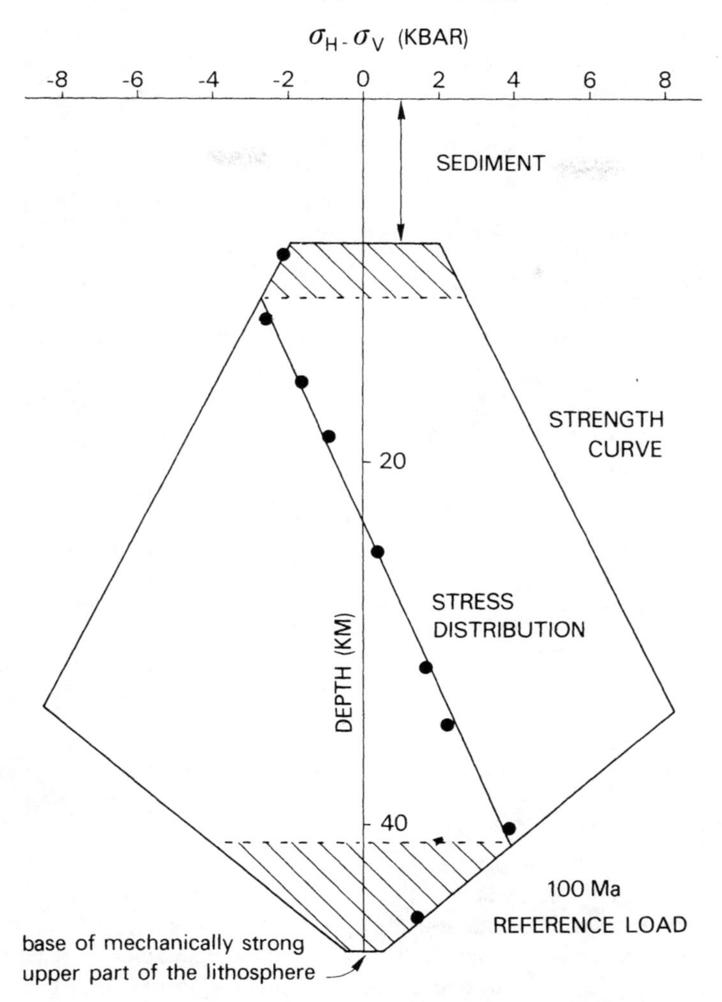

Figure 10. Comparison of stresses generated at 100 Ma old passive margin, under reference loading, with lithospheric strength. Strength envelope and results of stress calculation (solid dots) as a function of depth down to the base of MSL at the point of maximum flexure (see also figure 9). The stress distribution is given by the line inside the strength envelope connecting the solid dots. Differential stresses (σ_H - σ_V) are plotted against depth. Sign convention for the stresses: tension positive, compression negative. Zero-strength is assumed for the sediments. Differences between brittle strengths in compression and tension are ignored. Hatched areas in the upper and lower part of the MSL denote failure by brittle fracture and ductile flow, respectively.

debatable arguments are available for the estimation of the size of the proposed oceans. This applies in particular to the "type locality" of the Wilson cycle, the Iapetus Ocean, which is of unknown width (Windley, 1977). In fact, the best evidence for the Iapetus Ocean is the occurrence of ophiolite belts (Zwart and Dornsiepen, 1978). Several authors, (e.g. Dewey, 1976; Nicolas and Le Pichon, 1980; Spray, 1984) favour obduction of

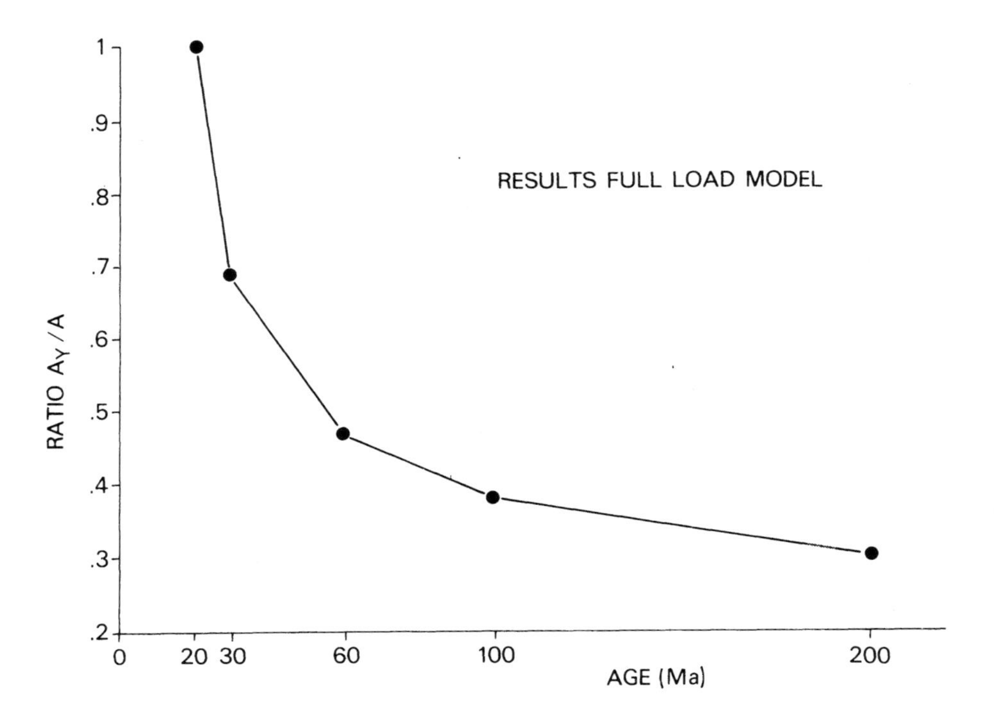

Figure 11. Ratio of A_y (the hatched area within the strength envelope, see figure 10) to A (total area of the envelope) as a function of lithospheric age for the full load model. For ages below 20 Ma lithospheric failure and subsequently initiation of subduction is induced.

thermally young immature oceanic lithosphere to account for the short time gap documented between ophiolite formation and ophiolite emplacement.

Similarly, near-horizontal subduction during closure of a small ocean basin provides an important element in the dynamical evolution of foreland thrust belts (Stockmal et al., 1986). Our findings also shed light on the far greater degree of basement fragmentation observed under foreland basins in collisional settings compared to their Andean analogs (Allen et al., 1986). Evolutionary frameworks of the Wilson cycle in terms of opening and closing of wide oceans often seem to be more inspired by an actualistic comparison with the present size of the Atlantic Ocean, than by a consideration of more pertinent geological observations. During its evolution, the Atlantic Ocean passed through a transition from a narrow to a wide ocean basin, without the formation of a system of subduction zones at its margins. Apparently, optimal loading conditions for transformation of passive into active margins were not reached at this stage, while further aging has not made the passive margins more susceptible to initiation of subduction. To rely too heavily on such an actualistic analogy might, therefore, be very misleading. Moreover, estimated widths of oceans inferred from palinspastic restorations are often routinely increased by an additional 1000 km meant to represent the lengths of slabs consumed in modern circum-Pacific subduction zones, associated with subduction of old oceanic lithosphere (e.g. Williams,

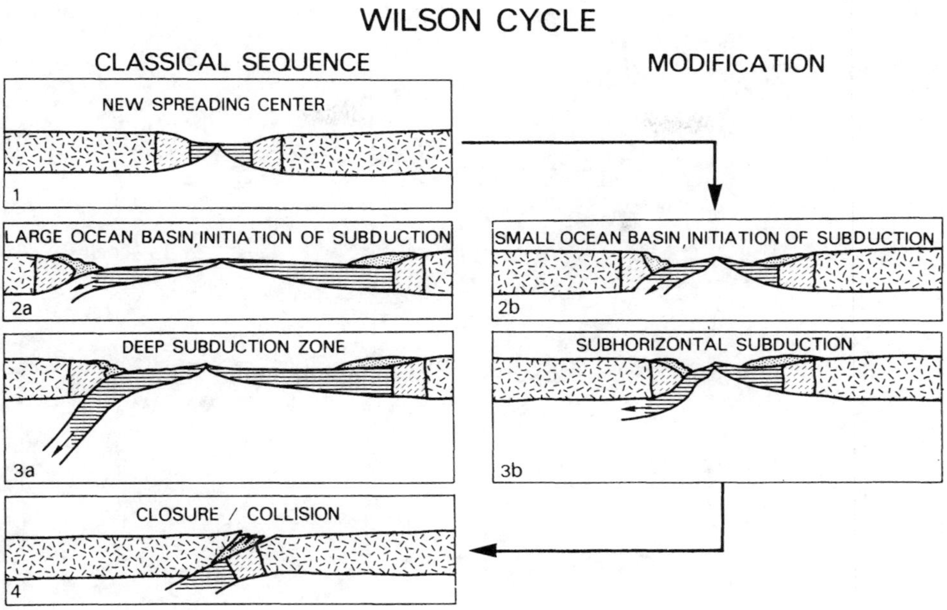

Figure 12. Scenarios for the Wilson Cycle. Left: classical scenario, in which rifting is followed by the formation of a large ocean basin. Initiation of subduction involves old oceanic lithosphere, which results in a deep subduction zone. Right: preferred scenario for Wilsonian orogenies, in which closure of the newly rifted basin occurs through initiation of subduction of young oceanic lithosphere leading to either shallow angle subduction or to obduction of gravitationally stable buoyant lithosphere. Implications for tectonics and volcanism strongly differ from those associated with deep subduction zone. Patterns indicate passive-margin sediments and continental, rift-stage, and oceanic lithosphere.

1980). Such reconstructions exclude a priori the possibility of the closing of a young oceanic basin and might even lead one to overlook possible interesting consequences of the subduction of young lithosphere. Therefore, we suggest that a more critical appraisal is made of the role large oceans play in the Wilson cycle concept.

4.2 The role of external forces

In the preceding section the flexural stresses induced by sediment loading were shown to be of the order of several kbars. As noted previously, the regional stress field caused by plate tectonic forces is dominated by concentration of slab-pull forces. For passive margins located in the interiors of plates not involved in subduction or collision processes, the local stress field induced by sediment loading dominates the stresses induced by the remaining plate-tectonic forces, which in general is of the order of a few hundred bars (e.g. Richardson et al., 1979). We have shown that under such conditions evolution of passive margins to maturity will not in itself lead to initiation of subduction.

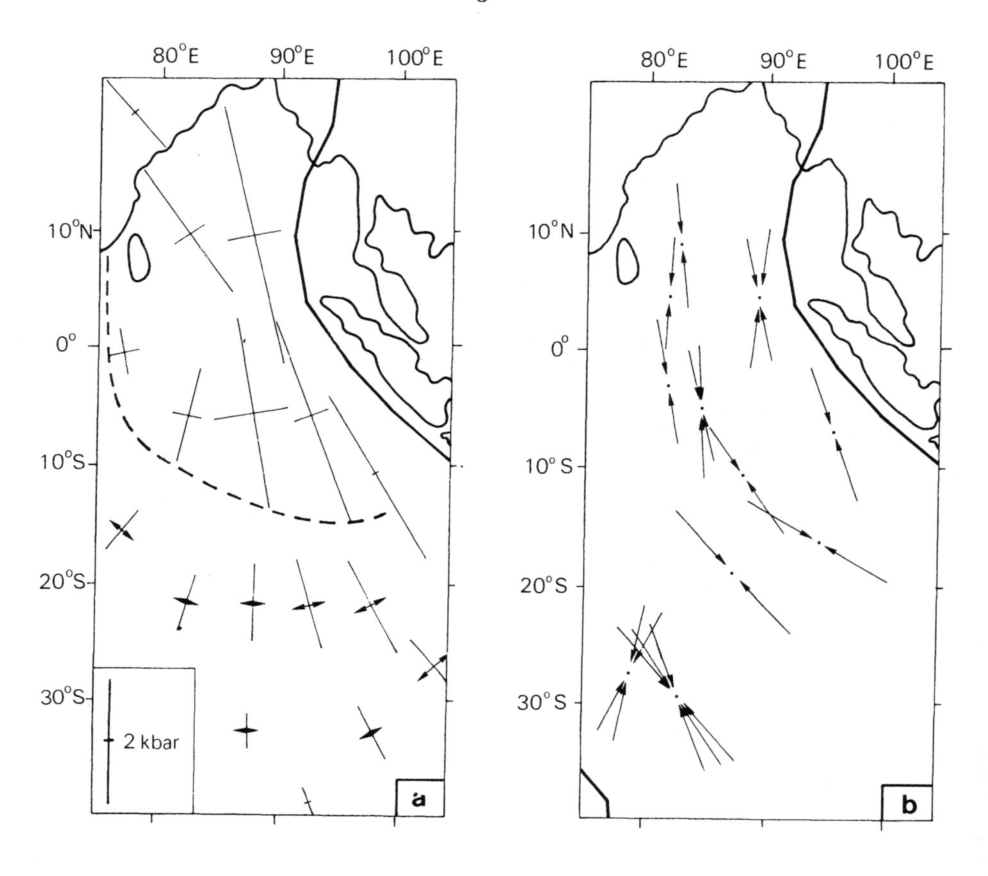

Figure 13. Regional stress field in the northeastern Indian Ocean. a (left). Calculated stress field after Cloetingh and Wortel (1986). The dashed line is the southern limit of the observed deformation in the northeastern Indian Ocean (Geller et al., 1983). Plotted are principal horizontal non-hydrostatic stresses averaged over a uniform elastic plate with a reference thickness of 100 km. b (right). The orientation of maximum horizontal compressive stress inferred from a focal mechanism study by Bergman and Solomon (1985).

Thus, in general, external forces in addition to sediment loading are required to cause rupture of the lithosphere. Figure 5 has given a comparison of the force required to induce complete lithospheric tensional failure and calculated intraplate tensional stress levels in the eastern Pacific. An example of a different tectonic setting is found in the northeastern Indian Ocean, where a high level of compressional stress (Figure 13) is induced by the focusing of resistive forces associated with the Himalayan collision zone and the subduction of young oceanic lithosphere in the northern part of the Sunda arc (Cloetingh and Wortel, 1985; 1986). The response of the oceanic lithosphere in the northeastern Indian Ocean to the compressional stresses is, however, not one of the lithosphere failure, but one of buckling (Weissel et al., 1980; McAdoo and Sandwell, 1985; Zuber, 1987). Figure 14 shows that the estimates of the forces required to buckle the ocean floor in the northeastern Indian Ocean given by McAdoo and Sandwell (1985) are in excellent agreement with the calculated stress levels shown in Figure 13. As noted by Cloetingh et al. (1983) the folding of the oceanic lithosphere caused by regional compressional stresses amplified by sediment loading might be a preparatory stage for the initiation of a new subduction zone in the northeastern Indian Ocean. More recently, strong independent evidence for the presence of a nascent diffuse plate boundary in the area has been presented by Wiens et al. (1985, 1986) and Wiens (1986) based on an inversion of current plate velocities and a study of the seismo-tectonics of the area.

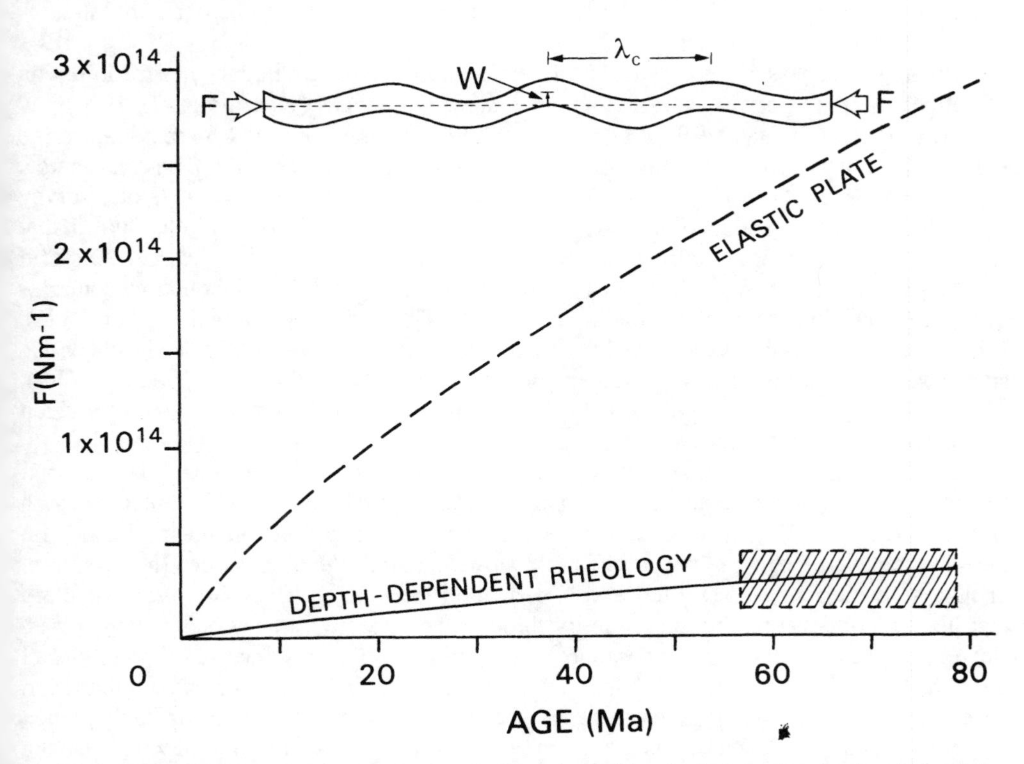

Figure 14. Buckling load versus age for oceanic lithosphere with a depth-dependent rheology inferred from rock-mechanics studies (Goetze and Evans, 1979) (solid curve) and for fully elastic oceanic lithosphere (dashed curve) (after McAdoo and Sandwell, 1985). Boxes indicate the stress levels calculated for the area in the northeastern Indian Ocean (Cloetingh and Wortel, 1986, see figure 13) where folding of oceanic lithosphere under the influence of compressional stresses has been observed (McAdoo and Sandwell, 1985; Geller et al., 1983).

5. Discussion

We have shown that concentration of slab-pull forces provides an effective mechanism for fragmentation of plates and the formation of new spreading centres. We have also demonstrated that evolution of passive margins to maturity will not in itself lead to initiation of subduction and shown that, in general, external forces in addition to sediment loading are required.

The action of external forces will be most effective when young passive margins are pre-stressed by thick sedimentary wedges. McKenzie (1977) demonstrated that a large slab pull is required to overcome the resistive forces active in the process of trench formation. The compressive stresses necessary to sustain the further development of subduction zones that involve young stable lithosphere (McKenzie, 1977; England and Wortel, 1980), may be provided by stress concentration of plate tectonic forces during plate reorganizations (Cloetingh and Wortel, 1985) and are an order of magnitude smaller than the stresses required to rupture the lithosphere.

Existing weakness zones located within plates might be more suitable sites for initiation of subduction than passive margins. As such, spreading ridges (Turcotte et al., 1977) and transform faults (Uyeda and Ben-Avraham, 1972) have been suggested. A spreading ridge may be thought of as the extreme case (0 Ma) of a young passive margin, and represents a rupture zone in the lithosphere. The former view is consistent with the results of a survey of recently initiated subduction zones in the Pacific (Karig, 1982), whichs shows that subduction zones initiated during the Neogene are frequently on the sites of transform faults. In other cases there has been rejuvenation of pre-existing subduction zones or geometrical adjustment of plate boundaries (Okal et al., 1986; Kroenke and Walker, 1986). It is widely held that the present Pacific coastline of North and South America has been the site of semi-continuous eastward subduction since the Late Proterozoic (Windley, 1977). The different plate configurations and thermal regimes at that time may have provided conditions more suitable for initiation of subduction.

Thus, in general, we do not expect initiation of subduction of oceanic lithosphere at passive margins to play a leading role in plate reorganizations such as documented by Rona and Richardson (1978). In an oceanic plate attached to a subduction zone the pull acting on the subducting slab can be concentrated to a sufficient stress level to induce the formation of new spreading centres (Wortel and Cloetingh, 1981; 1983). Therefore, we conjecture that plate reorganizations occur primarily through the formation of new spreading ridges, since stress relaxation in the lithosphere occurs much more easily via this process than through the formation of new subduction zones. During such a process new subduction zones might be subsequently created at the sites of already present transform faults, when the new spreading direction has a component perpendicular to the direction of the transform fault.

Better understanding of the mechanics of formation of new plate boundaries in general, and the processes underlying the transition of passive margins into active margins in particular, will enhance our insight into the evolution of sedimentary basins. We have shown (Cloetingh et al., 1985; Cloetingh, 1986) that the associated reorganizations of lithospheric stress field are recorded in the stratigraphic record of sedimentary basins. Specific short-term fluctuations (on time scales of a few Ma and longer) in apparent sea levels inferred from passive margins and intracratonic basins can now be associated quantitatively with particular plate-tectonic reorganizations. Alternatively, the seismostratigraphic record might provide a new source of information on paleo-stress fields (Cloetingh, 1986). Furthermore, the opening and closure of oceanic basins, with the associated changes in the area/age distribution of the ocean floor, has been shown (Heller and Angevine, 1985) to be the main controlling feature on the occurrence of long-term (time scales of tens of Ma) sea level cycles. Further work on modelling of paleo-stress field alongs the lines set out by Wortel and Cloetingh (1981) is required to more fully document differences in the roles of initiation of rifting and the initiation of subduction in the tectonic evolution of the plates' interiors.

6. Conclusions

Lateral variations in the age of the downgoing lithosphere at subduction zones are an effective mechanism for fragmentation of oceanic and continental plates and formation of new spreading centres. Aging of passive margins will not in itself lead to a spontaneous initiation of subduction. In general, the formation of new subduction zones at passive margins requires a focusing of external plate tectonic forces. The action of these external forces will be most effective when young passive margins are pre-stressed by thick sedimentary wedges. Plate reorganizations probably take place primarily through the formation of new spreading ridges, because stress relaxation in the lithosphere occurs much more effectively via this process than through the formation of new subduction zones.

References

Allen, P., Homewood, P. and Williams, G., 1986. Foreland basins: an introduction, *Spec. Publ. Intern. Assoc. Sediment.*, **8**, 8-15.

Artyushkov, E., 1973. Stress in the lithosphere caused by crustal thickness inhomogeneities, *J. Geophys. Res.*, **78**, 7675-7708.

Bergman, E.A. and Solomon, S.C., 1985. Earthquake source mechanisms from bodywave inversion and intraplate tectonics in the northern Indian Ocean, *Phys. Earth Planet. Int.*, **40**, 1-23.

Bodine, J.H., Steckler, M.S. and Watts, A.B., 1981. Observations of flexure and the rheology of the oceanic lithosphere, *J. Geophys. Res.*, **86**, 3695-3707.

Brace, J.H. and Kohlstedt, D.L., 1980. Limits on lithospheric stress imposed by laboratory experiments, *J. Geophys. Res.*, **85**, 6248-6252.

Byerlee, J.D., 1968. Brittle-ductile transition in rocks, *J. Geophys. Res.*, **73**, 4741-4750.

Byerlee, J.D., 1978. Friction of rocks, *Pure Appl. Geophys.*, **116**, 615-626.

Caldwell, J.G. and Turcotte, D.L., 1979. Dependence of the thickness of the elastic oceanic lithosphere on age, *J. Geophys. Res.*, **84**, 7572-7576.

Carter, N.L., Anderson, D.A., Hansen, F.D. and Kranz, R.L., 1981. Creep and creep rupture of granitic rocks, *Am. Geophys. Un. Geophys. Monogr.*, **24**, 61-82.

Chapple, W.M. and Tullis, T.E., 1977. Evaluation of the forces that drive the plates, *J. Geophys. Res.*, **82**, 1967-1984.

Church, W.R. and Stevens, R.K., 1971. Early Paleozoic ophiolite complexes of the Newfoundland Appalachians as mantle-oceanic crust sequences, *J. Geophys. Res.*, **76**, 1460-1466.

Cloetingh, S., 1982. *Evolution of passive continental margins and initiation of subduction zones*, Ph.D. Thesis Univ. Utrecht, 111 pp.

Cloetingh, S., 1986. Intraplate stresses: a new tectonic mechanism for fluctuations of relative sealevel, *Geology*, **14**, 617-620.

Cloetingh, S., McQueen, H. and Lambeck, K., 1985. On a tectonic mechanism for regional sealevel variations, *Earth Planet. Sci. Lett.*, **75**, 157-166.

Cloetingh, S. and Nieuwland, F., 1984. On the mechanics of lithospheric stretching and doming: a finite element analysis, *Geol. Mijnb.*, **63**, 315-322.

Cloetingh, S. and Wortel, R., 1985. Regional stress field of the Indian plate, *Geophys. Res. Lett.*, **12**, 77-80.

Cloetingh, S. and Wortel, R., 1986. Stress in the Indo-Australian plate, *Tectonophysics*, **132**, 49-67.

Cloetingh, S.A.P.L., Wortel, M.J.R. and Vlaar, N.J., 1982. Evolution of passive continental margins and initiation of subduction zones, *Nature*, **297**, 139-142.

Cloetingh, S.A.P.L., Wortel, M.J.R. and Vlaar, N.J., 1983. State of stress at passive margins and initiation of subduction zones, *Am. Ass. Petrol. Geol. Memoir*, **34**, 717-723.

Cloetingh, S.A.P.L., Wortel, M.J.R. and Vlaar, N.J., 1984. Passive margin evolution, initiation of subduction and the Wilson cycle, *Tectonophysics*, **109**, 147-163.

Cohen, C.R., 1982. Model for a passive to active continental margin transition: implications for hydrocarbon exploration, *Am. Ass. Petrol. Geol. Bull.*, **66**, 708-718.

Crough, S.T., 1983. Hotspot swells, *Ann. Rev. Earth Planet. Sci.*, **11**, 165-193.

Detrick, R.S., Von Herzen, R.P., Crough, S.T., Epp, D. and Fehn, U., 1981. Heat flow on the Hawaiian swell and lithospheric reheating, *Nature*, **292**, 142-143.

Dewey, J.F., 1969. Continental margins: a model for conversion of Atlantic-type to Andean type, *Earth Planet. Sci. Lett.*, **6**, 189-197.

Dewey, J.F., 1976. Ophiolite obduction, *Tectonophysics*, **31**, 93-120.

Dickinson, W.R. and Seely, D.R., 1979. Structure and stratigraphy of fore-arc regions, *Am. Ass. Petrol. Geol. Bull.*, **63**, 2-31.

Duncan, R.A. and McDougall, I., 1987. Time-space relationships for Cainozoic intraplate volcanism in eastern Australia, the Tasman Sea, and New Zealand, in: *Intraplate volcanism in eastern Australia and New Zealand*, R.W. Johnson and S.R. Taylor (eds.), Australian Academy of Sciences, in press.

England, P. and Wortel, R., 1980. Some consequences of the subduction of young slabs, *Earth Planet. Sci. Lett.*, **47**, 403-415.

Flinn, E.A., 1982. The International Lithosphere Program, *EOS Trans. Am. Geophys. Un.*, **63**, 209-210.

Forsyth, D.W. and Uyeda, S., 1975. On the relative importance of the driving forces of plate motion, *Geophys. J. R. Astr. Soc.*, **43**, 163-200.

Fuchs, K., Bonjer, K., Gajewski, D., Luschen, E., Prodehl, C., Sandmeier, K. J., Wenzel, F. and Wilhelm, H., 1987. Crustal evolution of the Rhinegraben area, I. Exploring the lower crust in the Rhinegraben rift by unified geophysical experiments, *Tectonophysics*, in press.

Geller, C.A., Weissel, J.K. and Anderson, R.N., 1983. Heat transfer and intraplate deformation in the central Indian Ocean, *J. Geophys. Res.*, **88**, 1018-1032.

Goetze, C., 1978. The mechanisms of creep in olivine, *Philos. Trans. Roy. Soc. Lond.*, Ser. A, **288**, 99-119.

Goetze, C. and Evans, B., 1979. Stress and temperature in the bending lithosphere as constrained by experimental rock mechanics, *Geophys. J. R. Astr. Soc.*, **59**, 463-478.

Hales, A.L., 1969. Gravitational sliding and continental drift, *Earth Planet. Sci. Lett.*, **6**, 31-34.

Handschumacher, D.W., 1976. Post-Eocene plate tectonics of the eastern Pacific, *Am. Geophys. Un. Geophys. Monogr.*, **19**, 177-202.

Heller, P.L. and Angevine, C.L., 1985. Sea-level cycles during the growth of Atlantic-type oceans, *Earth Planet. Sci. Lett.*, **75**, 417-426.

Hellinger, S.J. and Sclater, J.G., 1983. Some comments on two-layer extensional models for the evolution of sedimentary basins, *J. Geophys. Res.*, **88**, 8251-8270.

Hey, R., 1977. Tectonic evolution of the Cocos-Nazca spreading center, *Geol. Soc. Am. Bull.*, **88**, 1404-1420.

Houseman, G.A. and England, P.C., 1986. A dynamical model of lithosphere extension and sedimentary basin formation, *J. Geophys. Res.*, **91**, 719-729.

Hynes, A., 1982. Stability of the oceanic tectosphere-a model for early Proterozoic intracratonic orogeny, *Earth Planet. Sci. Lett.*, **61**, 333-345.

Karig, D.E., 1982. Initiation of subduction zones: implications for arc evolution and ophiolite development, *Geol. Soc. Lond. Spec. Publ.*, **10**, 563-576.

Karner, G.D. and Watts, A.B., 1982. On isostasy at Atlantic-type continental margins, *J. Geophys. Res.*, **87**, 2923-2948.

Kirby, S.H., 1980. Tectonic stresses in the lithosphere: constraints provided by the experimental deformation of rocks, *J. Geophys. Res.*, **85**, 6353-6363.

Kirby, S.H., 1983. Rheology of the lithosphere, *Rev. Geophys. Space. Phys.*, **21**, 1458-1487.

Kirby, S.H., 1985. Rock mechanics observations pertinent to the rheology of the continental lithosphere and the localization of strain along shear zones, *Tectonophysics*, **119**, 1-27.

Kobayashi, K. and Sacks, I.S., 1985. Preface to structures and processes in subduction zones, *Tectonophysics,* **112**, 561 pp.

Kroenke, L.W. and Walker, D.A., 1986. Evidence for the formation of a new trench in the western Pacific, *EOS Trans. Am. Geophys. Un.*, **67**, 145-146.

Kuznir, N. and Karner, G.D., 1985. Dependence of the flexural rigidity of the continental lithosphere on rheology and temperature, *Nature,* **316**, 138-142.

Lachenbruch, A.H. and Sass, J.H., 1978. Models of the extending lithosphere and heat flow in the Basin and Range Province, *Geol. Soc. Am. Memoir,* **152**, 209-250.

Lambeck, K., McQueen, H., Stephenson, R.A. and Denham, D., 1984. The state of stress within the Australian continent, *Ann. Geophys.,* **2**, 723-741.

Lister, C.R.B., 1975. Gravitational drive on oceanic plates caused by thermal contraction, *Nature,* **257**, 663-665.

Mammericks, J. and Sandwell, D., 1986. Rifting of old oceanic lithosphere, *J. Geophys. Res.*, **91**, 1975-1988.

McAdoo, D.C., Martin, C.F. and Poulouse, S., 1985. Seasat observation of flexure: evidence for a strong lithosphere, *Tectonophysics,* **116**, 209-222.

McAdoo, D.C. and Sandwell, D.T., 1985. Folding of oceanic lithosphere, *J. Geophys. Res.*, **90**, 8563-8568.

McKenzie, D.P., 1969. Speculation on the causes and consequences of plate motions, *Geophys. J. R. Astr. Soc.*, **18**, 1-32.

McKenzie, D.P., 1977. The initiation of trenches: a finite amplitude instability, *Maurice Ewing Series Am. Geophys. Un.*, **1**, 57-62.

Meissner, R. and Strehlau, J., 1982. Limits of stresses in continental crusts and their relation to the depth frequency of shallow earthquakes, *Tectonics,* **1**, 73-89.

Mitchell, A.H.G., 1984. Initiation of subduction of post-collision foreland thrusting and back-thrusting., *J. Geodyn.*, **1**, 103-120.

Molnar, P., Freedman, D. and Shih, J.S.F., 1979. Lengths of intermediate and deep seismic zones and temperatures in downgoing slabs of lithosphere, *Geophys. J. R. Astr. Soc.*, **56**, 41-54.

Morgan, P. and Baker, B.H., 1983. Processes of continental rifting, *Tectonophysics*, 680 pp.

Nicolas, A. and LePichon, X., 1980. Thrusting of young lithosphere in subduction zones with special reference to structures in ophiolite peridotites, *Earth Planet. Sci. Lett.*, **46**, 397-406.

Okal, E.A., Woods, D.F. and Lay, T., 1986. Intraplate deformation in the Samoa-Gilbert-Ralik area: a prelude to a change of plate boundaries in the southwest Pacific ?, *Tectonophysics*, **132**, 69-77.

Oxburgh, E.R. and Parmentier, E.M., 1977. Compositional and density stratification in oceanic lithosphere-causes and consequences, *J. Geol. Soc. Lond.*, **133**, 343-355.

Palmason, G., 1982. Continental and oceanic rifts, *Geodyn. Series Am. Geophys. Un.*, **8**, 309 pp.

Richardson, R.M., Solomon, S.C. and Sleep, N.H., 1979. Tectonic stress in the plates, *Rev. Geophys. Space. Phys.*, **17**, 981-1019.

Richter, F., 1973. Dynamical models for sea floor spreading, *Rev. Geophys. Space. Phys.*, **11**, 223-287.

Rona, P.A. and Richardson, E.S., 1978. Early Cenozoic global plate reorganization, *Earth Planet. Sci. Lett.*, **40**, 1-11.

Sawyer, D.S., 1985. Brittle failure in the upper mantle during extension of continental lithosphere, *J. Geophys. Res.*, **90**, 3021-3025.

Shelton, G. and Tullis, J., 1981. Experimental flow laws for crustal rocks, *EOS Trans. Am. Geophys. Un.*, **62**, 1111.

Shudofsky, G., Cloetingh, S., Stein, S. and Wortel, R., 1987. Unusually deep earthquakes in East Africa: constraints on the thermo-mechanical structure of a continental rift system, *Geophys. Res. Lett..*, **14**, 741-744.

Sibson, R.H., 1983. Continental fault structure and the shallow earthquake source, *J. Geol. Soc. Lond.*, **140**, 741-767.

Sleep, N.H., 1971. Thermal effects of the formation of Atlantic continental margins by continental break-up, *Geophys. J. R. Astr. Soc.*, **24**, 325-350.

Solomon, S.C., Sleep, N.H. and Richardson, R.M., 1975. On the forces driving plate motions : inferences from absolute plate velocities and intraplate stress, *Geophys. J. R. Astr. Soc.*, **42**, 769-801.

Southam, J.R. and Hay, W.W., 1981. Global sedimentary mass balance and sealevel changes, *In: C.Emiliani, editor, The oceanic lithosphere, The Sea,* **7**, 1617-1684.

Spray, J.G., 1984. Possible causes and consequences of upper mantle decoupling and ophiolite displacement, *Geol. Soc. Lond. Spec. Publ.*, **13**, 255-268.

Steckler, M.S. and Ten Brink, U.S., 1986. Lithospheric strength variations as a control on new plate boundaries: examples from the northern Red Sea region, *Earth Planet. Sci. Lett.*, **79**, 120-132.

Steckler, M.S. and Watts, A.B., 1981. Subsidence history and tectonic evolution of Atlantic type continental margins, *Geodyn. Series Am. Geophys. Un.*, **6**, 184-196.

Stein, C.A. and Cochran, J.R., 1985. The transition between the Sheba ridge and Owen Basin: rifting of old oceanic lithosphere, *Geophys. J. R. Astr. Soc.*, **81**, 47-74.

Stein, S., Cloetingh, S., Wiens, D.A. and Wortel, R., 1987. Why does near ridge extensional seismicity occur primarily in the Indian Ocean?, *Earth Planet. Sci. Lett.*, **82.**, 107-113.

Stockmal, G.S., Beaumont, C. and Boutilier, R., 1986. Geodynamic models of convergent margin tectonics: transition from rifted margin to overthrust belt and consequences for foreland-basin development, *Am. Ass. Petrol. Geol. Bull.*, **70**, 181-190.

Taylor, B. and Karner, G.D., 1983. On the evolution of marginal basins, *Rev. Geophys. Space. Phys.*, **21**, 1727-1741.

Torgersen, T. and Clarke, W.B., 1985. Helium accumulation in groundwater,I, an evaluation of sources and the continental flux of crustal ^{4}He in the Great Artesian Basin, Australia, *Geochim. Cosmochim. Act.*, **49**, 1211-1218.

Trumpy, R., 1982. Alpine paleogeography: a reappraisal, In: K.Hsu, editor, *Mountain Building Processes*, Academic Press, London, 149-156.

Turcotte, D.L. and Ahern, J.L., 1977. On the thermal and subsidence history of sedimentary basins, *J. Geophys. Res.*, **82**, 2762-3766.

Turcotte, D.L. and Emerman, S.H., 1983. Mechanisms of active and passive rifting, *Tectonophysics*, 39-50.

Turcotte, D.L., Haxby, W.F. and Ockendon, J.R., 1977. Lithospheric instabilities, *Maurice Ewing Series Am. Geophys. Un.*,**1**, 63-69.

Turcotte, D.L. and Oxburgh, E.R., 1976. Stress accumulation in the lithosphere, *Tectonophysics*, **35**, 183-199.

Turcotte, D.L. and Schubert, G., 1982. *Geodynamics: applications of continuum physics to geological problems*, Wiley, New York, 450 pp.

Uyeda, S. and Ben-Avraham, Z., 1972. Origin and development of the Philippine Sea, *Nature*, **240**, 176-178.

Vierbuchen, R.C., George, R.P. and Vail, P.R., 1983. A thermal-mechanical model of rifting with implications for outer highs on passive continental margins, *Am. Ass. Petrol. Geol. Memoir*, **34**, 765-778.

Vink, G.E., Morgan, W.J. and Zhao, W.-L., 1984. Preferential rifting of continents : a source of displaced terranes, *J. Geophys. Res.*, **89**, 10072-10076.

Vlaar, N.J., 1983. Thermal anomalies and magmatism due to lithospheric doubling and shifting, *Earth Planet. Sci. Lett.*, **65**, 322-330.

Vlaar, N.J. and Cloetingh, S.A.P.L., 1984. Orogeny and ophiolites: plate tectonics revisited with reference to the Alps, *Geol. Mijnb.*, **63**, 159-164.

Vlaar, N.J. and Wortel, M.J.R., 1976. Lithospheric aging instability and subduction, *Tectonophysics*, **32**, 331-351.

Warsi, W.E.K., Hilde, T.W.C. and Searle, R.C., 1983. Convergence structures of the Peru-Trench between 10 S and 14 S, *Tectonophysics*, **99**, 313-329.

Watkins, J.S. and Drake, C.L., 1983. Studies in Continental Margin.Geology, *Am. Ass. Petrol. Geol. Memoir*, **34**, 801 pp.

Watts, A.B., Bodine, J.H. and Ribe, N.M., 1980. Observations of flexure and the geologic evolution of the Pacific Ocean Basin, *Nature*, **283**, 532-537.

Weissel, J.K., Anderson, R.N. and Geller, C.A., 1980. Deformation of the Indo-Australian plate, *Nature*, **287**, 284-291.

Wiens, D.A., 1986. Historical seismicity near Chagos: deformation of the equitorial region of the Indo-Australian plate, *Earth Planet. Sci. Lett.*, **76**, 350-360.

Wiens, D.A., De Mets, C., Gordon, R.G., Stein, S., Angus, D., Engeln, J.F., Lundgren, P., Quible, D., Stein, C., Weinsten, S. and Woods, D.F., 1985. A diffuse plate boundary model for Indian ocean tectonics, *Geophys. Res. Lett.*, **12**, 429-432.

Wiens, D.A. and Stein, S., 1983. Age dependence of oceanic intraplate seismicity and implications for lithospheric evolution, *J. Geophys. Res.*, **88**, 6455-6468.

Wiens, D.A., Stein, S., DeMets, C., Gordon, R. and Stein, C., 1986. Plate tectonic models for Indian Ocean "intraplate" deformation, *Tectonophysics*, **132**, 37-48.

Williams, H., 1980. Structural telescoping across the Apalachian orogen and the minimum width of the Iapetus Ocean, *Geol. Assoc. Can. Spec. Paper*, **20**, 421-440.

Wilson, J.T., 1966. Did the Atlantic close and then re-open?, *Nature*, **211**, 676-681.

Windley, B.F., 1977. *The evolving continents*, Wiley, Chichester, 385 pp.

Winterer, E.L. and Bosellini, A., 1981. Subsidence and sedimentation of Jurassic passive continental margin Southern Alps, Italy, *Am. Ass. Petrol. Geol. Bull.*, **65**, 394-421.

Withjack, M., 1979. An analytical model of continental rift fault patterns, *Tectonophysics*, **59**, 59-81.

Wortel, R., 1980. *Age-dependent subduction of oceanic lithosphere*, Ph.D. Thesis Univ. Utrecht, 147 pp.

Wortel, M.J.R., 1984. Spatial and temporal variations in the Andean subduction zone, *J. Geol. Soc. Lond.*, **141**, 783-791.

Wortel, R. and Cloetingh, S., 1981. On the origin of the Cocos-Nazca spreading center, *Geology*, **9**, 425-430.

Wortel, R. and Cloetingh, S., 1983. A mechanism for fragmentation of oceanic plates, *Am. Ass. Petrol. Geol. Memoir*, **34**, 793-801.

Wortel, M.J.R. and Cloetingh, S.A.P.L., 1985. Accretion and lateral variations in tectonic structure along the Peru-Chile Trench, *Tectonophysics*, **112**, 443-462.

Wortel, M.J.R. and Vlaar, N.J., 1978. Age-dependent subduction of oceanic lithosphere beneath western South America, *Phys. Earth Planet. Int.*, **17**, 201-208.

Zuber, M.T., 1987. Compression of oceanic lithosphere: an analysis of intraplate deformation in the Central Indian Basin, *J. Geophys. Res.*, **92**, 4817-4825.

Zwart, H.J. and Dornsiepen, U.R., 1978. The tectonic framework of central and western Europe, *Geol. Mijnb.*, **57**, 627-654.

S.A.P.L. CLOETINGH and M.J.R. WORTEL, Department of Theoretical Geophysics, University of Utrecht, PO Box 80.021, 3508 TA Utrecht, The Netherlands.

Chapter 17

Supercontinent breakup: effect on eustatic sea level and the oceanic heat flux

C.L. Angevine, S.R. Linneman and P.L. Heller

First-order sea-level highstands have occured only twice during Phanerozoic time and in both cases have followed supercontinent breakups by 50—100 Ma. We propose that a major control on first order sea-level cycles is the formation of new ocean basins and destruction of preexisting ocean basins. During breakup the global plate generation rate increases as new spreading ridges are created. As older basins are consumed, in order to conserve total ocean area, plate generation rates gradually decrease. Taken together, these variations in the global plate generation cause sea level, as well as the ocean heat flux, to first rise and then fall as new ocean basins open. The eustatic highstand generally occurs later than the oceanic heat flux maximum which, in turn, follows the time of peak plate generation rates. In contrast to other models, our mechanism does not require anomalously high spreading rates during the Late Cretaceous highstand.

1. Introduction

The record of relative sea-level variation is well-documented for most of the Phanerozoic (Fig. 1). Sea level has peaked twice, once during the Ordovician and, more recently, during the Cretaceous. In both instances the breakup of a supercontinent preceded the sea level highstand by less than 100 Ma. The preferred driving mechanism for secular sea-level variations is changes in the volume of the mid-ocean ridge system (cf. Hays and Pitman, 1973) although other factors, such as hot spot activity (Schlanger et al., 1981) may also play a role. The mid-ocean ridge volume is affected by changes in spreading rate, the

N.J. Vlaar, G. Nolet, M.J.R. Wortel & S.A.P.L. Cloetingh (eds), Mathematical Geophysics, 389-399.

length of the ridge system, and the orientation of ridges with respect to subduction zones (Hays and Pitman, 1973; Pitman, 1977; Parsons, 1982). When ridge volumes increase seawater is displaced and sea level rises; sea level falls as ridge volumes diminish. A number of workers (Russell, 1977; Valentine and Moores, 1970; Worsley et al., 1984; Heller and Angevine, 1985) suggest that there may be a natural link between sea-level variations and supercontinent breakup. The explanation lies in the manner in which the volume of the ridge system varies during the course of the Wilson cycle. As new ridge segments are created between fractions of the former supercontinent the volume of the ridge system increases. If the new basins continue to open then older, pre-existing basins (including ridge segments) must be consumed in order to conserve total seafloor area and the volume of the ridge system must decrease. Thus it is expected that sea level should first rise and then fall after the breakup of a supercontinent in general agreement with the observed record.

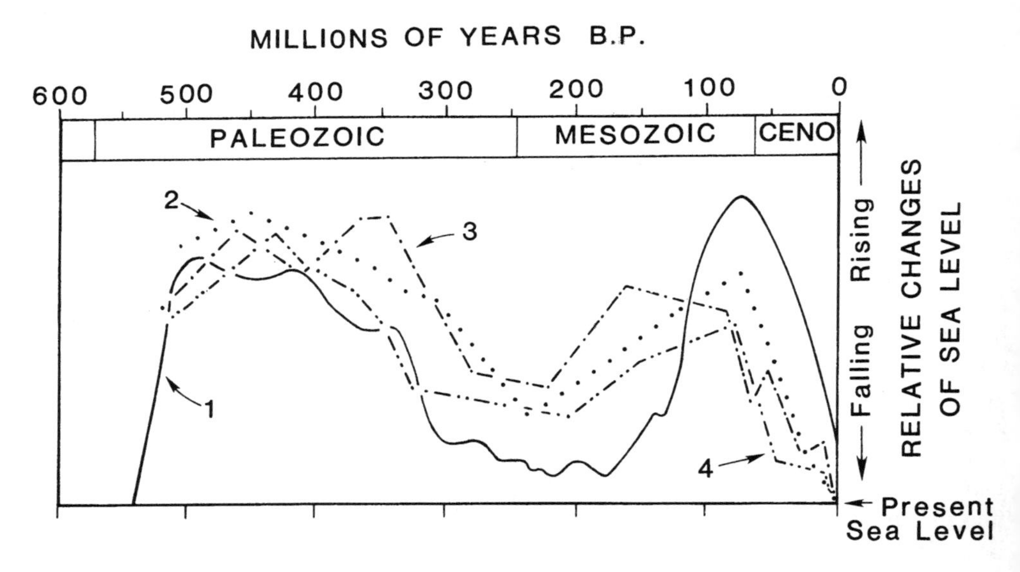

Figure 1. First-order relative sea-level histories for the Phanerozoic based on (1) the interpretation of seismic reflection data (Vail et al., 1977), and continental flooding histories for (2) the world, (3) the U.S.S.R., and (4) North America (Hallam ,1977).

Several methods can be used to obtain an absolute sea-level history for the Cenozoic and the late Mesozoic (Hallam, 1984). The oldest relies on continental flooding histories which are obtained by measuring the areal extent of marine sediments through time. Hypsometric curves show how much sea level must rise or fall to flood some percentage of the continent. A recent analysis of flooding data by Harrison et al. (1981) indicates the sea level has fallen by 173 m in the past 100 Ma. Similar data is interpreted by Bond (1979) to require a fall of only 110 m during the same interval. Because studies from individual continents tend to yield very different sea-level histories, the estimates given above represent weighted averages of the individual histories. Several workers (Southam and Hay, 1981; Wyatt,

1986) argue that hypsometric distributions were considerably deeper in the past, when larger land masses existed, and that estimates based on present distributions may underestimate the eustatic variation by as much as 100 m. This point aside, it is not clear why any such average should even approximate the true eustatic variation.

Sleep (1976) has taken a more direct approach to the problem, using the present elevation of Cretaceous marine sediments in Minnesota (assumed to be unaffected by epeirogenic displacements) to infer that eustatic sea level has fallen by 325 m since the Cenomanian. Sahagian (1987) has since revised this estimate downward to 275 m by taking into account the amount of post-glacial rebound yet to occur in Minnesota. To show decisively that this part of North America has not experienced significant amounts of epeirogenic uplift since the Cretaceous will, however, be a difficult task.

Eustatic variations have also been determined by analyzing the subsidence history of passive continental margins. Deviations from the subsidence predicted by thermal subsidence models are attributed to sea level fluctuations (Watts, 1982). Analyses of wells located off the eastern coasts of the United States and Canada (Watts and Steckler, 1979; Watts and Thorne, 1984) and isopach patterns in the North Sea (Thorne and Bell, 1983) indicate that sea level has fallen by approximately 100 m since the mid Cretaceous. An important drawback to this technique is that water depths, which are difficult to determine accurately, must be included in the models.

An independent estimate of eustatic sea-level variations can be obtained by determining how the total volume of the ocean basins, V, changes with time. Sea level variations, $\Delta h(t)$, measured from some time $t = 0$, are directly related to changes in basin volume according to:

$$\Delta h = \frac{\rho_m - \rho_w}{A_O \rho_m} \left[V(t) - V(0) \right] \tag{1}$$

where A_O is the area of seafloor, ρ is density, and the subscripts m and w refer to mantle and seawater, respectively. The density ratio in Eqn. 1 accounts for the isostatic deflection of the seafloor due to water loading. Equation 1 is incomplete in that is does not include a correction for continental flooding, but this correction is relatively small (Turcotte and Burke, 1978). The volume of the ocean basins can be calculated by convolving the depth-age relation, $d(\tau)$, for the seafloor with the area-age distribution of the seafloor, $dA/d\tau$ (Parsons, 1982; Heller and Angevine, 1985):

$$V(t) = \int_0^{t_{max}} d(\tau) \frac{dA}{d\tau} d\tau \tag{2}$$

where t_{max} is the age of the oldest seafloor. The area-age distribution simply specifies how much seafloor of a given age exists. At present the distribution may be determined using marine magnetic anomalies. Determining the area-age distribution for some time in the past is of course more difficult because seafloor is continually being subducted, destroying the record of spreading.

Hays and Pitman (1973) and Pitman (1978) made reconstructions of the area-age distributions by using fracture zone lineations and magnetic anomaly patterns to find stage-pole locations and calculate opening rates; they then work backward in time calculating changes in ocean basin volume. Hays and Pitman (1973) originally calculated a fall of 520 m since the Cretaceous highstand but this figure was subsequently revised downwards to 350 m (Pitman, 1978) on the basis of new magnetic anomaly data. Kominz (1984) emphasizes the many difficulties in making these reconstructions, including uncertainties in absolute dating of magnetic anomalies, errors in estimating the length of subducted ridges and inaccurate stage-pole locations. Kominz's (1984) estimates for the Cretaceous highstand, based on the time-scale of Larson et al. (1982), range from 90 to 310 m, with a preferred value of 200 m above present sea level. An additional 50 to 60 m must be added to these estimates to account for the overall increase in continental ice volume since the Eocene (Pitman, 1978). While Kominz's (1984) preferred estimate is significantly larger than the 100 to 200 m highstand obtained in most of the stratigraphic studies discussed above, it is in excellent agreement with the results of Sahagian (1987).

Because there is no consensus on the magnitude of the Cretaceous highstand it is important to consider independent approaches to the problem. In this chapter we will discuss an alternate technique (Berger and Winterer, 1974) for reconstructing the area-age distribution of the seafloor and derive sea level and oceanic heat flux histories from these reconstructions. We will show that the breakup of a supercontinent may result in sea-level variations of less than 10 %. When additional tectonoeustatic and glacioeustatic factors are considered we find that, during the Cretaceous, sea level may have been 220 m higher than present.

2. Area-age distributions

Berger and Winterer (1974) noted that the world's oceans have distinct area-age distributions. In oceans surrounded by subduction zones, such as the Pacific, the distribution is approximately triangular (Figure 2b) because there is much more young seafloor present than old seafloor. The Pacific's area-age distribution can be expressed as

$$\frac{dA_P}{d\tau} = \begin{cases} C_P\,(1 - \dfrac{t}{t_P}) & 0 \le t \le t_P \\ 0 & t > t_P \end{cases} \tag{3}$$

where C_P is the rate of plate generation and t_P is the age of the oldest seafloor in the Pacific. Berger and Winterer (1974) and Parsons (1982) have noted that the rate of consumption per unit age (C_P/t_P) must be constant, on average, to maintain a triangular distribution. Over time, equal areas of young and old seafloor must be consumed. In contrast an ocean such as the Atlantic, in which there is no subduction, will have a rectangular area-age distribution (Figure 2a) assuming a constant rate of plate generation, C_A:

$$\frac{dA_A}{d\tau} = \begin{cases} C_A & 0 \le t \le t_A \\ 0 & t > t_A \end{cases} \tag{4}$$

where T_A is the age of the oldest seafloor in the Atlantic. The total ocean area, given by the sum of the areas beneath each area-age distribution, can be expressed as

$$A_O = C_A t_A + \tfrac{1}{2} C_P t_P \tag{5}$$

If the area of the oceans is conserved as the Atlantic continues to open then the Pacific must be consumed, presumably through the subduction of young as well as old seafloor. It seems reasonable to assume that the slope of the area-age distribution (i.e. the rate of consumption per unit age) remains constant as the Pacific closes (Figure 2b). Thus, in the Pacific both the rate of plate generation as well as the age of the oldest seafloor will decrease with time.

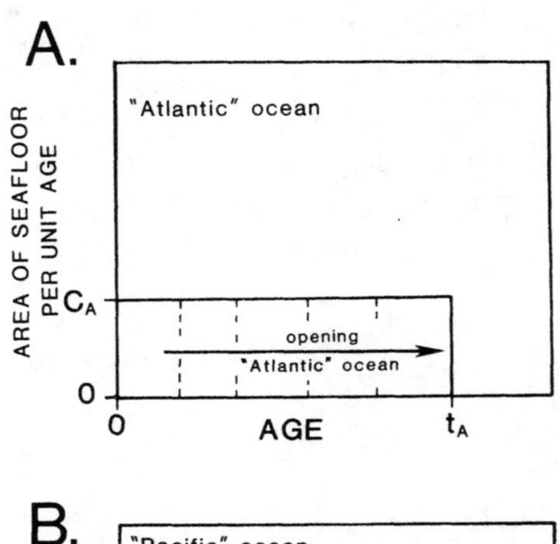

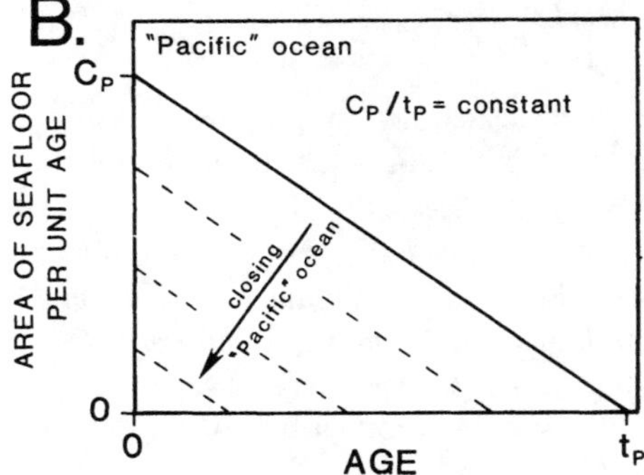

Figure 2. Idealized area-age distributions for the Atlantic (A) and Pacific (B) oceans. The Atlantic ocean is assumed to open a a constant rate C_A. Closure of the Pacific is accomplished by subduction of ridges and young seafloor as well as older seafloor, such that the slope of the distribution, C_P/t_P, remains constant.

Now, consider the evolution of the global area-age distribution after the breakup of a supercontinent and formation of a single " Atlantic " ocean. Just prior to breakup there would be, presumably, a " Pacific " ocean with a triangular distribution. Assuming a uniform rate of plate generation in the new Atlantic ocean, its area-age distribution will expand as shown in Figure 2a. The Pacific's plate generation rate and the age of its oldest seafloor will decrease as the Atlantic ages:

$$C_P = [2\beta(A_O - C_A t_A)]^{1/2} \tag{6}$$

$$t_P = [2(A_O - C_A t_A)/\beta]^{1/2} \tag{7}$$

where $\beta = C_P/t_P$ is the rate of consumption per unit age. Heller and Angevine (1985) discuss the evolution of the global area-age distribution in more detail.

3. Sea level variations

Having provided a model for the variations through time, we may obtain sea-level histories by determining the volumes of the Atlantic and Pacific oceans. The depth-age relations for the seafloor can be approximated by

$$d(\tau) = d_o - (d_o - d_r)\exp(-\tau/a) \tag{8}$$

where d_o = 6.4 km, d_r = 2.5 km and a = 63 Ma (Parson's and Sclater, 1977). This approximation overestimates the depth of young seafloor (Davis and Lister, 1974) but is adequate for determining variations in the volume of the ocean basins. The volume of the Atlantic ca be expressed as

$$V_A = C_A t_A \left\{ d_o - \frac{a}{t_A}(d_o - d_r)[1 - \exp(-t_A/a)] \right\} \tag{9}$$

similarly, the volume of the Pacific can be determined from

$$V_P = \tfrac{1}{2} C_P t_P \left\{ d_o - \frac{2a}{t_P}(d_o - d_r)\left[1 - \frac{a}{t_P} + \frac{a}{t_P}\exp(-t_P/a)\right]\right\} \tag{10}$$

We obtain the sea-level histories shown in Figure 3, using these volumes and three different opening rates for the Atlantic. In determining these histories we have assumed $\beta = 1.44 \times 10^{-8}$ km²/a and $A_O = 3.01 \times 10^8$ km².

In the three cases considered highstand occurs at 70 to 80 Ma after breakup, depending on the opening rate of the Atlantic. This is approximately the same time lag between the breakup of Pangea and the late Cretaceous highstand (Fig. 1). The magnitude of the sea-level variation is somewhat less than 100 m. For the more likely case of a protracted breakup, highstand can be delayed even longer but the magnitude of the variation is similar (Heller and Angevine, 1985). Worsley et al. (1984) obtain sea-level variations of 200 m using a similar model, but their method of calculating basin volumes is incorrect.

4. Heat-flux variations

As the oceanic area-age distribution varies so too will the total heat flux through the seafloor. Several-heat flux histories have been determined using the eustatic record.

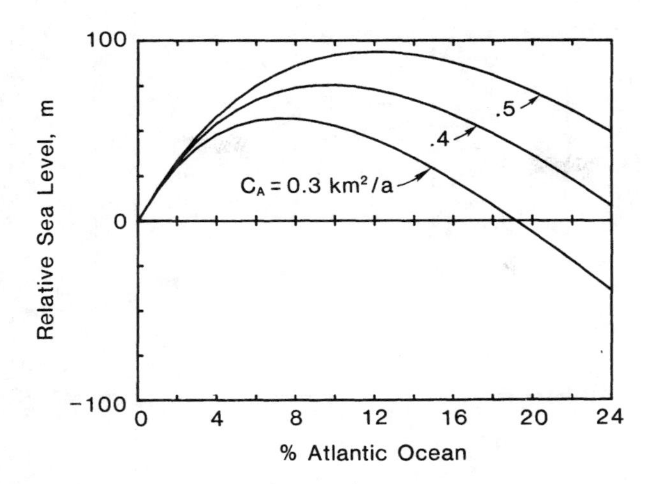

Figure 3. Relative sea level, as a function of area occupied by the Atlantic, for three different Atlantic opening rates. Highstand is reached at 70 to 75 Ma after supercontinent breakup. If the Atlantic began to open at 175 Ma ago, and now occupies 24 % of the total ocean area, then its average opening rate is 0.42 km^2/a. Our model predicts that a passive rearrangement of the oceanic area-age distribution, initiated by the breakup of a supercontinent, has caused sea level to fall by 70 m since the early Cretaceous.

Turcotte and Burke (1978) assume a rectangular area-age distribution and show a linear relation between sea level and oceanic heat flux. Harrison (1980) uses a triangular distribution with a constant plate generation rate and obtains an exponentional relation between sea level and heat flux. Both studies assume that sea-level fluctuations are due only to changes in the area-age distribution and predict that the oceanic heat flux is greatest at times of eustatic highstand.

We can determine the oceanic heat flux by convolving the heat flow-age relation, $q(\tau)$, with the area-age distribution of the seafloor:

$$Q = \int_0^{t_{max}} q(\tau) \frac{dA}{d\tau} d\tau \tag{11}$$

Heat flow is related to the age of the seafloor by (Sclater et al., 1980):

$$q(\tau) = k_m T_m / (\pi \kappa \tau)^{1/2} \tag{12}$$

where T_m is the asthenosphere temperature, k_m and κ are the thermal conductivity and diffusivity, respectively, of the lithosphere. The Atlantic heat flux can be written as

$$Q_A = 2 k_m T_m C_A (t_A / \pi \kappa)^{1/2} \tag{13}$$

while the Pacific heat flux is

$$Q_P = \frac{4}{3} k_m T_m C_P (t_P / \pi \kappa)^{1/2} \tag{14}$$

The total oceanic heat flux is simply the sum of these two contributions. Because we are primarily interested in how Q varies with time, we show the percent change in Q as a

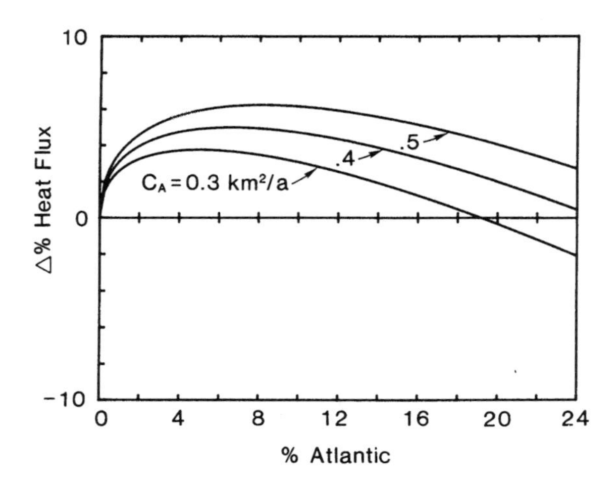

Figure 4. Variation in oceanic heat flux as a function of the fractional area occupied by the Atlantic, for three different Atlantic opening rates. Maximum heat flux occurs at 50 to 60 Ma after supercontinent breakup. The eustatic highstand occurs some 20 Ma after the heat flux peak.

function of the area occupied by the Atlantic (Figure 4).

The variation in the oceanic heat flux is broadly similar to the eustatic variation. An important difference is that the heat flux peaks at 50 to 60 Ma after breakup, some 20 Ma prior to the eustatic highstand. The magnitude of the heat flux variation is less than 10 %.

5. Discussion

Our discussion of the oceanic area-age distribution are based on the present distribution and pattern of consumption, rather than the partially-preserved spreading history, determined from marine magnetic anomalies and fracture zone geometries, preferred by other workers (Hays and Pitman, 1973; Pitman, 1978; Kominz, 1984). The two approaches give very different histories for the global plate generation rate, as shown in Figure 5. Our reconstructions, based on an average opening rate of 0.42 km^2/a for the Atlantic, predict that plate generation rates have never been much higher than the present value of 3.0 km^2/a. Neither does our model show a one-to one correlation between sea level and plate generation rates. Kominz's (1984) preferred estimate requires that plate generation rates during the Cretaceous highstand were almost twice the present rate.

The main problem with models that invoke high plate generation rates to explain the Cretaceous highstand is that they produce a surfeit of Cretaceous age seafloor which is not observed in the area-age distribution. To remove excess seafloor, consumption must be concentrated in that age range. For instance, Kominz's (1984) model predicts that 53.3 $\times 10^6$ km^2 of seafloor was created between 110 and 100 Ma age of which all but 17.5 $\times 10^6$ km^2 (Parsons, 1982; Fig. 2) has been consumed. The rate of consumption per unit age would have to average 3.41 $\times 10^{-8}$ km^2/a^2 over the past 105 Ma in order to remove this excess. Although this rate is not unreasonable, we note that the measured rate of consumption per unit age for seafloor in the age range 110 Ma - 100 Ma is only 2.4

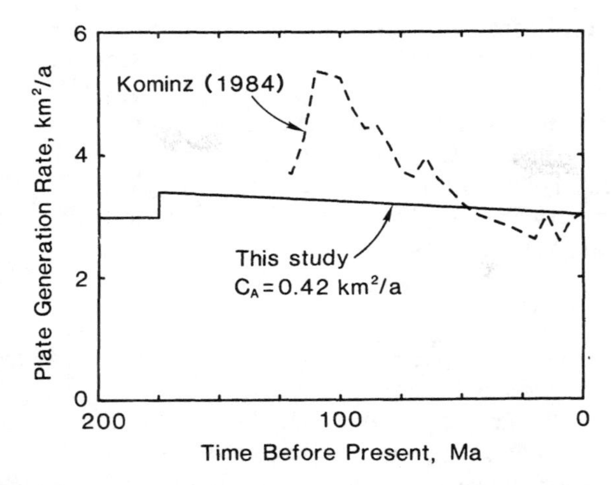

Figure 5. Global plate generation rate as a function of time. Our study (solid line) indicates that plate generation rates have decreased steadily since the Atlantic opened at 175 Ma and that a one-to-one correspondence between eustatic highstands and peak plate generation rates does not exist. In contrast reconstructions based on marine magnetic anomalies (cf. Kominz, 1984) predict much higher plate generation rates (dashed curve) during highstands. It may be necessary to turn to independent measures of seafloor generation rates, such as variations in seawater chemistry, to determine which model is more reasonable.

$\times 10^{-8}$ km^2/a^2 (Parsons, 1982).

The secular variation in the global rate of plate generation and oceanic heat flux, predicted by our model, results from the passive rearrangement of surface plates rather than global pulses of mantle convection. Episodic mantle convection is thought to be inhibited by the temperature-dependent rheology of silicates which strongly buffers the cooling rate of the mantle (Davies, 1979; Turcotte et al., 1979). Our prediction that the global plate generation rate was highest during the Jurassic and has decreased steadily since is consistent with the increase of $^{87}Sr/^{86}Sr$ in seawater (cf. Burke et al., 1982) over the same time, although climatic variations should also be taken into account.

Sea level can be affected in ways other than those considered above. If the continents were to increase in area as a result of continental growth (Reymer and Schubert, 1984) or extension (Heller and Angevine, 1985) then the area of the ocean basins would decrease and sea level would rise. The eustatic rise which began no later than early Jurassic time, and culminated in the Cretaceous highstand, may have been initiated by continental rifting during the early Jurassic (Smith and Noltimer, 1979) which preceded the actual breakup of Pangea. Similarly, collision and orogeny can reduce the area of the continents causing sea level to fall (Harrison et al., 1981). The collision of the Indian subcontinent with Eurasia proceeds at a rate of 0.15 km^2/a (Parsons, 1982) but its effect on the volume of the ocean basins may be considerably smaller due to the transport of sediment from the Himalayas to the Ganges and Indus fans. Schlanger et al. (1981) suggest that anomalous volcanism, occurring from 110 to 70 Ma ago, may have thermally uplifted a large area on the Pacific and Farallon plates, causing sea level to rise by 100 m or more above its present level. Changes in the volume of continental glaciers can also have a secular influence on sea level. If the present continental glaciers were to melt completely sea level would rise between 40 and 50 m (Pitman, 1978). It is likely that this ice began to accumulate no later than the end of the Eocene (Matthews and Poore, 1980). When the effects of volcanism and glaciation are included in our earlier estimates of sea level, we find that sea level has fallen by perhaps 220 m since the Cretaceous highstand. This value is intermediate between the estimates of Harrison et al. (1981) and Sahagian (1987). The timing of the Cretaceous highstand would be influenced by the history of volcanic uplift in the Pacific and the timing of supercontinent breakup.

6. Conclusions

As a supercontinent fragments and new ocean basins open the area-age distribution of the seafloor may be passively altered such that sea level first rises and the falls, reaching a highstand some 70 to 80 Ma after breakup. That two highstands occurred during the Phanerozoic could be largely a result of two supercontinent breakups during the Phanerozoic. Our most fundamental conclusion is that peak plate generation rates, the heat flux maximum, and the eustatic highstand need not occur at the same time. This is in strong contrast to prevailing opinion.

Acknowledgements

We have benefitted greatly from discussions with Mike Arthur. This work was supported in part by the donors to the Geophysics Development Fund.

References

Anderton, R., 1982. Dalradian deposition and the late Precambrian-Cambrian history of the N. Atlantic region: a review of the early history of the Iapetus ocean, *J. Geol. Soc. London*, **139**, 421-431.

Berger, W.H. and Winterer, E.L., 1974. Plate stratigraphy and the fluctuating carbonate line, *Int. Assoc. Sedimentol.*, Spec. Publ. 1, 11-48.

Bond, G.C., 1979. Speculations on real sea-level changes and vertical motions of continents at selected times in the Cretaceous and Tertiary periods, *Geology*, **6**, 247-250.

Burke, W.H., Denison, R.E., Hetherington, E.A., Koepnick, R.B., Nelson, H.F. and Otto, J.B., 1982. Variation of seawater $^{87}Sr/^{86}Sr$ throughout Phanerozoic time, *Geology*, **10**, 516-519.

Davies. G.F., 1979. Thermal histories of convective earth models and constraints on the radiogenic heat production in the earth, *J. Geophys. Res.*, **85**, 2517-2530.

Davis, E.E. and Lister, C.R.B., 1974. Fundamentals of ridge crest topography, *Earth Plan. Sci. Lett.*, **21**, 405-413.

Hallam, A., 1977. Secular changes in marine undulation of USSR and North America through the Phanerozoic, *Nature*, **269**, 769-772.

Hallam, A., 1984. Pre-Quaternary sea-level changes, *Ann. Rev. Earth Planet.*, **12**, 205-243.

Harrison, C.G.A., 1980. Spreading rates and heat flow, *Geophys. Res. Lett.*, **7**, 1041-1044.

Harrison, C.G.A., Brass,G.W., Saltzman, E., Sloan II, J., Southam, J. and Whitman, J.M., 1981. Sea-level variations, global sedimentary rates and the hypsographic curve, *Earth Plan. Sci. Lett.*, **54**, 1-16.

Hays, J.D. and Pitman III, W.C., 1973. Lithospheric plate motion, sea-level changes, and climatic and ecological consequences, *Nature*, **246**, 18-22.

Heller, P.L. and Angevine, C.L., 1985. Sea-level cycles during the growth of Atlantic-type oceans, *Earth Plan. Sci. Lett.*, **75**, 417-426.

Kominz, M.A., 1984. Oceanic ridge volumes and sea-level changes -an error analysis, in: *Interregional*

Unconformities and Hydrocarbon Accumulation, 109-127, ed. Schlee, J.S., Am. Assoc. Pet. Geol. Mem., **36**.

Larson, R.L., Golovchenko, X. and Pitman III, W.C., 1982. *Geomagnetic polarity time scale*, Am. Assoc. Pet. Geol. Plate-tectonic Map Circum-Pacific Region, scale 1:20,000,000.

Matthews, R.K. and Poore, R.Z., 1980. Tertiary $\delta^{18}O$ record and glacioeustatic sea-level fluctuations, *Geology*, **8**, 501-504.

Parsons, B. and Sclater, J.G., 1977. An analysis of the variation of ocean floor bathymetry and heat flow with age, *J. Geophys. Res.*, **82**, 803-827.

Parsons, B., 1982. Causes and consequences of the relation between area and age of the ocean floor, *J. Geophys. Res.*, **87**, 289-302.

Pitman III, W.C., 1978. Relationship between eustacy and stratigraphic sequences of passive margins, *Bull. Geol. Soc. Am.*, **89**, 1389-1403.

Reymer, A. and Schubert, G., 1984. Phanerozoic addition rates to the continental crust and crustal growth, *Tectonics*, **3**, 63-77.

Russell, K.L., 1987. Oceanic ridges and eustatic changes in sea level, *Nature*, **218**, 861-862.

Sahagian, D., 1987. Cretaceous shoreline deposits: application to epeirogeny and eustatic sea level, *J. Geophys. Res.*, **92**, 4895-4904.

Schlanger, S.O., Jenkyns, H.C. and Premoli-Silva, I., 1981. Volcanism and vertical tectonics in the Pacific basin related to global Cretaceous transgressions, *Earth Plan. Sci. Lett.*, **52**, 435-449.

Sclater, J.G., Jaupart, C. and Galson, D., 1980. The heat flow through oceanic and continental crust and the heat loss of the Earth, *Rev. Geophys. Space Phys.*, **18**, 269-311.

Sleep, N.H., 1976. Platform subsidence mechanisms and "eustatic" sea-level changes, *Tectonophysics*, **36**, 45-56.

Smith, T.E. and Noltimer, H.C., 1979. Paleomagnetism of the Newark Trend igneous rocks of the north central Appalachians and the opening of the central Atlantic Ocean, *Am. J. Sci.*, **279**, 778-807.

Southam, J.R. and Hay, W.W., 1981. Global sedimentary mass balance and sea-level changes, in: *The Sea*, 1617-1684, ed. Emiliani, C., Wiley, New York.

Thorne, J. and Bell, R., 1983. A eustatic sea-level curve from histograms of North Sea subsidence, *EOS*, **64**, 858.

Turcotte, D.L. and Burke, K., 1978. Global sea-level changes and the thermal structure of the Earth, *Earth Plan. Sci. Lett.*, **41**, 341-346.

Turcotte, D.L., Cook, F.A. and Willemann, R.J., 1979. Parameterized convection within the moon and terrestrial planets, *Proc. Lunar Plan. Sci. Conf.*, **10**, 2375-2392.

Vail, P.R., Mitchum, R.M. and Thompson III, S., 1977. Seismic stratigraphy and global changes of sea level, part 4: global cycles of relative changes of sea level, *Am. Assoc. Pet. Geol. Mem.*, **26**, 83-97.

Valentine, J.W. and Moores, E.M., 1970. Plate-tectonic regulation of faunal diversity and sea level, *Nature*, **228**, 657-659.

Watts, A.B., 1982. Tectonic subsidence, flexure and global changes in sea level, *Nature*, **297**, 469-474.

Watts, A.B. and Steckler, M.S., 1979. Subsidence and eustacy at the continental margin of eastern North America, *Am. Geophys. Union, Maurice Ewing Ser.*, **3**, 218-234.

Watts, A.B. and Thorne, J. 1984. Tectonics, global changes in sea level and their relationship to stratigraphical sequences at the US Atlantic continental margin, *Mar. Pet. Geol.*, **1**, 319-339.

Worsley, T.R., Nance, D. and Moody, J.B., 9184. Global tectonics and eustacy for the past two billion years, *Mar. Geol.*, **58**, 373-400.

Wyatt, A.R., 1986. Post-Triassic continental hypsometry and sea level, *J. Geol. Soc. London*, **143**, 907-910.

C.L. ANGEVINE, S.R. LINNEMAN and P.L. HELLER, Department of Geology and Geophysics, University of Wyoming, P.O. Box 3006, Laramie WY 82071, USA.

Index